PERGAMON INTERNATIONAL LIBRARY
of Science, Technology, Engineering and Social Studies

The 1000-volume original paperback library in aid of education, industrial training and the enjoyment of leisure

Publisher: Robert Maxwell, M.C.

L. D. LANDAU and E. M. LIFSHITZ

COURSE OF THEORETICAL PHYSICS

Volume 10

PHYSICAL KINETICS

Other Titles in Series

LANDAU and LIFSHITZ: COURSE OF THEORETICAL PHYSICS

Vol. 1 *Mechanics* (3rd edition)
Vol. 2 *The Classical Theory of Fields* (4th edition)
Vol. 3 *Quantum Mechanics – Non-relativistic Theory* (3rd edition)
Vol. 4 *Quantum Electrodynamics*
Vol. 5 *Statistical Physics*, Part 1 (3rd edition)
Vol. 6 *Fluid Mechanics*
Vol. 7 *Theory of Elasticity* (2nd edition)
Vol. 8 *Electrodynamics of Continuous Media*
Vol. 9 *Statistical Physics*, Part 2

PHYSICAL KINETICS

by

E. M. LIFSHITZ and L. P. PITAEVSKIĬ

Institute of Physical Problems, U.S.S.R. Academy of Sciences

Volume 10 of *Course of Theoretical Physics*

Translated from the Russian by

J. B. SYKES and R. N. FRANKLIN

PERGAMON PRESS

OXFORD · NEW YORK · TORONTO · SYDNEY · PARIS · FRANKFURT

UK	Pergamon Press Ltd., Headington Hill Hall, Oxford OX3 0BW, England
USA	Pergamon Press Inc., Maxwell House, Fairview Park, Elmsford, New York 10523, USA
CANADA	Pergamon Press Canada Limited, Suite 104, 150 Consumers Road, Willowdale, Ontario M2J 1P9, Canada
AUSTRALIA	Pergamon Press (Aust.) Pty. Ltd., P.O. Box 544, Potts Point, NSW 2011, Australia
FRANCE	Pergamon Press SARL, 24 rue des Ecoles, 75240 Paris, Cedex 05, France
FEDERAL REPUBLIC OF GERMANY	Pergamon Press GmbH, 6242 Kronberg-Taunus, Hammerweg 6, Federal Republic of Germany

First edition 1981

British Library Cataloguing in Publication Data

Lifshitz, E.M.
Physical kinetics. – (Course of theoretical physics; v. 10). – (Pergamon international library)
1. Plasma (Ionized gases)
I. Title II. Pitaevskii, L. P.
III. Series
530.4′4 QC718 80-42162

ISBN 0-08-020641-7 (Hard cover)
ISBN 0-08-026480-8 (Flexicover)

Translated from
Fizicheskaya kinetika,
Nauka, Moscow, 1979

Printed in Great Britain by
A. Wheaton and Co. Ltd., Exeter

CONTENTS

VIII. QUANTUM LIQUIDS

IX. METALS

X. THE DIAGRAM TECHNIQUE FOR NON-EQUILIBRIUM SYSTEMS

XI. SUPERCONDUCTORS

XII. KINETICS OF PHASE TRANSITIONS

PREFACE

THIS final volume of the *Course of Theoretical Physics* deals with physical kinetics, in the wide sense of the microscopic theory of processes in systems not in statistical equilibrium.

In contrast to the properties of systems that are in statistical equilibrium, the kinetic properties are much more closely related to the nature of the microscopic interactions in a particular physical object. This is the reason for the enormous variety in such properties and the considerably greater complexity of the relevant theory. The choice of topics to be included in a general course of theoretical physics thereby becomes less clear.

The scope of the book will be evident from the table of contents. Here we shall add only a few remarks.

Much attention is given to the theory of gases, as the simplest branch, in principle, of kinetic theory. Several chapters are concerned with plasma theory, not only because of the intrinsic physical significance of this department of kinetic theory, but also because many of the problems involved can be completely solved and furnish an instructive illustration of the general methods of the kinetic theory.

The kinetic properties of solids are especially multifarious. In the selection of material for the chapters in question, we naturally had to confine ourselves to the most general subjects which exhibit the basic physical kinetic phenomena and the methods of treating them. Here we must again emphasize that the book is part of a course of theoretical physics, and does not set out to be a textbook of solid state theory.

There are two evident omissions from the book: the kinetics of magnetic processes, and the theory of transport phenomena arising from the passage of fast particles through matter. These omissions are due to lack of time, and we resolved to accept them for the present edition, so as not to delay its publication any further. We trust that, although the book thus does not contain all that it might, everything in it will be found both interesting and useful.

This volume completes the programme laid down by Lev Davidovich Landau more than forty years ago. The entire *Course* comprises the following ten volumes:

Vol. 1 *Mechanics*
Vol. 2 *The Classical Theory of Fields*
Vol. 3 *Quantum Mechanics* (*Non-Relativistic Theory*)
Vol. 4 *Quantum Electrodynamics* (formerly *Relativistic Quantum Theory*)
Vol. 5 *Statistical Physics, Part 1*
Vol. 6 *Fluid Mechanics*

Vol. 7 *Theory of Elasticity*
Vol. 8 *Electrodynamics of Continuous Media*
Vol. 9 *Statistical Physics, Part 2*
Vol. 10 *Physical Kinetics*

The position of Vol. 9 results from the fact that it makes considerable use of material from fluid mechanics and macroscopic electrodynamics.

In the new series of Russian editions begun in 1973, Volumes 1, 2, 3, 5, 9 and 10 have so far appeared. Volume 7 can be reissued with only minor changes. Volume 4, previously published as *Relativistic Quantum Theory*, will lose the chapters on weak and strong interactions and shortly be reissued as *Quantum Electrodynamics*. Volumes 6 and 8, which have not been reissued for many years, require more substantial revision and expansion; we intend to proceed to this in the near future.

We should like to express our sincere thanks to A. F. Andreev, R. N. Gurzhi, V. L. Gurevich, Yu. M. Kagan, M. I. Kaganov and I. M. Lifshitz, with whom we have discussed matters treated in the book. We are also grateful to L. P. Gor'kov and A. A. Rukhadze, who read the manuscript and made a number of comments.

November 1978

E. M. LIFSHITZ
L. P. PITAEVSKIĬ

NOTATION

Particle distribution function f (Chapters I–VI); momentum distribution function always relative to d^3p.

Occupation numbers of quantum states $n(\mathbf{p})$ for electrons and $N(\mathbf{k})$ for phonons (Chapters VII and IX–XI); momentum distribution always relative to $d^3p/(2\pi\hbar)^3$.

Collision integral C; linearized collision integral I.

Thermodynamic quantities: temperature T, pressure P, chemical potential μ, particle number density N, total particle number $\mathcal{N}$, total volume $\mathcal{V}$.

Electric field $\mathbf{E}$, magnetic induction $\mathbf{B}$; unit electric charge e (electron charge $-e$).

In estimates: characteristic lengths L; atomic dimensions and lattice constant d; mean free path l; speed of sound u.

Averaging is denoted by angle brackets $\langle\ldots\rangle$ or by a bar over a letter.

Three-dimensional vector suffixes are denoted by Greek letters $\alpha, \beta, \ldots$

In Chapters III–VI:

Electron mass m, ion mass M.

Electron charge $-e$, ion charge ze.

Electron thermal velocity $v_{Te} = (T_e/m)^{1/2}$.

Ion thermal velocity $v_{Ti} = (T_i/M)^{1/2}$.

Plasma frequency $\Omega_e = (4\pi N_e e^2/m)^{1/2}$, $\Omega_i = (4\pi N_i z^2 e^2/M)^{1/2}$.

Debye length $a_e = (T_e/4\pi N_e e^2)^{1/2}$, $a_i = (T_i/4\pi N_i z^2 e^2)^{1/2}$, $a^{-2} = a_e^{-2} + a_i^{-2}$.

Larmor frequency $\omega_{Be} = eB/mc$, $\omega_{Bi} = zeB/Mc$.

References to other volumes in the *Course of Theoretical Physics*:

Mechanics = Vol. 1 (*Mechanics*, third English edition, 1976).
Fields = Vol. 2 (*The Classical Theory of Fields*, fourth English edition, 1975).
QM = Vol. 3 (*Quantum Mechanics*, third English edition, 1977).
RQT = Vol. 4 (*Relativistic Quantum Theory*, Part 1, English edition, 1971; Part 2, English edition, 1974); to be reissued (see Preface).
SP 1 = Vol. 5 (*Statistical Physics*, Part 1, third English edition, 1980).
FM = Vol. 6 (*Fluid Mechanics*, English edition, 1959).
TE = Vol. 7 (*Theory of Elasticity*, second English edition, 1970).
ECM = Vol. 8 (*Electrodynamics of Continuous Media*, English edition, 1960).
SP 2 = Vol. 9 (*Statistical Physics*, Part 2, English edition, 1980).

All are published by Pergamon Press.

CHAPTER I

KINETIC THEORY OF GASES

§1. The distribution function

THIS chapter deals with the kinetic theory of ordinary gases consisting of electrically neutral atoms or molecules. The theory is concerned with non-equilibrium states and processes in an ideal gas. An ideal gas, it will be recalled, is one so rarefied that each molecule in it moves freely at almost all times, interacting with other molecules only during close encounters with them. That is to say, the mean distance between molecules, $\bar{r} \sim N^{-1/3}$ (where N is the number of molecules per unit volume), is assumed large in comparison with their size, or rather in comparison with the range d of the intermolecular forces; the small quantity $Nd^3 \sim (d/\bar{r})^3$ is sometimes called the *gaseousness parameter.*

The statistical description of the gas is given by the *distribution function* $f(t, q, p)$ of the gas molecules in their phase space. It is, in general, a function of the generalized coordinates (chosen in some manner, and denoted jointly by q) and the corresponding generalized momenta (denoted jointly by p), and in a non-steady state also of the time t. Let $d\tau = dq\, dp$ denote a volume element in the phase space of the molecule; dq and dp conventionally denote the products of the differentials of all the coordinates and all the momenta respectively. The product $f\, d\tau$ is the mean number of molecules in a given element $d\tau$ which have values of q and p in given ranges dq and dp. We shall return later to this definition of the mean.

Although the function f will be everywhere understood as the distribution density in phase space, there is advantage in expressing it in terms of suitably chosen variables, which need not be canonically conjugate coordinates and momenta. Let us first of all decide on the choice to be made.

The translational motion of a molecule is always classical, and is described by the coordinates $\mathbf{r} = (x, y, z)$ of its centre of mass and by the components of the momentum $\mathbf{p}$ (or the velocity $\mathbf{v} = \mathbf{p}/m$) of its motion as a whole. In a monatomic gas, the motion of the particles, which are atoms, is purely translational. In polyatomic gases, the molecules also have rotational and vibrational degrees of freedom.

The rotational motion of a molecule in a gas is almost always classical too.† It is described in the first place by the angular momentum vector $\mathbf{M}$ of the molecule. For a diatomic molecule, this is sufficient. Such a molecule is a rotator turning in a plane perpendicular to $\mathbf{M}$. In actual physical problems, the distribution function

†The condition for the rotation to be classical is $\hbar^2/2I \ll T$, where I is the moment of inertia of the molecule and T the temperature of the gas. This condition can be violated in ordinary gases only for hydrogen and deuterium at low temperatures.

may be regarded as independent of the angle φ of rotation of the axis of the molecule in this plane, all orientations of the molecule in the plane being equally probable. This is because the angle φ changes rapidly as the molecule rotates, and the result may be understood as follows.

The rate of change of φ (the angular velocity of rotation of the molecule) is $\dot{\varphi} \equiv \Omega = M/I$. Its mean value $\bar{\Omega} \sim \bar{v}/d$, where d is the molecular dimension and $\bar{v}$ the mean linear speed. Different molecules have various values of Ω, distributed in some way about $\bar{\Omega}$. Thus molecules which initially had the same φ very soon acquire different values; there is a rapid "mixing" with regard to angles. Let the distribution of molecules in angle $\varphi = \varphi_0$ (in the range from 0 to 2π) and in Ω at the initial instant $t = 0$ be given by a function $f(\varphi_0, \Omega)$. We separate from it the mean value independent of φ:

$$f = \bar{f}(\Omega) + f'(\varphi_0, \Omega),$$

$$\bar{f}(\Omega) = \frac{1}{2\pi}\int_0^{2\pi} f(\varphi_0, \Omega)\, d\varphi_0,$$

so that $f'(\varphi_0, \Omega)$ is a function periodic in φ_0 with period 2π and zero mean. In the course of time, the free rotation of the molecules ($\varphi = \Omega t + \varphi_0$) changes the distribution function:

$$f(\varphi, \Omega, t) = \bar{f}(\Omega) + f'(\varphi - \Omega t, \Omega);$$

In the course of time, f' becomes a more and more rapidly oscillating function of Ω: the characteristic period of oscillation $\Delta\Omega \sim 2\pi/t$, and becomes small in comparison with $\bar{\Omega}$ even during the mean free time of the molecules between collisions. All observable physical quantities, however, involve some averaging of the distribution function with respect to Ω; the contribution of the rapidly oscillating function f' to such mean values is negligible. This enables us to replace the distribution $f(\varphi, \Omega)$ by the angle-averaged function $\bar{f}(\Omega)$.

The above arguments are, of course, general ones, and apply to any rapidly varying quantities (phases) which take values in finite ranges.

Returning to the rotational degrees of freedom of molecules, let us note that in polyatomic gases the distribution function may also depend on the angles which specify the fixed orientation of the axes of the molecules relative to the vector **M**. For example, in molecules of the symmetrical-top type this is the precession angle between **M** and the axis of the top, whereas the distribution function may again be regarded as independent of the rapidly varying angles of rotation of the top about its own axis and precession of this axis about **M**.†

The vibrational motion of the atoms within the molecule is practically always

†In the rotation of a spherical-top molecule, such as CH_4, the two angles remain constant which define the orientation of the molecule relative to **M** (i.e. the direction of the angular velocity $\mathbf{\Omega}$). In the rotation of an asymmetrical-top molecule, a combination of angles remains constant which represents the rotational energy $E_{\mathrm{rot}} = M_\xi^2/2I_1 + M_\eta^2/2I_2 + M_\zeta^2/2I_3$, where M_ξ, M_η, M_ζ are the components of the constant vector **M** along the rotating principal axes of inertia of the molecule.

quantized, so that the vibrational state of the molecule is specified by the appropriate quantum numbers. Under ordinary conditions (at not too high temperatures), however, the vibrations are not excited at all, and the molecule is at its ground vibrational level.

In this chapter we shall denote by Γ the set of all variables on which the distribution function depends, other than the coordinates of the molecule as a whole (and the time t). We separate from the phase volume element $d\tau$ the factor $dV = dx\,dy\,dz$, and denote by $d\Gamma$ the remaining factor in terms of the variables used (and integrated over the angles on which f does not depend). The quantities Γ have an important common property: they are integrals of the motion, and remain constant for each molecule during its free motion (in the absence of an external field) between successive collisions; but they are in general altered by each collision. The coordinates x, y, z of the molecule as a whole vary, of course, during its free motion.

For a monatomic gas, the quantities Γ are the three components of the momentum $\mathbf{p} = m\mathbf{v}$ of the atom, so that $d\Gamma = d^3p$. For a diatomic molecule, Γ includes not only the momentum $\mathbf{p}$ but also the angular momentum $\mathbf{M}$; accordingly, $d\Gamma$ may be expressed as

$$d\Gamma = 2\pi\, d^3p\, M\, dM\, do_{\mathbf{M}}, \tag{1.1}$$

where $do_{\mathbf{M}}$ is a solid-angle element for the direction of the vector $\mathbf{M}$†. For a symmetrical-top molecule, the quantities Γ include also the angle θ between $\mathbf{M}$ and the axis of the top; then

$$d\Gamma = 4\pi^2\, d^3p\, M^2\, dM\, do_{\mathbf{M}} d\cos\theta$$

(one factor of 2π comes from integration over the angle of rotation of the top about its axis, and another from integration over the angle of precessional rotation).

The integral

$$f(t, \mathbf{r}, \Gamma) d\Gamma = N(t, \mathbf{r})$$

is the spatial distribution density of gas particles; $N\,dV$ is the mean number of molecules in the volume element dV. Here the following comments are needed.

An infinitesimal volume element dV really means one that is not mathematically but physically small, i.e. a region of space which is very small in comparison with the characteristic dimensions L of the problem, but still large in comparison with molecular dimensions. The statement that a molecule is in a given volume element

†This expression can be derived by first writing

$$\begin{aligned} d\Gamma &= d^3p\,\delta(\mathbf{M}\,.\,\mathbf{n})\, d^3M\, do_{\mathbf{n}} \\ &= d^3p\,\delta(M\cos\theta) M^2\, dM\, do_{\mathbf{M}} d\cos\theta\, d\varphi, \end{aligned}$$

where $do_{\mathbf{n}} = d\cos\theta\, d\varphi$ is a solid-angle element for the direction of the molecule axis (θ being the angle between this axis and $\mathbf{M}$). The delta function expresses the fact that $\mathbf{M}$ has only two independent components (corresponding to the number of rotational degrees of freedom of a diatomic molecule): $\mathbf{M}$ is perpendicular to the molecule axis. Integration of this formula over $d\cos\theta\, d\varphi$ gives (1.1).

dV therefore defines its position, at best, only to within distances of the order of its dimensions. This is a very important point. If the coordinates of the gas particles were specified exactly, then the result of a collision between, say, two atoms of a monatomic gas moving in definite classical paths would also be entirely definite. If, however, the collision is between atoms in a given physically small volume (as always in the kinetic theory of gases), the uncertainty in the relative position of the atoms means that the result of the collision also is uncertain, and only the probability of one or another outcome can be considered.

We can now specify that the mean number density of particles refers to averaging over the volumes of physically infinitesimal elements thus defined, and correspondingly over times of the order of that taken by the particles to traverse such elements.

Since the dimensions of the volume elements used in defining the distribution function are large in comparison with the molecular dimensions d, the distances L over which this function varies considerably must always be large also, in comparison with d. The ratio between the size of the physically infinitesimal volume elements and the mean intermolecular distance $\bar{r}$ may in general have any value. There is, however, a difference in the nature of the density N determined by the distribution function, according to the value of that ratio. If the element dV is not large compared with $\bar{r}$, the density N is not a macroscopic quantity: the fluctuations of the number of particles present in dV are comparable with its mean value. The density N becomes a macroscopic quantity only if it is defined with respect to volumes dV containing many particles; the fluctuations in the number of particles in these volumes are then relatively small. It is, however, clear that such a definition is possible only if also the characteristic dimensions of the problem $L \gg \bar{r}$.

§2. The principle of detailed balancing

Let us consider collisions between two molecules, one of which has values of Γ in a given range $d\Gamma$, and the other in a range $d\Gamma_1$, and which acquire in the collision values in the ranges $d\Gamma'$ and $d\Gamma_1'$ respectively; for brevity, we shall refer simply to a collision of molecules with Γ and Γ_1, resulting in Γ' and Γ_1'. The total number of such collisions per unit time and unit volume of the gas may be written as a product of the number of molecules per unit volume, $f(t, \mathbf{r}, \Gamma)d\Gamma$, and the probability that any of them has a collision of the type concerned. This probability is always proportional to the number of molecules Γ_1 per unit volume, $f(t, \mathbf{r}, \Gamma_1)d\Gamma_1$, and to the ranges $d\Gamma'$ and $d\Gamma_1'$ of the values of Γ for the two molecules after the collision. Thus the number of collisions $\Gamma, \Gamma_1 \to \Gamma', \Gamma_1'$ per unit time and volume may be written as

$$w(\Gamma', \Gamma_1'; \Gamma, \Gamma_1) f f_1 \, d\Gamma \, d\Gamma_1 \, d\Gamma' \, d\Gamma_1'; \tag{2.1}$$

here and henceforward, the affixes to f correspond to those of their arguments Γ: $f_1 \equiv f(t, \mathbf{r}, \Gamma_1)$, $f' \equiv f(t, \mathbf{r}, \Gamma')$, and so on. The coefficient w is a function of all its arguments Γ.† The ratio of $w\,d\Gamma'\,d\Gamma_1'$ to the absolute value of the relative velocity $\mathbf{v}-\mathbf{v}_1$ of the colliding molecules has the dimensions of area, and is the effective collision cross-section:

$$d\sigma = \frac{w(\Gamma', \Gamma_1'; \Gamma, \Gamma_1)}{|\mathbf{v}-\mathbf{v}_1|}\,d\Gamma'\,d\Gamma_1'. \tag{2.2}$$

The function w can in principle be determined only by solving the mechanical problem of collision of particles interacting according to some given law. However, certain properties of this function can be elucidated from general arguments.‡

The collision probability is known to have an important property which follows from the symmetry of the laws of mechanics (classical or quantum) under time reversal; see *QM*, § 144. Let Γ^T denote the values of the quantities obtained from Γ by time reversal. This operation changes the signs of all linear and angular momenta; hence, if $\Gamma = (\mathbf{p}, \mathbf{M})$, then $\Gamma^T = (-\mathbf{p}, -\mathbf{M})$. Since time reversal interchanges the states that are "before" and "after" the collision we have

$$w(\Gamma', \Gamma_1'; \Gamma, \Gamma_1) = w(\Gamma^T, \Gamma_1{}^T; \Gamma'^T, \Gamma_1'^T). \tag{2.3}$$

This relation implies, in a state of statistical equilibrium, the *principle of detailed balancing*, according to which the number of collisions $\Gamma, \Gamma_1 \to \Gamma', \Gamma_1'$ is equal, in equilibrium, to the number $\Gamma'^T, \Gamma_1'^T \to \Gamma^T, \Gamma_1{}^T$. For, expressing these numbers in the form (2.1), we have

$$w(\Gamma', \Gamma_1'; \Gamma, \Gamma_1) f_0 f_{01}\,d\Gamma\,d\Gamma_1\,d\Gamma'\,d\Gamma_1' = w(\Gamma^T, \Gamma_1{}^T; \Gamma'^T, \Gamma_1'^T) f_0' f_{01}'\,d\Gamma^T\,d\Gamma_1{}^T\,d\Gamma'^T\,d\Gamma_1'^T,$$

where f_0 is the equilibrium (Boltzmann) distribution function. The product of phase volume elements $d\Gamma\,d\Gamma_1\,d\Gamma'\,d\Gamma_1'$ is unaltered by time reversal; the differentials on the two sides of the above equation may therefore be omitted. Next, when t is replaced by $-t$, the energy is unchanged: $\epsilon(\Gamma) = \epsilon(\Gamma^T)$, where $\epsilon(\Gamma)$ is the energy of the molecule as a function of the quantities Γ. Since the equilibrium distribution function (in a gas at rest as a whole) depends only on the energy,

$$f_0(\Gamma) = \text{constant} \times e^{-\epsilon(\Gamma)/T}, \tag{2.4}$$

where T is the gas temperature, we have $f_0(\Gamma) = f_0(\Gamma^T)$. Lastly, by the law of conservation of energy in the collision of two molecules $\epsilon + \epsilon_1 = \epsilon' + \epsilon_1'$. Hence

$$f_0 f_{01} = f_0' f_{01}', \tag{2.5}$$

and the above equation reduces to (2.3).

This assertion remains valid, of course, for a gas moving with a macroscopic

†The characteristics of the initial (i) and final (f) states in w are written from right to left, $w(f, i)$ as is customary in quantum mechanics.

‡It should be emphasized immediately that, although the free motion of molecules is assumed classical, this does not at all mean that their collision cross-section need not be determined quantum-mechanically; in fact, it usually must be so determined. The whole of the derivation of the transport equation given here is independent of the classical or quantum nature of the function w.

velocity **V**. The equilibrium distribution function is then

$$f_0(\Gamma) = \text{constant} \times \exp\left(-\frac{\epsilon(\Gamma) - \mathbf{p}\,.\,\mathbf{V}}{T}\right), \tag{2.6}$$

and equation (2.5) continues to be valid because of the conservation of momentum in collisions: $\mathbf{p} + \mathbf{p}_1 = \mathbf{p}' + \mathbf{p}_1'$.†

Note that (2.5) depends only on the form of the distribution (2.4) or (2.6) as a function of Γ; the parameters T and **V** may vary through the gas volume.

The principle of detailed balancing may also be expressed in a somewhat different form. To do so, we apply not only time reversal but spatial inversion, changing the sign of all coordinates. If the molecules are not sufficiently symmetrical, they become their stereoisomers on inversion, and they cannot be made to coincide with these by any rotation of the molecule as a whole.‡ In such cases, inversion would mean replacing the gas by an essentially different substance, and no new conclusions would be available as to its properties. If, however, the symmetry of the molecule does not allow stereoisomerism, the gas remains the same on inversion, and the quantities which describe the properties of a macroscopically homogeneous gas must remain unaltered.

Let Γ^{TP} denote the set of quantities obtained from Γ by simultaneous time reversal and inversion. Inversion changes the sign of all ordinary (polar) vectors, including the momentum **p**, but leaves unchanged the axial vectors, including the angular momentum **M**. Hence, if $\Gamma = (\mathbf{p}, \mathbf{M})$, then $\Gamma^{TP} = (\mathbf{p}, -\mathbf{M})$. As well as (2.3), we have the equation§

$$w(\Gamma', \Gamma_1'; \Gamma, \Gamma_1) = w(\Gamma^{TP}, \Gamma_1{}^{TP}; \Gamma'^{TP}, \Gamma_1'^{TP}). \tag{2.7}$$

Transitions corresponding to the functions w on the two sides of (2.3) are said to be mutually *time-reversed*. They are not strictly *direct* and *reverse* transitions, since Γ and Γ^T are not the same. For a monatomic gas, however, the principle of detailed balancing can also be expressed in relation to direct and reverse transitions. Since the quantities Γ are here just the three momentum components of the atom, $\Gamma = \Gamma^{TP} = \mathbf{p}$, and from (2.7)

$$w(\mathbf{p}', \mathbf{p}_1'; \mathbf{p}, \mathbf{p}_1) = w(\mathbf{p}, \mathbf{p}_1; \mathbf{p}', \mathbf{p}_1'). \tag{2.8}$$

This is detailed balancing in the literal sense: each microscopic collision process is balanced by the reverse process.

The function w satisfies one further general relation which does not depend on the symmetry under time reversal, and which can be most clearly derived in

†Equation (2.6) is obtained from (2.4) by transforming the energy of the molecule from the frame of reference K_0 in which the gas is at rest to the frame K in which it moves with velocity **V**: $\epsilon_0(\Gamma) = \epsilon(\Gamma) - \mathbf{p}\,.\,\mathbf{V} + \frac{1}{2}m\mathbf{V}^2$; see *Mechanics* (3.5).

‡Stereoisomers exist for molecules that have no centre of symmetry and no plane of symmetry.

§If the quantities Γ include also variables specifying the rotational orientation of the molecule, they too must be transformed in a certain way in going to Γ^T or Γ^{TP}. For instance, the precession angle of a symmetrical top is given by the product $\mathbf{M}\,.\,\mathbf{n}$, where **n** is the direction of the axis of the molecule; this quantity changes sign both under time reversal and under inversion.

quantum-mechanical terms, the transitions considered being between states forming a discrete series. These are states of a pair of molecules moving in a given finite volume. The probability amplitudes of various collision processes form a unitary matrix $\hat{S}$, the *scattering matrix* or *S-matrix*. The unitarity condition is $\hat{S}^+\hat{S} = 1$, or, in explicit form with the matrix suffixes which label the various states,

$$\sum_n S^+_{in} S_{nk} = \sum_n S^*_{ni} S_{nk} = \delta_{ik}.$$

In particular, when $i = k$,

$$\sum_n |S_{ni}|^2 = 1.$$

The square $|S_{ni}|^2$ gives the probability of a collision with the transition $i \to n$,† and the above equation is simply the normalization condition for probabilities: the sum of the probabilities for all possible transitions from a given initial state is unity. The unitarity condition may also be written as $\hat{S}\hat{S}^+ = 1$, with the opposite order of the factors $\hat{S}$ and $\hat{S}^+$. We then have $\Sigma_n S_{in} S^*_{kn} = \delta_{ik}$, and when $i = k$

$$\sum_n |S_{in}|^2 = 1,$$

so that the sum of the probabilities for all possible transitions to a given final state is unity. Subtracting from each sum the one term with $n = i$ (transition without change of state), we can write

$$\sum_n{}' |S_{ni}|^2 = \sum_n{}' |S_{in}|^2.$$

This is the required equation. In terms of the functions w, it becomes

$$\int w(\Gamma', \Gamma'_1; \Gamma, \Gamma_1)\, d\Gamma'\, d\Gamma'_1 = \int w(\Gamma, \Gamma_1; \Gamma', \Gamma'_1)\, d\Gamma'\, d\Gamma'_1. \tag{2.9}$$

§3. The Boltzmann transport equation

Let us now go on to derive the basic equation in the kinetic theory of gases, which is satisfied by the distribution function $f(t, \mathbf{r}, \Gamma)$.

If collisions between molecules were entirely negligible, each gas molecule would constitute a closed subsystem, and the distribution function of the molecules would obey Liouville's theorem, according to which

$$df/dt = 0; \tag{3.1}$$

†For large values of the time t, $|S_{ni}|^2$ is proportional to t, and division by t gives the transition probability per unit time; cf. *RQT*, §64. If the wave functions of the initial and final particles are normalized to one particle per unit volume, this "probability" has the same dimensions (volume/time) as the quantity $w\, d\Gamma\, d\Gamma_1$ defined by (2.1).

see *SP* 1, §3. The total derivative here corresponds to differentiation along the phase path of the molecule, which is determined by its equations of motion. Liouville's theorem applies to a distribution function defined as the density in phase space (i.e. in the space of variables that are canonically conjugate generalized coordinates and momenta). This of course does not prevent f itself from being subsequently expressed in terms of any other variables.

In the absence of an external field, the quantities Γ for a freely moving molecule remain constant, and only its coordinates $\mathbf{r}$ vary; then

$$df/dt = \partial f/\partial t + \mathbf{v} \cdot \nabla f. \tag{3.2}$$

If, on the other hand, the gas is in, for example, an external field $U(\mathbf{r})$ acting on the coordinates of the centre of mass of the molecule (a gravitational field, say), then

$$df/dt = \partial f/\partial t + \mathbf{v} \cdot \nabla f + \mathbf{F} \cdot \partial f/\partial \mathbf{p}, \tag{3.3}$$

where $\mathbf{F} = -\nabla U$ is the force exerted on the molecule by the field.

When collisions are taken into account, (3.1) is no longer valid, and the distribution function is no longer constant along the phase paths. Instead of (3.1), we have

$$df/dt = C(f), \tag{3.4}$$

where $C(f)$ denotes the rate of change of the distribution function by virtue of collisions: $dV\, d\Gamma\, C(f)$ is the change due to collisions, per unit time, in the number of molecules in the phase volume $dV\, d\Gamma$. Equation (3.4), in the form

$$\partial f/\partial t = -\mathbf{v} \cdot \nabla f + C(f),$$

with df/dt taken from (3.2), gives the total change in the distribution function at a given point in phase space; the term $dV\, d\Gamma\, \mathbf{v} \cdot \nabla f$ is the decrease per unit time in the number of molecules in this phase space element because of their free motion.

The quantity $C(f)$ is called the *collision integral*, and equations of the form (3.4) go by the general name of *transport equations*. Of course, the transport equation becomes meaningful only when the form of the collision integral has been established. We shall now discuss this topic.

When two molecules collide, their values of Γ are changed. Hence every collision undergone by a molecule transfers it out of a particular range $d\Gamma$; such collisions are referred to as "losses". The total number of collisions $\Gamma, \Gamma_1 \rightarrow \Gamma', \Gamma_1'$ with all possible values of $\Gamma_1, \Gamma', \Gamma_1'$ and given Γ, occurring in a volume dV per unit time, is equal to the integral

$$dV\, d\Gamma \int w(\Gamma', \Gamma_1'; \Gamma, \Gamma_1) f f_1\, d\Gamma_1\, d\Gamma'\, d\Gamma_1'.$$

There are also collisions ("gains") which bring into the range $d\Gamma$ molecules which originally had values outside that range. These are collisions $\Gamma', \Gamma_1' \rightarrow \Gamma, \Gamma_1$, again with all possible Γ_1, Γ', Γ_1' and given Γ. The total number of such collisions in the volume dV per unit time is

$$dV\, d\Gamma \int w(\Gamma, \Gamma_1; \Gamma', \Gamma_1') f' f_1'\, d\Gamma_1\, d\Gamma'\, d\Gamma_1'.$$

Subtracting the losses from the gains, we thus find that as a result of all collisions the relevant number of molecules is increased, per unit time, by

$$dV\, d\Gamma \int (w' f' f_1' - w f f_1) d\Gamma_1\, d\Gamma'\, d\Gamma_1',$$

where for brevity

$$w \equiv w(\Gamma', \Gamma_1'; \Gamma, \Gamma_1), \quad w' \equiv w(\Gamma, \Gamma_1; \Gamma', \Gamma_1'). \tag{3.5}$$

We therefore have the following expression for the collision integral:

$$C(f) = \int (w' f' f_1' - w f f_1)\, d\Gamma_1\, d\Gamma'\, d\Gamma_1'. \tag{3.6}$$

In the second term in the integrand, the integration over $d\Gamma'\, d\Gamma_1'$ relates only to w, since f and f_1 do not depend on these variables. This part of the integral can therefore be transformed by means of the unitarity relation (2.9). The collision integral then becomes

$$C(f) = \int w'(f' f_1' - f f_1)\, d\Gamma_1\, d\Gamma'\, d\Gamma_1', \tag{3.7}$$

in which both terms have the factor w'.†

Having established the form of the collision integral, we can write the transport equation as

$$\partial f/\partial t + \mathbf{v} \cdot \nabla f = \int w'(f' f_1' - f f_1)\, d\Gamma_1\, d\Gamma'\, d\Gamma_1'. \tag{3.8}$$

This integro-differential equation is also called the *Boltzmann equation*; it was first derived by Ludwig Boltzmann, the founder of the kinetic theory, in 1872.

The equilibrium statistical distribution must satisfy the transport equation identically. This condition is in fact fulfilled. The equilibrium distribution is stationary and (in the absence of an external field) uniform; the left-hand side of (3.8) is therefore identically zero. The collision integral also is zero, since the integrand vanishes by virtue of (2.5). The equilibrium distribution for a gas in an external field also satisfies the transport equation, of course. We need only recall that the left-hand side of the transport equation is the total derivative df/dt, which is

†The possibility of transforming the collision integral by means of (2.9) was noted by E. C. G. Stueckelberg (1952).

identically zero for any function f that depends only on integrals of the motion; and the equilibrium distribution is expressed solely in terms of the total energy $\epsilon(\Gamma)$ of the molecule, which is an integral of the motion.

In the above derivation of the transport equation, collisions were regarded as essentially instantaneous and occurring at a particular point in space. It is therefore clear that the equation allows us in principle to follow the variation of the distribution function only over times long compared with the duration of collisions, and over distances large compared with the size of the region in which a collision takes place. These distances are of the order of the range of action d of the molecular forces (and for neutral molecules, this is equal to their dimensions); the collision time is of the order of $d/\bar{v}$. Such values give the lower limit of distances and times that can be dealt with by means of the transport equation; the origin of these limitations will be considered in § 16. In practice, however, there is usually no need (and no possibility) for such a detailed account of the behaviour of the system, which would require, in particular, the specification of the initial conditions (coordinates and velocities of the gas molecules) with the same accuracy, which is impracticable. In actual physical problems, there are characteristic lengths L and times T imposed on the system by the conditions of the problem (characteristic gradient lengths for the macroscopic properties of the gas, wavelengths and periods of sound waves propagated in it, and so on). It is then sufficient to follow the behaviour of the system over distances and times small compared with these L and T. That is, the physically infinitesimal volume and time elements need be small only in comparison with L and T. The initial conditions of the problem are also averaged over such elements.

For a monatomic gas, the quantities Γ reduce to the three components of the momentum $\mathbf{p}$, and from (2.8) the functions w' in the collision integral can be replaced by $w = w'(\mathbf{p}', \mathbf{p}_1'; \mathbf{p}, \mathbf{p}_1)$. Then, expressing this function in terms of the differential collision cross-section $d\sigma$ by $w\, d^3p'\, d^3p_1' = v_{\text{rel}}\, d\sigma$ (where $v_{\text{rel}} = |\mathbf{v} - \mathbf{v}_1|$; see (2.2)), we find

$$C(f) = \int v_{\text{rel}}(f'f_1' - ff_1)\, d\sigma\, d^3p_1. \tag{3.9}$$

The function w, and therefore the cross-section $d\sigma$ defined by (2.2), contain delta-function factors which express the conservation laws for momentum and energy, as a result of which the variables $\mathbf{p}_1$, $\mathbf{p}'$ and $\mathbf{p}_1'$ (for a given $\mathbf{p}$) are not in fact independent. However, when the collision integral is expressed in the form (3.9), we can suppose that these delta functions have been removed by appropriate integrations; then $d\sigma$ will be the ordinary scattering cross-section, depending (for a given v_{rel}) only on the scattering angle.

For a qualitative treatment of transport phenomena in gases, the collision integral is roughly estimated by means of the *mean free path* l, an average distance traversed by a molecule between two successive collisions.† It has, of course, only qualitative significance; even its definition varies according to which transport phenomenon is under consideration.

†This concept is due to R. Clausius (1858).

The mean free path can be expressed in terms of the collision cross-section σ and the number density N of molecules in the gas. If a molecule travels a unit distance in its path, it collides with the molecules present in a volume σ (that of a cylinder with cross-sectional area σ and unit length), the number of which is σN. Hence

$$l \sim 1/N\sigma. \tag{3.10}$$

The collision cross-section $\sigma \sim d^2$, where d is the dimension of the molecule. With $N \sim 1/\bar{r}^3$, $\bar{r}$ being the mean distance between molecules, we find

$$l \sim \bar{r}(\bar{r}/d)^2 = d(\bar{r}/d)^3. \tag{3.11}$$

Since in a gas $\bar{r} \gg d$, the mean free path $l \gg \bar{r}$.

The ratio $\tau \sim l/\bar{v}$ is called the *mean free time*. For a rough estimate of the collision integral, we can put

$$C(f) \sim -(f - f_0)/\tau \sim -(\bar{v}/l)(f - f_0). \tag{3.12}$$

By writing the difference $f - f_0$ in the numerator we have taken account of the fact that the collision integral is zero for the equilibrium distribution function. The minus sign in (3.12) expresses the fact that collisions are the mechanism for reaching statistical equilibrium, i.e. they tend to reduce the deviation of the distribution function from its equilibrium form. In this sense, τ acts as a relaxation time for the establishment of equilibrium in each volume element of the gas.

§ 4. The *H* theorem

A gas left to itself, like any closed macroscopic system, will tend to reach a state of equilibrium. Accordingly, the time variation of a non-equilibrium distribution function in accordance with the transport equation must be accompanied by an increase in the entropy of the gas. We shall show that this is in fact so.

The entropy of an ideal gas in a non-equilibrium macroscopic state described by a distribution function f is

$$S = \int f \log(e/f)\, dV\, d\Gamma; \tag{4.1}$$

see *SP* 1, § 40. Differentiating this expression with respect to time, we have

$$\begin{aligned} \frac{dS}{dt} &= \int \frac{\partial}{\partial t}\left(f \log\frac{e}{f}\right) dV\, d\Gamma \\ &= -\int \log f \,\frac{\partial f}{\partial t}\, dV\, d\Gamma. \end{aligned} \tag{4.2}$$

Since the establishment of statistical equilibrium in the gas is brought about by collisions of molecules, the increase in the entropy must arise from the collisional part of the change in the distribution function. The change in this function due to the free motion of the molecules, on the other hand, cannot alter the entropy of the gas, since this part of the change in the distribution function is given (for a gas in an external field $U(\mathbf{r})$) by the first two terms on the right-hand side of the equation

$$\partial f/\partial t = -\mathbf{v}\,.\,\nabla f - \mathbf{F}\,.\,\partial f/\partial \mathbf{p} + C(f).$$

Their contribution to the derivative dS/dt is

$$-\int \log f[-\mathbf{v}\,.\,\partial f/\partial \mathbf{r} - \mathbf{F}\,.\,\partial f/\partial \mathbf{p}]\,dV\,d\Gamma = \int [\mathbf{v}\,.\,\partial/\partial \mathbf{r} + \mathbf{F}\,.\,\partial/\partial \mathbf{p}](f \log f/e)\,dV\,d\Gamma.$$

The integral over dV of the term involving the derivative $\partial/\partial \mathbf{r}$ is transformed by Gauss's theorem into a surface integral; it gives zero on integration through the whole volume of the gas, since $f = 0$ outside the region occupied by the gas. Similarly, the term involving the derivative $\partial/\partial \mathbf{p}$, on integration over d^3p, becomes an integral over an infinitely distant surface in momentum space, and likewise gives zero.

The change in the entropy is therefore expressed by

$$dS/dt = -\int \log f\,.\,C(f)\,d\Gamma\,dV. \tag{4.3}$$

This integral can be transformed by a device which, with a view to later applications, we shall formulate for the general integral $\int \varphi(\Gamma)C(f)\,d\Gamma$, where $\varphi(\Gamma)$ is any function of the quantities Γ. With the collision integral in the form (3.6), we write

$$\int \varphi(\Gamma)C(f)\,d\Gamma = \int \varphi w(\Gamma, \Gamma_1; \Gamma', \Gamma_1')f'f_1'\,d^4\Gamma - \int \varphi w(\Gamma', \Gamma_1'; \Gamma, \Gamma_1)ff_1\,d^4\Gamma,$$

where for brevity $d^4\Gamma = d\Gamma\,d\Gamma_1\,d\Gamma'\,d\Gamma_1'$. Since the integration here is over all the variables $\Gamma, \Gamma_1, \Gamma', \Gamma_1'$, we can, without altering the integral, rename the variables in any manner. Interchanging Γ, Γ_1 and Γ', Γ_1' in the second integral, we find

$$\int \varphi(\Gamma)C(f)d\Gamma = \int (\varphi - \varphi')w(\Gamma, \Gamma_1; \Gamma', \Gamma_1')f'f_1'd^4\Gamma.$$

Interchanging here Γ, Γ' and Γ_1, Γ_1', taking half the sum of the resulting integrals, and noting the obvious symmetry of w with respect to the two colliding particles, we obtain the transformation rule

$$\int \varphi(\Gamma)C(f)\,d\Gamma = \frac{1}{2}\int (\varphi + \varphi_1 - \varphi' - \varphi_1')w'f'f_1'\,d^4\Gamma. \tag{4.4}$$

In particular, $\int C(f)\,d\Gamma = 0$: with $C(f)$ here in the form (3.7), we have

$$\int C(f)\,d\Gamma = \int w'(f'f_1' - ff_1)\,d^4\Gamma = 0. \tag{4.5}$$

Applied to the integral (4.3), the rule (4.4) gives

$$dS/dt = \frac{1}{2}\int w'f'f'_1 \log (f'f'_1/ff_1)\, d^4\Gamma\, dV$$

$$= \frac{1}{2}\int w'ff_1 x \log x\, d^4\Gamma\, dV,$$

where $x = f'f'_1/ff_1$. Subtracting from this equation half of the zero integral (4.5), we convert it to

$$dS/dt = \frac{1}{2}\int w'ff_1(x \log x - x + 1)\, d^4\Gamma\, dV. \tag{4.6}$$

The function in the parentheses in the integrand is non-negative for all $x > 0$; it is zero when $x = 1$, increasing on either side of that point. By definition, the factors w', f and f_1 in the integrand are also positive. We thus obtain the required result,

$$dS/dt \geq 0, \tag{4.7}$$

expressing the law of increase of entropy; the equality occurs at equilibrium.†

Note that, since the integrand in (4.6) (and therefore in (4.3)) is non-negative, not only the whole integral (4.3) over $d\Gamma\, dV$ but also that over $d\Gamma$ alone is positive. Thus collisions increase the entropy in each volume element of the gas. This does not, of course, imply that the entropy itself increases in every volume element, since it can be transferred from one region to another by the free motion of the molecules.

§ 5. The change to macroscopic equations

The Boltzmann transport equation gives a microscopic description of the way in which the state of the gas varies with time. We shall show how the transport equation can be converted into the usual equations of fluid mechanics, which give a less detailed, macroscopic description of this time variation. The description is valid when the macroscopic properties (temperature, density, velocity, etc.) of the gas vary sufficiently slowly through its volume: the distances L over which they change appreciably must be much greater than the mean free path l of the molecules.

It has already been mentioned that the integral

$$N(t, \mathbf{r}) = \int f(t, \mathbf{r}, \Gamma)\, d\Gamma \tag{5.1}$$

is the spatial distribution density of gas molecules; the product $\rho = mN$ is correspondingly the mass density of the gas. The macroscopic velocity of the gas is

†The proof of the law of increase of entropy by means of the transport equation is due to Boltzmann, and was the first microscopic proof of that law. As applied to gases, the law is often called the *H theorem*, since Boltzmann used the symbol H for the entropy.

denoted by $\mathbf{V}$ (in contrast to the microscopic velocities $\mathbf{v}$ of the molecules); it is defined as the mean

$$\mathbf{V} = \bar{\mathbf{v}} = \frac{1}{N}\int \mathbf{v} f \, d\Gamma. \tag{5.2}$$

Collisions do not alter either the number of colliding particles or their total energy and momentum. It is therefore clear that the collisional part of the change in the distribution function also cannot affect the macroscopic quantities in each volume element of the gas—its density, internal energy, and macroscopic velocity $\mathbf{V}$: the collisional parts of the change in the total number, energy and momentum of the molecules in unit volume of the gas are given by the zero integrals

$$\int C(f)\, d\Gamma = 0, \quad \int \epsilon C(f)\, d\Gamma = 0, \quad \int \mathbf{p} C(f)\, d\Gamma = 0. \tag{5.3}$$

These equations are easily derived by applying to the integrals the transformation (4.4) with $\varphi = 1$, ϵ and $\mathbf{p}$ respectively; the first integral is zero identically, the other two are zero by virtue of the conservation of energy and momentum in collisions.

Let us now take the transport equation

$$\frac{\partial f}{\partial t} + \frac{\partial}{\partial x_\alpha}(v_\alpha f) = C(f) \tag{5.4}$$

and integrate over $d\Gamma$ after first multiplying by m, p_β or ϵ. In every case, the right-hand side is zero, and we have the equations

$$\partial \rho / \partial t + \operatorname{div} \rho \mathbf{V} = 0, \tag{5.5}$$

$$\partial (\rho V_\alpha)/\partial t + \partial \Pi_{\alpha\beta}/\partial x_\beta = 0, \tag{5.6}$$

$$\partial (N\bar{\epsilon})/\partial t + \operatorname{div} \mathbf{q} = 0. \tag{5.7}$$

The first of these is the usual continuity equation of fluid mechanics, expressing the conservation of mass of the gas. The second equation expresses the conservation of momentum; the tensor $\Pi_{\alpha\beta}$ is defined as

$$\Pi_{\alpha\beta} = \int m v_\alpha v_\beta f \, d\Gamma \tag{5.8}$$

and is the momentum flux tensor; its component $\Pi_{\alpha\beta}$ is the α-component of the momentum transferred in unit time by molecules across unit area perpendicular to the x_β-axis. Lastly, (5.7) is the equation of conservation of energy; the vector $\mathbf{q}$ is defined as

$$\mathbf{q} = \int \epsilon \mathbf{v} f \, d\Gamma, \tag{5.9}$$

and is the energy flux in the gas.

To reduce (5.6) and (5.7) to the usual equations of fluid mechanics, however, we have still to express $\Pi_{\alpha\beta}$ and $\mathbf{q}$ in terms of macroscopic quantities. It has already been mentioned that the macroscopic description of the gas presupposes sufficiently small gradients of its macroscopic properties. We can then suppose, as a first approximation, that in each separate region of the gas thermal equilibrium is reached, whereas the gas as a whole is not in equilibrium. Thus the distribution function f in each volume element is assumed to be a local equilibrium function, equal to the equilibrium function f_0 for the density, temperature and macroscopic velocity that prevail in that volume element. This approximation implies the neglect of all dissipative processes (viscosity and thermal conduction) in the gas. Equations (5.6) and (5.7) then naturally reduce to those for an ideal fluid; this may be proved as follows.

The equilibrium distribution in a region of the gas moving as a whole with velocity $\mathbf{V}$ differs from that in a gas at rest only by a Galilean transformation; on changing to a frame of reference K' that moves with the gas, we obtain the ordinary Boltzmann distribution. The velocities $\mathbf{v}'$ of the molecules in this frame are related to those in the original frame K by $\mathbf{v} = \mathbf{v}' + \mathbf{V}$. We write

$$\begin{aligned}\Pi_{\alpha\beta} &= mN\langle v_\alpha v_\beta\rangle \\ &= mN\langle (V_\alpha + v'_\alpha)(V_\beta + v'_\beta)\rangle \\ &= mN(V_\alpha V_\beta + \langle v'_\alpha v'_\beta\rangle);\end{aligned}$$

the terms $V_\alpha v'_\beta$ and $V_\beta v'_\alpha$ give zero on averaging over the directions of $\mathbf{v}'$, since all directions of the velocity of a molecule in the frame K' are equally probable. For the same reason,

$$\langle v'_\alpha v'_\beta\rangle = \tfrac{1}{3}\langle v'^2\rangle \delta_{\alpha\beta}; \tag{5.10}$$

the mean square of the thermal velocity is $\langle v'^2\rangle = 3T/m$, where T is the temperature of the gas. Finally, since NT is equal to the gas pressure P, we find

$$\Pi_{\alpha\beta} = \rho V_\alpha V_\beta + \delta_{\alpha\beta} P, \tag{5.11}$$

the familiar expression for the momentum flux tensor in an ideal fluid; with this tensor, equation (5.6) is equivalent to Euler's equation in fluid mechanics (*FM*, §7).

In order to transform the integral (5.9), we note that the energy ϵ of a molecule in the frame K is related to its energy ϵ' in the frame K' by

$$\epsilon = \epsilon' + m\mathbf{V}\,.\,\mathbf{v}' + \tfrac{1}{2}mV^2.$$

Substituting this and $\mathbf{v} = \mathbf{v}' + \mathbf{V}$ in $\mathbf{q} = N\overline{\epsilon\mathbf{v}}$, we have

$$\begin{aligned}\mathbf{q} &= N\mathbf{V}[\tfrac{1}{2}mV^2 + \tfrac{1}{3}m\overline{v'^2} + \overline{\epsilon'}] \\ &= \mathbf{V}(\tfrac{1}{2}\rho V^2 + P + N\overline{\epsilon'}),\end{aligned}$$

using (5.10) in averaging the product $\mathbf{v}'(\mathbf{V}\,.\,\mathbf{v}')$. But $N\overline{\epsilon'}$ is the thermodynamic internal energy of the gas per unit volume; the sum $N\overline{\epsilon'}+P$ is the heat function W of the gas per unit volume. Thus

$$\mathbf{q} = \mathbf{V}(\tfrac{1}{2}\rho V^2 + W), \tag{5.12}$$

in agreement with the known expression for the energy flux in the dynamics of an ideal fluid (*FM*, §6).

Lastly, let us consider the law of conservation of angular momentum in the transport equation. This law should apply exactly only to the total angular momentum of the gas, made up of the orbital angular momentum of the molecules in their translational motion and their intrinsic rotational angular momenta **M**; the total angular momentum density is given by the sum

$$\int \mathbf{r}\times\mathbf{p}\, f\, d\Gamma + \int \mathbf{M}\, f\, d\Gamma. \tag{5.13}$$

These two terms, however, have different orders of magnitude. The orbital angular momentum of the relative motion of two molecules at a mean distance $\bar{r}$ apart is of the order of $m\bar{v}\bar{r}$, but the intrinsic angular momentum $M \sim m\bar{v}d$, which is small in comparison, since we always have $d \ll \bar{r}$.

Naturally, therefore, the Boltzmann transport equation, which corresponds to the first non-vanishing approximation with respect to the small quantity $d/\bar{r}$, cannot take account of the small changes in the orbital angular momentum due to the exchange between the two parts of the total angular momentum (5.13). This has the result that the Boltzmann equation conserves the total orbital angular momentum of the gas: the equation $\int \mathbf{p}\, C(f)\, d\Gamma = 0$ which expresses the conservation of momentum necessarily implies that

$$\int \mathbf{r}\times\mathbf{p}\, C(f)\, d\Gamma = \mathbf{r}\times\int \mathbf{p}\, C(f)\, d\Gamma = 0. \tag{5.14}$$

The reason for this property is evident: since, in the Boltzmann equation, collisions are regarded as taking place at a point, the sum of the orbital angular momenta of the colliding molecules is conserved, as well as the sum of their momenta. In order to derive an equation for the change in the orbital angular momentum, it would be necessary to take account of terms of the next higher order in $d/\bar{r}$, arising from the fact that the molecules are at a finite distance apart at the time of collision.

However, the actual process of angular momentum exchange between the translational and rotational degrees of freedom can be described in terms of the Boltzmann equation by a relation of the form

$$d\mathfrak{M}/dt = \int \mathbf{M}\, C(f)\, d\Gamma, \tag{5.15}$$

where $\mathfrak{M}$ is the intrinsic angular momentum density of the molecules. Since the sum of the intrinsic angular momenta of two molecules need not be conserved in a collision, the integral on the right of (5.15) is in general not zero, and gives the rate of change of $\mathfrak{M}$. If a non-zero angular momentum density is created in the gas by some means, its subsequent relaxation is described by (5.15).

§6. The transport equation for a slightly inhomogeneous gas

In order to take account of dissipative processes (thermal conduction and viscosity) in a slightly inhomogeneous gas, we must go to the next approximation beyond that treated in §5. Instead of regarding the distribution function in each region of the gas as just the local-equilibrium function f_0, we shall now allow for a slight deviation of f from f_0, putting

$$f = f_0 + \delta f, \quad \delta f = -(\partial f_0/\partial \epsilon)\chi(\Gamma) = f_0\chi/T, \tag{6.1}$$

where δf is a small correction ($\ll f_0$). The latter is conveniently represented in the above form, with the factor $-\partial f_0/\partial\epsilon$ separated; for the Boltzmann distribution, this derivative differs from f_0 itself only by a factor $1/T$. The correction δf must in principle be determined by solving the transport equation linearized with respect to the correction.†

The function χ must satisfy not only the transport equation itself but also certain additional conditions. The reason is that f_0 is the equilibrium distribution function corresponding to given values (in the volume element concerned) of the gas particle number, energy and momentum densities, i.e. to given values of the integrals

$$\int f_0\, d\Gamma, \quad \int \epsilon f_0\, d\Gamma, \quad \int \mathbf{p} f_0\, d\Gamma. \tag{6.2}$$

The non-equilibrium distribution function (6.1) must yield the same values of these quantities, i.e. the integrals with f and f_0 must be the same. The function χ must therefore satisfy the contitions

$$\int f_0\chi\, d\Gamma = 0, \quad \int f_0\chi\epsilon\, d\Gamma = 0, \quad \int f_0\chi\mathbf{p}\, d\Gamma = 0. \tag{6.3}$$

It must be emphasized that even the concept of the temperature in a non-equilibrium gas becomes determinate only when specific values are assigned to the integrals (6.2). The concept becomes entirely rigorous only when the gas as a whole is in complete equilibrium; to define the temperature in a non-equilibrium gas, a further condition is necessary, which may be the specification of these values.

Let us first of all transform the collision integral in the transport equation (3.8). When the functions (6.1) are substituted, the terms not containing the small correction χ cancel, since the equilibrium distribution function makes the collision integral zero. The first-order terms give

$$C(f) = f_0 I(\chi)/T, \tag{6.4}$$

†This method of solving the transport equation is due to D. Enskog (1917).

where $I(\chi)$ denotes the linear integral operator

$$I(\chi) = \int w' f_{01}(\chi' + \chi_1' - \chi - \chi_1)\, d\Gamma_1\, d\Gamma'\, d\Gamma_1'. \tag{6.5}$$

Here we have used the equation $f_0 f_{01} = f_0' f_{01}'$; the factor f_0 can be taken outside the integral, since there is no integration over $d\Gamma$.

The integral (6.5) is identically zero for the functions

$$\chi = \text{constant}, \quad \chi = \text{constant} \times \epsilon, \quad \chi = \mathbf{p} \cdot \delta\mathbf{V}, \tag{6.6}$$

where $\delta\mathbf{V}$ is a constant vector; this result for the second and third functions follows from the conservation of energy and momentum in each collision. The functions (6.6), which are independent of time and coordinates, therefore satisfy the transport equation itself.

The origin of these solutions is simple. The transport equation is identically satisfied by the equilibrium distribution function with any (constant) particle density and temperature. It is therefore necessarily satisfied also by the small correction

$$\delta f = (\partial f_0/\partial N)\delta N = f_0 \delta N/N,$$

which arises when the density changes by δN; this gives the first solution (6.6). Similarly, the equation is satisfied by the increment

$$\delta f = (\partial f_0/\partial T)\delta T,$$

which arises when T changes by a small constant amount δT. The derivative $\partial f_0/\partial T$ is made up of a term constant $\times f_0$ (arising from differentiation of the normalization factor in f_0) and a term proportional to ϵf_0; this gives the second solution (6.6). The third solution expresses Galileo's relativity principle: the equilibrium distribution function must satisfy the transport equation in any other inertial frame. When we change to a frame moving relative to the original one with a small constant velocity $\delta\mathbf{V}$, the velocities $\mathbf{v}$ of the molecules become $\mathbf{v} + \delta\mathbf{V}$, and the distribution function therefore receives the increment

$$\delta F = (\partial f_0/\partial \mathbf{v}) \cdot \delta\mathbf{V} = -(f_0/T)\mathbf{p} \cdot \delta\mathbf{V},$$

corresponding to the third solution (6.6). The "extra" solutions (6.6) are excluded by applying the three conditions (6.3).

We shall transform the left-hand side of the transport equation in a general manner, which covers both thermal conduction and viscosity. That is, we allow the presence of gradients of all macroscopic properties of the gas, including the macroscopic velocity $\mathbf{V}$.

The equilibrium distribution function in a gas at rest ($\mathbf{V} = 0$) is the Boltzmann distribution, which we write as

$$f_0 = \exp\left(\frac{\mu - \epsilon(\Gamma)}{T}\right), \tag{6.7}$$

where μ is the chemical potential of the gas. The distribution in a moving gas differs from (6.7) only by a Galilean transformation of the velocity, as already noted in §5. In order to write this function explicitly, we separate from the total energy $\epsilon(\Gamma)$ of the molecule the kinetic energy of its translational motion:

$$\epsilon(\Gamma) = \tfrac{1}{2}mv^2 + \epsilon_{\text{int}}; \tag{6.8}$$

the internal energy ϵ_{int} includes the energy of rotation of the molecule and the vibrational energy. Replacing $\mathbf{v}$ by $\mathbf{v}-\mathbf{V}$, we find the Boltzmann distribution in a moving gas:

$$f_0 = \exp\left(\frac{\mu - \epsilon_{\text{int}}}{T}\right) \exp\left(\frac{m(\mathbf{v}-\mathbf{V})^2}{2T}\right). \tag{6.9}$$

In a slightly inhomogeneous gas, f_0 depends on the coordinates and the time, as a result of the variation through the gas (and in the course of time) of its macroscopic properties: the velocity $\mathbf{V}$, the temperature T and the pressure P (and therefore μ). Since the gradients of these quantities are assumed small, it is sufficient (in this approximation) to replace f by f_0 on the left of the transport equation.

The calculations can be somewhat simplified by noting the obvious fact that the kinetic coefficients, our real subject of interest, do not depend on the velocity $\mathbf{V}$. It is therefore sufficient to consider any one point in the gas, and to choose the point where $\mathbf{V}$ (but not, of course, its derivatives) is zero.

Differentiating the expression (6.9) with respect to time and then putting $\mathbf{V} = 0$, we obtain

$$\frac{T}{f_0}\frac{\partial f_0}{\partial t} = \left[\left(\frac{\partial \mu}{\partial T}\right)_P - \frac{\mu - \epsilon(\Gamma)}{T}\right]\frac{\partial T}{\partial t} + \left(\frac{\partial \mu}{\partial P}\right)_T \frac{\partial P}{\partial t} + m\mathbf{v}\cdot\frac{\partial \mathbf{V}}{\partial t}.$$

By the familiar formulae of thermodynamics,

$$(\partial\mu/\partial T)_P = -s, \quad (\partial\mu/\partial P)_T = 1/N, \quad \mu = w - Ts,$$

where w, s and $1/N$ are the heat function, entropy and volume per gas particle. Hence

$$\frac{T}{f_0}\frac{\partial f_0}{\partial t} = \frac{\epsilon(\Gamma) - w}{T}\frac{\partial T}{\partial t} + \frac{1}{N}\frac{\partial P}{\partial t} + m\mathbf{v}\cdot\frac{\partial \mathbf{V}}{\partial t}. \tag{6.10}$$

Similarly,

$$\frac{T}{f_0}\mathbf{v}\cdot\nabla f_0 = \frac{\epsilon(\Gamma) - w}{T}\mathbf{v}\cdot\nabla T + \frac{1}{N}\mathbf{v}\cdot\nabla P + mv_\alpha v_\beta V_{\alpha\beta}, \tag{6.11}$$

where for brevity

$$V_{\alpha\beta} = \frac{1}{2}\left(\frac{\partial V_\alpha}{\partial x_\beta} + \frac{\partial V_\beta}{\partial x_\alpha}\right), \quad V_{\alpha\alpha} = \text{div}\,\mathbf{V}; \tag{6.12}$$

in the last term in (6.11), we have made the identical substitution

$$v_\alpha v_\beta \,\partial V_\beta/\partial x_\alpha = v_\alpha v_\beta V_{\alpha\beta}.$$

The left-hand side of the transport equation is found by adding the expressions (6.10) and (6.11). All derivatives of macroscopic quantities with respect to time can be expressed in terms of their spatial gradients by means of the equations of an ideal (non-viscous and thermally non-conducting) medium; the inclusion of dissipative terms here would lead to quantities of a higher order of smallness. At the point where $\mathbf{V} = 0$, Euler's equation gives

$$\partial \mathbf{V}/\partial t = -(1/\rho)\nabla P = -(1/Nm)\nabla P. \tag{6.13}$$

At this same point, the equation of continuity gives $\partial N/\partial t = -N \operatorname{div} \mathbf{V}$, or

$$\frac{1}{N}\frac{\partial N}{\partial t} = \frac{1}{P}\frac{\partial P}{\partial t} - \frac{1}{T}\frac{\partial T}{\partial t} = -\operatorname{div} \mathbf{V}, \tag{6.14}$$

with the equation of state for an ideal gas, $N = P/T$. Lastly, the equation of conservation of entropy, $\partial s/\partial t + \mathbf{V} . \nabla s = 0$, gives $\partial s/\partial t = 0$, or

$$\frac{c_p}{T}\frac{\partial T}{\partial t} - \frac{1}{P}\frac{\partial P}{\partial t} = 0, \tag{6.15}$$

with the use of the thermodynamic formulae

$$(\partial s/\partial T)_P = c_p/T, \quad (\partial s/\partial P)_T = -1/P,$$

c_p being the specific heat, again per molecule; the second of these formulae relates to an ideal gas. Equations (6.14) and (6.15) give

$$\frac{1}{T}\frac{\partial T}{\partial t} = -\frac{1}{c_v}\operatorname{div} \mathbf{V}, \quad \frac{1}{P}\frac{\partial P}{\partial t} = -\frac{c_p}{c_v}\operatorname{div} \mathbf{V} \tag{6.16}$$

since for an ideal gas $c_p - c_v = 1$.

A straightforward calculation leads to the result

$$\frac{\partial f_0}{\partial t} + \mathbf{v} . \nabla f_0 = \frac{f_0}{T}\left\{\frac{\epsilon(\Gamma) - w}{T}\mathbf{v} . \nabla T + m v_\alpha v_\beta V_{\alpha\beta} + \frac{w - Tc_p - \epsilon(\Gamma)}{c_v}\operatorname{div} \mathbf{V}\right\}. \tag{6.17}$$

It must be emphasized that no specific assumption has so far been made about the temperature dependence of the thermodynamic quantities; only the general equation of state of an ideal gas has been used. For a gas with a classical rotation of molecules, and vibrations not excited, the specific heat is independent of temperature, and the heat function is†

$$w = c_p T. \tag{6.18}$$

†The energy $\epsilon(\Gamma)$ of the molecule is assumed to be measured from its lowest value; accordingly the temperature-independent additive constant in w is omitted.

The last term in (6.17) can then be simplified; equating (6.17) and (6.4), we write the transport equation in the final form

$$\frac{\epsilon(\Gamma)-c_pT}{T}\mathbf{v}\,.\,\nabla t+\left[mv_\alpha v_\beta-\delta_{\alpha\beta}\frac{\epsilon(\Gamma)}{c_v}\right]V_{\alpha\beta}=I(\chi). \tag{6.19}$$

In §§7 and 8, this equation will be further studied with reference to thermal conduction and viscosity.

From the law of increase of entropy, it follows that a pressure gradient (in the absence of temperature and velocity gradients) does not bring about dissipative processes; cf. *FM*, §49. In the transport equation, this condition is necessarily satisfied, as is shown by the absence of the pressure gradient on the left of (6.19).

§7. Thermal conduction in the gas

To calculate the thermal conductivity of the gas, we have to solve the transport equation with a temperature gradient. Retaining only the first term on the left of (6.19), we have

$$\frac{\epsilon(\Gamma)-c_pT}{T}\mathbf{v}\,.\,\nabla T=I(\chi). \tag{7.1}$$

The solution is to be sought in the form

$$\chi=\mathbf{g}\,.\,\nabla T, \tag{7.2}$$

where the vector $\mathbf{g}$ depends only on the quantities Γ, since a factor ∇T results on both sides of (7.1) when this substitution is made. Since the equation must be valid for any vector ∇T, the coefficients of this on the two sides must be equal, and so we obtain for $\mathbf{g}$ the equation

$$\mathbf{v}\frac{\epsilon(\Gamma)-c_pT}{T}=I(\mathbf{g}), \tag{7.3}$$

which does not involve ∇T (nor therefore any explicit dependence on the coordinates).

The function χ must also satisfy the conditions (6.3). With χ in the form (7.2), the first two of these are necessarily satisfied, as is evident from the fact that (7.3) contains no vector parameters which might give the direction of the constant vector integrals $\int f_0\mathbf{g}\,d\Gamma$ and $\int f_0\epsilon\mathbf{g}\,d\Gamma$. The third condition imposes on the solution of (7.3) the further condition

$$\int f_0\mathbf{v}\,.\,\mathbf{g}\,d\Gamma=0. \tag{7.4}$$

If the transport equation has been solved and the function χ is known, the thermal conductivity can be determined by calculating the energy flux, or rather its dissipative part that is not due simply to convective energy transfer, which we shall denote by $\mathbf{q}'$. In the absence of macroscopic motion in the gas, $\mathbf{q}'$ is equal to the total energy flux $\mathbf{q}$ given by the integral (5.9). When $f = f_0$, this integral is zero identically, because of the integration over the directions of $\mathbf{v}$. On substitution of f from (6.1), there thus remains

$$\mathbf{q} = \frac{1}{T}\int \mathbf{v} f_0 \chi \epsilon \, d\Gamma$$

$$= \frac{1}{T}\int f_0 \epsilon \mathbf{v}(\mathbf{g} \cdot \nabla T)\, d\Gamma,$$

or in components

$$q_\alpha = -\kappa_{\alpha\beta}\, \partial T/\partial x_\beta, \quad \kappa_{\alpha\beta} = -\frac{1}{T}\int f_0 \epsilon v_\alpha g_\beta \, d\Gamma. \tag{7.5}$$

Since a gas in equilibrium is isotropic, there are no preferred directions in it, and the tensor $\kappa_{\alpha\beta}$ can only be expressible in terms of the unit tensor $\delta_{\alpha\beta}$, i.e. it reduces to a scalar:

$$\kappa_{\alpha\beta} = \kappa \delta_{\alpha\beta}, \quad \kappa = \tfrac{1}{3}\kappa_{\alpha\alpha}.$$

Thus the energy flux is

$$\mathbf{q} = -\kappa \nabla T, \tag{7.6}$$

where the scalar *thermal conductivity* is

$$\kappa = -\frac{1}{3T}\int f_0 \epsilon \mathbf{v} \cdot \mathbf{g} \, d\Gamma. \tag{7.7}$$

The transport equation necessarily makes this quantity positive (see §9): the flux $\mathbf{q}$ must be in the opposite direction to the temperature gradient.

In monatomic gases, the velocity $\mathbf{v}$ is the only vector on which the function $\mathbf{g}$ depends; it is therefore clear that this function must have the form

$$\mathbf{g} = (\mathbf{v}/v) g(v). \tag{7.8}$$

In polyatomic gases, $\mathbf{g}$ depends on two vectors: the velocity $\mathbf{v}$ and the angular momentum $\mathbf{M}$. If the symmetry of the molecules does not allow stereoisomerism, the collision integral, and therefore equation (7.3), are invariant under inversion; the solution χ must be similarly invariant. In other words, $\chi = \mathbf{g} \cdot \nabla T$ must be a true scalar, and, since the gradient ∇T is a true vector, so must be the function $\mathbf{g}$. For instance, in a diatomic gas, where the quantities Γ are just the vectors $\mathbf{v}$ and $\mathbf{M}$, the function $\mathbf{g}(\Gamma)$ has the form

$$\mathbf{g} = \mathbf{v} g_1 + \mathbf{M}(\mathbf{v} \cdot \mathbf{M}) g_2 + (\mathbf{v} \times \mathbf{M}) g_3, \tag{7.9}$$

where g_1, g_2, g_3 are scalar functions of the scalar arguments $\mathbf{v}^2$, $\mathbf{M}^2$, $(\mathbf{v}.\mathbf{M})^2$; this is the most general form of a true vector that can be constructed from the true vector $\mathbf{v}$ and the pseudovector $\mathbf{M}$.†

If, however, the substance is stereoisomeric, there is no invariance under inversion: as already mentioned in §2, inversion then "transforms" the gas into what is essentially a different substance. Accordingly, the function χ may also contain pseudoscalar terms, and the function $\mathbf{g}$ may contain pseudovector terms, e.g. one of the form $g_4\mathbf{M}$.

The condition for the above method of solving the transport equation (based on the assumption that f is close to f_0) to be valid can be ascertained by estimating the collision integral from (3.12). The mean energy of a molecule is $\bar{\epsilon} \sim T$, and so an estimate of the two sides of (7.3) gives $\bar{v} \sim g/\tau \sim g\bar{v}/l$, whence $g \sim l$. The condition $\chi/T \sim g|\nabla T|/T \ll 1$ (equivalent to $\delta f \ll f_0$) therefore signifies that the distances L over which the temperature undergoes a considerable change ($|\nabla T| \sim T/L$) must be large in comparison with l. That is, a function having the form (6.1) constitutes the leading terms in an expansion of the solution of the transport equation in powers of the small ratio l/L.

An estimate of (7.7) with $g \sim l$ gives

$$\kappa \sim cNl\bar{v}, \tag{7.10}$$

where c is the specific heat per molecule of the gas. This is a well-known elementary formula in the kinetic theory of gases (cf. the last footnote to §11). Putting $l \sim 1/N\sigma$, $c \sim 1$ and $\bar{v} \sim \sqrt{(T/m)}$, we have

$$\kappa \sim (1/\sigma)\sqrt{(T/m)}. \tag{7.11}$$

In this estimate, the cross-section σ relates to the mean thermal speed of the molecules, and in that sense is to be regarded as a function of temperature. As the speed increases, the cross-section in general decreases; accordingly, σ is a decreasing function of the temperature. When the temperature is not too low, the gas molecules behave qualitatively as hard elastic particles which interact only when they actually collide. This type of interaction corresponds to a collision cross-section varying only slightly with the speed (and therefore with the temperature). Under such conditions, κ is approximately proportional to $\sqrt{T}$.

At a given temperature, the thermal conductivity is seen from (7.11) to be independent of the gas density, i.e. of the gas pressure. It must be emphasized that this important property is not related to the assumptions used in making the estimate, but is an exact consequence of the Boltzmann transport equation. It arises because this equation takes account only of collisions between pairs of molecules (for which reason the mean free path is inversely proportional to the gas density).

†The solution of the Boltzmann equation for a gas of rotating molecules was first discussed by Yu. M. Kagan and A. M. Afanas'ev (1961).

§8. Viscosity in the gas

The viscosity of a gas is calculated by means of the transport equation in the same way as the thermal conductivity. The only difference is that the deviation from equilibrium is due not to the temperature gradient but to the non-uniformity of the gas flow as regards the macroscopic velocity $\mathbf{V}$. It is again assumed that the characteristic dimensions of the problem $L \gg l$.

There are, as we know, two kinds of viscosity, the corresponding coefficients being usually denoted by η and ζ. They are defined as the coefficients in the viscous stress tensor $\sigma'_{\alpha\beta}$ which forms part of the momentum flux tensor:

$$\Pi_{\alpha\beta} = P\delta_{\alpha\beta} + \rho V_\alpha V_\beta - \sigma'_{\alpha\beta}, \tag{8.1}$$

$$\sigma'_{\alpha\beta} = 2\eta(V_{\alpha\beta} - \tfrac{1}{3}\delta_{\alpha\beta}\,\mathrm{div}\,\mathbf{V}) + \zeta\delta_{\alpha\beta}\,\mathrm{div}\,\mathbf{V}, \tag{8.2}$$

where $V_{\alpha\beta}$ is defined by (6.12); see *FM*, §15. In an incompressible fluid, only the viscosity η occurs. The "second viscosity" ζ appears in motion such that $\mathrm{div}\,\mathbf{V} \neq 0$. It is convenient to calculate the two coefficients separately.

Omitting the temperature-gradient term from the general transport equation (6.19), we can write

$$mv_\alpha v_\beta(V_{\alpha\beta} - \tfrac{1}{3}\delta_{\alpha\beta}\,\mathrm{div}\,\mathbf{V}) + [\tfrac{1}{3}mv^2 - \epsilon(\Gamma)/c_v]\,\mathrm{div}\,\mathbf{V} = I(\chi), \tag{8.3}$$

where the terms containing the first and second viscosities have been separated on the left-hand side. In calculating the first viscosity, we have to assume that $\mathrm{div}\,\mathbf{V} = 0$. The resulting equation can be identically rewritten as

$$m(v_\alpha v_\beta - \tfrac{1}{3}\delta_{\alpha\beta}v^2)V_{\alpha\beta} = I(\chi), \tag{8.4}$$

where the two tensor factors on the left have zero trace.

The solution of this equation is sought in the form

$$\chi = g_{\alpha\beta}V_{\alpha\beta}, \tag{8.5}$$

where $g_{\alpha\beta}(\Gamma)$ is a symmetric tensor; since the trace $V_{\alpha\alpha} = 0$, by adding a term in $\delta_{\alpha\beta}$ to $g_{\alpha\beta}$ we can always ensure that $g_{\alpha\alpha} = 0$, without altering χ. The equation for $g_{\alpha\beta}$ is

$$m(v_\alpha v_\beta - \tfrac{1}{3}\delta_{\alpha\beta}v^2) = I(g_{\alpha\beta}). \tag{8.6}$$

The extra conditions (6.3) are necessarily satisfied.

The momentum flux is calculated from the distribution function as the integral (5.8). The required part of this, namely the viscous stress tensor, is

$$\sigma'_{\alpha\beta} = -(m/T)\int v_\alpha v_\beta f_0\chi\,d\Gamma = \eta_{\alpha\beta\gamma\delta}V_{\gamma\delta}, \tag{8.7}$$

$$\eta_{\alpha\beta\gamma\delta} = -(m/T)\int f_0 v_\alpha v_\beta g_{\gamma\delta}\,d\Gamma. \tag{8.8}$$

The quantities $\eta_{\alpha\beta\gamma\delta}$ form a tensor of rank four, symmetric in the suffixes α, β and γ, δ and giving zero on contraction with respect to the pair γ, δ. Because the gas is isotropic, this tensor can only be expressed in terms of the unit tensor $\delta_{\alpha\beta}$. An expression satisfying these conditions is

$$\eta_{\alpha\beta\gamma\delta} = \eta[\delta_{\alpha\gamma}\delta_{\beta\delta} + \delta_{\alpha\delta}\delta_{\beta\gamma} - \tfrac{2}{3}\delta_{\alpha\beta}\delta_{\gamma\delta}].$$

Then $\sigma'_{\alpha\beta} = 2\eta V_{\alpha\beta}$, so that η is the required scalar viscosity coefficient. It is determined by contracting the tensor with respect to the pairs of suffixes α, γ and β, δ:

$$\eta = -(m/10T)\int v_\alpha v_\beta g_{\alpha\beta} f_0 \, d\Gamma. \tag{8.9}$$

In a monatomic gas, $g_{\alpha\beta}$ is a function only of the vector $\mathbf{v}$. The general form of such a symmetric tensor with zero trace is

$$g_{\alpha\beta} = (v_\alpha v_\beta - \tfrac{1}{3}\delta_{\alpha\beta}v^2)g(v), \tag{8.10}$$

with a single scalar function $g(v)$. In polyatomic gases, the tensor $g_{\alpha\beta}$ is composed of a large number of variables, including the two vectors $\mathbf{v}$ and $\mathbf{M}$. In the absence of stereoisomerism, $g_{\alpha\beta}$ can include only true tensor terms; in a stereoisomeric gas, pseudotensor terms also are possible.

An estimate of the viscosity coefficient, similar to (7.10) for the thermal conductivity, gives a well-known elementary formula in the kinetic theory of gases,

$$\eta \sim m\bar{v}Nl; \tag{8.11}$$

see the last footnote to §11. The thermometric conductivity and the kinematic viscosity are found to be of the same order:

$$\kappa/Nc_p \sim \eta/Nm \sim \bar{v}l. \tag{8.12}$$

Putting in (8.11) $l \sim 1/N\sigma$ and $\bar{v} \sim (T/m)^{1/2}$, we obtain

$$\eta \sim \sqrt{(mT)}/\sigma. \tag{8.13}$$

The description of the pressure and temperature dependence of κ in §7 is entirely valid for the viscosity η also.

In order to calculate the second viscosity coefficient, we must take the second term on the left of the transport equation (8.3) to be non-zero:

$$[\tfrac{1}{3}mv^2 - \epsilon(\Gamma)/c_v] \operatorname{div} \mathbf{V} = I(\chi). \tag{8.14}$$

We shall seek the solution in the form

$$\chi = g \operatorname{div} \mathbf{V} \tag{8.15}$$

and obtain for g the equation

$$\tfrac{1}{3}mv^2 - \epsilon(\Gamma)/c_v = I(g). \tag{8.16}$$

Calculation of the stress tensor and comparison with the expression $\zeta\delta_{\alpha\beta}\,\mathrm{div}\,\mathbf{V}$ gives the viscosity coefficient as

$$\zeta = -(m/3T)\int v^2 g f_0\, d\Gamma. \tag{8.17}$$

In monatomic gases $\epsilon(\Gamma) = \frac{1}{2}mv^2$, $c_v = 3/2$, and the left-hand side of (8.16) is zero. The equation $I(g) = 0$ then shows that $g = 0$, and therefore $\zeta = 0$. We reach, therefore, the interesting conclusion that the second viscosity of monatomic gases is zero.†

PROBLEM

Show that a gas of ultra-relativistic particles has zero second viscosity (I. M. Khalatnikov, 1955).

SOLUTION. The energy ϵ of a relativistic particle in a frame of reference K in which the gas moves with a (non-relativistic) velocity $\mathbf{V}$ is related to its energy ϵ' in the frame K' in which the gas is at rest by $\epsilon' = \epsilon - \mathbf{p}.\mathbf{V}$, where $\mathbf{p}$ is the momentum of the particle in the frame K; this is the Lorentz transformation formula with the terms above the first order in $\mathbf{V}$ omitted. The distribution function in the frame K is $f_0(\epsilon - \mathbf{p}.\mathbf{V})$, where $f_0(\epsilon')$ is the Boltzmann distribution.

Considering only the viscosity, we can immediately assume that the gradients of all macroscopic quantities are zero except that of the velocity $\mathbf{V}$; then $\partial\mathbf{V}/\partial t = 0$, and the last term in (6.10) vanishes.‡ In (6.11), the first two terms are also absent, and the third becomes

$$\mathbf{v}.\nabla(\mathbf{p}.\mathbf{V}) = v_\alpha p_\beta\, \partial V_\beta/\partial x_\alpha = v_\alpha p_\beta V_{\alpha\beta};$$

the directions of $\mathbf{v}$ and $\mathbf{p}$ are the same, and so $p_\alpha v_\beta = p_\beta v_\alpha$. The equations of continuity and entropy conservation in the form used in §6 remain valid in the motion of a relativistic gas (with small velocities $\mathbf{V}$). The formulae (6.16) therefore remain valid also. The transport equation thus becomes

$$(v_\alpha p_\beta - \delta_{\alpha\beta}\epsilon/c_v)V_{\alpha\beta} = I(\chi).$$

In the second-viscosity problem, we must put $V_{\alpha\beta} = \frac{1}{3}\delta_{\alpha\beta}\,\mathrm{div}\,\mathbf{V}$, and then

$$(\tfrac{1}{3}vp - \epsilon/c_v)\,\mathrm{div}\,\mathbf{V} = I(\chi).$$

In an ultra-relativistic gas, $v \approx c$, $\epsilon = cp$, and the specific heat $c_v = 3$ (see *SP* 1, §44, Problem); the left-hand side of the equation, and therefore χ, are then zero.

§9. Symmetry of the kinetic coefficients

The thermal conductivity and the viscosity are among the quantities which govern relaxation processes in systems slightly departing from equilibrium. These *kinetic coefficients* satisfy *Onsager's symmetry principle*, which may be established

†It must be emphasized that these gases are being treated in the approximation with respect to the gaseousness parameter Nd^3 which corresponds to the Boltzmann equation (and in which η is independent of the density). In higher approximations (the subsequent terms in the "virial expansion"; see §18), a non-zero viscosity ζ does appear. Another important point is the quadratic dependence of the particle energy on the momentum; in a relativistic "monatomic" gas, the second viscosity is not zero (although it vanishes in another limiting case, the ultra-relativistic case; see the Problem).

‡To avoid misunderstandings, it may be mentioned that in a relativistic gas the pressure gradient makes a contribution to the thermal-conduction energy flux; see *FM*, §126.

in a general form without discussing specific mechanisms of relaxation. However, in a specific calculation of kinetic coefficients from the transport equations, the symmetry principle does not yield any extra conditions to be imposed on the solution of the equations. In such a calculation the requirements of the principle are necessarily satisfied. It is useful to see how this occurs.

In the general formulation of Onsager's principle (see *SP* 1, § 120), there appears a set of quantities x_a which describe the deviation of the system from equilibrium, and a set of quantities "thermodynamically conjugate" to these, $X_a = -\partial S/\partial x_a$ (where S is the entropy of the system). The relaxation process of a system slightly departing from equilibrium is described by equations which determine the rates of change of the x_a as linear functions of the X_a:

$$\dot{x}_a = -\sum_b \gamma_{ab} X_b, \tag{9.1}$$

where the γ_{ab} are the kinetic coefficients. According to Onsager's principle, if x_a and x_b behave in the same way under time reversal, then

$$\gamma_{ab} = \gamma_{ba}. \tag{9.2}$$

The rate of change of the entropy is given by the quadratic form

$$\dot{S} = -\sum_a X_a \dot{x}_a = \sum_{a,b} \gamma_{ab} X_a X_b. \tag{9.3}$$

The first of these expressions is often convenient for establishing the correspondence between the $\dot{x}_a$ and the X_a.

For thermal conductivity, we take as the rates $\dot{x}_a$ the components q'_α of the dissipative heat flux vector (at any given point in the medium); the suffix a is then the same as the vector suffix α. The corresponding quantities X_a are the derivatives $T^{-2}\partial T/\partial x_a$; cf. *SP* 2, § 88. Equations (9.1) correspond to $q'_\alpha = -\kappa_{\alpha\beta}\partial T/\partial x_\beta$, so that the kinetic coefficients γ_{ab} are the quantities $T^2\kappa_{\alpha\beta}$. According to Onsager's principle, we should have $\kappa_{\alpha\beta} = \kappa_{\beta\alpha}$.

Similarly, for the viscosity, we take as the $\dot{x}_a$ the components $\sigma'_{\alpha\beta}$ of the viscous momentum flux tensor; the corresponding X_a are $-V_{\alpha\beta}/T$ (the suffix a here answering to the pair of tensor suffixes $\alpha\beta$). Equations (9.1) correspond to $\sigma'_{\alpha\beta} = \eta_{\alpha\beta\gamma\delta}V_{\gamma\delta}$, and the kinetic coefficients are $T\eta_{\alpha\beta\gamma\delta}$. According to Onsager's principle, we must have $\eta_{\alpha\beta\gamma\delta} = \eta_{\gamma\delta\alpha\beta}$.

In the problems of thermal conduction and viscosity of gases, considered in §§7 and 8, the symmetry of the tensors $\kappa_{\alpha\beta}$ and $\eta_{\alpha\beta\gamma\delta}$ was a necessary consequence of the isotropy of the medium, independent of the solution of the transport equation. We shall show, however, that it would also follow from this solution, independently of the isotropy of the gas.

The procedure for problems of thermal conduction and viscosity in a slightly inhomogeneous gas was to seek the correction to the equilibrium distribution function in the form

$$\chi = \sum_a g_a(\Gamma) X_a, \tag{9.4}$$

obtaining for the functions g_a equations of the form

$$L_a = I(g_a). \tag{9.5}$$

The quantities L_a are components of the vector

$$T[\epsilon(\Gamma) - c_p T]v_\alpha$$

for thermal conduction, or the tensor

$$-T\left[mv_\alpha v_\beta - \frac{\epsilon(\Gamma)}{c_v}\delta_{\alpha\beta}\right]$$

for viscosity; cf. (6.19). The solutions of equations (9.5) must satisfy the further conditions

$$\int f_0 g_a \, d\Gamma = 0, \quad \int f_0 g_a \epsilon \, d\Gamma = 0, \quad \int f_0 g_a \mathbf{p} \, d\Gamma = 0.$$

With these conditions, the kinetic coefficients γ_{ab} can be written as the integrals

$$T^2 \gamma_{ab} = -\int f_0 L_a g_b \, d\Gamma. \tag{9.6}$$

The proof of the symmetry $\gamma_{ab} = \gamma_{ba}$ thus reduces to that of the equation

$$\int f_0 L_a g_b \, d\Gamma = \int f_0 L_b g_a \, d\Gamma. \tag{9.7}$$

It is based on the property that the linearized operator I is "self-conjugate", which may be arrived at as follows.

Let us consider the integral

$$\int f_0 \varphi I(\psi) \, d\Gamma = \int f_0 f_{01} w' \varphi(\psi' + \psi_1' - \psi - \psi_1) \, d^4\Gamma,$$

where $\psi(\Gamma)$ and $\varphi(\Gamma)$ are any two functions of the variables Γ. Since the integration is over all the variables $\Gamma, \Gamma_1, \Gamma', \Gamma_1'$, we can rename these in any way (as was done in §4) without affecting the value of the integral. We make the change $\Gamma, \Gamma' \leftrightarrow \Gamma_1, \Gamma_1'$, and then in each of the two resulting forms the further change $\Gamma, \Gamma_1 \leftrightarrow \Gamma', \Gamma_1'$. The sum of all four expressions gives

$$\int f_0 \varphi I(\psi) \, d\Gamma = \frac{1}{4}\int f_0 f_{01}[w'(\varphi + \varphi_1) - w(\varphi' + \varphi_1')][(\psi' + \psi_1') - (\psi + \psi_1)] \, d^4\Gamma; \tag{9.8}$$

the notation w and w' is as in (3.5). Let us now consider a similar integral in which $\psi(\Gamma)$ and $\varphi(\Gamma)$ are replaced by $\varphi(\Gamma^T)$ and $\psi(\Gamma^T)$ respectively (without changing w

and w'). With the change $\Gamma^T, \Gamma_1^T, \ldots \rightarrow \Gamma, \Gamma_1, \ldots$ in this integral, and the principle of detailed balancing (2.3), we have

$$\int f_0 \psi^T I(\varphi^T)\, d\Gamma$$

$$= \frac{1}{4} \int f_0 f_{01}[w(\psi + \psi_1) - w'(\psi' + \psi_1')][(\varphi' + \varphi_1') - (\varphi + \varphi_1)]\, d^4\Gamma, \tag{9.9}$$

where the equation $f_0(\Gamma^T) = f_0(\Gamma)$ has also been used. Expanding the square brackets in (9.8) and (9.9), and comparing corresponding terms, we see that the two integrals are equal. In making the comparison, it is necessary to take account of the unitarity relation (2.9), which gives, for example,

$$\int f_0 f_{01} w(\psi + \psi_1)(\varphi + \varphi_1)\, d^4\Gamma = \int f_0 f_{01} w'(\psi + \psi_1)(\varphi + \varphi_1)\, d^4\Gamma;$$

the relation (2.9) is applied here to the integration over the variables Γ' and Γ_1', on which only w and w' depend in the integrand.

Thus we reach the equation

$$\int f_0 \varphi I(\psi)\, d\Gamma = \int f_0 \psi^T I(\varphi^T)\, d\Gamma. \tag{9.10}$$

If the principle of detailed balancing is valid in its simple form (2.8), $w = w'$, then (9.10) reduces to a literal self-conjugacy of the operator I:

$$\int f_0 \varphi I(\psi)\, d\Gamma = \int f_0 \psi I(\varphi)\, d\Gamma, \tag{9.11}$$

where both integrals contain functions φ and ψ of the same variables Γ; this is immediately evident when $w = w'$, from the expression (9.8).

Returning to the kinetic coefficients, we make in the first integral (9.7) the change $\Gamma \rightarrow \Gamma^T$, and note that

$$L_a(\Gamma^T) = \pm L_a(\Gamma), \tag{9.12}$$

the upper and lower signs relating to viscosity and thermal conduction respectively. We now use the relations (9.5) and (9.10). In the latter, we can integrate over Γ^T in place of Γ; this clearly does not affect the value of the integral. We have

$$\begin{aligned}\int f_0 g_b L_a\, d\Gamma &= \pm \int f_0 g_b{}^T I(g_a)\, d\Gamma^T \\ &= \pm \int f_0 g_a{}^T I(g_b)\, d\Gamma^T \\ &= \pm \int f_0 g_a{}^T L_b(\Gamma)\, d\Gamma^T.\end{aligned}$$

Now, changing $\Gamma^T \to \Gamma$ on the right-hand side, and using (9.12), we have the required result (9.7).

The kinetic coefficients must also satisfy conditions which follow from the law of increase of entropy; in particular, the "diagonal" coefficients γ_{aa} must be positive. Since the transport equation guarantees the increase of entropy, these conditions are of course necessarily satisfied when the kinetic coefficients are calculated from that equation.

The increase of entropy is expressed by the inequality

$$-\int \log f \, C(f) \, d\Gamma > 0;$$

see §4. Substituting

$$f = f_0(1 + \chi/T), \quad C(f) = (f_0/T)I(\chi),$$

we have

$$-\int \log f_0 \, C(f) \, d\Gamma - \frac{1}{T}\int f_0 \log(1 + \chi/T) I(\chi) \, d\Gamma > 0.$$

The first integral is identically zero; in the second integral, since χ is small, $\log(1 + \chi/T) \approx \chi/T$, and so we find

$$-\int f_0 \chi I(\chi) \, d\Gamma > 0. \tag{9.13}$$

This inequality ensures the necessary properties of the kinetic coefficients. In particular, when $\chi = g_a$ it expresses the fact that γ_{aa} is positive.

§10. Approximate solution of the transport equation

Because of the complexity of the law of interaction of molecules (especially polyatomic ones), which determines the function w in the collision integral, the Boltzmann equation cannot really be even written down in an exact form for specific gases. However, even with linearization and some simple assumptions about the nature of the molecular interaction, the complexity of the mathematical structure of the transport equation makes it generally impossible to solve in an exact analytical form. Fairly efficient methods for the approximate solution of the Boltzmann equation are therefore of particular significance in the kinetic theory of gases. The principle as applied to a monatomic gas is as follows (S. Chapman 1916).

Let us first take the problem of thermal conduction. For a monatomic gas, the specific heat $c_p = 5/2$, and the linearized equation (7.3) becomes

$$-\mathbf{v}\left(\frac{5}{2} - \beta v^2\right) = I(\mathbf{g}), \tag{10.1}$$

where $\beta = m/2T$; the linear integral operator $I(\mathbf{g})$ is defined by

$$I(\mathbf{g}) = \int\int v_{\text{rel}} f_{01}(\mathbf{g}' + \mathbf{g}_1' - \mathbf{g} - \mathbf{g}_1)\, d^3p_1\, d\sigma, \tag{10.2}$$

corresponding to the collision integral (3.9), and the equilibrium distribution function is†

$$f_0(v) = (N\beta^{3/2}/m^3\pi^{3/2})e^{-\beta v^2}. \tag{10.3}$$

An efficient method of approximately solving equation (10.1) is based on expanding the required functions in terms of a complete set of mutually orthogonal functions, which may with especial advantage be taken as the Sonine polynomials (D. Burnett 1935). These are defined by‡

$$S_r^s(x) = \frac{1}{s!} e^x x^{-r} \frac{d^s}{dx^s}(e^{-x}x^{r+s}), \tag{10.4}$$

where r is any number and s is a positive integer or zero. In particular,

$$S_r^0 = 1, \quad S_r^1(x) = r + 1 - x. \tag{10.5}$$

The orthogonality property of these polynomials for a given r and different s is

$$\int_0^\infty e^{-x}x^r S_r^s(x) S_r^{s'}(x)\, dx = \Gamma(r + s + 1)\delta_{ss'}/s!. \tag{10.6}$$

We shall seek the solution of (10.1) as the expansion

$$\mathbf{g}(\mathbf{v}) = (\beta/N)\mathbf{v} \sum_{s=1}^{\infty} A_s S_{3/2}^s(\beta v^2). \tag{10.7}$$

By omitting the term with $s = 0$, we automatically satisfy the condition (7.4), the integral being zero because the polynomials with $s = 0$ and $s \neq 0$ are orthogonal. The expression in parentheses on the left of (10.1) is the polynomial $S_{3/2}^1(\beta v^2)$, and this equation therefore becomes

$$-\mathbf{v}S_{3/2}^1(\beta v^2) = (\beta/N) \sum_{s=1}^{\infty} A_s I(\mathbf{v}S_{3/2}^s). \tag{10.8}$$

Multiplying both sides scalarly by $\mathbf{v}f_0(v)S_{3/2}^l(\beta v^2)$ and integrating over d^3p, we

†The distribution function is everywhere taken to be defined in momentum space. This, however, does not prevent it from being expressed for convenience in terms of the velocity $\mathbf{v} = \mathbf{p}/m$.

‡They differ only in normalization and affix numbering from the generalized Laguerre polynomials:

$$S_r^s(x) = \frac{(-1)^r}{(r+s)!} L_{r+s}^r(x).$$

obtain a set of algebraic equations

$$\sum_{s=1}^{\infty} a_{ls}A_s = \frac{15}{4}\delta_{l1}, \quad l = 1, 2, \ldots, \tag{10.9}$$

with

$$\begin{aligned} a_{ls} &= -(\beta^2/N^2)\int f_0\mathbf{v} \,.\, S^l_{3/2} I(\mathbf{v}S^s_{3/2})d^3p \\ &= (\beta^2/4N^2)\{\mathbf{v}S^l_{3/2}, \mathbf{v}S^s_{3/2}\}, \end{aligned} \tag{10.10}$$

the notation being

$$\left.\begin{aligned} \{F, G\} &= \int f_0(v)f_0(v_1)|\mathbf{v}-\mathbf{v}_1|\Delta(F)\Delta(G)\, d^3p\, d^3p_1\, d\sigma, \\ \Delta(F) &= F(\mathbf{v}') + F(\mathbf{v}_1') - F(\mathbf{v}) - F(\mathbf{v}_1). \end{aligned}\right\} \tag{10.11}$$

There is no equation with $l = 0$ in (10.9), since $a_{0s} = 0$ because of the conservation of momentum: $\Delta(\mathbf{v}S^0_{3/2}) = \Delta(\mathbf{v}) = 0$. The thermal conductivity is calculated by substituting (10.7) in the integral (7.7). The condition (7.4) shows that this integral (with $\epsilon = \frac{1}{2}mv^2$) can be put in the form

$$\kappa = -\frac{1}{3}\int f_0 S^1_{3/2}(\beta v^2)\mathbf{v}\,.\,\mathbf{g}\, d^3p$$

and the result is

$$\kappa = 5A_1/4. \tag{10.12}$$

The advantage of expanding in Sonine polynomials is shown by the simplicity of the right-hand side of equations (10.9) and the expression (10.12).

The calculations are entirely similar for the viscosity. The solution of (8.6) is sought in the form

$$g_{\alpha\beta} = -(\beta^2/N^2)(v_\alpha v_\beta - \tfrac{1}{3}v^2\delta_{\alpha\beta})\sum_{s=0}^{\infty} B_s S^s_{5/2}(\beta v^2). \tag{10.13}$$

Substitution in (8.6), multiplication by

$$f_0(v)S^l_{5/2}(\beta v^2)(v_\alpha v_\beta - \tfrac{1}{3}v^2\delta_{\alpha\beta}),$$

and integration over d^3p leads to the set of equations

$$\sum_{s=0}^{\infty} b_{ls}B_s = 5\delta_{l0}, \quad l = 0, 1, 2, \ldots, \tag{10.14}$$

where

$$b_{ls} = (\beta^3/N^2)\{(v_\alpha v_\beta - \tfrac{1}{3}v^2\delta_{\alpha\beta})S^l_{5/2}, \quad (v_\alpha v_\beta - \tfrac{1}{3}v^2\delta_{\alpha\beta})S^s_{5/2}\}. \tag{10.15}$$

The viscosity is found from (8.9) as

$$\eta = \tfrac{1}{4} m B_0. \tag{10.16}$$

The infinite set of equations (10.9) or (10.14) is approximately solved by retaining only the first few terms in the expansion (10.7) or (10.13), i.e. by artificially terminating the set. The approximation converges extremely rapidly as the number of terms increases: in general, retaining just one term gives the value of κ or η with an accuracy of 1–2%.†

We shall show that the approximate solution of the linearized transport equation for monatomic gases by the above method gives values of the kinetic coefficients that are certainly less than would follow from the exact solution of the equation.

The transport equation may be written in the symbolic form

$$I(g) = L, \tag{10.17}$$

where the functions g and L are vectors in the thermal conduction problem, and tensors of rank two in the viscosity problem. The corresponding kinetic coefficient is determined from the function g as a quantity proportional to the integral

$$-\int f_0 g I(g)\, d^3p; \tag{10.18}$$

see §9. The approximate function g, however, satisfies not equation (10.17) itself but only the integral relation

$$\int f_0 g I(g)\, d^3p = \int f_0 L g\, d^3p, \tag{10.19}$$

as is evident from the way in which the coefficients in the expansions of g are determined.

The statement made above follows immediately from the "variational principle" whereby the solution of (10.17) gives a maximum of the functional (10.18) within the class of functions that satisfy the condition (10.19). The validity of this principle is easily shown by considering the integral

$$-\int f_0 (g - \varphi) I(g - \varphi)\, d^3p,$$

where g is the solution of (10.17), and φ any trial function that satisfies the condition (10.19). This integral is positive, by the general property (9.13) of the operator I. Expanding the parentheses, we write

$$-\int f_0 \{ g I(g) + \varphi I(\varphi) - \varphi I(g) - g I(\varphi) \}\, d^3p.$$

†The convergence is, however, somewhat less good in problems of diffusion, and especially of thermal diffusion.

Since for a monatomic gas the principle of detailed balancing is valid in the form (2.8), the operator I has the self-conjugacy property (9.11).† Hence the integrals of the last two terms in the braces are equal. Then substitution of $I(g) = L$ gives

$$-\int f_0\{gI(g)+\varphi I(\varphi)-2\varphi I(g)\}\,d^3p = -\int f_0\{gI(g)+\varphi I(\varphi)-2L\varphi\}\,d^3p > 0.$$

Finally, transforming the integral of the last term by means of (10.19), we find

$$-\int f_0 gI(g)\,d^3p > -\int f_0\varphi I(\varphi)\,d^3p,$$

as was to be proved.

There is a case that is of formal interest though having no direct physical significance, namely a gas of particles interacting according to $U = \alpha/r^4$.‡ This has the property that the collision cross-section for such particles (determined by classical mechanics) is inversely proportional to the relative speed v_{rel}, and so the product $v_{rel}d\sigma$ which appears in the collision integral depends only on the scattering angle θ, not on v_{rel}. The property in question is easily proved by dimensional arguments: the cross-section depends only on three parameters, namely the constant α, the particle mass m, and the velocity v_{rel}, and from these we can form no dimensionless combination, and only one combination $v_{rel}^{-1}(\alpha/m)^{1/2}$ having the dimensions of area, which must therefore be proportional to the cross-section. This property of the cross-section greatly simplifies the structure of the collision integral, and it becomes possible to find exact solutions of the linearized transport equations for the thermal conduction and viscosity problems. These solutions are found to be just the first terms in the expansions (10.7) and (10.13).§

PROBLEMS‖

PROBLEM 1. Find the thermal conductivity of a monatomic gas, retaining only the first term in the expansion (10.7).

SOLUTION. With one term of the expansion, equations (10.9) reduce to $A_1 = 15/4a_{11}$. To calculate the integral (10.10) with $l = s = 1$, we express $\mathbf{v}$, $\mathbf{v}_1$, $\mathbf{v}'$, $\mathbf{v}_1'$ in terms of the velocity of the centre of mass and the relative velocities of the two atoms:

$$\mathbf{V} = \tfrac{1}{2}(\mathbf{v}+\mathbf{v}_1) = \tfrac{1}{2}(\mathbf{v}'+\mathbf{v}_1'),$$

$$\mathbf{v}_{rel} = \mathbf{v}-\mathbf{v}_1, \quad \mathbf{v}_{rel}' = \mathbf{v}'-\mathbf{v}_1',$$

$$v^2+v_1^2 = 2V^2+\tfrac{1}{2}v_{rel}^2,$$

$$d^3p\,d^3p_1 = m^6\,d^3V\,d^3v_{rel}.$$

†It must be emphasized that the variational principle as stated above is dependent on this, and is not valid when the principle of detailed balancing has only its most general form (2.3).

‡The transport properties of this gas model were first discussed by J. C. Maxwell (1866).

§A detailed account of the theory for this case is given in §§38–40 of L. Waldmann's article in *Handbuch der Physik* **12**, 295, 1958.

‖Formulae (1)–(6) are due to Chapman and Enskog.

A simple calculation gives

$$\Delta(\mathbf{v}S^{1}_{3/2}) = \Delta(\beta v^2\mathbf{v}) = \beta[(\mathbf{V}\,.\,\mathbf{v}'_{\text{rel}})\mathbf{v}'_{\text{rel}} - (\mathbf{V}\,.\,\mathbf{v}_{\text{rel}})\mathbf{v}_{\text{rel}}].$$

Squaring and averaging over the directions of $\mathbf{V}$, we obtain

$$\tfrac{2}{3}\beta^2[v^4_{\text{rel}} - (\mathbf{v}_{\text{rel}}\,.\,\mathbf{v}'_{\text{rel}})^2]V^2 = \tfrac{2}{3}\beta^2 v^4_{\text{rel}} V^2 \sin^2\theta.$$

Integration over $4\pi V^2\, dV$ and over the directions of $\mathbf{v}_{\text{rel}}$ (the latter reducing to a multiplication by 4π) gives finally

$$a_{11} = \frac{1}{4}\beta^4(\beta/2\pi)^{1/2}\int_0^{\pi}\int_0^{\infty}\exp(-\tfrac{1}{2}\beta v^2_{\text{rel}})v^7_{\text{rel}}\sin^2\theta\,\frac{d\sigma}{d\theta}\,dv_{\text{rel}}\,d\theta; \tag{1}$$

the thermal conductivity is

$$\kappa = 75/16a_{11}. \tag{2}$$

PROBLEM 2. The same as Problem 1, but for the viscosity.

SOLUTION. We find in a similar manner

$$B_0 = 5/b_{00}, \quad \eta = 5m/4b_{00}.$$

In the integral (10.15) with $l = s = 0$,

$$\Delta(v_\alpha v_\beta - \tfrac{1}{3}v^2\delta_{\alpha\beta}) = \tfrac{1}{2}(v_{\text{rel},\alpha}v_{\text{rel},\beta} - v'_{\text{rel},\alpha}v'_{\text{rel},\beta}).$$

The square of this is

$$\tfrac{1}{2}v^4_{\text{rel}}\sin^2\theta.$$

Integration over d^3V and over the directions of $\mathbf{v}_{\text{rel}}$ shows that $b_{00} = a_{11}$, so that

$$\eta = 4m\kappa/15. \tag{3}$$

For a monatomic gas, the specific heat $c_p = 5/2$; hence the ratio of the kinematic viscosity $\nu = \eta/Nm$ to the thermometric conductivity $\chi = \kappa/Nc_p$, called the *Prandtl number*, is, in this approximation

$$\nu/\chi = 2/3 \tag{4}$$

whatever the law of interaction of the atoms.†

PROBLEM 3. In the same approximation, find the thermal conductivity and viscosity of a monatomic gas when the atoms are regarded as hard elastic spheres with diameter d.

SOLUTION. The scattering of one sphere by another is equivalent to that of a point particle by an impenetrable sphere of radius d; the cross-section is therefore $d\sigma = (\tfrac{1}{2}d)^2 do$. Calculation of the integral (1) gives the results‡

$$\kappa = \frac{75}{64\sqrt{\pi}d^2}\sqrt{\frac{T}{m}} = \frac{0.66}{d^2}\sqrt{\frac{T}{m}}, \tag{5}$$

$$\eta = \frac{5}{16\sqrt{\pi}d^2}\sqrt{(mT)} = 0.18\frac{\sqrt{(mT)}}{d^2}. \tag{6}$$

†For a gas with the interaction law $U = \alpha/r^4$, formulae (1)–(4) become exact, and lead to the values

$$\kappa = 3.04T(m\alpha)^{-1/2}, \quad \eta = 0.81T(m/\alpha)^{1/2}.$$

‡To illustrate the rapidity with which successive approximations converge, it may be mentioned that the inclusion of the second and third terms in the expansions (10.7) and (10.13) multiplies the expressions (5) and (6) by $(1 + 0.015 + 0.001)$ and $(1 + 0.023 + 0.002)$ respectively.

§ 11. Diffusion of a light gas in a heavy gas

The phenomenon of diffusion in a mixture of two gases will be studied here for some particular cases which allow a fairly extensive theoretical analysis.

Let N_1 and N_2 denote the particle number densities of the two components of the mixture, and let the concentration of the mixture be expressed by $c = N_1/N$, where $N = N_1 + N_2$. The total number density of particles is related to the pressure and temperature by $N = P/T$. The gas pressure is constant throughout the volume; let the concentration and the temperature vary along the x-axis (by allowing a temperature variation, we include thermal diffusion in the problem).

Let us consider diffusion in a mixture of gases of which one (the "heavy" gas) consists of molecules whose mass is much larger than that of the particles of the other (the "light" gas). The latter will be assumed monatomic. Since the mean thermal energy of translational motion is the same for all particles (at a given temperature), the mean speed of the heavy molecules is much less than that of the light ones, and they can be approximately regarded as being at rest. When a light and a heavy particle collide, the latter may be assumed to remain fixed, while the velocity of the light particle changes direction but remains unaltered in magnitude.

In this section we shall take the case where the concentration of the light gas (gas 1) in the mixture is small. Then collisions between its atoms are relatively rare and we may suppose that the light particles collide only with the heavy ones.†

In the general case of an arbitrary gas mixture, a separate transport equation has to be set up for the distribution function of the particles of each component, the right-hand side containing the sum of the collision integrals between the particles of each component and those of that and every other component. In the particular case under discussion, however, it is convenient to derive the simplified transport equation *ab initio.*

The required equation is to determine the distribution function for the particles of the light gas, which we denote by $f(\mathbf{p}, x)$. With the assumptions made, collisions between light and heavy particles do not affect the distribution of the latter, and in the diffusion problem this distribution can be taken as given.

Let θ be the angle between the direction of the momentum $\mathbf{p} = m_1\mathbf{v}$ of a light particle and the x-axis. It is evident from the symmetry of the conditions of the problem that the distribution function will depend only on θ (and on the variables p and x). Let $d\sigma = F(p, \alpha)\, do'$ denote the cross-section for collisions in which a light particle with momentum $\mathbf{p}$ acquires a momentum $\mathbf{p}' = m\mathbf{v}'$ directed into the solid-angle element do'; α is the angle between the vectors $\mathbf{p}$ and $\mathbf{p}'$ (whose magnitudes are equal). The probability per unit path length that the particle undergoes such a collision is $N_2 d\sigma$, where N_2 is the number density of heavy particles; the probability per unit time is found by multiplying by the speed of the particle: $N_2 v\, d\sigma$.

Let us consider particles in a given unit of volume having momenta in a given range dp of magnitudes and directed into the solid-angle element do. The number of such particles is $f\, d^3p = f(p, \theta, x)p^2\, dp\, do$. Of these,

$$f(p, \theta, x)p^2\, dp\, do\, .\, N_2 v F(p, \alpha)\, do'$$

†The kinetic theory of this model was first developed by H. A. Lorentz (1905).

particles per unit time acquire by collisions a momentum $\mathbf{p}'$ directed into do'. Thus the total number of particles whose momentum changes direction is

$$d^3p \int N_2 v f(p, \theta, x) F(p, \alpha)\, do'.$$

Conversely, of the particles in $d^3p' = p'^2\, dp'\, do'$,

$$f(p', \theta', x) p'^2\, dp'\, do' \,.\, N_2 v' F(p', \alpha)\, do$$

acquire a velocity directed into do. Since $p' = p$, the total number of particles that acquire a velocity in d^3p as a result of collisions is

$$d^3p \int N_2 v f(p, \theta', x) F(p, \alpha) do'.$$

Thus the change in the number of particles in d^3p is the difference

$$d^3p \,.\, N_2 v \int F(p, \alpha)[f(p, \theta', x) - f(p, \theta, x)] do'.$$

This must equal the total time derivative

$$d^3p(df/dt) = d^3p \mathbf{v} \,.\, \nabla f = d^3p(\partial f/\partial x) v \cos\theta.$$

Equating the two expressions gives the required transport equation

$$v \cos\theta\, \partial f/\partial x = N_2 v \int F(p, \alpha)[f(p, \theta', x) - f(p, \theta, x)]\, do' \equiv C(f). \qquad (11.1)$$

The right-hand side is zero for any function f that does not depend on the direction of $\mathbf{p}$, and not only for the Maxwellian function f_0 as in the case of the Boltzmann equation. This is because of the assumption that the magnitude of the momentum is unchanged in the scattering of light particles by heavy ones: such collisions evidently leave steady any energy distribution of light particles. In reality, equation (11.1) corresponds only to the zero-order approximation with respect to the small quantity m_1/m_2, and energy relaxation occurs in the next approximation.

If the concentration and temperature gradients are not too large (these quantities varying only slightly over distances of the order of the mean free path), f may be sought as the sum

$$f = f_0(p, x) + \delta f(p, \theta, x),$$

where δf is a small correction to the local-equilibrium distribution function f_0 and is linear in the gradients of c and T. In turn, we seek δf in the form

$$\delta f = \cos\theta \,.\, g(p, x). \qquad (11.2)$$

where g is a function of p and x only. In substituting in (11.1), it is sufficient to retain the f_0 term on the left-hand side; in the collision integral, the f_0 term disappears:

$$C(f) = gN_2v \int F(p, \alpha)(\cos\theta' - \cos\theta)\, do';$$

the function g, which is independent of the angles, has been taken outside the integral.

This integral may be simplified as follows. We take the direction of the momentum $\mathbf{p}$ as the polar axis for the measurement of angles. Let φ and φ' be the azimuths of the x-axis and the momentum $\mathbf{p}'$ relative to this polar axis. Then

$$\cos\theta' = \cos\theta\cos\alpha + \sin\theta\sin\alpha\cos(\varphi - \varphi').$$

The solid-angle element $do' = \sin\alpha\, d\alpha\, d\varphi'$, since α is the polar angle for the momentum $\mathbf{p}'$. The integral of the term in $\cos(\varphi - \varphi')$ gives zero from the integration over $d\varphi'$. The result is

$$C(f) = -N_2\sigma_t(p)vg\cos\theta = -N_2\sigma t(p)v\delta f, \tag{11.3}$$

where

$$\sigma_t(p) = 2\pi \int F(p, \alpha)(1 - \cos\alpha)\sin\alpha\, d\alpha$$

$$= \int (1 - \cos\alpha)\, d\sigma \tag{11.4}$$

is called the *transport cross-section* for collisions.

From (11.1), we now find

$$g(p, x) = -\frac{1}{N_2\sigma_t}\frac{\partial f_0}{\partial x}. \tag{11.5}$$

The diffusion flux $\mathbf{i}$ is, by definition, the flux of molecules of one component of the mixture (in this case, the light component). It is calculated from the distribution function as the integral

$$\mathbf{i} = \int f\mathbf{v}\, d^3p, \tag{11.6}$$

or, since the vector $\mathbf{i}$ is along the x-axis,

$$i = \int \cos\theta \,.\, fv\, d^3p = \int \cos^2\theta \,.\, gv\, d^3p; \tag{11.7}$$

the f_0 term disappears on integration over angles. Substitution of (11.5) gives

$$i = -\frac{1}{N_2}\frac{\partial}{\partial x}\int \frac{f_0 v \cos^2\theta}{\sigma_t(p)}\, d^3p$$
$$= -\frac{1}{3N_2}\frac{\partial}{\partial x}\int \frac{f_0 v}{\sigma_t}\, d^3p.$$

This expression may be written

$$i = -\frac{1}{3N_2}\frac{\partial}{\partial x}\{N_1\langle v/\sigma_t\rangle\},$$

where the averaging is over the Maxwellian distribution. Lastly, we use the concentration $c = N_1/N \approx N_1/N_2$ (since by hypothesis $N_2 \gg N_1$), and replace N_2 approximately by $N = P/T$. The pressure being constant, we find the result

$$i = -\tfrac{1}{3}T\frac{\partial}{\partial x}\left\{\frac{c}{T}\langle v/\sigma_t\rangle\right\}$$
$$= -\tfrac{1}{3}\langle v/\sigma_t\rangle\frac{\partial c}{\partial x} - \tfrac{1}{3}cT\frac{\partial}{\partial T}\left[\frac{1}{T}\langle v/\sigma_t\rangle\right]\frac{\partial T}{\partial x}. \tag{11.8}$$

This is to be compared with the phenomenological expression for the diffusion flux,

$$i = -ND\left(\nabla c + \frac{k_T}{T}\nabla T\right), \tag{11.9}$$

which defines the *diffusion coefficient D* and the *thermal diffusion ratio* k_T; the product $D_T = Dk_T$ is the *thermal diffusion coefficient* (see *FM*, § 58).† Thus we find

$$D = (T/3P)\langle v/\sigma_t\rangle, \tag{11.10}$$

$$k_T = cT\frac{\partial}{\partial T}\log\frac{\langle v/\sigma_t\rangle}{T}. \tag{11.11}$$

In diffusion equilibrium in a non-uniformly heated gas, a concentration distribution is set up in which the diffusion flux $\mathbf{i} = 0$. Equating to a constant the expression in the braces in (11.8), we obtain

$$c = \text{constant} \times \frac{T}{\langle v/\sigma_t\rangle}. \tag{11.12}$$

Assuming that the cross-section σ_t is independent of the velocity, and noting that $\langle v\rangle \sim (T/m_1)^{1/2}$, we find that, in diffusion equilibrium of a mixture with a low

†The phenomenon of thermal diffusion was predicted by Enskog (1911) for precisely this model of a gas mixture.

concentration of the light gas, that concentration is proportional to $\sqrt{T}$, i.e. the light gas is concentrated in the regions where the temperature is high.

The diffusion coefficient is, in order of magnitude,

$$D \sim \bar{v} l, \tag{11.13}$$

where $\bar{v}$ is the mean thermal speed of the light-gas molecules and $l \sim 1/N\sigma$ the mean free path. There is a well-known elementary derivation of this formula. The number of molecules of gas 1 passing across unit area perpendicular to the x-axis from left to right per unit time is equal in order of magnitude to the product $N_1 \bar{v}$, where the density N_1 must be taken at a distance l to the left of the area, i.e. at the points from which the molecules reach that area without undergoing collisions. We similarly find the number of molecules crossing the same area from right to left, and the difference between the two numbers gives the diffusion flux:

$$i \sim N_1(x-l)\bar{v} - N_1(x+l)\bar{v} \sim -l\bar{v}\, dN_1/dx,$$

which gives (11.13).†

§ 12. **Diffusion of a heavy gas in a light gas**

Let us now consider the opposite limiting case, where the concentration of the heavy gas in the mixture is small. In this case, the diffusion coefficient may be calculated indirectly without using the transport equation, by finding the *mobility* of the heavy-gas particles, regarding this gas as being in an external field. The mobility b is related to the diffusion coefficient of the same particles by the familiar Einstein's relation

$$D = bT; \tag{12.1}$$

see *FM*, § 59.

The mobility is, by definition, the proportionality coefficient between the mean velocity $\mathbf{V}$ acquired by a gas particle in the external field, and the force $\mathbf{f}$ exerted on the particle by the field:

$$\mathbf{V} = b\mathbf{f}. \tag{12.2}$$

The velocity $\mathbf{V}$ is determined from the condition that the force $\mathbf{f}$ balances the resistance $\mathbf{f}_r$ exerted on the moving heavy particle by the light particles; collisions between heavy particles may be neglected, because there are relatively few of

†Diffusion, thermal conduction and viscosity are brought about by the same mechanism, namely direct molecular transport. The thermal conduction may be regarded as a "diffusion of energy" and the viscosity as a "diffusion of momentum". We may therefore assert that the diffusion coefficient D, the thermometric conductivity $\chi = \kappa/Nc_p$ and the kinematic viscosity $\nu = \eta/Nm$ are of the same order of magnitude; this leads to the formulae (7.10) for the thermal conductivity and (8.11) for the viscosity.

them. The distribution function of the light particles is Maxwellian:

$$f_0 = \frac{N_1}{(2\pi m_1 T)^{3/2}} \exp\left(-\frac{m_1 v^2}{2T}\right),$$

where m_1 is the mass of a light particle.

Let us consider one particular heavy particle with velocity $\mathbf{V}$, and take coordinates moving with that particle; let $\mathbf{v}$ denote the velocities of the light particles in these coordinates. The distribution function of the light particles in these coordinates is $f_0(\mathbf{v}+\mathbf{V})$; cf. (6.9). Assuming that $\mathbf{V}$ is small, we can write

$$f_0(\mathbf{v}+\mathbf{V}) \approx f_0(v)(1 - m_1\mathbf{v}\cdot\mathbf{V}/T). \qquad (12.3)$$

The required resistance $\mathbf{f}_r$ can be calculated as the total momentum transferred to the heavy particle by light particles colliding with it per unit time. The frame of reference is unchanged in a collision. The light particle carries momentum $m_1\mathbf{v}$; after the collision, in which its momentum is turned through an angle α, it carries away an average momentum $m_1\mathbf{v}\cos\alpha$. The average momentum transferred to the heavy particle in such a collision is therefore $m_1\mathbf{v}(1-\cos\alpha)$. Multiplying this by the flux of light particles with velocity $\mathbf{v}$ and by the cross-section $d\sigma$ for such a collision, and integrating, we obtain the total momentum transferred to the heavy particle:

$$\mathbf{f}_r = m_1 \int f_0(\mathbf{v}+\mathbf{V}) v \mathbf{v} \sigma_t \, d^3p,$$

again with the notation (11.4). When $f_0(\mathbf{v}+\mathbf{V})$ is substituted in the form (12.3), the first term gives zero in the integration over directions of $\mathbf{v}$, leaving

$$\mathbf{f}_r = -\frac{m_1^2}{T} \int f_0(v) \mathbf{V}\cdot\mathbf{v}\, \mathbf{v} v \sigma_t \, d^3p,$$

or, averaging over directions of $\mathbf{v}$,

$$\mathbf{f}_r = -\frac{m_1^2}{3T} \mathbf{V} \int f_0(v) \sigma_t v^3 \, d^3p$$

$$= -N_1 \frac{m_1^2}{3T} \mathbf{V} \langle \sigma_t v^3 \rangle,$$

where the angle brackets again denote averaging over the ordinary Maxwellian distribution. Lastly, since in this case $N_1 \gg N_2$, we write $N_1 \approx N = P/T$, so that

$$\mathbf{f}_r = -\frac{m_1^2 P}{3T^2} \langle \sigma_t v^3 \rangle \mathbf{V}.$$

Equating to zero the sum of the resistance $\mathbf{f}_r$ and the external force $\mathbf{f}$, we find from

(12.2) the mobility b, and thence the diffusion coefficient

$$D = bT = 3T^3/m_1^2P\langle\sigma_t v^3\rangle. \tag{12.4}$$

To calculate the thermal diffusion in this case, it would be necessary to know the distribution function of the light-gas particles in the presence of a temperature gradient. The thermal diffusion coefficient therefore cannot be calculated in a general form here.

In order of magnitude $D \sim \bar{v}/N\sigma$, where $\bar{v} \sim \sqrt{(T/m_1)}$ is, as in (11.13), the mean thermal speed of the light-gas molecules. Thus the order of magnitude of the diffusion coefficient is the same in each case:

$$D \sim T^{3/2}/\sigma P m_1^{1/2}. \tag{12.5}$$

PROBLEM

Determine the diffusion coefficient in a mixture of two gases (one light and one heavy), regarding their particles as hard elastic spheres with diameters d_1 and d_2.

SOLUTION. The collision cross-section $d\sigma = \pi(d_1 + d_2)^2\, do/16\pi$, and so the transport cross-section $\sigma_t = \frac{1}{4}\pi(d_1 + d_2)^2$, equal in this case to the total cross-section σ. The diffusion coefficient is

$$D = AT^{3/2}/(d_1 + d_2)^2 P m_1^{1/2},$$

where m_1 is the mass of a light particle and A is a numerical factor. When the concentration of the light gas is small, a calculation from (11.10) gives

$$A = \tfrac{4}{3}(2/\pi)^{3/2} = 0.68.$$

When the concentration of the heavy gas is small, (12.4) gives

$$A = 3/2\sqrt{(2\pi)} = 0.6.$$

Note that the values of A in the two limiting cases are almost equal.

§ 13. Transport phenomena in a gas in an external field

The rotational degrees of freedom of molecules provide the mechanism whereby an external magnetic or electric field can affect transport phenomena in a gas.† The effect is of the same nature in the magnetic and electric cases; we shall first discuss a gas in a magnetic field.

A rotating molecule has in general a magnetic moment, whose average value (in the quantum-mechanical sense) will be denoted by $\boldsymbol{\mu}$. The magnetic field will be assumed so weak that μB is small in comparison with the intervals in the fine structure of molecular levels.‡ We can then neglect the influence of the field on the

†This mechanism was pointed out by Yu. M. Kagan and L. A. Maksimov (1961), who also derived the results given in this section.

‡In macroscopic electrodynamics, the mean value (over physically infinitesimal volumes) of the magnetic field is called the *magnetic induction* and denoted by **B**. When the density of the medium is low, as in a gas, the magnetization is negligible, and the vector **B** then coincides with the macroscopic field **H**.

state of the molecule, so that the magnetic moment is calculated for the unperturbed state. For fairly high temperatures, the case we shall consider, μB is small in comparison with T also; this enables us to neglect the influence of the field on the equilibrium distribution function of the gas molecules.

The magnetic moment is parallel to the rotational angular momentum $\mathbf{M}$ of the molecule, and may be written

$$\boldsymbol{\mu} = \gamma \mathbf{M}. \tag{13.1}$$

Classical rotation of the molecule corresponds to large rotational quantum numbers; we can then neglect in $\mathbf{M}$ the difference between the total angular momentum (including spin) and the rotational angular momentum. The value of the constant coefficient γ depends on the nature of the molecule and the nature of its magnetic moment. For example, with a diatomic molecule having non-zero spin S,

$$\gamma \approx (2\sigma/M)\mu_B, \tag{13.2}$$

where μ_B is the Bohr magneton, and the number $\sigma = J - K$ is the difference between the quantum numbers J of the total angular momentum and K of the rotational angular momentum (σ takes the values $S, S-1, \ldots, -S$); in the denominator, the difference between J and K is not significant: $M \approx \hbar J \approx \hbar K$. In (13.2) it is assumed that the spin–axis interaction in the molecule is small in comparison with the intervals in the rotational structure of the levels (Hund's case *b*).†

In a magnetic field $\mathbf{B}$, the molecule is subjected to a torque $\boldsymbol{\mu} \times \mathbf{B}$. The vector $\mathbf{M}$ is then no longer constant during the "free" motion of the molecule, but varies according to

$$d\mathbf{M}/dt = \boldsymbol{\mu} \times \mathbf{B} = -\gamma \mathbf{B} \times \mathbf{M}; \tag{13.3}$$

the vector $\mathbf{M}$ precesses about the direction of the field with angular velocity $-\gamma \mathbf{B}$. The left-hand side of the transport equation thus has an added term $(\partial f/\partial \mathbf{M}) \cdot \dot{\mathbf{M}}$, and the equation becomes

$$\frac{\partial f}{\partial t} + \mathbf{v} \cdot \frac{\partial f}{\partial \mathbf{r}} + \gamma \mathbf{M} \times \mathbf{B} \cdot \frac{\partial f}{\partial \mathbf{M}} = C(f). \tag{13.4}$$

The variables Γ on which the distribution function depends must also include the discrete variable σ, which determines the value of the magnetic moment, if there is such a variable, as in (13.2).

In problems of thermal conduction and viscosity, we again take a distribution close to the equilibrium one, and express it as

$$f = f_0(1 + \chi/T). \tag{13.5}$$

†Formula (13.2) follows from the exact formula for case *b*, derived in *QM*, § 113, Problem 3, on taking the limit of large J and K with a fixed difference $J - K$. The contribution of the orbital angular momentum Λ is then negligible, being of the next order of smallness in $1/J$.

We shall first show that a term in $\partial f_0/\partial \mathbf{M}$ does not occur in the transport equation. Since f_0 depends only on the energy $\epsilon(\Gamma)$ of the molecule, and $\partial\epsilon/\partial\mathbf{M}$ is equal to the angular velocity $\mathbf{\Omega}$, we have

$$\gamma\mathbf{M}\times\mathbf{B}\,.\,\partial f_0/\partial\mathbf{M} = \gamma\mathbf{M}\times\mathbf{B}\,.\,\mathbf{\Omega}\;\partial f_0/\partial\epsilon. \tag{13.6}$$

For molecules of the rotator and spherical-top types, $\mathbf{M}$ and $\mathbf{\Omega}$ are parallel, and the expression (13.6) is zero identically. In other cases, it becomes zero after averaging over the rapidly varying phases, the necessity for which has been explained in §1. When molecules of the symmetrical-top or asymmetrical-top type rotate, there is a rapid variation both of the direction of the axes of the molecule itself and of that of its angular velocity $\mathbf{\Omega}$. After the averaging mentioned, $\mathbf{\Omega}$ can retain only the component $\mathbf{\Omega}_M$ along the constant vector $\mathbf{M}$, and for this component the product $\mathbf{M}\,.\,\mathbf{B}\times\mathbf{\Omega}_M = 0$.

The remaining terms in the transport equation are transformed in the same way as in §7 or §8. For instance, in the thermal conduction problem we find the equation

$$\frac{\epsilon(\Gamma)-c_pT}{T}\mathbf{v}\,.\,\nabla T = -\gamma\mathbf{M}\times\mathbf{B}\cdot\frac{\partial\chi}{\partial\mathbf{M}}+I(\chi). \tag{13.7}$$

The solution of this equation is again to be sought in the form $\chi = \mathbf{g}\,.\,\nabla T$, but there are now three vectors $\mathbf{v}, \mathbf{M}, \mathbf{B}$, not two, available to construct the vector function $\mathbf{g}(\Gamma)$. The external field creates a distinctive direction in the gas. The process of thermal conduction therefore becomes anisotropic, and the scalar coefficient κ has to be replaced by a thermal conductivity tensor $\kappa_{\alpha\beta}$, which determines the heat flux by

$$q_\alpha = -\kappa_{\alpha\beta}\,\partial T/\partial x_\beta. \tag{13.8}$$

The tensor $\kappa_{\alpha\beta}$ is calculated from the distribution function as the integral

$$\kappa_{\alpha\beta} = -\frac{1}{T}\int f_0\epsilon v_\alpha g_\beta\,d\Gamma; \tag{13.9}$$

cf. (7.5).

The general form of a tensor of rank two depending on the vector $\mathbf{B}$ is

$$\kappa_{\alpha\beta} = \kappa\delta_{\alpha\beta}+\kappa_1 b_\alpha b_\beta+\kappa_2 e_{\alpha\beta\gamma}b_\gamma, \tag{13.10}$$

where $\mathbf{b} = \mathbf{B}/B$, $e_{\alpha\beta\gamma}$ is the antisymmetric unit tensor, and κ, κ_1, κ_2 are scalars depending on the field strength B. The tensor (13.10) obviously has the property†

$$\kappa_{\alpha\beta}(\mathbf{B}) = \kappa_{\beta\alpha}(-\mathbf{B}). \tag{13.11}$$

†This property expresses the symmetry of the kinetic coefficients in the presence of a magnetic field. In the present case, it necessarily follows from the existence of only the one vector $\mathbf{b}$ from which the tensor $\kappa_{\alpha\beta}$ can be constructed.

The expression (13.10) corresponds to the heat flux

$$\mathbf{q} = -\kappa\nabla T - \kappa_1\mathbf{b}(\mathbf{b}\,.\,\nabla T) - \kappa_2\nabla T \times \mathbf{b}. \tag{13.12}$$

The last term is what is called an odd effect, changing sign with the field.

The integral term $I(\chi)$ on the right of (13.7) is given by (6.5). The integrand contains the function f_0, which is proportional to the gas density N. Separating this factor and dividing both sides of the equation by it, we find that N appears only in the combinations $\mathbf{B}/N$ with the field and $\nabla T/N$ with the temperature gradient. It is therefore clear that the function $f_0\chi = f_0\mathbf{g}\,.\,\nabla T$ will depend on the parameters N and B only through the ratio B/N; the integrals (13.9) will also depend only on this quantity, and therefore so will the coefficients κ, κ_1, κ_2 in (13.12). The density N is proportional (at a given temperature) to the gas pressure P. Thus the thermal conductivity of a gas in a magnetic field depends on the field and the pressure only through the ratio B/P.†

When B increases, the first term on the right of (13.7) increases, but the second term is unchanged. It is therefore clear that as $B \to \infty$ the solution of the equation must be a function depending only on the direction (not the magnitude) of the field, and this function must make identically zero the term $\mathbf{M}\times\mathbf{B}\,.\,\partial\chi/\partial\mathbf{M}$ in the equation; accordingly, the coefficients κ, κ_1, κ_2 tend to constant limits independent of B, as $B \to \infty$.

The treatment of the viscosity of a gas in a magnetic field is similar. The corresponding transport equation is

$$\left(mv_\alpha v_\beta - \frac{\epsilon(\Gamma)}{c_v}\delta_{\alpha\beta}\right)V_{\alpha\beta} = I(\chi) - \gamma\mathbf{M}\times\mathbf{B}\cdot\frac{\partial\chi}{\partial\mathbf{M}}; \tag{13.13}$$

cf. (6.19). The solution is to be sought in the form $\chi = g_{\alpha\beta}V_{\alpha\beta}$. Instead of the two viscosity coefficients η and ζ, we must now use a tensor $\eta_{\alpha\beta\gamma\delta}$ of rank four which determines the viscous stress tensor

$$\sigma'_{\alpha\beta} = \eta_{\alpha\beta\gamma\delta}V_{\gamma\delta}; \tag{13.14}$$

by definition, the tensor $\eta_{\alpha\beta\gamma\delta}$ is symmetric in the pairs of suffixes α, β and γ, δ. With the known function χ, its components are calculated as

$$\eta_{\alpha\beta\gamma\delta} = -\int mv_\alpha v_\beta f_0 g_{\gamma\delta}\,d\Gamma. \tag{13.15}$$

The viscosity tensor thus found will necessarily satisfy the condition

$$\eta_{\alpha\beta\gamma\delta}(\mathbf{B}) = \eta_{\gamma\delta\alpha\beta}(-\mathbf{B}), \tag{13.16}$$

which expresses the symmetry of the kinetic coefficients.

With the vector $\mathbf{b} = \mathbf{B}/B$ (and the unit tensors $\delta_{\alpha\beta}$ and $e_{\alpha\beta\gamma}$), we can construct the

†The change in the thermal conductivity of a gas in a magnetic field is called the *Senftleben effect*.

following independent tensor combinations having the symmetry properties of $\eta_{\alpha\beta\gamma\delta}$:

$$\left.\begin{aligned}
&(1)\ \delta_{\alpha\gamma}\delta_{\beta\delta}+\delta_{\alpha\delta}\delta_{\beta\gamma},\\
&(2)\ \delta_{\alpha\beta}\delta_{\gamma\delta},\\
&(3)\ \delta_{\alpha\gamma}b_\beta b_\delta+\delta_{\beta\gamma}b_\alpha b_\delta+\delta_{\alpha\delta}b_\beta b_\gamma+\delta_{\beta\delta}b_\alpha b_\gamma,\\
&(4)\ \delta_{\alpha\beta}b_\gamma b_\delta+\delta_{\gamma\delta}b_\alpha b_\beta,\\
&(5)\ b_\alpha b_\beta b_\gamma b_\delta,\\
&(6)\ b_{\alpha\gamma}\delta_{\beta\delta}+b_{\beta\gamma}\delta_{\alpha\delta}+b_{\alpha\delta}\delta_{\beta\gamma}+b_{\beta\delta}\delta_{\alpha\gamma},\\
&(7)\ b_{\alpha\gamma}b_\beta b_\delta+b_{\beta\gamma}b_\alpha b_\delta+b_{\alpha\delta}b_\beta b_\gamma+b_{\beta\delta}b_\alpha b_\gamma,
\end{aligned}\right\}\qquad(13.17)$$

where $b_{\alpha\beta}=-b_{\beta\alpha}=e_{\alpha\beta\gamma}b_\gamma$. In all these combinations except (4), the property (13.16) follows automatically from the symmetry with respect to the pairs of suffixes α,β and γ,δ; in (4), the two terms are combined in order to satisfy the condition (13.16).†

In accordance with the number of tensors (13.17), a gas in a magnetic field in general has seven independent viscosity coefficients. These may be defined as the coefficients in the following expression for the viscous stress tensor:

$$\begin{aligned}
\sigma'_{\alpha\beta}&=2\eta(V_{\alpha\beta}-\tfrac{1}{3}\delta_{\alpha\beta}\operatorname{div}\mathbf{V})+\zeta\delta_{\alpha\beta}\operatorname{div}\mathbf{V}\\
&\quad+\eta_1(2V_{\alpha\beta}-\delta_{\alpha\beta}\operatorname{div}\mathbf{V}+\delta_{\alpha\beta}V_{\gamma\delta}b_\gamma b_\delta-2V_{\alpha\gamma}b_\gamma b_\beta\\
&\quad-2V_{\beta\gamma}b_\gamma b_\alpha+b_\alpha b_\beta\operatorname{div}\mathbf{V}+b_\alpha b_\beta V_{\gamma\delta}b_\gamma b_\delta)\\
&\quad+2\eta_2(V_{\alpha\gamma}b_\gamma b_\beta+V_{\beta\gamma}b_\gamma b_\alpha-2b_\alpha b_\beta V_{\gamma\delta}b_\gamma b_\delta)\\
&\quad+\eta_3(V_{\alpha\gamma}b_{\beta\gamma}+V_{\beta\gamma}b_{\alpha\gamma}-V_{\gamma\delta}b_{\alpha\gamma}b_\beta b_\delta-V_{\gamma\delta}b_{\beta\gamma}b_\alpha b_\delta)\\
&\quad+2\eta_4(V_{\gamma\delta}b_{\alpha\gamma}b_\beta b_\delta+V_{\gamma\delta}b_{\beta\gamma}b_\alpha b_\delta)+\zeta_1(\delta_{\alpha\beta}V_{\gamma\delta}b_\gamma b_\delta+b_\alpha b_\beta\operatorname{div}\mathbf{V});
\end{aligned}\qquad(13.18)$$

$V_{\alpha\beta}$ is defined in (6.12). This is so constructed that $\eta,\eta_1,\ldots,\eta_4$ are coefficients of tensors which give zero on contraction with respect to the suffixes α,β; ζ and ζ_1 are coefficients of tensors with non-zero trace, and may be called second viscosity coefficients. Note that they contain not only the scalar div **V** but also $V_{\gamma\delta}b_\gamma b_\delta$. The first two terms in (13.18) correspond to the usual expression for the stress tensor, so that η and ζ are the ordinary viscosity coefficients.

The tensors $\kappa_{\alpha\beta}$ and $\eta_{\alpha\beta\gamma\delta}$ must be true tensors, since they satisfy the condition of symmetry under inversion. The abandonment of this condition (for a gas of stereoisomeric material) would therefore not lead to the presence of any new terms.

Such abandonment would, however, bring about new effects, with a heat flux $q^{(V)}$ due to the velocity gradients and viscous stresses $\sigma'^{(T)}$ due to the temperature gradient. These *cross-effects* are described by the formulae

$$q_\gamma^{(V)}=c_{\gamma,\alpha\beta}V_{\alpha\beta},\quad \sigma'^{(T)}_{\alpha\beta}=-a_{\alpha\beta,\gamma}\partial T/\partial x_\gamma,\qquad(13.19)$$

†It is unnecessary to write down combinations of terms with two factors $b_{\alpha\beta}$: since the product of two tensors $e_{\alpha\beta\gamma}$ reduces to products of tensors $\delta_{\alpha\beta}$, such combinations would reduce to those already included in (13.17).

where $c_{\gamma,\alpha\beta}$ and $a_{\alpha\beta,\gamma}$ are tensors of rank three symmetric in the pair of suffixes separated by the comma. With $\dot{x}_a$ and X_a chosen as in § 9, the kinetic coefficients γ_{ab} and γ_{ba} are $Tc_{\gamma,\alpha\beta}$ and $T^2 a_{\alpha\beta,\gamma}$. Onsager's principle thus shows that in the presence of a magnetic field we must have

$$Ta_{\alpha\beta,\gamma}(\mathbf{B}) = c_{\gamma,\alpha\beta}(-\mathbf{B}). \tag{13.20}$$

The general form of such tensors is

$$a_{\alpha\beta,\gamma} = a_1 b_\alpha b_\beta b_\gamma + a_2 b_\gamma \delta_{\alpha\beta} + a_3(b_\alpha \delta_{\beta\gamma} + b_\beta \delta_{\alpha\gamma}) + a_4(b_{\alpha\gamma} b_\beta + b_{\beta\gamma} b_\alpha). \tag{13.21}$$

All the terms here are pseudotensors, and so the relations (13.19) with these coefficients are not invariant under inversion.

Let us now briefly consider transport phenomena in a gas in an electric field. We take a gas consisting of polar molecules (i.e. having a dipole moment **d**) of the symmetrical-top type. In an electric field, a polar molecule is acted on by a torque $\mathbf{d}\times\mathbf{E}$, so that the transport equation contains a term

$$\dot{\mathbf{M}} \cdot \partial f/\partial \mathbf{M} = \mathbf{d}\times\mathbf{E} \cdot \partial f/\partial \mathbf{M}.$$

The direction of **d** is along the axis of the molecule and is unrelated to that of the rotational angular momentum **M**. However, as a result of averaging with respect to the rapid precession of the top's axis about the direction of the constant vector **M**, there remains in the above term only the component d along **M**, and it becomes

$$\gamma\mathbf{M}\times\mathbf{E} \cdot \partial f/\partial \mathbf{M}, \tag{13.22}$$

where $\gamma = \sigma d/M$; the variable σ (the cosine of the angle between **d** and **M**) now takes a continuous series of values from -1 to $+1$. The expression (13.22) differs from the corresponding term in the magnetic case only in that **B** is replaced by **E**. Thus all the above transport equations and the conclusions drawn from them remain valid.†

There is, however, a difference arising from the fact that the electric field **E** is a true vector, not a pseudovector, and is unaffected by time reversal. For this reason, Onsager's principle for the thermal conductivity and viscosity tensors is here expressed by

$$\kappa_{\alpha\beta}(\mathbf{E}) = \kappa_{\beta\alpha}(\mathbf{E}), \quad \eta_{\alpha\beta\gamma\delta}(\mathbf{E}) = \eta_{\gamma\delta\alpha\beta}(\mathbf{E}), \tag{13.23}$$

instead of (13.11) and (13.16). Correspondingly, $\kappa_2 \equiv 0$ and $\eta_3 = \eta_4 \equiv 0$ in (13.10) and (13.18) (where now $\mathbf{b} = \mathbf{E}/E$).‡ On the other hand, cross-effects are possible not only in a stereoisomeric gas, for which (13.21) is fully valid, but also in a gas of non-stereoisomeric molecules: the expression (13.21) with $a_4 \equiv 0$ is then a true tensor.

†Diatomic molecules rotate in a plane perpendicular to **M**; hence $\sigma = 0$ for a diatomic polar molecule. In such a case the effect of the electric field on the motion of the molecules appears in the transport equation only in the quadratic approximation with respect to the field.

‡In a gas of non-stereoisomeric molecules, the absence of the terms in κ_2, η_3, η_4 in an electric field is also required by the condition of invariance under inversion.

§ 14. **Phenomena in slightly rarefied gases**

The dynamical equations of motion of a gas, including thermal conduction and internal friction, contain the heat flux $\mathbf{q}'$ (the dissipative part of the energy flux $\mathbf{q}$) and the viscous stress tensor $\sigma'_{\alpha\beta}$ (the dissipative part of the momentum flux $\Pi_{\alpha\beta}$). These equations acquire real meaning when $\mathbf{q}'$ and $\sigma'_{\alpha\beta}$ have been expressed in terms of the temperature and velocity gradients in the gas. However, the usual expressions linear in these gradients are just the first terms of expansions in powers of the small ratio l/L of the mean free path to the characteristic dimensions of the problem (called the *Knudsen number* K). If this ratio is not very small, it may be reasonable to make corrections based on the terms of the next order of smallness in l/L. Such corrections arise both in the equations of motion themselves and in the boundary conditions on these equations at the surfaces of bodies in the gas flow.

The successive terms in the expansions of the fluxes $\mathbf{q}'$ and $\sigma'_{\alpha\beta}$ are expressed by means of the spatial derivatives of temperature, pressure and velocity, of various orders and raised to various powers. These terms must in principle be calculated by going to further approximations in the solution of the transport equation. The zero-order approximation corresponds to the local-equilibrium distribution function f_0 and the dynamical equations of an ideal fluid. The first-order approximation corresponds to the distribution function $f = f_0(1 + \chi^{(1)}/T)$ considered in §§6–8, and the Navier–Stokes equations of fluid dynamics, and the equation of thermal conduction. In the second-order approximation, the distribution function is to be sought in the form

$$f = f_0\left[1 + \frac{1}{T}\chi^{(1)} + \frac{1}{T}\chi^{(2)}\right] \qquad (14.1)$$

and the transport equation is to be linearized with respect to the second-order correction $\chi^{(2)}$. The resulting equation is

$$\frac{T}{f_0}\left(\frac{\partial_0}{\partial t} + \mathbf{v}\cdot\nabla\right)\frac{f_0\chi^{(1)}}{T} + \frac{T}{f_0}\frac{\partial_1}{\partial t}f_0$$
$$-\frac{1}{T^2}\int w' f_{01}[\chi^{(1)\prime}\chi_1^{(1)\prime} - \chi^{(1)}\chi_1^{(1)}]\,d\Gamma_1\,d\Gamma'\,d\Gamma_1' = \frac{1}{T}I(\chi^{(2)}), \qquad (14.2)$$

where I is again the linear integral operator (6.5). The symbol $\partial_0/\partial t$ signifies that the time derivatives of macroscopic quantities which appear as a result of differentiating $f_0\chi^{(1)}/T$ are to be expressed in terms of spatial derivatives by means of the zero-order equations of fluid dynamics (Euler's equations). The symbol $\partial_1/\partial t$ signifies that the time derivatives are to be eliminated by means of the first-order terms in the Navier–Stokes equations and the equation of thermal conduction (the terms containing η, ζ and κ).

We shall not write out all the numerous terms in $\mathbf{q}'$ and $\sigma'_{\alpha\beta}$ that arise in the second approximation and are called *Burnett terms* (D. Burnett, 1935). In many cases these terms make a contribution to the solution that is small in comparison with the corrections in the boundary conditions, to be discussed below. In such cases, the inclusion of corrections in the equations themselves would be an unjustifiable exaggeration of the attainable accuracy. We shall merely consider some typical correction terms and make estimates of them for motions of various kinds.

First of all, let us note that the small parameter $\mathrm{K} = l/L$ is related in a certain way to two parameters which describe the fluid motion, namely the Reynolds number R and the Mach number M. The Reynolds number is defined as $\mathrm{R} \sim VL/\nu$, where V is the characteristic scale of velocity of the flow and ν the kinematic viscosity; the Mach number $\mathrm{M} \sim V/u$, where u is the speed of sound. In a gas, the order of magnitude of the speed of sound is the same as the mean thermal speed $\bar{v}$ of the molecules, and the kinematic viscosity $\nu \sim l\bar{v}$. Hence $\mathrm{R} \sim VL/l\bar{v}$, $\mathrm{M} \sim V/\bar{v}$ and the Knudsen number

$$\mathrm{K} \sim \mathrm{M}/\mathrm{R}. \tag{14.3}$$

Hence it is clear that the condition $\mathrm{K} \ll 1$ for the flow to be governed by the linear equations of fluid dynamics imposes a limitation on the relative order of magnitude of R and M. Let us first consider "slow" motions, with

$$\mathrm{R} \lesssim 1, \quad \mathrm{M} \ll 1. \tag{14.4}$$

Let us take any of the Burnett terms in the viscous stress tensor containing the product of two first derivatives of the velocity, for instance

$$\rho l^2 \frac{\partial V_\alpha}{\partial x_\gamma} \frac{\partial V_\beta}{\partial x_\gamma}; \tag{14.5}$$

the coefficient ρl^2 (where ρ is the gas density) is an order-of-magnitude estimate. This term gives a contribution $\sigma^{(2)} \sim \rho l^2 V^2/L^2$ to $\sigma'_{\alpha\beta}$. The order of magnitude of the principal (Navier–Stokes) terms in the viscous stresses is

$$\sigma^{(1)} \sim \eta\, \partial V/\partial x \sim \rho l \bar{v} V/L,$$

and the ratio

$$\sigma^{(2)}/\sigma^{(1)} \sim lV/L\bar{v} \sim l^2\mathrm{R}/L^2.$$

Since $\mathrm{R} \lesssim 1$, we see that the terms (14.5) give a correction to the viscous stresses whose relative order is $\lesssim (l/L)^2$; the correction in the boundary conditions (see below) gives much larger corrections ($\sim l/L$) to the motion.

The corrections are even smaller that arise from terms of the form†

$$\frac{\rho l^2}{m^2 \bar{v}^2} \frac{\partial T}{\partial x_\alpha} \frac{\partial T}{\partial x_\beta}, \tag{14.6}$$

if the temperature gradients are those which result from the motion itself; this follows because the characteristic temperature differences $\Delta T \sim TV^2/u^2$. If,

†Terms of this kind in the viscous stresses were first discussed by Maxwell (1879).

however, temperature differences are imposed from outside (e.g. by heated bodies immersed in the gas), the Burnett terms of the form (14.6) may cause a steady motion with characteristic velocities determined by the equilibrium equation

$$\frac{\partial}{\partial x_\beta}(\sigma^{(1)}_{\alpha\beta} + \sigma^{(2)}_{\alpha\beta}) = \frac{\partial P}{\partial x_\alpha}.$$

An estimate of the speed of this motion is

$$V \sim l(\Delta T)^3/Lm\bar{v}T^2 \tag{14.7}$$

(M. N. Kogan, V. S. Galkin and O. G. Fridlender 1970). In making the estimate, it must be remembered that the Laplacian of the temperature can be expressed, by means of the thermal conduction equation div $(\kappa\nabla T) = 0$, in terms of the square of the temperature gradient, and that the motion is caused only by the non-potential part $\partial\sigma^{(2)}_{\alpha\beta}/\partial x_\beta$ of the force; the potential part is balanced by the pressure.

Similar considerations apply to the correction terms in the heat flux $\mathbf{q}'$. It is impossible to construct a second-order correction term from the derivatives of the temperature alone; the first such correction term (after $-\kappa\nabla T$) is constant $\times\, \nabla\triangle T$ (where $\triangle$ is the Laplacian operator), and thus is of the third order. The terms which include velocity derivatives as well as temperature derivatives, such as

$$(\rho l^2/m)\,\mathrm{div}\,\mathbf{V}\,.\,\nabla T,$$

again give corrections of relative order l^2/L^2.

Let us now go on to "fast" motions, with

$$\mathrm{R} \gg 1, \quad \mathrm{M} \lesssim 1. \tag{14.8}$$

In such cases, the gas motion takes place in two regions: the main volume, where the viscous terms in the equations of motion are unimportant, and a thin boundary layer, in which the gas velocity decreases rapidly.

Let us consider, for example, the flow of gas past a flat plate, taking the direction of flow as the x-axis. The thickness δ of the boundary layer on the plate is

$$\delta \sim (x\nu/V)^{1/2} \sim (xl\bar{v}/V)^{1/2},$$

where x is the distance from the leading edge; see *FM*, §39. The characteristic dimension for the variation of the velocity in the x-direction is given by the coordinate x itself, and that in the y-direction, perpendicular to the plate, is given by the thickness δ of the boundary layer. Here, by the equation of continuity, $V_y \sim V_x\delta/x$. The principal term in the Navier–Stokes viscous stress tensor is

$$\sigma'_{xy} \sim \rho\nu\,\partial V_x/\partial y \sim \rho\bar{v}lV/\delta.$$

Among the Burnett terms in σ'_{xy}, however, there is none containing $(\partial V_x/\partial y)^2$; it is easily seen that the derivatives $\partial V_\alpha/\partial x_\beta$ do not yield a tensor of rank two quadratic in them whose xy-component contains that square. The largest terms in $\sigma_{xy}^{(2)}$ can only be those of the form

$$\rho l^2(\partial V_x/\partial y)\,\mathrm{div}\,\mathbf{V} \sim \rho l^2 V^2/x\delta.$$

Their ratio to $\sigma_{xy}^{(1)}$ is $\sigma^{(2)}/\sigma^{(1)} \sim lV/x\bar{v} \sim (l/\delta)^2$, which is again of the second order.

We shall now show that the correction terms in the conditions at gas–solid boundaries yield effects of the first order in l/L. It follows that appreciable consequences of the rarefaction of the gas occur near solid surfaces.

In non-rarefied gases, the boundary condition at the surface of a solid is that the temperatures of the gas and the solid are equal. In reality, however, this is an approximate condition, and applies only if the mean free path may be regarded as infinitesimal. When the finite mean free path at the surface of contact between a solid and a non-uniformly heated gas is taken into account, there is a difference of temperatures, which falls to zero, in general, only when there is complete thermal equilibrium and the gas temperature is constant.†

Near a solid surface (at distances from it that are small, but not too small), the temperature gradient of the gas may be assumed constant, so that the temperature varies linearly with the distance. In the immediate neighbourhood of the wall, however, at distances $\sim l$, the temperature variation is in general more complex and its gradient is not constant. The continuous curve in Fig. 1 shows the approximate form of the gas temperature near the surface.

However, this true form of the temperature in the vicinity of the wall, which relates to distances comparable with the mean free path, is not important when considering the temperature distribution throughout the gas. As regards the temperature distribution near a solid wall, we are mainly concerned with only the straight part of the curve in Fig. 1, which extends to distances large compared with

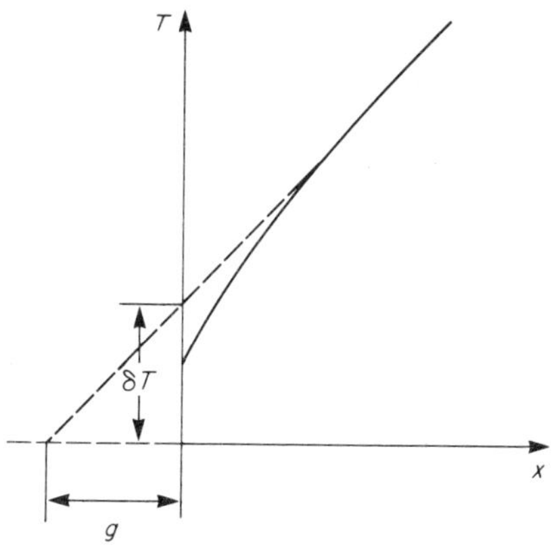

FIG. 1.

†In referring to the temperature of a gas in regions whose size is of the order of the mean free path, it is necessary, strictly speaking, to define what is meant by temperature. In the present case it will be defined in terms of the mean energy of the molecules at a given point in the gas, the function which determines the temperature from that mean energy being taken as the same as for large volumes of gas.

the mean free path. The equation of this straight line is determined by its slope and by the intercept on the ordinate axis. We are thus concerned not with the actual discontinuity of temperature at the wall, but with the discontinuity that results when the temperature gradient is assumed constant near the wall at all distances down to zero, as shown by the broken line in Fig. 1. Let δT denote this extrapolated temperature discontinuity, defined as the gas temperature minus the wall temperature (the latter being arbitrarily taken as zero in Fig. 1).

When the temperature gradient is zero, so is the discontinuity δT. Hence, for fairly small temperature gradients,

$$\delta T = g\, \partial T/\partial n; \tag{14.9}$$

the derivative is taken along the normal to the surface into the gas. The coefficient g may be called the *temperature discontinuity coefficient.* If the gas temperature increases into the volume ($\partial T/\partial n > 0$), we must also have $\delta T > 0$, and so the coefficient g is positive.

Similar effects occur at the boundary between a solid wall and a moving gas. Instead of "sticking" completely to the surface, a rarefied gas maintains a small but finite velocity near it, and slips along the surface. As in (14.9), we have as the speed of slip

$$v_0 = \xi\, \partial V_t/\partial n, \tag{14.10}$$

where V_t is the tangential component of the gas velocity near the wall. Like g, the *slip coefficient* ξ is positive. The same comments apply to v_0 as were made regarding the temperature discontinuity δT given by (14.9). This speed is, strictly speaking, not the actual speed of the gas at the wall itself, but the speed extrapolated on the assumption of a constant gradient $\partial V_t/\partial n$ in the layer of gas along the wall.

The coefficients g and ξ have the dimensions of length, and are of the same order of magnitude as the mean free path:

$$g \sim l, \quad \xi \sim l. \tag{14.11}$$

The temperature discontinuity and the slip speed themselves are consequently quantities of the first order in l/L. To calculate the coefficients g and ξ, it would be necessary to solve the transport equation for the distribution function of the gas molecules near the surface. This equation would have to take account of collisions between the gas molecules and the wall, and it would therefore be necessary to know the law governing their scattering in such collisions.

If the broken line in Fig. 1 is continued to intersect the abscissa axis, it makes an intercept of length g. Thus we can say that the temperature distribution in the presence of a temperature discontinuity is the same as if there were no discontinuity but the wall were moved back a distance g. The same applies to the slip, with the wall moved back a distance ξ. Of course, with these changes only the first-order terms in g or ξ should be retained in the solutions of problems in fluid mechanics. Since taking account of the temperature or velocity discontinuities is equivalent to moving the boundaries by distances of the order of l, the resulting corrections in the solutions are of the order of $l\partial/\partial x \sim l/L$, i.e. of the first order in l/L.

As well as the above corrections to the boundary conditions, there are other effects of the same order in l/L, which in many instances are more important, since some qualitatively new phenomena occur.

One of these is a movement of gas near a non-uniformly heated solid surface, called *thermal slip*. It bears some analogy to thermal diffusion in a mixture of gases. Just as, in the presence of a temperature gradient in a gas mixture, collisions with molecules of the other gas create a flux of particles, so in this case a flux results from collisions with the non-uniformly heated wall by molecules in a thin layer of gas at the wall, whose thickness $\sim l$.

Let $\mathbf{V}_t$ denote the tangential velocity acquired by the gas near the wall as a result of thermal slip, and $\nabla_t T$ the tangential component of the temperature gradient. In the first approximation, we can suppose that $\mathbf{V}_t$ is proportional to $\nabla_t T$, i.e. for an isotropic surface

$$\mathbf{V}_t = \mu \nabla_t T. \tag{14.12}$$

The coefficient μ must be proportional to the mean free path, since it is due to particles in a gas layer of that thickness. Then clearly, from dimensional arguments, $\mu \sim l/m\bar{v}$. Expressing the mean free path in terms of the collision cross-section and the gas density, we have $l \sim 1/N\sigma \sim T/\sigma P$, and, finally,

$$\mu \sim \frac{1}{\sigma P}\sqrt{\frac{T}{m}}. \tag{14.13}$$

The sign of μ is not determined by thermodynamic requirements; experimental results show that usually $\mu > 0$.

One further first-order effect is the presence in a moving gas of an additional surface heat flux (i.e. restricted to a layer at the wall with thickness $\sim l$) $\mathbf{q}'_{\text{surf}}$, proportional to the normal gradient of the tangential velocity:

$$\mathbf{q}'_{\text{surf}} = \varphi \partial \mathbf{V}_t/\partial n, \tag{14.14}$$

with the dimensions energy/length × time.

The coefficients μ and φ are connected by a relation which follows from Onsager's principle. To derive this, let us consider the "surface" part of the rate of increase of entropy $\dot{S}_{\text{surf}}$, due to the motion of the gas at the wall and taken per unit area of the wall surface. This quantity consists of two parts. The presence of the heat flux $\mathbf{q}'_{\text{surf}}$ contributes $-T^{-2}\mathbf{q}'_{\text{surf}} \cdot \nabla T$; cf. the corresponding expression for the rate of increase of entropy due to a bulk heat flux (*FM*, § 49; *SP* 2, § 88). Secondly, the wall past which the gas is flowing is subject to a frictional force $-\eta\partial\mathbf{V}_t/\partial n$ per unit area. The energy dissipated per unit time is equal to the work done by this force, $-\eta\partial\mathbf{V}_t/\partial n \cdot \mathbf{V}_t$, and division by T gives the contribution to the rate of increase of entropy. Thus we have

$$\dot{S}_{\text{surf}} = -\frac{1}{T^2}\mathbf{q}'_{\text{surf}} \cdot \nabla T - \frac{1}{T}\eta \mathbf{V}_t \cdot \frac{\partial \mathbf{V}_t}{\partial n}. \tag{14.15}$$

We now take as the X_a, in the general statement of Onsager's principle (§9), the vectors

$$\mathbf{X}_1 = \frac{1}{T^2}\nabla_t T, \quad \mathbf{X}_2 = \frac{1}{T}\frac{\partial \mathbf{V}_t}{\partial n}.$$

A comparison of (14.15) with the expressions (9.3) shows that the corresponding quantities $\dot{x}_a$ are the vectors

$$\dot{\mathbf{x}}_1 = \mathbf{q}'_{\text{surf}}, \quad \dot{\mathbf{x}}_2 = \eta \mathbf{V}_t.$$

The "equations of motion" (9.1) are the relations (14.12) and (14.14); writing these as

$$\dot{\mathbf{x}}_1 = T\varphi \mathbf{X}_2, \quad \dot{\mathbf{x}}_2 = \eta\mu T^2 \mathbf{X}_1,$$

we obtain the required relation

$$\varphi = T\eta\mu \tag{14.16}$$

(L. Waldmann 1967).

PROBLEMS

PROBLEM 1. Two vessels containing a gas at different temperatures T_1 and T_2 are connected by a long tube. As a result of thermal slip, a pressure difference is established between the gases in the two vessels (the *thermo-mechanical effect*). Determine this difference.

SOLUTION. The boundary condition at the surface of the tube for Poiseuille flow under the influence of the pressure and temperature gradients, with allowance for thermal slip, is $v = \mu\, dT/dx$ at $r = R$ (where R is the tube radius and the x-axis is along the length of the tube). We find in the usual way (see *FM*, §17) the velocity distribution over the tube cross-section:

$$v = -\frac{1}{4\eta}\frac{dP}{dx}(R^2 - r^2) + \mu\frac{dT}{dx}.$$

The mass of gas flowing through a cross-section of the tube per unit time is

$$Q = -\frac{\rho\pi R^4}{8\eta}\frac{dP}{dx} + \rho\mu\pi R^2\frac{dT}{dx}, \tag{1}$$

where ρ is the gas density. In mechanical equilibrium $Q = 0$, whence

$$\frac{dP}{dx} = \frac{8\eta\mu}{R^2}\frac{dT}{dx}.$$

Integration over the whole length of the tube gives the pressure difference:

$$P_2 - P_1 = (8\eta\mu/R^2)(T_2 - T_1)$$

(if $T_2 - T_1$ is fairly small, η and μ may be taken as constants). An estimate of the order of magnitude of the effect by means of (14.13) and (8.11) gives

$$\delta P/P \sim (l^2/R^2)\delta T/T.$$

The velocity distribution over the tube cross-section when $Q = 0$ is

$$v = \mu\left(\frac{2r^2}{R^2} - 1\right)\frac{dT}{dx}.$$

The gas flows along the walls in the direction of the temperature gradient ($v > 0$), and near the axis of the tube it flows in the opposite direction ($v < 0$).

PROBLEM 2. Two tubes of length L and different radii ($R_1 < R_2$) are joined at their ends; the junctions are maintained at different temperatures ($T_2 > T_1$), the difference being small. As a result of thermal slip, a circulatory motion of gas is established in the tubes. Find the total gas flow through the tube cross-sections.

SOLUTION. Dividing (1) in Problem 1 by R^4 and integrating along a closed contour formed by the two tubes, we have

$$Q = \frac{\rho\mu\pi}{L}(T_2 - T_1)(R_2^2 - R_1^2)\frac{R_1^2 R_2^2}{R_2^4 + R_1^4}.$$

The flow takes place in the direction shown in Fig. 2.

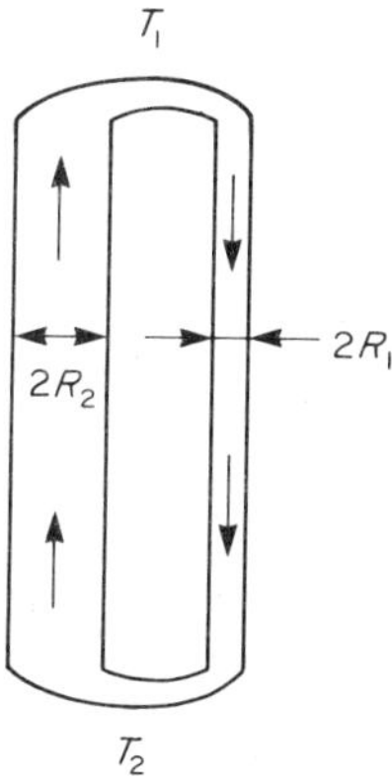

FIG. 2.

PROBLEM 3. Determine the force **F** acting on a sphere of radius R immersed in a gas where a constant temperature gradient $\nabla T = \mathbf{A}$ is maintained.

SOLUTION. The temperature distribution within the sphere is given by

$$T = \frac{3\kappa_2}{\kappa_1 + 2\kappa_2} Ar \cos\theta,$$

where κ_1 and κ_2 are the thermal conductivities of the sphere and the gas; r and θ are spherical polar coordinates with the origin at the centre of the sphere and the polar axis along **A** (see *FM*, § 50, Problem 2). Hence we find for the temperature gradient along the surface of the sphere

$$\frac{1}{R}\frac{\partial T}{\partial\theta} = -\frac{3\kappa_2}{\kappa_1 + 2\kappa_2} A \sin\theta.$$

The laminar flow of the gas resulting from the thermal slip is determined only by the one vector **A**. The corresponding solution of the Navier–Stokes equation may therefore be sought in the same form as in the problem of liquid flow past a sphere moving in it (see *FM*, § 20):

$$\mathbf{v} = -a\frac{\mathbf{A} + \mathbf{n}(\mathbf{A}\,.\,\mathbf{n})}{r} + b\frac{3\mathbf{n}(\mathbf{A}\,.\,\mathbf{n}) - \mathbf{A}}{r^3},$$

where $\mathbf{n} = \mathbf{r}/r$; the additive constant in $\mathbf{v}$ is omitted, since we must have $v = 0$ as $r \to \infty$. The constants a and b are found from the conditions

$$v_r = 0, \quad v_\theta = (\mu/R)\partial T/\partial\theta \quad \text{at} \quad r = R;$$

their values are

$$a = b/R^2 = -3\kappa_2 R\mu/2(\kappa_1 + 2\kappa_2).$$

The force on the sphere is

$$\mathbf{F} = 8\pi a\eta \mathbf{A} = -12\pi\eta\mu R\kappa_2 \nabla T/(\kappa_1 + 2\kappa_2).$$

For the surface effects considered in these Problems to be in fact small compared with the volume effects, the temperature must vary only slightly over the radius of the tube in Problems 1 and 2, and over the radius of the sphere in Problem 3.

PROBLEM 4. Two vessels joined by a long tube contain gas at the same temperature and at pressures P_1 and P_2. Determine the heat flux between the vessels which accompanies Poiseuille flow in the tube (the *mechano-caloric effect*).

SOLUTION. According to (14.14) and (14.16), the heat flux along the walls of the tube is

$$q' = 2\pi R q'_{\text{surf}} = 2\pi R T\eta\mu\, dV/dr.$$

From the condition of mechanical equilibrium of the liquid in a steady flow, we have

$$2\pi R\eta\, dV/dr = \pi R^2\, dP/dx = \pi R^2(P_2 - P_1)/L.$$

Hence, finally,

$$q' = \pi R^2 T\mu(P_2 - P_1)/L.$$

§ 15. Phenomena in highly rarefied gases

The phenomena discussed in § 14 are no more than correction effects associated with higher powers of the ratio of the mean free path l to the characteristic dimensions L of the problem; this ratio was supposed still small. If the gas is so rarefied, or the dimensions L are so small, that $l/L \gtrsim 1$, the equations of fluid dynamics become completely inapplicable, even with corrected boundary conditions.

In the general case of any l/L, it is in principle necessary to solve the transport equation with specified boundary conditions on solid surfaces in contact with the gas. These conditions depend on the interaction between the gas molecules and the surface, and relate the distribution function for particles incident on the surface to that for particles leaving it. If this interaction amounts to scattering of molecules without chemical transformation, ionization, or absorption by the surface, it is described by the probability $w(\Gamma', \Gamma)d\Gamma'$ that a molecule with given values of Γ strikes the surface and is reflected into a given range $d\Gamma'$; the function w is normalized by the condition

$$\int w(\Gamma', \Gamma)\, d\Gamma' = 1. \tag{15.1}$$

With this function, the boundary condition for the distribution function $f(\Gamma)$ becomes

$$\int_{\mathbf{n}\cdot\mathbf{v}<0} w(\Gamma', \Gamma)\mathbf{n}\cdot\mathbf{v} f(\Gamma) d\Gamma = -\mathbf{n}\cdot\mathbf{v}' f(\Gamma') \quad \text{with} \quad \mathbf{n}\cdot\mathbf{v}' > 0. \tag{15.2}$$

The integral on the left multiplied by $d\Gamma'$ is the number of molecules incident on unit area of the surface per unit time and scattered into a given range $d\Gamma'$; the integration is taken over the range of values of Γ that corresponds to molecules moving towards the surface (**n** being a unit vector along the outward normal to the surface of the body). The expression on the right of (15.2) is the number of molecules leaving unit area of the surface per unit time. The values of Γ' on each side of the equation must correspond to molecules moving away from the surface.

In equilibrium, when the temperature of the gas is the same as that of the body, the distribution function must have the Boltzmann form for both the incident and the reflected particles. Hence it follows that the function w must satisfy identically the equation

$$\int_{\mathbf{n}\,.\,\mathbf{v}<0} w(\Gamma', \Gamma)\mathbf{n}\,.\,\mathbf{v}\, e^{-\epsilon/T_1}\, d\Gamma = -\mathbf{n}\,.\,\mathbf{v}'\, e^{-\epsilon'/T_1}, \tag{15.3}$$

which is obtained by substituting in (15.2) $f(\Gamma) = \text{constant} \times \exp(-\epsilon/T_1)$, with T_1 the temperature of the body.

In the general formulation described, the solution of the problem of highly rarefied gas flow is of course very difficult. The problem can, however, be more simply stated in the limiting case where the gas is so highly rarefied that $l/L \gg 1$.

A large class of such problems relate to situations where a considerable mass of gas occupies a volume large compared with the dimensions L of solid bodies immersed in the gas, and also compared with the mean free path l. Then collisions of molecules with solid surfaces are comparatively rare, and are unimportant relative to collisions between molecules. If the gas itself is in equilibrium, with temperature T_2, we can assume under these conditions that the equilibrium is not destroyed by the immersed body. There may be any temperature difference between the gas and the body. The same is true of the macroscopic velocities.

Let $\tau = T_2 - T_1$ be the difference between the temperature of the gas and that of some part df of the surface of the body, and $\mathbf{V}$ the velocity of the gas relative to the body. For non-zero τ and $\mathbf{V}$ there is heat exchange between the gas and the body, and a force is exerted on the body by the gas. Let q be the dissipative heat flux from the gas to the body, and let $\mathbf{F} - P\mathbf{n}$ denote the force per unit area acting along the outward normal $\mathbf{n}$ at each point on the surface of the body. The second term here is the ordinary gas pressure; $\mathbf{F}$ is the additional force under consideration, due to τ and $\mathbf{V}$. The quantities q and $\mathbf{F}$ are functions of τ and $\mathbf{V}$, and are zero when these are zero.

If τ and $\mathbf{V}$ are sufficiently small (τ with respect to the temperatures themselves of the gas and the solid, $\mathbf{V}$ with respect to the thermal velocity of the gas molecules), then q and $\mathbf{F}$ can be expanded in powers of τ and $\mathbf{V}$ as far as the linear terms. Let F_n and V_n denote the components of $\mathbf{F}$ and $\mathbf{V}$ along the normal $\mathbf{n}$; $\mathbf{F}_t$ and $\mathbf{u}_t$ their tangential parts, which are vectors having two independent components. Then the expansions mentioned are

$$q = \alpha\tau + \beta V_n, \quad F_n = \gamma\tau + \delta V_n, \quad \mathbf{F}_t = \theta \mathbf{V}_t, \tag{15.4}$$

where $\alpha, \beta, \gamma, \delta, \theta$ are constants (or rather functions of temperature and pressure), characteristic of any given gas and solid material. The "scalar" quantities q and F_n

cannot, by symmetry, contain terms linear in the vector $\mathbf{V}_t$. For the same reason, the expansion of the vector $\mathbf{F}_t$ does not contain terms linear in the "scalars" τ and V_n.

The quantities α, δ and θ are positive. For example, if the gas temperature exceeds the body temperature ($\tau > 0$), heat will pass from the gas to the body, and the corresponding part of the flux q will be positive; hence $\alpha > 0$. Next, the forces F_n and $\mathbf{F}_t$ due to the gas flow relative to the body must be in the same direction as V_n and $\mathbf{V}_t$; hence $\delta > 0$ and $\theta > 0$. The sign of the coefficients β and γ does not follow from general thermodynamic considerations, although in practice they seem to be usually positive. There is a simple relation between them which is a consequence of the symmetry of the kinetic coefficients.

To derive this relation, we calculate the time derivative of the total entropy of the system comprising the gas and the body in it. A quantity of heat $q\,df$ is gained by the body from the gas in unit time through each surface element df. The increment in the entropy S_1 of the body is

$$\dot{S}_1 = \oint (q/T_1)\, df,$$

where the integration is over the whole surface of the body.

To calculate the increase in the entropy of the gas, we take coordinates such that the gas is at rest at the position of the body; then the velocity of each point on the surface is $-\mathbf{V}$. In order to demonstrate the required relation, we shall suppose that the shape of the body may vary during its motion; then the velocities $\mathbf{V}$ of various points on its surface are arbitrary independent variables. From the thermodynamic relation $dE = T\,dS - P\,d\mathcal{V}$, the change in the entropy of the gas per unit time is

$$\dot{S}_2 = (\dot{E}_2 + P_2\dot{\mathcal{V}}_2)/T_2,$$

quantities with the suffix 2 relating to the gas. The derivative $\dot{E}_2$ is, by the conservation of the total energy of the system, minus the change in the energy of the body. This change is made up of the quantity of heat $\oint q\,df$ and the work $\oint -\mathbf{V}\,.\,(\mathbf{F} - P\mathbf{n})\,df$ done on the body. Thus we find as the change in the energy of the gas

$$\dot{E}_2 = \oint (-q + F_n V_n + \mathbf{F}_t \cdot \mathbf{V}_t - P_2 V_n)\, df.$$

The change in the gas volume is equal to minus the change in the volume of the body:

$$\dot{\mathcal{V}}_2 = \oint V_n\, df.$$

The change in the entropy of the gas is therefore

$$\dot{S}_2 = \frac{1}{T_2}\oint (-q + F_n V_n + \mathbf{F}_t\,.\,\mathbf{V}_t)\, df.$$

Adding the derivatives of S_1 and S_2, and then putting (for small τ) $T_1 \approx T_2 \equiv T$, we finally have as the rate of change of the total entropy of the system

$$\dot{S} = \int \left[\frac{q\tau}{T^2} + \frac{F_n V_n}{T} + \frac{\mathbf{F}_t \cdot \mathbf{V}_t}{T}\right] df. \tag{15.5}$$

We take as the quantities $\dot{x}_1, \dot{x}_2, \dot{x}_3, \dot{x}_4$ in the general formulation of Onsager's principle (§ 9) respectively q, F_n and the two components of the vector $\mathbf{F}_t$ at any given point on the surface of the body. To find the corresponding quantities X_a, we compare (15.5) with the general expression (9.3) for the rate of change of the entropy, and see that X_1, X_2, X_3, X_4 are respectively $-\tau/T^2$, $-V_n/T$ and the two components of the vector $-\mathbf{V}_t/T$ at the same point. The kinetic coefficients (i.e. those in the relations (9.1)) are

$$\gamma_{11} = \alpha T^2, \quad \gamma_{22} = \delta T, \quad \gamma_{33} = \gamma_{44} = \theta T,$$

$$\gamma_{12} = \beta T, \quad \gamma_{21} = \gamma T^2.$$

The symmetry $\gamma_{12} = \gamma_{21}$ thus gives the required relation:

$$\beta = \gamma T. \tag{15.6}$$

Moreover, from the condition that the quadratic form (9.3) is positive ($\dot{S} > 0$), we have the inequalities $\alpha, \beta, \theta > 0$ already mentioned, and also the inequality

$$T\alpha\delta > \beta^2.$$

To calculate the coefficients in (15.4), we need to know the specific form of the law of scattering of gas molecules by the surface of the body, expressed by the function $w(\Gamma', \Gamma)$ defined above. As an example, let us derive a formula which in principle allows α to be calculated.

The energy flux from the gas to the body is given by the integral

$$q = \int (\epsilon - \epsilon')|v_x| w(\Gamma', \Gamma) f(\Gamma)\, d\Gamma\, d\Gamma', \tag{15.7}$$

taken over the ranges $v_x < 0$, $v'_x > 0$, since an amount of energy $\epsilon - \epsilon'$ is transferred to the wall at each collision of a molecule with the wall.

Let us transform this expression by means of the principle of detailed balancing, according to which, in equilibrium, the number of transitions $\Gamma \to \Gamma'$ in the scattering of molecules by the wall is equal to the number of transitions $\Gamma'^T \to \Gamma^T$. This means that

$$w(\Gamma', \Gamma)|v_x| \exp\left(\frac{\mu - \epsilon}{T_1}\right) = w(\Gamma^T, \Gamma'^T)|v'_x| \exp\left(\frac{\mu - \epsilon'}{T_1}\right); \tag{15.8}$$

in equilibrium, the temperatures of the gas and the wall are equal.

In (15.7) we rename the variables of integration: $\Gamma \to \Gamma'^T$, $\Gamma' \to \Gamma^T$. Half the sum of the two resulting expressions gives

$$q = \frac{1}{2} \int (\epsilon - \epsilon') e^{\mu/T_2} [w(\Gamma', \Gamma)|v_x| e^{-\epsilon/T_2} - w(\Gamma^T, \Gamma'^T)|v_x'| e^{-\epsilon'/T_2}] \, d\Gamma \, d\Gamma'.$$

Lastly, substituting $w(\Gamma^T, \Gamma'^T)$ from (15.8) and then expanding the integrand in powers of the small difference $\tau = T_2 - T_1$, we find that $q = \alpha\tau$, where

$$\alpha = \frac{1}{2T^2} \int (\epsilon - \epsilon')^2 |v_x| w(\Gamma', \Gamma) \exp\left(\frac{\mu - \epsilon(\Gamma)}{T}\right) d\Gamma \, d\Gamma' \quad (v_x < 0, \quad v_x' > 0); \qquad (15.9)$$

the subscript is omitted from the temperature $T_1 \approx T_2$.

The distribution function for molecules scattered from the wall depends on the specific nature of their interaction with the wall. There is said to be *complete accommodation* if the molecules reflected from each surface element of the body have (whatever the magnitude and direction of their velocity before the impact) the same distribution as in a beam leaving a small aperture in a vessel containing gas at a temperature equal to that of the body. Thus, with complete accommodation, the gas scattered by the wall reaches thermal equilibrium with it. The values of the coefficients in (15.4) may reasonably be compared with those for complete accommodation. In particular, energy exchange between the gas molecules and the solid wall is usually described by the accommodation coefficient, defined as the ratio α/α_0, where α_0 corresponds to complete accommodation. In actual cases, complete accommodation is not usually achieved, and the accommodation coefficient is less than unity.

The fact that α_0 is in fact the greatest possible value is easily shown as follows. Let us view the entropy S in (15.5) somewhat differently: not as the total entropy of the body and the gas together, but as the entropy of the body together with just the gas molecules that reach the surface of the body in a time Δt. For this system, reflection of the molecules with complete accommodation denotes a transition to a state of complete equilibrium, and its entropy therefore takes the maximum possible value. Accordingly, the change of entropy $\Delta S = \dot{S}\Delta t$ accompanying this transition will also be a maximum.† That is, for complete accommodation the quadratic form (9.3) must be a maximum for any given values of the X_a (i.e. of τ, V_n and $\mathbf{V}_t$). Denoting the corresponding values of the coefficients γ_{ab} by the suffix zero, we can write this condition as

$$\frac{\alpha_0 - \alpha}{T^2} \tau^2 + \frac{2(\beta_0 - \beta)}{T^2} \tau V_n + \frac{\delta_0 - \delta}{T} V_n^2 + \frac{\theta_0 - \theta}{T} \mathbf{V}_t^2 > 0.$$

From this, there follow the inequalities

$$\left.\begin{aligned} \alpha_0 > \alpha, \quad \delta_0 > \delta, \quad \theta_0 > 0, \\ T(\alpha_0 - \alpha)(\delta_0 - \delta) > (\beta_0 - \beta)^2. \end{aligned}\right\} \qquad (15.10)$$

†Important points in this argument are that the body (which acts as a "heat reservoir") may be regarded as in equilibrium throughout the process, and that the entropy of an ideal gas depends only on the distribution law for its molecules, not on the law of interaction between them.

Let us consider the outflow of a highly rarefied gas from a small orifice with linear dimensions L. In the limit $l/L \gg 1$, this process is a very simple one. The molecules will leave the vessel independently, forming a molecular beam in which each molecule moves at the speed with which it reached the orifice. The number of molecules leaving the orifice per unit time is equal to the number of collisions per unit time between molecules and a surface with area s equal to that of the orifice. The number of collisions per unit wall area is $P/(2\pi mT)^{1/2}$, where P is the gas pressure and m the mass of a molecule; see *SP* 1, § 39. Thus the mass of gas leaving per unit time is

$$Q = sP\sqrt{(m/2\pi T)}. \tag{15.11}$$

If two vessels containing gas are joined by an orifice, for $l \ll L$ in mechanical equilibrium the pressures P_1 and P_2 of the gases in the two vessels are equal, whatever their temperatures T_1 and T_2. If $l \gg L$, the condition of mechanical equilibrium is that the numbers of molecules passing through the orifice in each direction are equal. By (15.11), this gives

$$P_1/\sqrt{T_1} = P_2/\sqrt{T_2}. \tag{15.12}$$

Thus the pressures of rarefied gases in two communicating vessels will be different, and proportional to the square roots of the temperatures (the *Knudsen effect*).

So far we have discussed phenomena in a large mass of highly rarefied gas in equilibrium by itself. Let us now briefly consider phenomena of another type, where the gas itself is not in equilibrium, for instance in heat transfer between two solid plates heated to different temperatures and immersed in a rarefied gas, the distance between them being small compared with the mean free path. Molecules moving in the space between the plates undergo almost no collisions with one another; after reflection from one plate, they move freely until they strike the other. When scattered by the hotter plate, the molecules gain some energy from it, and then transfer some of their energy to the cooler plate when they reach it. The heat transfer mechanism in this case thus differs essentially from that of ordinary conduction in a non-rarefied gas. It may be described by a heat transfer coefficient κ, defined (by analogy with the ordinary conductivity) so that

$$q = \kappa(T_2 - T_1)/L, \tag{15.13}$$

where q is the amount of heat transferred per unit area of the plates per unit time, T_1 and T_2 the temperatures of the plates and L the distance between them. The value of κ may be estimated in order of magnitude by means of (7.10). Since collisions between molecules are now replaced by collisions of molecules with the plates, the mean free path l must be replaced by the distance L between the plates. Thus

$$\kappa \sim L\bar{v}N \sim PL/\sqrt{(mT)}. \tag{15.14}$$

The heat transfer coefficient in a highly rarefied gas is proportional to the pressure, in contrast to the conductivity of a non-rarefied gas, which is independent of the pressure. It should be emphasized, however, that κ here is not a property of the gas alone: it depends also on the specific conditions of the problem, namely the distance L between the plates.

A similar effect is the "viscosity" of a highly rarefied gas, which occurs, for example, in the relative motion of two plates in it (again with $L \ll l$). The viscosity coefficient η must here be defined so that

$$F = \eta V/L, \tag{15.15}$$

where F is the friction force per unit area on the moving plate and V the relative speed of the plates. Replacing the mean free path l in (8.11) by the distance L, we have

$$\eta \sim m\bar{v}NL \sim LP\sqrt{(m/T)}, \tag{15.16}$$

i.e. the viscosity of a rarefied gas is likewise proportional to the pressure.

PROBLEMS

PROBLEM 1. At the initial instant $t = 0$, a gas occupies the half-space $x < 0$. Neglecting collisions, determine the density distribution at subsequent instants.

SOLUTION. If collisions are neglected, the transport equation reduces to

$$\partial f/\partial t + \mathbf{v} \,.\, \partial f/\partial \mathbf{r} = 0,$$

the general solution being $f = f(\mathbf{r} - \mathbf{v}t, \mathbf{v})$. With the given initial condition, we have

$$f_0 = f_0(v) \quad \text{for} \quad v_x > x/t, \quad f = 0 \quad \text{for} \quad v_x < x/t,$$

where f_0 is the Maxwellian distribution. The gas density is

$$N(t, x) = \int_{-\infty}^{\infty}\int_{-\infty}^{\infty}\int_{x/t}^{\infty} f_0(v) m^3 \, dv_x \, dv_y \, dv_z$$

$$= \tfrac{1}{2} N_0\left[1 - \Phi\left(\frac{x}{t}\sqrt{\frac{m}{2T}}\right)\right],$$

where

$$\Phi(\xi) = \frac{2}{\sqrt{\pi}}\int_0^{\xi} e^{-y^2}\, dy,$$

and N_0 is the initial density. Since collisions have been neglected, these formulae are actually valid only in the range $|x| \ll l$.

PROBLEM 2. Determine the force acting on a sphere of radius R moving in a rarefied gas with velocity $\mathbf{V}$.

SOLUTION. The total resistance to the motion of the sphere is

$$\mathbf{F} = -(4\pi/3)\mathbf{V}R^2(\delta + 2\theta).$$

PROBLEM 3. Determine the speed of movement, in a rarefied gas, of a light plane disc whose sides are heated to different temperatures T_1 and T_2.

SOLUTION. The speed V of the disc (in the direction perpendicular to its plane) is found from the condition that the total forces acting on the two sides of zero. It moves with the cooler side forwards at a speed given (when $T_2 > T_1$) by

$$V = \gamma(T_2 - T_1)/2\delta.$$

PROBLEM 4. Calculate the value α_0 of the coefficient α corresponding to complete accommodation.

SOLUTION. The amount of energy contributed per unit time by molecules colliding with unit area of the surface of a body is $\int f_2 v_x \epsilon \, d\Gamma$, where f_2 is the Boltzmann distribution function with the temperature T_2 of the gas, ϵ is the energy of a molecule and the x-axis is perpendicular to the surface. The amount of energy carried away by the same molecules is found (in the case of complete accommodation) simply by replacing T_2 by the temperature T_1 of the body. The heat flux is

$$q = \int (f_2 - f_1)\epsilon v_x \, d\Gamma,$$

the integration over v_x being from 0 to ∞. The energy of the molecule is written as $\epsilon = \epsilon_{\text{int}} + \frac{1}{2}mv^2$, where ϵ_{int} is the internal energy. The value given by calculation for each integral is

$$\int f\epsilon v_x \, d\Gamma = \nu(\bar{\epsilon}_{\text{int}} + 2T) = \nu(\bar{\epsilon} + \tfrac{1}{2}T) = \nu T(c_v + \tfrac{1}{2}),$$

where $\bar{\epsilon} = c_v T$ is the mean energy of a molecule and $\nu = P/\sqrt{(2\pi mT)}$ the number of molecules striking unit area of the surface per unit time. The heat q is equal to the difference between the energies of the molecules arriving and leaving in equal numbers, i.e. for the same ν. The value obtained for the coefficient in $q = \alpha(T_2 - T_1)$ is

$$\alpha_0 = \frac{P}{\sqrt{(2\pi mT)}}(c_v + \tfrac{1}{2});$$

the difference $T_2 - T_1$ is assumed small, and so we put $T_1 \approx T_2 \equiv T$.

PROBLEM 5. The same as Problem 4, but for the coefficients β and γ.

SOLUTION. The normal component of the momentum contributed per unit time by the molecules striking unit area of the surface of the body is half the gas pressure. Expressing the pressure in terms of ν, we have

$$\tfrac{1}{2}P = \nu\sqrt{(\tfrac{1}{2}\pi mT)}.$$

The difference between the values of this quantity at the temperature T_1 and T_2 for the same ν gives the additional force F_n caused by the temperature difference. If $T_2 - T_1$ is small, we find

$$\gamma_0 = P/4T.$$

For β, in accordance with (15.6), $\beta_0 = P/4$.

PROBLEM 6. The same as Problem 4, but for the coefficients δ and θ.

SOLUTION. We take coordinates in which the body is at rest and the gas moves with velocity $\mathbf{V}$, the x-axis being normal to the surface and the xy-plane containing $\mathbf{V}$. The distribution function in these coordinates is

$$f = \text{constant} \times \exp\left\{-\frac{\epsilon_{\text{int}}}{T} - \frac{m}{2T}[(v_x - V_x)^2 + (v_y - V_y)^2 + v_z^2]\right\}.$$

With complete accommodation, the reflected molecules have a distribution function with $\mathbf{V} = 0$; τ is assumed to be zero.

To calculate the tangential force F_y, we put $V_x = 0$. The total y-component of momentum contributed by molecules reaching the surface of the body is

$$\int m v_y v_x f \, d\Gamma = mV_y \int v_x f \, d\Gamma = mV_y \nu,$$

the integration over v_x being always from 0 to ∞. The y-component of momentum carried away by these molecules is zero. Thus $F_y = m\nu V_y$, and so

$$\theta_0 = \nu m = P\sqrt{(m/2\pi T)}.$$

Now let $V_x \neq 0$, $V_y = 0$. As far as the first order in V_x we have

$$f = f_0 + V_x(mv_x/T)f_0,$$

where f_0 is the distribution function with $\mathbf{V} = 0$. The number of molecules colliding with unit area of the surface per unit time is

$$\nu = \int f v_x \, d\Gamma = \frac{P}{\sqrt{(2\pi m T)}} + \frac{P V_x}{2T}.$$

The x-component of momentum contributed by these molecules is

$$\int m v_x^2 f \, d\Gamma = \tfrac{1}{2}P + P V_x \sqrt{(2m/\pi T)}.$$

The molecules reflected from the bounding surface have the distribution function with $V_x = 0$, normalized so that the integral $\int f v_x \, d\Gamma$ is equal to the number ν of incident molecules determined above. The x-component of momentum carried away by these molecules is

$$-\tfrac{1}{2}\nu\sqrt{(2\pi m T)} = -\tfrac{1}{2}P - \tfrac{1}{2}P V_x \sqrt{(\pi m/2T)}.$$

The normal force additional to the pressure is $F_x = \delta_0 V_x$, where

$$\delta_0 = P\sqrt{\frac{m}{2\pi T}}\,(2 + \tfrac{1}{2}\pi) = \tfrac{1}{2}\theta_0(4 + \pi).$$

PROBLEM 7. Assuming complete accommodation, determine the temperature of a plate moving in its own plane with speed V in a rarefied gas.

SOLUTION. Proceeding as in Problem 4, we have for the energy contributed $\nu(c_v T_2 + \frac{1}{2}T_2 + \frac{1}{2}mV^2)$, and for that carried away $\nu T_1(c_v + \frac{1}{2})$. Equating these gives

$$T_1 - T_2 = mV^2/(2c_v + 1).$$

PROBLEM 8. Determine the quantity of gas flowing per unit time through the cross-section of a cylindrical tube of radius R as a result of pressure and temperature gradients. The gas is so rarefied that the mean free path $l \gg R$.† There is complete accommodation in collisions of molecules with the tube walls.

SOLUTION. The speed distribution of the molecules reflected from the wall with complete accommodation is $v_x f \, d^3p$, where f is the Maxwellian distribution function and the x-axis is perpendicular to the surface. If ϑ is the angle between the velocity of a molecule and the x-axis, we find that the distribution of the reflected molecules with respect to their directions of motion (whatever their speed) is

$$(\nu/\pi)\cos\vartheta \, do,$$

this function being normalized so as to give ν on integration over all solid angles on one side of the plane.

We take the z-axis along the axis of the tube, and the origin in the cross-section considered. Molecules last reflected from various parts of the tube surface pass through this cross-section. Of those scattered by an element df of the wall surface at a distance z, the ones that pass through the cross-section concerned are those reflected in directions lying in the solid angle subtended by this cross-section at the relevant point on the surface of the tube; their number is thus $df \,.\, \nu \int \cos\vartheta \, do/\pi$, with integration over the angle range mentioned.

This integral is evidently the same for all points lying at the same distance from the cross-section concerned. The total number of molecules passing through this cross-section per unit time is therefore obtained by replacing df by the annular surface element $2\pi R \, dz$ and integrating along the whole length of the tube; multiplying also by the mass m of a molecule, we get the mass flow rate of the gas through a cross-section of the tube:

$$Q = 2mR \int \nu\left(\int \cos\vartheta \, do\right) dz.$$

The number ν, being a function of pressure and temperature, varies along the tube. If the lengthwise gradients of pressure and temperature are not too great, we can write

$$\nu(z) = \nu(0) + z[d\nu/dz]_{z=0}.$$

†Gas flow of this type is called *free-molecular flow*.

The integral containing $\nu(0)$ is evidently zero, and so

$$Q = 2\pi R[d\nu/dz]_{z=0} \int\int z \cos\vartheta \, do \, dz.$$

To carry out the integration, we take coordinates r and φ in the plane of the cross-section considered, r being the distance of a variable point A' from a fixed point O on the circumference of the cross-section, and φ the angle between OA' and the radius of the cross-section (Fig. 3). A molecule

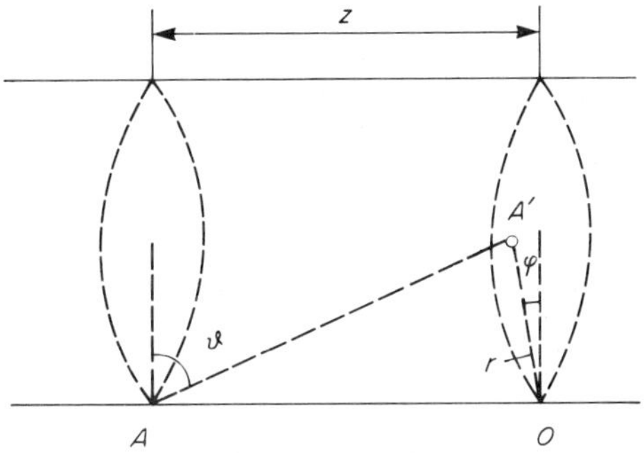

FIG. 3.

reflected from the wall at a point A on the same generator as O and then passing through A' must have a velocity at an angle ϑ to the normal to the tube surface at A such that

$$\cos\vartheta = \frac{r\cos\varphi}{\sqrt{(r^2+z^2)}}.$$

The solid-angle element may be written

$$do = \frac{r\,dr\,d\varphi}{r^2+z^2}\frac{z}{\sqrt{(r^2+z^2)}};$$

the area $r\,dr\,d\varphi$ is projected on the plane perpendicular to the line AA', and the result is divided by the square of the length of that line. The integration is carried over the region $-\frac{1}{2}\pi \leqslant \varphi \leqslant \frac{1}{2}\pi$, $0 \leqslant r \leqslant 2R\cos\varphi$, $-\infty \leqslant z \leqslant \infty$, and the result is

$$Q = (8\pi R^3/3)d\nu/dz.$$

Finally, putting $\nu = P/\sqrt{(2\pi mT)}$, we obtain

$$Q = \frac{4\pi R^3}{3L}\sqrt{(2\pi m)}\left(\frac{P_2}{\sqrt{T_2}} - \frac{P_1}{\sqrt{T_1}}\right),$$

where the difference in parentheses is between the values of $P/\sqrt{T}$ over a length L of the tube; the replacement of the derivative by the difference is allowable because Q, and therefore this derivative, are constant along the tube.

PROBLEM 9. Assuming complete accommodation, find the frictional force between two solid planes at a distance apart $L \ll l$, moving at relative speed V and having temperatures T_1 and T_2.

SOLUTION. Let plane 1 (at temperature T_1) be at rest, and plane 2 be moving at speed V in the x-direction, and let the y-direction be from plane 1 to plane 2. Molecules with speeds $v_y > 0$ and $v_y < 0$ are reflected from planes 1 and 2 respectively; with complete accommodation, their distribution functions are

$$f = \frac{2N_1}{(2\pi mT_1)^{3/2}}\exp\left(-\frac{mv^2}{2T_1}\right) \quad \text{for} \quad v_y > 0,$$

$$f = \frac{2N_2}{(2\pi mT_2)^{3/2}}\exp\left(-\frac{m(\mathbf{v}-\mathbf{V})^2}{2T_2}\right) \quad \text{for} \quad v_y < 0,$$

where N_1 and N_2 are the corresponding number densities of particles; the total density $N = N_1 + N_2$. The condition of zero total flux in the y-direction gives

$$N_1\sqrt{T_1} = N_2\sqrt{T_2}.$$

A pressure $P = N_1T_1 + N_2T_2$ acts on each plane, and the frictional force per unit area is

$$F_2 = -F_1 = mV\int_{v_y>0} v_y f\, d^3p$$
$$= VN_2\sqrt{(2mT_2/\pi)}$$
$$= VN\sqrt{(2m/\pi)}(T_1T_2)^{1/2}/(T_1^{1/2} + T_2^{1/2}).$$

If $T_1 = T_2 \equiv T$, then

$$F_2 = -F_1 = VP\sqrt{(m/2\pi T)},$$

in agreement with (15.15) and (15.16).

PROBLEM 10. Assuming complete accommodation, determine the heat transfer coefficient κ between two plates with almost equal temperatures T_1 and T_2.

SOLUTION. With complete accommodation, the molecules incident on plate 1 have an equilibrium distribution with temperature T_2. The energy flux from plate 1 to plate 2 is therefore $q = \alpha_0(T_2 - T_1)$. Taking α_0 from Problem 4 and determining κ from (15.13), we find

$$\kappa = \alpha_0 L = \frac{PL}{\sqrt{(2\pi mT)}}(c_v + \tfrac{1}{2}),$$

in accordance with the estimate (15.14).

PROBLEM 11. Determine the gas density on the axis behind a circular disc of radius $R \ll l$, moving in a gas with a velocity $-\mathbf{V}$ much greater than the mean thermal speed v_T of the atoms.

SOLUTION. When $V \gg v_T$, the particles reflected from the rear surface of the disc are unimportant (except for a narrow region near that surface; see below). The problem is a matter of the "shadow" of the disc in the incident flow. In coordinates for which the disc is at rest (and the gas is moving with velocity $\mathbf{V}$), in the absence of the disc the distribution function would be

$$f_0(\mathbf{v}) = \frac{N_0}{(2\pi mT)^{3/2}}\exp\left\{-\frac{m(\mathbf{v}-\mathbf{V})^2}{2T}\right\}.$$

In the presence of the disc, the number density of gas particles on the z-axis (Fig. 4) is

$$N(z) = 2\pi\int_0^\infty\int_{\vartheta_0}^\pi f_0(\mathbf{v})p^2\sin\vartheta\, d\vartheta\, dp,$$

where ϑ is the angle between $\mathbf{v}$ and the z-axis, and ϑ_0 the angle subtended by the radius of the disc at the point of observation on the z-axis ($\tan\vartheta_0 = R/z$; particles with $\vartheta < \vartheta_0$ are cut off by the disc). Integration, with the condition $V \gg v_T$, gives

$$N(z) = \frac{N_0}{V}\left(\frac{m}{2\pi T}\right)^{1/2}\int_0^\infty \exp\left\{-\frac{m}{2T}[(v - V\cos\vartheta_0)^2 + V^2\sin^2\theta_0]\right\}v\, dv$$
$$\approx N_0\cos\vartheta_0\exp\left\{-\frac{mV^2}{2T}\sin^2\vartheta_0\right\}$$
$$= N_0\frac{z}{\sqrt{(R^2+z^2)}}\exp\left\{-\frac{mV^2}{2T}\frac{R^2}{R^2+z^2}\right\},$$

where N_0 is the gas density far from the disc. The integration over dp is carried out with the assumption that $\cos\vartheta_0 \gg v_T/V$; it can be shown that this inequality is also the condition for particles reflected from the rear face to be negligible.

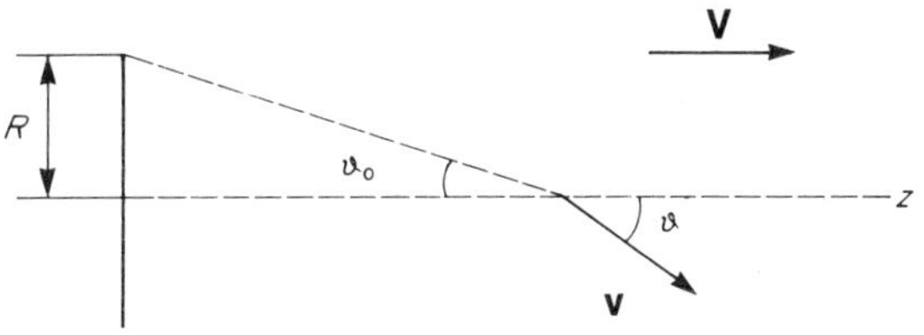

FIG. 4.

§ 16. Dynamical derivation of the transport equation

Although the derivation of the transport equation given in § 3 is satisfactory from the physical point of view, there is considerable interest in ascertaining how the equation can be derived analytically from the mathematical formalism of the theory, i.e. from the equations of motion of the gas particles. Such a derivation has been given by N. N. Bogolyubov (1946). The value of the method lies also in the fact that it affords a regular procedure for deriving in principle not only the Boltzmann equation but also the corrections to it, i.e. the terms of higher orders in the small "gaseousness parameter"—the ratio $(d/\bar{r})^3$, where d is the molecular dimension (range of action of molecular forces) and $\bar{r}$ the mean distance between the molecules. The derivation given below relates to a monatomic gas in purely classical terms, i.e. on the assumption that not only free motion but also collisions of the gas particles are describable by classical mechanics.

We start from Liouville's theorem regarding the distribution function for the gas as a whole, as a system of $\mathcal{N}$ particles. This function, in $6\mathcal{N}$-dimensional phase space, is denoted by $f^{(\mathcal{N})}(t, \tau_1, \tau_2, \ldots, \tau_{\mathcal{N}})$, where τ_a is the set of coordinates and momentum components for the ath particle: $\tau_a = (\mathbf{r}_a, \mathbf{p}_a)$. The function is assumed normalized to unity:

$$\int f^{(\mathcal{N})}(t, \tau_1, \tau_2, \ldots, \tau_{\mathcal{N}})\, d\tau_1 \ldots d\tau_{\mathcal{N}} = 1, \quad d\tau_a = d^3x_a\, d^3p_a.$$

The "one-particle" distribution function which appears in the Boltzmann equation is obtained by integrating $f^{(\mathcal{N})}$ over all $d\tau_a$ but one:

$$f^{(1)}(t, \tau_1) = \int f^{(\mathcal{N})}\, d\tau_2 \ldots d\tau_{\mathcal{N}}; \tag{16.1}$$

the function $f^{(1)}$ also is normalized to unity, and we shall retain the notation f (without superscript) for the distribution function normalized to the total number of particles: $f = \mathcal{N} f^{(1)}$.

It has been noted in *SP* 1, § 3, that Liouville's theorem arises as a consequence of the equation of continuity in phase space which must be satisfied by the distribution function for a closed system:

$$\frac{\partial f^{(\mathcal{N})}}{\partial t} + \sum_{a=1}^{\mathcal{N}} \left\{ \frac{\partial}{\partial \mathbf{r}_a} (f^{(\mathcal{N})} \dot{\mathbf{r}}_a) + \frac{\partial}{\partial \mathbf{p}_a} (f^{(\mathcal{N})} \dot{\mathbf{p}}_a) \right\} = 0. \tag{16.2}$$

With Hamilton's equations

$$\dot{\mathbf{r}}_a = \partial H/\partial \mathbf{p}_a, \quad \dot{\mathbf{p}}_a = -\partial H/\partial \mathbf{r}_a, \tag{16.3}$$

this gives

$$\frac{\partial f^{(\mathcal{N})}}{\partial t} + \sum_{a=1}^{\mathcal{N}} \left\{ \frac{\partial f^{(\mathcal{N})}}{\partial \mathbf{r}_a} \cdot \dot{\mathbf{r}}_a + \frac{\partial f^{(\mathcal{N})}}{\partial \mathbf{p}_a} \cdot \dot{\mathbf{p}}_a \right\} = \frac{df^{(\mathcal{N})}}{dt} = 0, \tag{16.4}$$

where the $\dot{\mathbf{r}}_a \equiv \mathbf{v}_a$ and $\dot{\mathbf{p}}_a$ are assumed to be expressed in terms of $\tau_1, \tau_2, \ldots$ by means of equations (16.3). Equation (16.4) expresses the content of Liouville's theorem.

We write the Hamiltonian function for a monatomic gas in the form

$$H = \sum_{a \le \mathcal{N}} \frac{p_a^2}{2m} + \sum_{b<a\le \mathcal{N}} U(|\mathbf{r}_a - \mathbf{r}_b|). \tag{16.5}$$

Here it is assumed that there is no external field, and that the interaction between the gas particles reduces to the sum of their pair interactions.† Equation (16.4) then becomes

$$\frac{\partial f^{(\mathcal{N})}}{\partial t} + \sum_{a=1}^{\mathcal{N}} \left\{ \frac{\partial f^{(\mathcal{N})}}{\partial \mathbf{r}_a} \cdot \mathbf{v}_a - \frac{\partial f^{(\mathcal{N})}}{\partial \mathbf{p}_a} \cdot \sum_{b<a} \frac{\partial U_{ab}}{\partial \mathbf{r}_a} \right\} = 0, \tag{16.6}$$

where U_{ab} $(a \neq b)$ denotes $U(|\mathbf{r}_a - \mathbf{r}_b|)$.

Let us now integrate this equation over $d\tau_2 \ldots d\tau_{\mathcal{N}}$. Then, of all the terms in the sum in (16.6), only those remain which involve differentiation with respect to $\mathbf{p}_1$ or $\mathbf{r}_1$; the integrals of the other terms are transformed into integrals over infinite surfaces in momentum or coordinate space, and are zero. Thus we have

$$\frac{\partial f^{(1)}(t, \tau_1)}{\partial t} + \mathbf{v}_1 \cdot \frac{\partial f^{(1)}(t, \tau_1)}{\partial \mathbf{r}_1} = \mathcal{N} \int \frac{\partial U_{12}}{\partial \mathbf{r}_1} \cdot \frac{\partial f^{(2)}(t, \tau_1, \tau_2)}{\partial \mathbf{p}_1} \, d\tau_2, \tag{16.7}$$

where $f^{(2)}$ is the two-particle distribution function normalized to unity, i.e. the integral

$$f^{(2)}(t, \tau_1, \tau_2) = \int f^{(\mathcal{N})} \, d\tau_3 \ldots d\tau_{\mathcal{N}}; \tag{16.8}$$

the factor $\mathcal{N}$ in (16.7) takes account of terms that differ only in the nomenclature of the variables of integration; strictly speaking, the number of such terms is $\mathcal{N} - 1$, but this is very large and may be replaced by $\mathcal{N}$.

Similarly, integrating (16.6) over $d\tau_3 \ldots d\tau_{\mathcal{N}}$, we obtain

$$\frac{\partial f^{(2)}}{\partial t} + \mathbf{v}_1 \cdot \frac{\partial f^{(2)}}{\partial \mathbf{r}_1} + \mathbf{v}_2 \cdot \frac{\partial f^{(2)}}{\partial \mathbf{r}_2} - \frac{\partial U_{12}}{\partial \mathbf{r}_1} \cdot \frac{\partial f^{(2)}}{\partial \mathbf{p}_1} - \frac{\partial U_{12}}{\partial \mathbf{r}_2} \cdot \frac{\partial f^{(2)}}{\partial \mathbf{p}_2}$$

$$= \mathcal{N} \int \left[\frac{\partial f^{(3)}}{\partial \mathbf{p}_1} \cdot \frac{\partial U_{13}}{\partial \mathbf{r}_1} + \frac{\partial f^{(3)}}{\partial \mathbf{p}_2} \cdot \frac{\partial U_{23}}{\partial \mathbf{r}_2} \right] d\tau_3, \tag{16.9}$$

where $f^{(3)}(t, \tau_1, \tau_2, \tau_3)$ is the three-particle distribution function.

†The latter assumption constitutes a model, but it does not affect the result in the first approximation (which corresponds to the Boltzmann equation): in this approximation, only pair collisions of particles occur, in which other (non-pair) interactions play no part.

Continuing in this way, we should obtain an almost infinite ($\mathcal{N}$ being very large) sequence of equations, each expressing $f^{(n)}$ in terms of $f^{(n+1)}$. All these equations are exact in the sense that no assumption has been made in them as to the rarefaction of the gas. To obtain a closed set of equations, the series has to be terminated in some way by making use of the condition that the gas is rarefied. In particular, the first approximation in this method corresponds to terminating the series already at the first equation, (16.7), in which the two-particle function $f^{(2)}$ is expressed approximately in terms of $f^{(1)}$. This is done by using the rarefaction of the gas, by means of equation (16.9).

Returning to this equation, we shall first of all show that the integral on the right-hand side is small. The function $U(r)$ is noticeably different from zero only within the range of action of the forces, i.e. when $r \lesssim d$. Hence, in both parts of the integral in (16.9), the integrations over coordinates are in practice only over the region $|\mathbf{r}_3 - \mathbf{r}_1| \lesssim d$ or $|\mathbf{r}_3 - \mathbf{r}_2| \lesssim d$, i.e. over a volume $\sim d^3$. Since in an integration over the whole volume of the gas, $\mathcal{V} \sim \mathcal{N}\bar{r}^3$, we should have $\int f^{(3)} d\tau_3 = f^{(2)}$, we obtain the estimate

$$\mathcal{N} \int \frac{\partial f^{(3)}}{\partial \mathbf{p}_1} \cdot \frac{\partial U_{13}}{\partial \mathbf{r}_1} d\tau_3 \sim \frac{\partial U(r)}{\partial r} \frac{\partial f^{(2)}}{\partial p_1} \frac{d^3}{\bar{r}^3}.$$

From this we see that the right-hand side of (16.9) is small in the ratio $(d/\bar{r})^3$ relative to the terms containing $\partial U/\partial \mathbf{r}$ on the left-hand side, and may therefore be neglected. The terms on the left constitute the total derivative $df^{(2)}/dt$, in which $\mathbf{r}_1$, $\mathbf{r}_2$, $\mathbf{p}_1$, $\mathbf{p}_2$ are regarded as functions of time which satisfy the equations of motion (16.3) with the two-body Hamiltonian

$$H = \frac{\mathbf{p}_1^2}{2m} + \frac{\mathbf{p}_2^2}{2m} + U(|\mathbf{r}_1 - \mathbf{r}_2|).$$

Thus we have

$$df^{(2)}(t, \tau_1, \tau_2)/dt = 0. \tag{16.10}$$

So far, all the transformations of the equations have been purely mechanical ones. To derive the transport equation, of course, some statistical assumption is also necessary. This may be formulated as the statistical independence of each pair of colliding particles, which has essentially been assumed in deriving the transport equation in § 3 (where the collision probability was written in the form (2.1), proportional to the product ff_1). In the method under consideration, this statement acts as the initial condition for the differential equation (16.10). It creates the asymmetry in relation to the two directions of time, and as a result the irreversible transport equation is derived from the equations of mechanics invariant under time reversal. The correlation between the positions and the momenta of the gas particles arises only as a result of their collisions and extends to distances $\sim d$. Thus the assumption of the statistical independence of colliding particles is also the source of the fundamental limitations as regards the distances and time intervals allowed by the transport equation, already discussed in § 3.

Let t_0 be some instant before the collision, when the two particles are still far apart ($|\mathbf{r}_{10}-\mathbf{r}_{20}| \gg d$, where the suffix zero denotes the values of quantities at that instant). The statistical independence of colliding particles means that at such an instant t_0 the two-particle distribution function is the product of two one-particle functions $f^{(1)}$. Hence the integration of (16.10) from t_0 to t gives

$$f^{(2)}(t, \tau_1, \tau_2) = f^{(1)}(t_0, \tau_{10}) f^{(1)}(t_0, \tau_{20}). \tag{16.11}$$

Here $\tau_{10} = (\mathbf{r}_{10}, \mathbf{p}_{10})$ and $\tau_{20} = (\mathbf{r}_{20}, \mathbf{p}_{20})$ are to be understood as those values of the coordinates and momenta which the particles must have at the instant t_0 in order to acquire the necessary values $\tau_1 = (\mathbf{r}_1, \mathbf{p}_1)$ and $\tau_2 = (\mathbf{r}_2, \mathbf{p}_2)$ at the instant t; in this sense, τ_{10} and τ_{20} are functions of τ_1, τ_2 and $t - t_0$ (only $\mathbf{r}_{10}$ and $\mathbf{r}_{20}$ depend on $t - t_0$; the values of $\mathbf{p}_{10}$ and $\mathbf{p}_{20}$ relate to particles moving freely before the collision, and do not depend on the choice of $t - t_0$).

Let us now return to (16.7), which is to become the transport equation. The left-hand side already has the required form; we shall now be concerned with the integral on the right, which is ultimately to become the collision integral in the Boltzmann equation. Substituting in this integral $f^{(2)}$ from (16.11) and changing on both sides from $f^{(1)}$ to $f = \mathcal{N} f^{(1)}$, we write

$$\frac{\partial f(t, \tau_1)}{\partial t} + \mathbf{v}_1 \cdot \frac{\partial f(t, \tau_1)}{\partial \mathbf{r}_1} = C(f),$$

where

$$C(f) = \int \frac{\partial U_{12}}{\partial \mathbf{r}_1} \cdot \frac{\partial}{\partial \mathbf{p}_1} \{ f(t_0, \tau_{10}) f(t_0, \tau_{20}) \} d\tau_2. \tag{16.12}$$

Only the range $|\mathbf{r}_2 - \mathbf{r}_1| \sim d$, i.e. the region in which the collision occurs, is important in the integral (16.12). In this range, however, we can neglect (in the first approximation, which is being considered here) the coordinate dependence of f, which varies appreciably only over distances L, the characteristic dimensions of the problem, which are certainly large in comparison with d. The final form of the collision integral will therefore be unaltered if, in order to simplify somewhat the analysis and the formulae, we take the case of spatial homogeneity, i.e. assume that f is independent of the coordinates. It may be noted immediately that the explicit time dependence through $\mathbf{r}_{10}(t)$ and $\mathbf{r}_{20}(t)$ then disappears from the functions $f(t_0, \mathbf{p}_{10})$ and $f(t_0, \mathbf{p}_{20})$.

We can transform the integrand in (16.12) by using the fact that the expression in the braces is an integral of the motion (and appeared as such in (16.11)); independently of this, it is obvious that $\mathbf{p}_{10}$ and $\mathbf{p}_{20}$, the values of the momenta at a fixed instant t_0, are by definition integrals of the motion. Using also the fact mentioned above that they contain no explicit dependence on the time t, we have

$$\frac{d}{dt} f(t_0, \mathbf{p}_{10}) f(t_0, \mathbf{p}_{20})$$

$$= \left(\mathbf{v}_1 \cdot \frac{\partial}{\partial \mathbf{r}_1} + \mathbf{v}_2 \cdot \frac{\partial}{\partial \mathbf{r}_2} - \frac{\partial U_{12}}{\partial \mathbf{r}_1} \cdot \frac{\partial}{\partial \mathbf{p}_1} - \frac{\partial U_{12}}{\partial \mathbf{r}_2} \cdot \frac{\partial}{\partial \mathbf{p}_2} \right) f(t_0, \mathbf{p}_{10}) f(t_0, \mathbf{p}_{20}) = 0. \tag{16.13}$$

From this, we express the derivative with respect to $\mathbf{p}_1$ in terms of those with respect to $\mathbf{r}_1$, $\mathbf{r}_2$ and $\mathbf{p}_2$, and substitute in (16.12). The term containing the derivative $\partial/\partial\mathbf{p}_2$ disappears when the integral is transformed to a surface integral in momentum space. We then find

$$C(f(t,\mathbf{p}_1)) = \int \mathbf{v}_{\text{rel}} \cdot \frac{\partial}{\partial \mathbf{r}} \{f(t_0, \mathbf{p}_{10}) f(t_0, \mathbf{p}_{20})\}\, d^3x\, d^3p_2, \tag{16.14}$$

with the relative velocity of the particles $\mathbf{v}_{\text{rel}} = \mathbf{v}_1 - \mathbf{v}_2$, taking into account the fact that $\mathbf{p}_{10}$ and $\mathbf{p}_{20}$ (and therefore the whole expression in the braces) depend on $\mathbf{r}_1$ and $\mathbf{r}_2$ only through the difference $\mathbf{r} = \mathbf{r}_1 - \mathbf{r}_2$. Replacing $\mathbf{r} = (x, y, z)$ by cylindrical polar coordinates z, ρ, φ with the z-axis along $\mathbf{v}_{\text{rel}}$, we have $\mathbf{v}_{\text{rel}} \cdot \partial/\partial\mathbf{r} = v_{\text{rel}}\partial/\partial z$, and the integration over dz converts (16.14) into†

$$C(f(t,\mathbf{p}_1)) = \int [f(t_0, \mathbf{p}_{10}) f(t_0, \mathbf{p}_{20})]_{z=-\infty}^{z=\infty} v_{\text{rel}}\rho\, d\rho\, d\varphi\, d^3p_2. \tag{16.15}$$

We now use the fact that $\mathbf{p}_{10}$ and $\mathbf{p}_{20}$ are the initial (at time t_0) momenta of particles which at the final instant t have momenta $\mathbf{p}_1$ and $\mathbf{p}_2$. If at the final instant $z = z_1 - z_2 = -\infty$, it is clear that at the initial instant the particles were "even further" apart, i.e. there has been no collision. In this case, therefore, the initial and final momenta are the same:

$$\mathbf{p}_{10} = \mathbf{p}_1, \quad \mathbf{p}_{20} = \mathbf{p}_2 \quad \text{for} \quad z = -\infty.$$

If $z = +\infty$, $\mathbf{p}_{10}$ and $\mathbf{p}_{20}$ act as the initial momenta for the collision which gives the particles momenta $\mathbf{p}_1$ and $\mathbf{p}_2$; in this case, we write

$$\mathbf{p}_{10} = \mathbf{p}_1'(\rho), \quad \mathbf{p}_{20} = \mathbf{p}_2'(\rho) \quad \text{for} \quad z = +\infty.$$

These are functions of the coordinate ρ, which acts as the impact parameter for the collision. The product

$$\rho\, d\rho\, d\varphi = d\sigma$$

is the classical collision cross-section.

Lastly, it is to be noted that the explicit dependence of the functions $f(t_0, \mathbf{p}_{10})$ and $f(t_0, \mathbf{p}_{20})$ on t_0 can be replaced in this approximation by a similar dependence on t. The validity of (16.11) requires only the inequality $t - t_0 \gg d/\bar{v}$ to be satisfied: at the instant t_0, the distance between the particles must be large in comparison with the range d of the forces. The difference $t - t_0$, however, may be so chosen as to satisfy also the condition $t - t_0 \ll l/\bar{v}$, where l is the mean free path; the ratio $l/\bar{v}$, which is the mean free time, is just the characteristic quantity that determines the

†The limits $z = \pm\infty$ are to be understood as distances large compared with d, but small compared with the mean free path l; if they were taken literally, the result would be zero, since $f \equiv 0$ outside the region occupied by the gas. This has arisen because in going from (16.12) to (16.14) we used equation (16.13), which is valid only until the particles in question undergo their next collisions.

periods of possible time variation of the distribution function. The change in this function during the time $t - t_0$ will then be relatively small and may be neglected.

From these considerations, we obtain the final form of the integral (16.15):

$$C(f(t, \mathbf{p}_1)) = \int \{f(t, \mathbf{p}_1')f(t, \mathbf{p}_2') - f(t, \mathbf{p}_1)f(t, \mathbf{p}_2)\}v_{rel}\, d\sigma\, d^3p_2, \tag{16.16}$$

which agrees with the Boltzmann collision integral (3.9).

§ 17. The transport equation including three-particle collisions

To find the first correction terms to the Boltzmann equation, we must go back to the points in § 16 where terms were neglected, and increase the accuracy of the calculations by one further order of magnitude relative to the gaseousness parameter. First of all, terms containing the triple correlation $f^{(3)}$ were omitted in (16.9), and three-atom collisions were thereby left out of consideration. Moreover, in converting the collision integral (16.12) to the final form (16.16), we neglected the variation of the distribution function over distances $\sim d$ and times $\sim d/\bar{v}$; the pair collisions were thereby regarded as "local" events occurring at a single point. We must now take both these corrections into account: three-particle collisions, and the "non-localness" of pair collisions.

In the first approximation, the sequence of equations was terminated at the second equation, which relates $f^{(2)}$ and $f^{(3)}$. In the second approximation, we must go to the third equation, which relates $f^{(3)}$ and $f^{(4)}$, omitting the $f^{(4)}$ terms in the same way as the $f^{(3)}$ terms were omitted in (16.9) in the first approximation. The equation then becomes

$$df^{(3)}(t, \tau_1, \tau_2, \tau_3)/dt = 0, \tag{17.1}$$

corresponding to the earlier equation (16.10) for $f^{(2)}$; the variables τ_1, τ_2, τ_3 in (17.1) are assumed to vary with time according to the equations of motion in the three-body problem; a pair interaction between particles is again assumed.† With the statistical independence of the particles before the collision, the solution of (17.1) is

$$f^{(3)}(t, \tau_1, \tau_2, \tau_3) = f^{(1)}(t_0, \tau_{10})f^{(1)}(t_0, \tau_{20})f^{(1)}(t_0, \tau_{30}). \tag{17.2}$$

The quantities t_0, τ_{a0} $(a = 1, 2, 3)$ here have the same sense as in (16.11); $\tau_{a0} = \tau_{a0}(t, t_0, \tau_1, \tau_2, \tau_3)$ are the values of the coordinates and momenta which the particles must have at the instant t_0 in order to reach the specified points τ_1, τ_2, τ_3 in phase space at the instant t. The only difference from (16.11) is that $\tau_{a0} = (\mathbf{r}_{a0}, \mathbf{p}_{a0})$ are now

†In contrast to the first approximation (cf. the first footnote to § 16), this assumption now places some limitations on the generality of the treatment, since in three-body collisions there could be an effect of three-body interactions, i.e. terms of the form $U(r_2 - r_1, r_3 - r_1)$ in the Hamiltonian, which do not reduce to pair interactions.

the initial coordinates and momenta in a *three*-body problem, which will be supposed solved in principle.†

To write down and transform the subsequent formulae, it is convenient to define an operator $\hat{S}_{123}$ whose effect on a function of the variables τ_1, τ_2, τ_3 (pertaining to the three particles in the three-body problem) is to change these variables according to

$$\left.\begin{aligned} \mathbf{r}_a &\to \tilde{\mathbf{r}}_a = \mathbf{r}_{a0} + (\mathbf{p}_{a0}/m)(t - t_0), \\ \mathbf{p}_a &\to \tilde{\mathbf{p}}_a = \mathbf{p}_{a0}. \end{aligned}\right\} \tag{17.3}$$

Similarly, the operator $\hat{S}_{12}$ will make this change in functions of the variables τ_1 and τ_2 which pertain to the two particles in the two-body problem. An important property of the transformation (17.3) is that for times $t - t_0 \gg d/\bar{v}$ it is no longer time-dependent: for such $t - t_0$, the particles are far apart and move freely with constant velocities $\mathbf{v}_{a0} = \mathbf{p}_{a0}/m$, the values of the $\mathbf{r}_{a0}$ vary with time as constant $-\mathbf{v}_{a0}(t - t_0)$, and the time dependence in (17.3) disappears. Moreover, if there were no interaction between the particles, the transformation (17.3) would reduce to an identity: in motion that is free at all times, the right-hand sides are identically equal to the left-hand sides. For the same reason, if one of the particles, say particle 1, does not interact with particles 2 and 3, then $\hat{S}_{123} \equiv \hat{S}_{23}$; the operators $\hat{S}_{12}$ and $\hat{S}_{13}$ then reduce to unity. It is therefore evident that the operator

$$\hat{G}_{123} = \hat{S}_{123} - \hat{S}_{12} - \hat{S}_{13} - \hat{S}_{23} + 2 \tag{17.4}$$

is zero if any one of the three particles does not interact with the other two, i.e. this operator separates from the functions the part that is due to the interaction of all three particles (whereas the three-body problem also includes, as particular cases, pair interactions, with the third particle in free motion).

With the operator $\hat{S}_{123}$, (17.2) becomes

$$f^{(3)}(t, \tau_1, \tau_2, \tau_3) = \hat{S}_{123}\tilde{f}^{(1)}(t, t_0, \tau_1)\tilde{f}^{(1)}(t, t_0, \tau_2)\tilde{f}^{(1)}(t, t_0, \tau_3), \tag{17.5}$$

where

$$\tilde{f}^{(1)}(t, t_0, \tau) = f^{(1)}(t_0, \mathbf{r} - \mathbf{p}(t - t_0)/m, \mathbf{p}); \tag{17.6}$$

the shift of the argument $\mathbf{r}$ in $f^{(1)}$ compensates that due to the operator $\hat{S}_{123}$.

The two-particle distribution $f^{(2)}$ is obtained by integrating the function $f^{(3)}$ with respect to the variables τ_3, and integration with respect to τ_2 and τ_3 gives the distribution function $f^{(1)}$:

$$f^{(2)}(t, \tau_1, \tau_2) = \int f^{(3)}(t, \tau_1, \tau_2, \tau_3)\, d\tau_3, \tag{17.7}$$

$$f^{(1)}(t, \tau_1) = \int f^{(3)}(t, \tau_1, \tau_2, \tau_3)\, d\tau_2\, d\tau_3. \tag{17.8}$$

†In practice, of course, an analytical solution of the three-body problem can be given only in a few cases such as that of hard spheres.

The object of the subsequent calculation is to eliminate $\tilde{f}^{(1)}$ from these two equations, with $f^{(3)}$ from (17.5), and so express $f^{(2)}$ with the necessary accuracy in terms of $f^{(1)}$. Then, substituting this expression in (16.7), which is itself exact, we arrive at the transport equation sought.

To carry out this programme, we first of all transform the integral (17.8), expressing the operator $\hat{S}_{123}$ in (17.5) in terms of $\hat{G}_{123}$ by (17.4). With the equations

$$\int \tilde{f}^{(1)}(t, t_0, \tau)\, d\tau = \int f^{(1)}(t_0, \tau)\, d\tau = 1,$$

$$\int \hat{S}_{12}\tilde{f}^{(1)}(t, t_0, \tau_1)\tilde{f}^{(1)}(t, t_0, \tau_2)\, d\tau_1\, d\tau_2 = 1,$$

which are obvious from the conservation of the total number of molecules, we obtain

$$f^{(1)}(t, \tau_1) = \tilde{f}^{(1)}(t, t_0, \tau_1) + 2\int \{(\hat{S}_{12} - 1)\tilde{f}^{(1)}(t, t_0, \tau_1)\tilde{f}^{(1)}(t, t_0, \tau_2)\}d\tau_2$$
$$+ \int \{\hat{G}_{123}\tilde{f}^{(1)}(t, t_0, \tau_1)\tilde{f}^{(1)}(t, t_0, \tau_2)\tilde{f}^{(1)}(t, t_0, \tau_3)\}\, d\tau_2\, d\tau_3. \qquad (17.9)$$

This equation can be solved for $\tilde{f}^{(1)}$ by successive approximation, bearing in mind that $\hat{S}_{12} - 1$ is of the first order of smallness, and $\hat{G}_{123}$ of the second order; compare the estimate of the right-hand side of (16.9). In the zero-order approximation, $\tilde{f}^{(1)}(t, t_0, \tau_1) = f^{(1)}(t, \tau_1)$. In the next two approximations,

$$\tilde{f}^{(1)}(t, t_0, \tau_1) = f^{(1)}(t, \tau_1) - 2\int \{(\hat{S}_{12} - 1)f^{(1)}(t, \tau_1)f^{(1)}(t, \tau_2)\}d\tau_2$$
$$- \int \{\hat{G}_{123} - 4(\hat{S}_{12} - 1)(\hat{S}_{13} + \hat{S}_{23} - 2)f^{(1)}(t, \tau_1)f^{(1)}(t, \tau_2)f^{(1)}(t, \tau_3)\}\, d\tau_2\, d\tau_3.$$

It now remains to substitute this expression in (17.5) and then in (17.7), retaining only the terms that are not above the second order of smallness, $\sim (\hat{S}_{12} - 1)^2$ or $\sim \hat{G}_{123}$. The final result is

$$f^{(2)}(t, \tau_1, \tau_2) = \hat{S}_{12}f^{(1)}(t, \tau_1)f^{(1)}(t, \tau_2) + \int \{\hat{R}_{123}f^{(1)}(t, \tau_1)f^{(1)}(t, \tau_2)f^{(1)}(t, \tau_3)\}\, d\tau_3, \qquad (17.10)$$

where

$$\hat{R}_{123} = \hat{S}_{123} - \hat{S}_{12}\hat{S}_{13} - \hat{S}_{12}\hat{S}_{23} + \hat{S}_{12}. \qquad (17.11)$$

It should be emphasized that the order of the S operators in their products is significant. The operator $\hat{S}_{12}\hat{S}_{23}$, for example, first changes the variables $\tau_1, \tau_2, \tau_3 \to \tau_1, \tilde{\tau}_2(\tau_2, \tau_3), \tilde{\tau}_3(\tau_2, \tau_3)$, the functions $\tilde{\tau}_{2,3}(\tau_2, \tau_3)$ being determined from the equations of motion of the interacting particles 2 and 3; the variables τ_1, τ_2, τ_3 are then subjected to the transformation $\tau_1, \tau_2, \tau_3 \to \tilde{\tau}_1(\tau_1, \tau_2), \tilde{\tau}_2(\tau_1, \tau_2), \tau_3$, where now the functions $\tilde{\tau}_{1,2}(\tau_1, \tau_2)$ are determined by the problem of the motion of a pair of interacting particles 1 and 2.

Next, substituting (17.10) in (16.7) and changing everywhere from the functions $f^{(1)}$ to $f = \mathcal{N}f^{(1)}$, we have the transport equation in the form†

$$\frac{\partial f(t, \tau_1)}{\partial t} + \mathbf{v}_1 \cdot \frac{\partial f(t, \tau_1)}{\partial \mathbf{r}_1} = C^{(2)}(f) + C^{(3)}(f), \qquad (17.12)$$

where

$$C^{(2)}(f(t, \tau_1)) = \int \frac{\partial U_{12}}{\partial \mathbf{r}_1} \cdot \frac{\partial}{\partial \mathbf{p}_1} \{\hat{S}_{12} f(t, \tau_1) f(t, \tau_2)\}\, d\tau_2, \qquad (17.13)$$

$$C^{(3)}(f(t, \tau_1)) = \frac{1}{\mathcal{N}} \int \frac{\partial U_{12}}{\partial \mathbf{r}_1} \cdot \frac{\partial}{\partial \mathbf{p}_1} \{\hat{R}_{123} f(t, \tau_1) f(t, \tau_2) f(t, \tau_3)\}\, d\tau_2\, d\tau_3. \qquad (17.14)$$

The first of these is the pair collision integral, and the second is the three-body collision integral. Let us consider their structure in more detail.

In both integrals, the integrands involve functions f taken at different points in space. In the pair collision integral, the effect of this "non-localness" is to be separated as a correction to the ordinary (Boltzmann) integral. To do so, we expand the functions f, which vary only slowly (over distances $\sim d$), in powers of $\mathbf{r}_2 - \mathbf{r}_1$.

Since these functions in the integrand are preceded by the operator $\hat{S}_{12}$, let us first consider the quantities $\hat{S}_{12}\mathbf{r}_1$ and $\hat{S}_{12}\mathbf{r}_2$ into which that operator transforms the variables $\mathbf{r}_1$ and $\mathbf{r}_2$. The centre of mass $\frac{1}{2}(\mathbf{r}_1 + \mathbf{r}_2)$ of the two particles moves uniformly in the two-body problem; the operator $\hat{S}_{12}$ therefore leaves this sum unchanged. We can thus write

$$\begin{aligned} \hat{S}_{12}\mathbf{r}_1 &= \hat{S}_{12}(\tfrac{1}{2}(\mathbf{r}_1 + \mathbf{r}_2) + \tfrac{1}{2}(\mathbf{r}_1 - \mathbf{r}_2)) \\ &= \mathbf{r}_1 + \tfrac{1}{2}(\mathbf{r}_2 - \mathbf{r}_1) - \tfrac{1}{2}\hat{S}_{12}(\mathbf{r}_2 - \mathbf{r}_1), \\ \hat{S}_{12}\mathbf{r}_2 &= \mathbf{r}_1 + \tfrac{1}{2}(\mathbf{r}_2 - \mathbf{r}_1) + \tfrac{1}{2}\hat{S}_{12}(\mathbf{r}_2 - \mathbf{r}_1). \end{aligned}$$

Now, expanding the functions

$$\hat{S}_{12} f(t, \mathbf{r}_1, \mathbf{p}_1) = f(t, \hat{S}_{12}\mathbf{r}_1, \mathbf{p}_{10}),$$
$$\hat{S}_{12} f(t, \mathbf{r}_2, \mathbf{p}_2) = f(t, \hat{S}_{12}\mathbf{r}_2, \mathbf{p}_{20})$$

in powers of $\mathbf{r}_2 - \mathbf{r}_1$ as far as the first-order terms, we obtain

$$C^{(2)}(f) = C_0^{(2)}(f) + C_1^{(2)}(f), \qquad (17.15)$$

where

$$C_0^{(2)}(f(t, \mathbf{r}_1, \mathbf{p}_1)) = \int \frac{\partial U_{12}}{\partial \mathbf{r}_1} \cdot \frac{\partial}{\partial \mathbf{p}_1} \{f(t, \mathbf{r}_1, \mathbf{p}_{10}) f(t, \mathbf{r}_2, \mathbf{p}_{20})\}\, d\tau_2, \qquad (17.16)$$

$$\begin{aligned} C_1^{(2)}(f(t, \mathbf{r}_1, \mathbf{p}_1)) = \frac{1}{2} \int \frac{\partial U_{12}}{\partial \mathbf{r}_1} \cdot \frac{\partial}{\partial \mathbf{p}_1} \Big\{ &(\mathbf{r}_2 - \mathbf{r}_1) \cdot \frac{\partial}{\partial \mathbf{r}_1} f(t, \mathbf{r}_1, \mathbf{p}_{10}) f(t, \mathbf{r}_2, \mathbf{p}_{20}) \\ &+ \Big[f(t, \mathbf{r}_1, \mathbf{p}_{10}) \frac{\partial}{\partial \mathbf{r}_1} f(t, \mathbf{r}_1, \mathbf{p}_{20}) - f(t, \mathbf{r}_1, \mathbf{p}_{20}) \frac{\partial}{\partial \mathbf{r}_1} f(t, \mathbf{r}_1, \mathbf{p}_{10}) \Big] \cdot \hat{S}_{12}(\mathbf{r}_2 - \mathbf{r}_1) \Big\}\, d\tau_2; \end{aligned} \qquad (17.17)$$

the differentiation with respect to $\mathbf{r}_1$ is taken at constant $\mathbf{p}_{10}$ or $\mathbf{p}_{20}$.

†The way to derive the correction terms to the Boltzmann equation was pointed out by N. N. Bogolyubov (1946). These terms were first brought to their final form by M. S. Green (1956).

The integral (17.16) is the same as (16.12);† it has been shown in § 16 how this integral is reducible to the ordinary Boltzmann form by carrying out one of the three integrations with respect to spatial coordinates.

Let us now consider the three-body collision integral (17.14). To include "non-localness" in this integral would be to go beyond the assumed accuracy, since the integral itself is a small correction. Hence, in the arguments of the three functions f, all the radius vectors $\mathbf{r}_1, \mathbf{r}_2, \mathbf{r}_3$ are to be taken as the same $\mathbf{r}_1$, and moreover we must assume that the operator $\hat{R}_{123}$ does not act on these variables at all:‡

$$C^{(3)}(f(t, \mathbf{r}_1, \mathbf{p}_1)) = \frac{1}{\mathcal{N}} \int \frac{\partial U_{12}}{\partial \mathbf{r}_1} \cdot \frac{\partial}{\partial \mathbf{p}_1} \{\hat{R}_{123} f(t, \mathbf{r}_1, \mathbf{p}_1) f(t, \mathbf{r}_1, \mathbf{p}_2) f(t, \mathbf{r}_1, \mathbf{p}_3)\}\, d\tau_2\, d\tau_3. \quad (17.18)$$

Let us next examine in somewhat more detail the structure of the operator $\hat{R}_{123}$, in order to elucidate the nature of the collision processes covered by the integral (17.18).

First of all, the operator $\hat{R}_{123}$, like $\hat{G}_{123}$ in (17.4), is zero if any one of the three particles does not interact with the others. However, the processes for which $\hat{R}_{123} \neq 0$ include not only three-body collisions in the literal sense, but also combinations of several pair collisions.

In genuine three-body collisions, three particles come simultaneously into the "sphere of interaction", as shown diagrammatically in Fig. 5a. But the operator $\hat{R}_{123}$ is also different from zero for "three-body interactions" which consist of three successive pair collisions, one pair colliding twice; Fig. 5b shows diagrammatically an example of such a process, for which $\hat{S}_{13} = 1$, so that $\hat{R}_{123} = \hat{S}_{123} - \hat{S}_{12}\hat{S}_{23}$.§ Moreover, the operator $\hat{R}_{123}$ also takes account of cases where one (or more) of the three collisions is "imaginary", i.e. occurs only if the influence of one of the real collisions on the path of the particles is ignored. An example is shown in Fig. 5c,

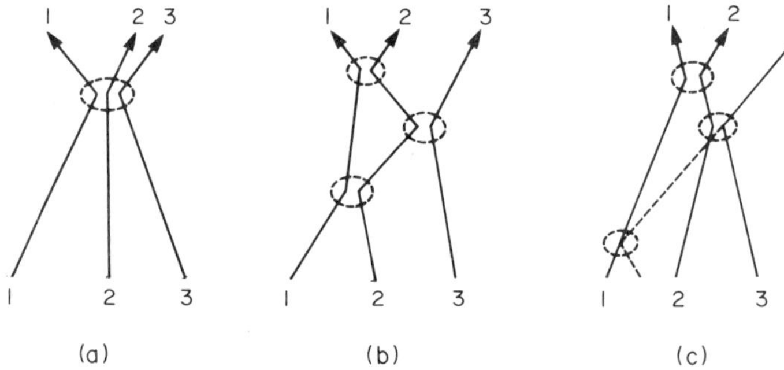

Fig. 5.

†It differs in that t_0 is replaced by t in the arguments of the functions f, but the right-hand equation (16.13) is then still valid, since the dependences on $\mathbf{r}_1, \mathbf{r}_2, \mathbf{p}_1, \mathbf{p}_2$ enter only through $\mathbf{p}_{10}$ and $\mathbf{p}_{20}$, which are integrals of the motion.

‡It should be stressed, to avoid misunderstanding, that these simplifications do not imply that the integrand no longer depends on $\mathbf{r}_2$ and $\mathbf{r}_3$; a dependence on these variables still occurs through the S operators, which transform the momenta $\mathbf{p}_a$ into functions $\tilde{\mathbf{p}}_a(\mathbf{r}_1, \mathbf{r}_2, \mathbf{r}_3, \mathbf{p}_1, \mathbf{p}_2, \mathbf{p}_3)$.

§The operator $\hat{R}_{123}$, unlike $\hat{G}_{123}$, is zero for a sequence of two collisions. For instance, in a process consisting of collisions 2–3 and 1–2, we should have $\hat{S}_{123} = \hat{S}_{12}\hat{S}_{23}$, $\hat{S}_{13} = 1$, and so $\hat{R}_{123} = 0$.

where the collision 1–3 would occur only if the path of particle 3 were unaffected by its collision with particle 2;† for this process, $\hat{S}_{123} = \hat{S}_{12}\hat{S}_{23}$ but $\hat{S}_{13} \neq 1$, so that $\hat{R}_{123} = -\hat{S}_{12}\hat{S}_{13} + \hat{S}_{12}$.

In the same kind of way as the integral $C_0^{(2)}$ was transformed in § 16, one of the six integrations with respect to coordinates in the three-body collision integral can be carried out; the interaction potential U_{12} then no longer appears explicitly in the formulae.‡

§ 18. The virial expansion of the kinetic coefficients

It was shown in §§ 7 and 8 that the thermal conductivity and the viscosity were independent of the gas density (or pressure) because only pair collisions of molecules were taken into account. For such collisions, the collision frequency, i.e. the number of collisions undergone by a given molecule per unit time, is proportional to the density N, the mean free path $l \propto 1/N$, and since η and κ are proportional to Nl they are independent of N. The values η_0 and κ_0 thus obtained are, of course, only the first terms in expansions of these quantities in powers of the density, called *virial expansions*. In the next approximation, there is already a density dependence in the form

$$\kappa = \kappa_0(1+\alpha N d^3), \quad \eta = \eta_0(1+\beta N d^3), \tag{18.1}$$

where d is a parameter of the order of molecular dimensions, and α and β are dimensionless constants. These first corrections have a twofold origin reflected in the correction terms $C^{(3)}$ and $C_1^{(2)}$ in the transport equation. Three-body collisions (whose frequency is proportional to N^2) decrease the mean free path. The non-localness of the pair collisions makes possible a transfer of momentum and energy across a certain surface without its actually being crossed by the colliding particles: the particles approach to a distance $\sim d$ and then separate, remaining on opposite sides of the surface. This effect increases the momentum and energy fluxes.

The solution of the problem of thermal conduction or viscosity with the more accurate transport equation (17.12) is to be based on the procedure as already described in §§ 6–8. We seek the distribution function in the form $f = f_0(1+\chi/T)$, where f_0 is the local-equilibrium function, and $\chi/T \sim l/L$ is a small correction. The three-body collision integral $C^{(3)}$, like $C_0^{(2)}$, is zero for the function f_0. We must therefore retain the χ term in it, and so the integral $C^{(3)}$ is, relative to the Boltzmann integral $C^{(2)}$, a correction of relative order $\sim(d/\bar{r})^3$. In the integral $C_1^{(2)}$, however, which contains spatial derivatives of the distribution function, it is sufficient to take $f = f_0$, and in this sense the term $C_1^{(2)}$ should be taken to the left-hand side of the equation, where it gives a correction of the same relative order $\sim(d/\bar{r})^3$. Thus the two additional terms $C^{(3)}$ and $C_1^{(2)}$ in the transport equation give contributions of the same order.§

†Having regard to the sense of action of the S operators, we must follow the paths of the particles *backwards* in time.

‡The transformation is carried out in a paper by M. S. Green (*Physical Review* **136**, A905, 1964).

§This argument clears up any misapprehension which might arise because the integral $C_1^{(2)}$ contains derivatives $\partial f/\partial \mathbf{r} \sim f/L$, which are not found in $C^{(3)}$, as a result of which the two terms might appear to give corrections of different orders of magnitude.

For reference, the results of solving the more accurate transport equation for the thermal conductivity and the viscosity of the gas, with the model of hard spheres (diameter d) are

$$\kappa = \kappa_0(1 + 1.2Nd^3), \quad \eta = \eta_0(1 + 0.35Nd^3), \tag{18.2}$$

where κ_0 and η_0 are the values obtained in § 10, Problem 3 (J. V. Sengers 1966).†

By making further corrections to the transport equation (arising from four-body collisions, etc.), it would in principle be possible to determine also the subsequent terms in the virial expansion of the kinetic coefficients. It is important to note, however, that these terms will involve non-integral powers of N; the functions $\kappa(N)$ and $\eta(N)$ are found to be non-analytic at the point $N = 0$. To elucidate the origin of this behaviour, let us consider the convergence of the integrals occurring in the theory (E. G. D. Cohen, J. R. Dorfman and J. Weinstock 1963).

We take first the integral in (17.10), which determines the contribution of three-body collisions to the two-particle distribution function. The type of convergence of the integral is different for the different kinds of collision process covered by the operator $\hat{R}_{123}$. Let us use as an example the process as in Fig. 5b.

The integration is over the phase volume $d\tau_3$ with given phase points τ_1 and τ_2. As the variable in the last integration we leave the distance r_3 of particle 3 (at time t) from the point where the collision 2–3 occurred. Before this last integration, the integrand will contain the following factors: (1) the volume element $r_3^2\, dr_3$ for the variable r_3; (2) if we follow the motion of particle 3 backwards in time, it will be clear that the direction of its momentum $\mathbf{p}_3$ must lie in a certain solid-angle element for the collision 3–2 to occur, namely the angle subtended by the region of collision at the distance r_3, giving a factor d^2/r_3^2; (3) another such factor arises from the further limitation on the possible directions of the momentum $\mathbf{p}_3$ imposed by the condition that the "recoiling" particle 2 enters the sphere of collision with particle 1. Thus we get an integral of the form $\int dr_3/r_3^2$, which is to be taken from $r_3 \sim d$ to ∞, and we see that it converges. Similarly, it can be shown that for collision processes of other types the convergence of the integral is even more rapid.

The contribution of four-body collisions would be expressed in (17.10) by an integral of similar form, taken over the phase space of particles 3 and 4, again for given τ_1 and τ_2.

Let us consider a four-body collision of the kind shown in Fig. 6. We leave as the last variable of integration the distance r_4 from particle 4 (at time t) to the 4–3 collision point. The difference from the preceding estimate arises because the phase point τ_3 (at time t) is not specified, unlike the point τ_2 in the integral corresponding to Fig. 5b. The position of the collision 4–3 is therefore also not fixed; it may occur anywhere in a cylindrical region with diameter $\sim d$ and axis along $\mathbf{p}_3$ (shown by the broken lines in Fig. 6). Accordingly, the solid angle subtended by this region at a distance r_4 is $\sim d/r_4$, instead of d^2/r_3^2 as in the previous case. The integral is therefore of the form $\int dr_4/r_4$, and so diverges logarithmically at the upper limit. Cutting off the integral at some distance Λ, we obtain a contribution to the function

†The calculations, which are exceedingly laborious, are given by Sengers in *Lectures in Theoretical Physics*, Vol. IXC, *Kinetic Theory* (ed. by W. E. Brittin), Gordon & Breach, New York, 1967.

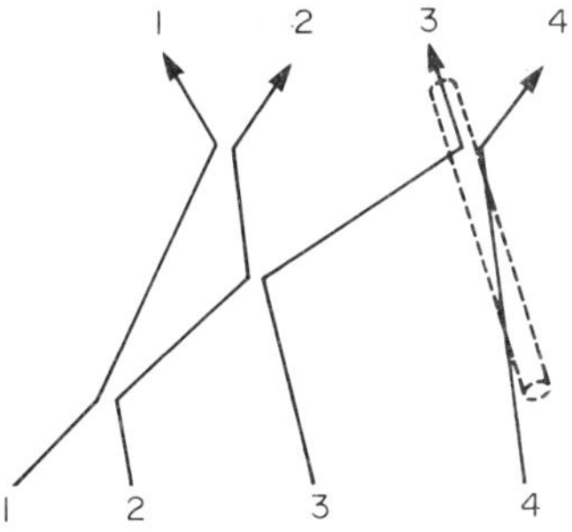

FIG. 6.

$f^{(2)}$ which contains the large logarithm $\log(\Lambda/d)$. This logarithm appears correspondingly in the correction to the transport coefficients, which is proportional not to $(Nd^3)^2$ but to $(Nd^3)^2 \log(\Lambda/d)$.

The presence of divergent terms signifies that the four-body collisions cannot be treated separately from those of all higher orders (five-body, etc.). For the divergence shows that large r_4 are important, but even when $r_4 \sim l$ particle 4 can collide with some particle 5, and so on. The way to remove the divergence is thus clear: in the expression for $f^{(2)}(t, \tau_1, \tau_2)$ we must take account of terms relating to collisions of all orders, retaining in each order the most rapidly divergent integrals. Such a summation can be carried out, and has a foreseeable result: the arbitrarily large parameter Λ in the logarithm is replaced by the order of magnitude of the mean free path, $l \sim 1/Nd^2$.†

Thus the expansion of the transport coefficients has the form

$$\kappa = \kappa_0[1 + \alpha_1 Nd^3 + \alpha_2 (Nd^3)^2 \log(1/Nd^3) + \dots], \tag{18.3}$$

and similarly for η.

§ 19. Fluctuations of the distribution function in an equilibrium gas

The distribution function determined by the transport equation, denoted in §§ 19 and 20 by $\bar{f}$, gives the mean numbers of molecules in the phase volume element $d^3x\, d\Gamma$; for a gas in statistical equilibrium, $\bar{f}(\Gamma)$ is the Boltzmann distribution function f_0 (6.7), independent of time and (if there is no external field) of the coordinates $\mathbf{r}$. It is natural to consider the fluctuations of the exact microscopic distribution function $f(t, \mathbf{r}, \Gamma)$ as it varies with time in the motion of the gas particles under their exact equations of motion.‡

We define the *correlation function* of the fluctuations as

$$\langle \delta f(t_1, \mathbf{r}_1, \Gamma_1) \delta f(t_2, \mathbf{r}_2, \Gamma_2) \rangle, \tag{19.1}$$

†See K. Kawasaki and I. Oppenheim, *Physical Review* **139**, A1763, 1965.
‡This topic was first discussed by B. B. Kadomtsev (1957).

where $\delta f = f - \bar{f}$. In an equilibrium gas, this function depends only on the time difference $t = t_1 - t_2$; the averaging is taken with respect to one of the times t_1 and t_2, with a fixed value of their difference. Since the gas is homogeneous, the coordinates $\mathbf{r}_1$ and $\mathbf{r}_2$ also occur in the correlation function as the difference $\mathbf{r} = \mathbf{r}_1 - \mathbf{r}_2$. We can therefore arbitrarily take t_2 and $\mathbf{r}_2$ as zero, and write the correlation function as

$$\langle \delta f(t, \mathbf{r}, \Gamma_1)\delta f(0, 0, \Gamma_2)\rangle. \tag{19.2}$$

Since the gas is isotropic, the dependence of this function on $\mathbf{r}$ in fact reduces to a dependence on the magnitude r.

If the function (19.2) is known, integration of it gives the correlation function of the particle number density:

$$\langle \delta N(t, \mathbf{r})\delta N(0, 0)\rangle = \int \langle \delta f(t, \mathbf{r}, \Gamma_1)\delta f(0, 0, \Gamma_2)\rangle \, d\Gamma_1 \, d\Gamma_2. \tag{19.3}$$

For distances r that are large compared with the mean free path l, the density correlation function may be calculated by the hydrodynamic theory of fluctuations (see *SP* 2, § 88), but at distances $\lesssim l$ a kinetic treatment is needed.

It is immediately evident from the definition (19.1) that

$$\langle \delta f(t, \mathbf{r}, \Gamma_1)\delta f(0, 0, \Gamma_2)\rangle = \langle \delta f(-t, -\mathbf{r}, \Gamma_2)\delta f(0, 0, \Gamma_1)\rangle. \tag{19.4}$$

The correlation function also has a more profound symmetry which corresponds to that of the equilibrium state of the system under time reversal. The latter process replaces a later time t by an earlier one $-t$, and also replaces the values of Γ by the time-reversed ones Γ^T. The symmetry in question is therefore expressed by

$$\langle \delta f(t, \mathbf{r}, \Gamma_1)\delta f(0, 0, \Gamma_2)\rangle = \langle \delta f(-t, \mathbf{r}, \Gamma_1{}^T)\delta f(0, 0, \Gamma_2{}^T)\rangle. \tag{19.5}$$

When $t = 0$, the function (19.2) relates the fluctuations at different points in phase space at the same instant. But the correlations between simultaneous fluctuations are propagated only to distances of the order of the range of molecular forces, whereas in the theory under consideration such distances are regarded as zero, so that the simultaneous-correlation function vanishes. It should be emphasized that this result is due to the equilibrium nature of the state relative to which the fluctuations are considered. We shall see in § 20 that simultaneous fluctuations also are correlated in the non-equilibrium case.

In the absence of correlation at non-zero distances, the simultaneous-correlation function reduces to delta functions, whose coefficient is the mean square fluctuation at one point in phase space; cf. *SP* 2, § 88. In an ideal equilibrium gas, the mean square fluctuation of the distribution function is equal to the mean value of the function itself (see *SP* 1, § 113); thus

$$\langle \delta f(0, \mathbf{r}, \Gamma_1)\delta f(0, 0, \Gamma_2)\rangle = \bar{f}(\Gamma_1)\delta(\mathbf{r})\delta(\Gamma_1 - \Gamma_2). \tag{19.6}$$

The non-simultaneous correlation between fluctuations at different points occurs even in the theory which neglects molecular dimensions. That this correlation necessarily arises is evident from the fact that particles which participate at a certain instant in fluctuations at some point in phase space will already be at other points at any subsequent instant.

The problem of calculating the correlation function for $t \neq 0$ cannot be solved in a general form, but can be reduced to the solution of particular equations. To do so, a proposition is needed from the general theory of quasi-steady fluctuations; see *SP* 1, §§118 and 119.

Let $x_a(t)$ be fluctuating quantities (with zero mean values). It is assumed that, if the system is in a non-equilibrium state with values of x_a beyond the limits of their mean fluctuations (but still small), the process of relaxation of the system to equilibrium is described by linear "equations of motion"

$$\dot{x}_a = -\sum_b \lambda_{ab} x_b \tag{19.7}$$

with constant coefficients λ_{ab}. Then we can say that the correlation functions of the x_a satisfy similar equations

$$\frac{d}{dt}\langle x_a(t)x_c(0)\rangle = -\sum_b \lambda_{ab}\langle x_b(t)x_c(0)\rangle, \quad t > 0, \tag{19.8}$$

with c a free suffix. Solving these equations for $t > 0$, we then find the values of the functions for $t < 0$ from the symmetry property

$$\langle x_a(t)x_b(0)\rangle = \langle x_b(-t)x_a(0)\rangle, \tag{19.9}$$

which follows from the definition of the correlation functions.

In the present case, the equations of motion (19.7) are represented by the linearized Boltzmann equation for the small addition δf to the equilibrium distribution function $\bar{f}$. Thus the correlation function of the distribution function must satisfy the integro-differential equation

$$\left(\frac{\partial}{\partial t} + \mathbf{v}_1 \cdot \frac{\partial}{\partial \mathbf{r}} - \hat{I}_1\right)\langle \delta f(t, \mathbf{r}, \Gamma_1)\delta f(0, 0, \Gamma_2)\rangle = 0 \quad \text{for} \quad t > 0, \tag{19.10}$$

where $\hat{I}_1$ is a linear integral operator acting on the variables Γ_1 in the function following it:

$$\hat{I}_1 g(\Gamma_1) = \int w(\Gamma_1, \Gamma; \Gamma_1', \Gamma')[\bar{f}_1' g_1' + \bar{f}' g' - \bar{f}_1 g_1 - \bar{f} g]\, d\Gamma\, d\Gamma_1\, d\Gamma'. \tag{19.11}$$

The variables Γ_2 in (19.10) are free variables. The initial condition for the equation is the value (19.6) of the correlation function for $t = 0$; that for $t < 0$ is then given by (19.4), the condition (19.5) being automatically satisfied by the result. The formulae (19.10), (19.11) and (19.4) constitute a set of equations sufficient in principle for a complete determination of the correlation function.

What is usually of interest is not the correlation function itself but its Fourier transform with respect to coordinates and time, denoted by $(\delta f_1 \delta f_2)_{\omega\mathbf{k}}$, where the suffixes 1 and 2 refer to the arguments Γ_1 and Γ_2:

$$(\delta f_1 \delta f_2)_{\omega\mathbf{k}} = \int_{-\infty}^{\infty} dt \int \langle \delta f(t, \mathbf{r}, \Gamma_1) \delta f(0, 0, \Gamma_2) \rangle e^{-i(\mathbf{k}\cdot\mathbf{r} - \omega t)} \, d^3x, \tag{19.12}$$

the *spectral function* or *spectral correlation function* of the fluctuations. If a fluctuating function is expanded as a Fourier integral with respect to time and coordinates, the mean value of the products of its Fourier components is related to the spectral correlation function by

$$\langle \delta f_{\omega\mathbf{k}}(\Gamma_1) \delta f_{\omega'\mathbf{k}'}(\Gamma_2) \rangle = (2\pi)^4 \delta(\omega + \omega') \delta(\mathbf{k} + \mathbf{k}') (\delta f_1 \delta f_2)_{\omega\mathbf{k}}; \tag{19.13}$$

cf. *SP* 1, § 122.

It is easy to derive an equation which in principle allows a determination of the spectral function of the fluctuations without previous calculation of the space–time correlation function. Dividing the range of integration with respect to t in (19.12) into two parts, from $-\infty$ to 0 and from 0 to ∞, and using (19.4), we have

$$(\delta f_1 \delta f_2)_{\omega\mathbf{k}} = (\delta f_1 \delta f_2)^{(+)}_{\omega\mathbf{k}} + (\delta f_2 \delta f_1)^{(+)}_{-\omega-\mathbf{k}}, \tag{19.14}$$

where

$$(\delta f_1 \delta f_2)^{(+)}_{\omega\mathbf{k}} = \int_0^{\infty} dt \int \langle \delta f(t, \mathbf{r}, \Gamma_1) \delta f(0, 0, \Gamma_2) \rangle e^{-i(\mathbf{k}\cdot\mathbf{r} - \omega t)} \, d^3x. \tag{19.15}$$

To the equation (19.10) we apply the one-sided Fourier transformation (19.15). The terms containing derivatives with respect to t and $\mathbf{r}$ are integrated by parts, using the facts that the correlation function must tend to zero as $\mathbf{r} \to \infty$ and as $t \to \infty$, and must be given by (19.6) when $t = 0$. The required equation is then found to be

$$[i(\mathbf{k}\cdot\mathbf{v}_1 - \omega) - \hat{I}_1](\delta f_1 \delta f_2)^{(+)}_{\omega\mathbf{k}} = \bar{f}(\Gamma_1) \delta(\Gamma_1 - \Gamma_2). \tag{19.16}$$

If we are interested in the fluctuations of the gas density, and not in those of the distribution function itself, it is appropriate to integrate equation (19.16) over $d\Gamma_2$:

$$[i(\mathbf{k}\cdot\mathbf{v} - \omega) - \hat{I}](\delta f(\Gamma) \delta N)^{(+)}_{\omega\mathbf{k}} = \bar{f}(\Gamma). \tag{19.17}$$

The spectral function $(\delta N^2)_{\omega\mathbf{k}}$ sought is found from the solution of this equation by a single integration, not a double one as in (19.3).

Another method of finding $(\delta N^2)_{\omega\mathbf{k}}$ is based on the relation between the density correlation function and the generalized susceptibility with respect to a weak external field of the form

$$U(t, \mathbf{r}) = U_{\omega\mathbf{k}} e^{i(\mathbf{k}\cdot\mathbf{r} - \omega t)}; \tag{19.18}$$

see *SP* 2, § 86.† If this field causes a density change

$$\delta N_{\omega\mathbf{k}} = \alpha(\omega, \mathbf{k}) U_{\omega\mathbf{k}}, \tag{19.19}$$

†This relation exists *only* in the equilibrium case.

then from *SP* 2 (86.20) the spectral correlation function of the density is, in the classical limit,

$$(\delta N^2)_{\omega\mathbf{k}} = (2T/\omega)\,\mathrm{im}\,\alpha(\omega, \mathbf{k}). \tag{19.20}$$

Let $\delta f(t, \mathbf{r})$ be the change in the distribution function due to the same field; it satisfies the transport equation

$$\frac{\partial}{\partial t}\delta f + \mathbf{v}\cdot\frac{\partial}{\partial \mathbf{r}}\delta f - \frac{\partial U}{\partial \mathbf{r}}\cdot\frac{\partial \bar{f}}{\partial \mathbf{v}} = \hat{I}\,\delta f.$$

The Fourier components of $\delta f(t, \mathbf{r}, \Gamma)$ are written

$$f_{\omega\mathbf{k}}(\Gamma) = \chi_{\omega\mathbf{k}}(\Gamma) U_{\omega\mathbf{k}},$$

in which the external field is separated as a factor. Then the equation for $\chi_{\omega\mathbf{k}}$ is

$$[i(\mathbf{k}\,.\,\mathbf{v} - \omega) - \hat{I}]\chi_{\omega\mathbf{k}}(\Gamma) = i\mathbf{k}\,.\,\partial\bar{f}/\partial\mathbf{v}. \tag{19.21}$$

The solution of this equation gives the required spectral correlation function by a single integration:

$$(\delta N^2)_{\omega\mathbf{k}} = (2T/\omega)\,\mathrm{im}\int \chi_{\omega\mathbf{k}}(\Gamma)\,d\Gamma. \tag{19.22}$$

PROBLEMS

PROBLEM 1. Determine the density correlation function in a monatomic gas in equilibrium, neglecting collisions.

SOLUTION. For a monatomic gas, the quantities Γ are the three components of the momentum $\mathbf{p}$. The solution of (19.10) for $\hat{I}_1 = 0$ is

$$\langle \delta f(t, \mathbf{r}, \mathbf{p}_1)\delta f(0, 0, \mathbf{p}_2)\rangle = \bar{f}(\mathbf{p}_1)\delta(\mathbf{r} - \mathbf{v}_1 t)\delta(\mathbf{p}_1 - \mathbf{p}_2),$$

and its Fourier component is

$$(\delta f_1 \delta f_2)_{\omega\mathbf{k}} = 2\pi\bar{f}(\mathbf{p}_1)\delta(\mathbf{p}_1 - \mathbf{p}_2)\delta(\omega - \mathbf{k}\,.\,\mathbf{v}_1).$$

Integration of these expressions with the Maxwellian function $\bar{f}$ gives as the density correlation function

$$\langle \delta N(t, \mathbf{r})\delta N(0, 0)\rangle = \bar{N}(m/2\pi T)^{3/2} t^{-3}\exp(-mr^2/2Tt^2), \tag{1}$$

$$(\delta N^2)_{\omega\mathbf{k}} = (\bar{N}/k)(2\pi m/T)^{1/2}\exp(-m\omega^2/2Tk^2). \tag{2}$$

PROBLEM 2. The same as Problem 1, but for a collision integral in the form $\hat{I}_1 g = -g/\tau$ with a constant time τ.

SOLUTION. Equation (19.16) reduces to an algebraic equation, from which we determine $(\delta f_1 \delta f_2)^{(+)}_{\omega\mathbf{k}}$, and then find from (19.14)

$$(\delta f_1 \delta f_2)_{\omega\mathbf{k}} = \frac{2\tau\bar{f}(\mathbf{p}_1)}{1 + \tau^2(\mathbf{k}\,.\,\mathbf{v}_1 - \omega)^2}\delta(\mathbf{p}_1 - \mathbf{p}_2). \tag{3}$$

The presence of even a small number of collisions changes the asymptotic behaviour of the spectral correlation function of the density at high frequencies, $\omega \gg k\bar{v}$, i.e. for fluctuations with a phase velocity much greater than the thermal speed of the molecules: in this limit,

$$(\delta N^2)_{\omega \mathbf{k}} = 2\bar{N}/\tau\omega^2, \tag{4}$$

i.e. the correlation function decreases with increasing frequency according to a power law, instead of exponentially as in (2).

§ 20. Fluctuations of the distribution function in a non-equilibrium gas

Let a gas be in a steady but non-equilibrium state with some distribution function $\bar{f}(\mathbf{r}, \Gamma)$ which satisfies the transport equation

$$\mathbf{v} \,.\, \partial \bar{f}/\partial \mathbf{r} = C(\bar{f}); \tag{20.1}$$

the function $\bar{f}$ may deviate greatly from the equilibrium distribution function f_0, and so the collision integral $C(\bar{f})$ is not assumed linearized with respect to the difference $\bar{f} - f_0$. The steady non-equilibrium state has to be maintained in the gas by external interactions: the gas may, for example, contain a temperature gradient supported by external sources, or it may execute a steady motion (which does not consist in a motion of the gas as a whole).

Let us seek to calculate the fluctuations of the distribution $f(t, \mathbf{r}, \Gamma)$ relative to $\bar{f}(\mathbf{r}, \Gamma)$. These fluctuations will again be described by a correlation function (19.1), in which the averaging is carried out in the usual way with respect to time for a given difference $t = t_1 - t_2$, and the correlation function depends only on t. Since the distribution $\bar{f}(\mathbf{r}, \Gamma)$ is not uniform, however, the correlation function now depends on the coordinates $\mathbf{r}_1$ and $\mathbf{r}_2$ separately, and not only on their difference. The property (19.4) becomes

$$\langle \delta f_1(t) \delta f_2(0) \rangle = \langle \delta f_2(-t) \delta f_1(0) \rangle, \tag{20.2}$$

where $f_1(t) \equiv f(t, \mathbf{r}_1, \Gamma_1)$, $f_2(0) \equiv f(0, \mathbf{r}_2, \Gamma_2)$.

The relation (19.5) involving time reversal does not apply in general in the non-equilibrium case.

The correlation function of the distribution function again satisfies the equation (19.10):

$$\left(\frac{\partial}{\partial t} + \mathbf{v}_1 \cdot \frac{\partial}{\partial \mathbf{r}_1} - \hat{I}_1\right) \langle \delta f_1(t) \delta f_2(0) \rangle = 0, \tag{20.3}$$

where $\hat{I}_1$ is the linear integral operator (19.11), which acts on the variables Γ_1.† The problem of the initial condition on this equation, i.e. the form of the single-time correlation function, is considerably more complex than in the equilibrium case, where it was given simply by (19.6). In a non-equilibrium gas, the single-time correlation function is itself determined from a transport equation whose form can

†The use of this equation in the non-equilibrium case is due to M. Lax (1966).

be established by using the relation between the correlation function and the two-particle distribution function $\bar{f}^{(2)}$ defined in § 16. In a steady state the function $\bar{f}^{(2)}(\mathbf{r}_1, \Gamma_1; \mathbf{r}_2, \Gamma_2)$, like $\bar{f}(\mathbf{r}, \Gamma)$, does not depend explicitly on the time.

To derive this relation, we note that, since the volume $d\tau = d^3x\, d\Gamma$ is infinitesimal, it cannot contain more than one particle at a time.† Hence the mean number $\bar{f}\, d\tau$ is also the probability that a particle is in the element $d\tau$ (the probability that there are two particles in it at once being a quantity of a higher order of smallness). It follows that the mean value of the product of the numbers of particles in the two elements $d\tau_1$ and $d\tau_2$ is equal to the probability of simultaneously finding one particle in each of them. For a given pair of particles this is the product $\bar{f}_{12}^{(2)}\, d\tau_1\, d\tau_2$, by the definition of the two-particle distribution function. Since a pair of particles can be chosen from the (very large) total number of particles in $\mathcal{N}(\mathcal{N}-1) \approx \mathcal{N}^2$ ways, we have

$$\langle f_1 d\tau_1 \,.\, f_2 d\tau_2 \rangle = \mathcal{N}^2 \bar{f}_{12}^{(2)}\, d\tau_1\, d\tau_2.$$

The equation $\langle f_1 f_2 \rangle = \mathcal{N}^2 \bar{f}_{12}^{(2)}$ thus obtained relates, however, only to different points in phase space. The passage to the limit $\mathbf{r}_1, \Gamma_1 \to \mathbf{r}_2, \Gamma_2$ makes it necessary to take into account that, if $d\tau_1$ and $d\tau_2$ coincide, an atom in $d\tau_1$ is also in $d\tau_2$. A relation which allows for this is

$$\langle f_1 f_2 \rangle = \mathcal{N}^2 \bar{f}_{12}^{(2)} + \bar{f}_1 \delta(\mathbf{r}_1 - \mathbf{r}_2)\delta(\Gamma_1 - \Gamma_2): \tag{20.4}$$

when it is multiplied by $d\tau_1 d\tau_2$ and integrated over some small volume $\Delta\tau$, the first term on the right gives a small quantity of the second order, $\propto (\Delta\tau)^2$, and the term containing the delta functions gives $\bar{f}\Delta\tau$, a first-order quantity. Thus we have

$$\left\langle \left(\int_{\Delta\tau} f\, d\tau \right) \right\rangle = \bar{f}\Delta\tau,$$

as it should be, since as far as first-order quantities there can only be either no particle or one particle in the small volume $\Delta\tau$.

Substituting (20.4) in the definition of the single-time correlation function

$$\langle \delta f_1(0) \delta f_2(0) \rangle = \langle f_1(0) f_2(0) \rangle - \bar{f}_1 \bar{f}_2,$$

we obtain the required relation between it and the two-particle distribution function:

$$\langle \delta f_1(0) \delta f_2(0) \rangle = \mathcal{N}^2 \bar{f}_{12}^{(2)} - \bar{f}_1 \bar{f}_2 + \bar{f}_1 \delta(\mathbf{r}_1 - \mathbf{r}_2)\delta(\Gamma_1 - \Gamma_2). \tag{20.5}$$

In an ideal gas in equilibrium, the two-particle distribution function reduces to the product $\bar{f}_{12}^{(2)} = \bar{f}_1 \bar{f}_2 / \mathcal{N}^2$, and (20.5) reduces to (19.6). In any case, $\bar{f}_{12}^{(2)}$ tends to this product as the distance between the points 1 and 2 increases, so that

$$\langle \delta f_1(0) \delta f_2(0) \rangle \to 0 \quad \text{as} \quad |\mathbf{r}_1 - \mathbf{r}_2| \to \infty. \tag{20.6}$$

†The derivation which follows is a paraphrase of the argument in *SP* 1, § 116.

The two-particle distribution function satisfies a transport equation analogous to the Boltzmann equation, which could be derived from equation (16.9) for $\bar{f}^{(2)}$ in the same way as the equation for the single-particle function was derived from (16.7).† Here, however, we shall give a derivation of the equation for $\bar{f}^{(2)}$ analogous to that of the Boltzmann equation in § 3, based on intuitive physical arguments.

We take as the unknown function not $\bar{f}^{(2)}$ itself but the difference

$$\varphi(\mathbf{r}_1, \Gamma_1; \mathbf{r}_2, \Gamma_2) = \mathcal{N}^2 \bar{f}^{(2)}_{12} - \bar{f}_1 \bar{f}_2, \tag{20.7}$$

which tends to zero as $|\mathbf{r}_1 - \mathbf{r}_2| \to \infty$; it is the correlation function (20.5) without the last term. This quantity is small in the usual sense of fluctuation theory, namely of the order of $1/\mathcal{N}$ in comparison with $\bar{f}_1 \bar{f}_2$.

In the absence of collisions, the function φ satisfies an equation which simply expresses Liouville's theorem—the constancy of $\bar{f}^{(2)}$ along the phase trajectory of a pair of particles:

$$\frac{d\bar{f}^{(2)}}{dt} = \frac{d\varphi}{dt} = \mathbf{v}_1 \cdot \frac{\partial \varphi}{\partial \mathbf{r}_1} + \mathbf{v}_2 \cdot \frac{\partial \varphi}{\partial \mathbf{r}_2} = 0. \tag{20.8}$$

The change in φ as a result of collisions is due to processes of two kinds.

Collisions of particles 1 and 2 with any other particles, but not with each other, cause the appearance, on the right of (20.8), of terms $\hat{I}_1 \varphi + \hat{I}_2 \varphi$, where $\hat{I}_1$ and $\hat{I}_2$ are the linear integral operators (19.11) acting on the variables Γ_1 and Γ_2 respectively.

Collisions between the particles 1 and 2 play a special role, causing a simultaneous "jump" of both particles from one pair of points in phase space to another pair. Exactly the same arguments as were used in the derivation of (3.7) give on the right of (20.8) a term $\delta(\mathbf{r}_1 - \mathbf{r}_2) C_{12}(\bar{f})$, where

$$C_{12}(\bar{f}) = \int w(\Gamma_1, \Gamma_2; \Gamma_1', \Gamma_2')(\bar{f}_1' \bar{f}_2' - \bar{f}_1 \bar{f}_2)\, d\Gamma_1'\, d\Gamma_2'; \tag{20.9}$$

in this integral, fluctuations may be neglected. The factor $\delta(\mathbf{r}_1 - \mathbf{r}_2)$ expresses the fact that particles undergoing collisions are at the same point in space.‡

Thus we have finally the equation

$$\mathbf{v}_1 \cdot \frac{\partial \varphi}{\partial \mathbf{r}_1} + \mathbf{v}_2 \cdot \frac{\partial \varphi}{\partial \mathbf{r}_2} - \hat{I}_1 \varphi - \hat{I}_2 \varphi = \delta(\mathbf{r}_1 - \mathbf{r}_2) C_{12}(\bar{f}). \tag{20.10}$$

Solution of this equation gives, in accordance with (20.5), the function which acts as the initial condition for equation (20.3) at $t = 0$.§

†In § 17 equation (16.9) was used only for a specific purpose, namely to eliminate $\bar{f}^{(2)}$ from the equation for $\bar{f}$.

‡A further integration of (20.9) over $d\Gamma_2$ yields the ordinar Boltzmann collision integral.

§This result is due to S. V. Gantsevich, V. L. Gurevich and R. Katilyus (1969) and to Sh. M. Kogan and A. Ya. Shul'man (1969).

Without the right-hand side, the homogeneous equation (20.10) has the solution

$$\left.\begin{aligned} \varphi &= f_{01}\Delta f_{02} + f_{02}\Delta f_{01}, \\ \Delta f_0 &= \frac{\partial f_0}{\partial \mathcal{N}}\Delta\mathcal{N} + \frac{\partial f_0}{\partial T}\Delta T + \frac{\partial f_0}{\partial \mathbf{v}}\cdot\Delta\mathbf{V}, \end{aligned}\right\} \tag{20.11}$$

corresponding to arbitrary small changes in the number of particles, the temperature, and the macroscopic velocity in the equilibrium distribution f_0.

This "spurious" solution is, however, excluded by the condition that $\varphi \to 0$ as $|\mathbf{r}_1 - \mathbf{r}_2| \to \infty$. Hence, in the equilibrium case, when the integral C_{12} is identically zero, equation (20.10) gives $\varphi = 0$, and we return to the initial condition (19.6).

The right-hand side of (20.10), i.e. the pair collisions between particles in given states Γ_1 and Γ_2, may thus be regarded as the source of the single-time correlation of fluctuations in a non-equilibrium gas. By causing a simultaneous change in the occupation numbers of the two states, pair collisions generate a correlation between these numbers. In the equilibrium state, owing to the exact compensation of direct and reverse pair collisions, this mechanism has no effect and there are no single-time correlations.

If the distribution $\bar{f}$ is independent of the coordinates $\mathbf{r}$ (as may happen when the deviation from equilibrium is maintained by an external field), we can consider fluctuations of the distribution function averaged over the whole volume of the gas, i.e. of the function

$$f(t, \Gamma) = \frac{1}{\mathcal{V}}\int f(t, \mathbf{r}, \Gamma)\, d^3x, \tag{20.12}$$

which we denote by the same letter f but without the argument $\mathbf{r}$. The corresponding correlation function satisfies an equation that differs from (20.3) in not having a term containing the derivative with respect to the coordinates:

$$\left(\frac{\partial}{\partial t} + \mathbf{F}_1\cdot\frac{\partial}{\partial \mathbf{p}_1} - \hat{I}_1\right)\langle \delta f(t, \Gamma_1)\delta f(0, \Gamma_2)\rangle = 0 \quad \text{for} \quad t > 0; \tag{20.13}$$

on the left-hand side, a term has been added which arises from the force $\mathbf{F}$ acting on the particles in the external field. The single-time correlation function

$$\begin{aligned} \langle \delta f(0, \Gamma_1)\delta f(0, \Gamma_2)\rangle &= \mathcal{N}^2\bar{f}^{(2)}(\Gamma_1, \Gamma_2) - \bar{f}(\Gamma_1)\bar{f}(\Gamma_2) + \frac{\bar{f}(\Gamma_1)}{\mathcal{V}}\delta(\Gamma_1 - \Gamma_2) \\ &\equiv \varphi(\Gamma_1, \Gamma_2) + \frac{\bar{f}(\Gamma_1)}{\mathcal{V}}\delta(\Gamma_1 - \Gamma_2) \end{aligned} \tag{20.14}$$

satisfies the equation

$$\left[\mathbf{F}_1\cdot\frac{\partial}{\partial \mathbf{p}_1} + \mathbf{F}_2\cdot\frac{\partial}{\partial \mathbf{p}_2} - (\hat{I}_1 + \hat{I}_2)\right]\varphi(\Gamma_1, \Gamma_2) = C_{12}(\varphi(\Gamma_1, \Gamma_2)). \tag{20.15}$$

If the gas is in a closed vessel, this equation is to be solved with the additional condition that expresses a fixed value (without fluctuations) of the total number of particles in the gas:

$$\int \langle \delta f(0, \Gamma_1)\delta f(0, \Gamma_2)\rangle \, d\Gamma_1 = \int \langle \delta f(0, \Gamma_1)\delta f(0, \Gamma_2)\rangle \, d\Gamma_2 = 0. \tag{20.16}$$

This condition must be satisfied in the equilibrium case also, but it is not satisfied by the expression $[\bar{f}(\Gamma_1)/\mathcal{V}]\delta(\Gamma_1 - \Gamma_2)$ which corresponds to the correlation function (19.6). The correct expression is obtained by making use of the arbitrary choice (20.11); with the appropriate value of the parameter $\Delta\mathcal{N}$,

$$\langle \delta f(0, \Gamma_1)\delta f(0, \Gamma_2)\rangle = \frac{1}{\mathcal{V}}\bar{f}(\Gamma_1)\delta(\Gamma_1 - \Gamma_2) - \frac{1}{\mathcal{N}}\bar{f}(\Gamma_1)\bar{f}(\Gamma_2). \tag{20.17}$$

This correlation function includes a term which does not contain a delta function.

CHAPTER II

THE DIFFUSION APPROXIMATION

§21. The Fokker–Planck equation

A CONSIDERABLE class of transport phenomena is constituted by processes in which the mean changes of quantities (on which the distribution function depends) in each event are small in comparison with their characteristic values. The relaxation times for such processes are long in comparison with the times of the individual events which constitute their microscopic mechanism; in this sense, they may be called slow processes.

A typical instance is the problem of momentum relaxation of a small admixture of a heavy gas in a light one, the latter being regarded as itself in equilibrium. Because of the low concentration of heavy particles, their collisions with one another may be neglected, and only those with the particles of the light gas considered. When a heavy particle collides with light ones, however, its momentum undergoes only a relatively small change.

We shall refer to this specific example, and derive the transport equation satisfied in such cases by the momentum distribution function $f(t, \mathbf{p})$ of the impurity particles.

Let $w(\mathbf{p}, \mathbf{q})d^3q$ denote the probability per unit time of a change $\mathbf{p} \to \mathbf{p} - \mathbf{q}$ in the momentum of a heavy particle in an individual collision with a light particle. Then the transport equation for the function $f(t, \mathbf{p})$ is

$$\partial f(t, \mathbf{p})/\partial t = \int \{w(\mathbf{p}+\mathbf{q}, \mathbf{q})f(t, \mathbf{p}+\mathbf{q}) - w(\mathbf{p}, \mathbf{q})f(t, \mathbf{p})\}d^3q, \qquad (21.1)$$

where the right-hand side is the difference between the numbers of particles per unit time that enter and leave a given momentum space element d^3p. According to the hypotheses used, the function $w(\mathbf{p}, \mathbf{q})$ decreases rapidly with increasing $\mathbf{q}$, and so the most important values of $\mathbf{q}$ in the integral are those which are small compared with the mean momentum of the particles. This allows the following expansion to be used in the integrand:

$$w(\mathbf{p}+\mathbf{q}, \mathbf{q})f(t, \mathbf{p}+\mathbf{q}) \approx w(\mathbf{p}, \mathbf{q})f(t, \mathbf{p}) + \mathbf{q}\cdot\frac{\partial}{\partial \mathbf{p}}w(\mathbf{p}, \mathbf{q})f(t, \mathbf{p})$$
$$+ \tfrac{1}{2} q_\alpha q_\beta \frac{\partial^2}{\partial p_\alpha \partial p_\beta} w(\mathbf{p}, \mathbf{q})f(t, \mathbf{p}).$$

The transport equation then becomes

$$\frac{\partial f}{\partial t} = \frac{\partial}{\partial p_\alpha}\left\{\tilde{A}_\alpha f + \frac{\partial}{\partial p_\beta}(B_{\alpha\beta} f)\right\}, \qquad (21.2)$$

where

$$\tilde{A}_\alpha = \int q_\alpha w(\mathbf{p}, \mathbf{q})\, d^3q, \quad B_{\alpha\beta} = \frac{1}{2}\int q_\alpha q_\beta w(\mathbf{p}, \mathbf{q})\, d^3q. \tag{21.3}$$

Thus the transport equation is now a differential (not integro-differential) equation. The quantities $\tilde{A}_\alpha$ and $B_{\alpha\beta}$ may be symbolically written as follows to show their significance more clearly:

$$\tilde{A}_\alpha = \sum q_\alpha/\delta t, \quad B_{\alpha\beta} = \sum q_\alpha q_\beta/2\delta t, \tag{21.4}$$

where the summations are over the (large) number of collisions that occur in the time δt.

The expression on the right of (21.2) is the divergence in momentum space, $-\partial s_\alpha/\partial p_\alpha$, of the vector

$$\left.\begin{aligned} s_\alpha &= -\tilde{A}_\alpha f - \frac{\partial}{\partial p_\beta}(B_{\alpha\beta} f) \\ &= -A_\alpha f - B_{\alpha\beta}\partial f/\partial p_\beta, \\ A_\alpha &= \tilde{A}_\alpha + \partial B_{\alpha\beta}/\partial p_\beta. \end{aligned}\right\} \tag{21.5}$$

Thus equation (21.2) is, as it should be, an equation of continuity in momentum space, and so the number of particles is automatically conserved in the process. The vector $\mathbf{s}$ is the particle flux density in momentum space.

According to equations (21.4), the coefficients in the transport equation are expressible in terms of the average characteristics of the collisions, and in this sense their calculation is a purely mechanical problem. In fact, however, there is no need to calculate the A_α and $B_{\alpha\beta}$ individually; they can be expressed in terms of one another by means of the condition for the flux to be zero in statistical equilibrium. In the present case, the equilibrium distribution function is

$$f = \text{constant} \times \exp(-\mathbf{p}^2/2MT),$$

where M is the mass of the heavy gas particles and T the temperature of the light gas. Substitution of this expression in the equation $\mathbf{s} = 0$ gives

$$MTA_\alpha = B_{\alpha\beta} p_\beta. \tag{21.6}$$

Thus the transport equation becomes

$$\frac{\partial f(t, \mathbf{p})}{\partial t} = \frac{\partial}{\partial p_\alpha}\left[B_{\alpha\beta}\left(\frac{p_\beta}{MT} f + \frac{\partial f}{\partial p_\beta}\right)\right]. \tag{21.7}$$

The coefficients in the first two terms of the expansion of the collision integral are of the same order of magnitude; the reason is that the averaging of the first powers of the quantities q_α in (21.4), whose sign is variable, involves a greater degree of cancellation than the averaging of quadratic expressions. The later terms of the expansion are all small in comparison with the first two.

The only vector on which the coefficients $B_{\alpha\beta}$ can depend is the momentum $\mathbf{p}$ of the heavy particles. If the velocities $\mathbf{p}/M$ of these particles are on average small compared with those of the light particles, they may be regarded as immobile in collisions; in this approximation, the $B_{\alpha\beta}$ are independent of $\mathbf{p}$. Thus the tensor $B_{\alpha\beta}$ reduces to a constant scalar B:

$$B_{\alpha\beta} = B\delta_{\alpha\beta}, \quad B = \frac{1}{6}\int q^2 w(0, \mathbf{q})\, d^3q, \tag{21.8}$$

and equation (21.7) becomes

$$\frac{\partial f}{\partial t} = B\frac{\partial}{\partial \mathbf{p}}\cdot\left(\frac{\mathbf{p}}{MT}f + \frac{\partial f}{\partial \mathbf{p}}\right). \tag{21.9}$$

There is a formal similarity between (21.7) and the equation of diffusion in an external field, which is especially evident from the form (21.9). The diffusion equation is

$$\partial c/\partial t = \operatorname{div}(D\nabla c - bc\mathbf{F}),$$

where c is the impurity concentration, $\mathbf{F}$ the force exerted on the impurity particles by the external field, D the diffusion coefficient, and b the mobility. The processes described by (21.9) may be referred to as diffusion in momentum space, with B acting as the diffusion coefficient; the relation between the coefficients in the two terms on the right of (21.9) is analogous to the familiar Einstein relation $D = bT$ between the diffusion coefficient and the mobility (*FM*, §59).

The transport equation in the form (21.2), with the coefficients defined in terms of the averaged characteristics of the elementary events by means of (21.4), is called the *Fokker–Planck equation* (A. D. Fokker 1914; M. Planck 1917). The specific properties of the variables p_α as the momenta of the particles have played no part in the derivation given. It is therefore clear that an equation of the same form is valid also for the distribution function f with respect to other variables, provided that the conditions on which the proof is based are satisfied: the relative smallness of the change in the quantities in each event, and the linearity with respect to f of the integral operator which expresses the change in the function resulting from these events.

As an example, let us take the case of a light gas forming a slight impurity in a heavy gas. In collisions with heavy particles, the momentum of a light particle changes greatly in direction but only slightly in magnitude. Although equation (21.7) is invalid under these conditions for the distribution function of the impurity gas particles with respect to the momentum vector $\mathbf{p}$, a similar equation can be set up for the distribution with respect to the magnitude p only. If the distribution function is, as before, taken relative to the momentum space element d^3p, so that the number of particles having the value of p in a range dp is $f(t, p).\ 4\pi p^2\, dp$, the Fokker–Planck equation is valid for the function $4\pi p^2 f$ in relation to dp:

$$\frac{\partial(fp^2)}{\partial t} = \frac{\partial}{\partial p}\left\{fp^2 A + B\frac{\partial}{\partial p}fp^2\right\},$$

or

$$\frac{\partial f}{\partial t} = \frac{1}{p^2}\frac{\partial}{\partial p} p^2\left\{fA + \frac{B}{p^2}\frac{\partial}{\partial p} fp^2\right\}, \tag{21.10}$$

where

$$B = \frac{1}{2}\sum(\delta p)^2/\delta t. \tag{21.11}$$

The expression in the braces is the radial flux s in momentum space. It must reduce to zero the case of the equilibrium distribution

$$f = \text{constant} \times \exp(-p^2/2mT),$$

where m is the mass of a light particle and T the temperature of the heavy gas. This condition gives a relation between A and B, and the transport equation (21.10) thus becomes

$$\frac{\partial f}{\partial t} = -\frac{1}{p^2}\frac{\partial (p^2 s)}{\partial p}, \quad s = -B\left(\frac{p}{mT} f + \frac{\partial f}{\partial p}\right). \tag{21.12}$$

PROBLEMS

PROBLEM 1. Determine the diffusion coefficient in momentum space (B in equation (21.9)) for an admixture of a heavy gas in a light one, assuming the speeds of the heavy particles to be small compared with those of the light ones.

SOLUTION. As shown in the text, under the conditions stated the momentum transfer may be calculated by assuming the heavy particle to be fixed and neglecting the change in its energy in the collision. The change in the momentum of the heavy particle is then calculated as the equal change in that of the light particle: $(\Delta\mathbf{p})^2 = 2p'^2(1-\cos\alpha)$, where p' is the momentum of the light particle and α the angle through which it is scattered. Hence

$$\sum(\Delta\mathbf{p})^2 = \delta t \int 2p'^2(1-\cos\alpha)Nv'\,d\sigma,$$

where N is the number density of light gas particles, and we have finally

$$B = (N/3m)\langle p'^3\sigma_t\rangle,$$

where $\sigma_t = \int(1-\cos\alpha)\,d\sigma$ is the transport cross-section, and the averaging is taken over the light gas particle distribution.

PROBLEM 2. Use the Fokker–Planck equation to determine the mobility of a heavy particle in a light gas.

SOLUTION. When an external field is present, a term $\mathbf{F}\,.\,\partial f/\partial\mathbf{p}$ is added to the left-hand side of (21.9), $\mathbf{F}$ being the force acting on the particle. Assuming this force to be small, we seek a steady solution of the equation in the form $f = f_0 + \delta f$, where f_0 is the Maxwellian distribution and $\delta f \ll f_0$. The equation for δf is then

$$B\frac{\partial}{\partial\mathbf{p}}\cdot\left(\frac{\partial\delta f}{\partial\mathbf{p}} + \frac{\mathbf{p}}{MT}\delta f\right) = \mathbf{F}\cdot\frac{\partial f_0}{\partial\mathbf{p}}.$$

Hence

$$B\left(\frac{\partial\delta f}{\partial\mathbf{p}} + \frac{\mathbf{p}}{MT}\delta f\right) = \mathbf{F}f_0,$$

and so $\delta f = f_0 \mathbf{F} \cdot \mathbf{p}/B$. The mobility b is the coefficient in the equation

$$\bar{\mathbf{v}} = \int \delta f \cdot \mathbf{v}\, d^3p = b\mathbf{F}.$$

A calculation of the integral gives

$$b = T/B = 3mT/N\langle\sigma_t p'^3\rangle,$$

in agreement with (12.4).

§ 22. A weakly ionized gas in an electric field

Let us consider an ionized gas in a uniform electric field **E**. This field perturbs the equilibrium distribution of free electrons in the gas and causes an electric current in it. We shall derive the transport equation which governs the electron distribution.† If the ionization is weak, the electron (and ion) density in the gas is small. Hence only collisions between electrons and neutral molecules are important; those between electrons and ions or other electrons may be neglected. We shall also assume that the mean energy acquired by the electrons in the electric field is (even in a strong field; see below) insufficient to excite or ionize the molecules; the collisions between electrons and molecules may then be regarded as elastic.

Because of the large difference between the electron mass m and the molecule mass M, the mean speed of the electrons is much greater than that of the molecules. For the same reason, the electron momentum changes its direction greatly in a collision, but its magnitude only slightly. Under these conditions, the collision integral in the transport equation falls into the sum of two parts which represent the changes in the number of particles in a given element of momentum space due to the change in the magnitude and in the direction of the momentum; the first part may be expressed in the Fokker–Planck differential form.

Because of the symmetry about the direction of the field, the distribution function depends on only two variables (apart from the time): the magnitude p of the momentum, and the angle θ between $\mathbf{p} = m\mathbf{v}$ and the direction of **E** (which we take as the z-axis). The transport equation for the function $f(t, p, \theta)$ has the form‡

$$\frac{\partial f}{\partial t} - e\mathbf{E} \cdot \frac{\partial f}{\partial \mathbf{p}} = -\frac{1}{p^2}\frac{\partial}{\partial p}(p^2 s) + Nv\int [f(t, p, \theta') - f(t, p, \theta)]\, d\sigma, \tag{22.1}$$

where

$$s = -B\left(\frac{v}{T}f + \frac{\partial f}{\partial p}\right), \quad B = \tfrac{1}{2}\sum(\Delta p)^2/\delta t.$$

The first term on the right of (22.1) corresponds to the right-hand side of the Fokker–Planck equation (21.12). The second term is the collision integral with

†The theory given in § 22 is due to B. I. Davydov (1936). The limiting expression (22.18) was earlier derived by M. J. Druyvesteyn (1930).

‡In this book, e always denotes a positive quantity, the absolute magnitude of the unit charge. The charge on the electron is therefore $-e$.

respect to the change in direction of the momentum. In this integral, the molecules may be regarded as immovable (N being their number density); the number of collisions undergone by an electron per unit time and changing the direction of the momentum from θ to θ' (or vice versa) is $Nv\,d\sigma$, where $d\sigma$ is the cross-section for scattering of an electron by a molecule at rest, which depends on p and on the angle α between $\mathbf{p}$ and $\mathbf{p}'$; we assume that the cross-section is already averaged over the orientations of the molecule.

We shall consider a steady state with a time-independent distribution function, and accordingly the term $\partial f/\partial t$ in (22.1) will be omitted.

To calculate B, we use the equation

$$(\mathbf{v}-\mathbf{V})^2 = (\mathbf{v}'-\mathbf{V}')^2,$$

which expresses the constant magnitude of the relative velocity of the two particles in an elastic collision; $\mathbf{v}$, $\mathbf{V}$ and $\mathbf{v}'$, $\mathbf{V}'$ are the initial and final velocities of the electron and the molecule. The change in the velocity of the molecule is small compared with that of the electron ($\Delta\mathbf{V} = -m\Delta\mathbf{v}/M$); hence, after expanding the above equation, we can put $\mathbf{V} = \mathbf{V}'$. Then

$$2\mathbf{V}\,.\,(\mathbf{v}-\mathbf{v}') = v^2 - v'^2 \approx 2v\Delta v,$$

where $\Delta v = v - v'$ is a small quantity. Thus

$$(\Delta p)^2 = m^2(\Delta v)^2 = (m^2/v^2)[(\mathbf{V}\,.\,\mathbf{v})^2 + (\mathbf{V}\,.\,\mathbf{v}')^2 - 2(\mathbf{V}\,.\,\mathbf{v})(\mathbf{V}\,.\,\mathbf{v}')].$$

The averaging of this expression is carried out in two stages. First, we average over the (Maxwellian) distribution of the molecular velocities $\mathbf{V}$. Because of the isotropy of this distribution, $\langle V_\alpha V_\beta\rangle = \frac{1}{3}\delta_{\alpha\beta}\langle V^2\rangle$, and $\langle V^2\rangle = 3T/M$. We therefore have

$$(\Delta p)^2 = (m^2T/Mv^2)(v^2 + v'^2 - 2\mathbf{v}\,.\,\mathbf{v}') \approx (2m^2T/M)(1-\cos\alpha). \qquad (22.2)$$

We must now average over the collisions undergone by the electron per unit time; this is done by integrating over $Nv\,d\sigma$. The result is

$$B = Nm^2v\sigma_t T/M = pmT/Ml, \qquad (22.3)$$

where $\sigma_t = \int(1-\cos\alpha)\,d\sigma$ is the transport cross-section, and l the mean free path defined as

$$l = 1/N\sigma_t, \qquad (22.4)$$

in general a function of p. The flux in (22.1) is therefore

$$s = -\frac{mp}{Ml}\left(vf + T\frac{\partial f}{\partial p}\right). \qquad (22.5)$$

Note that, according to (22.2), the change in the electron energy in the collision is $\Delta\epsilon \sim \bar{v}\Delta p \sim T(m/M)^{1/2} \sim \bar{\epsilon}(m/M)^{1/2}$. Hence an appreciable change in this energy occurs only after $\sim M/m$ collisions, whereas the direction of the electron momentum alters considerably in even one collision. That is, the electron energy relaxation time $\tau_\epsilon \sim \tau_p M/m$, where $\tau_p \sim l/\bar{v}$ is the momentum direction relaxation time.

The left-hand side of (22.1) is also to be transformed to the variables p and θ:

$$e\mathbf{E}\cdot\frac{\partial f}{\partial \mathbf{p}} = eE\frac{\partial f}{\partial p_z} = eE\left[\cos\theta\,\frac{\partial f}{\partial p} + \frac{\sin^2\theta}{p}\,\frac{\partial f}{\partial\cos\theta}\right]. \tag{22.6}$$

The solution of the transport equation thus derived may be sought as an expansion in Legendre polynomials:

$$f(p,\theta) = \sum_{n=0}^{\infty} f_n(p)P_n(\cos\theta). \tag{22.7}$$

We shall see later that the successive terms in this expansion decrease rapidly in order of magnitude. It is therefore sufficient in practice to take only the first two terms in the expansion:

$$f(p,\theta) = f_0(p) + f_1(p)\cos\theta. \tag{22.8}$$

The integral in (22.1), with (22.8) substituted, gives

$$\int [f(p,\theta') - f(p,\theta)]\,d\sigma = -f_1\sigma_t\cos\theta;$$

compare the transformation of a similar integral in (11.1). The transport equation then becomes

$$-eE[f_0'\cos\theta + f_1'\cos^2\theta + (f_1/p)\sin^2\theta] + (1/p^2)(s_0p^2)' + (v/l)f_1\cos\theta = 0,$$

where the prime denotes differentiation with respect to p. The term $p^{-2}(s_1p^2)'\cos\theta$ has been omitted; it is certainly small (in the ratio $\sim m/M$) in comparison with the term $(vf_1/l)\cos\theta$ (s_0 and s_1 are the expressions (22.5) with f_0 or f_1 in place of f). Multiplying this equation by $P_0 = 1$ or $P_1 = \cos\theta$ and integrating over $d\cos\theta$, we obtain two equations:

$$(1/p^2)(p^2S)' = 0, \quad S = -\frac{1}{lM}(p^2f_0 + mpTf_0') - \tfrac{1}{3}eEf_1, \tag{22.9}$$

$$f_1 = (eEl/v)f_0'. \tag{22.10}$$

The expression S represents the particle flux in momentum space as modified by the electric field. It follows from (22.9) that $S = \text{constant}/p^2$. The flux S must, however, be finite for all p, and the constant is therefore zero. Now substituting f_1 from (22.10) in the equation $S = 0$, we find an equation determining $f_0(p)$:

$$\left[pT + \frac{(eEl)^2M}{3p}\right]f_0' + \frac{p^2}{m}f_0 = 0. \tag{22.11}$$

So far we have made no assumptions as to the form of the function $l(p)$, and the integral of the first-order equation (22.11) can be written for any $l(p)$. In order to obtain more specific results, we shall assume that l = constant, which is equivalent to assuming that the cross-section σ_t is independent of the momentum.† The integration of (22.11) gives

$$f_0(p) = \text{constant} \times \left(\frac{\epsilon}{T} + \frac{\gamma^2}{6}\right)^{\gamma^2/6} e^{-\epsilon/T}, \tag{22.12}$$

where

$$\gamma = (eEl/T)\surd(M/m). \tag{22.13}$$

For the function $f_1(p)$, we have from (22.10) and (22.12)

$$f_1 = -f_0\sqrt{\left(\frac{m}{M}\right)}\frac{\gamma\epsilon/T}{\epsilon/T + \gamma^2/6}. \tag{22.14}$$

The quantity γ is the parameter that describes the extent to which the field affects the electron distribution. The limiting case of weak fields corresponds to $\gamma \ll 1$. In the first approximation, $f_0(p)$ then reduces to the unperturbed Maxwellian distribution: $f_0 \propto e^{-\epsilon/T}$, $\bar{\epsilon} = 3T/2$, and

$$f_1 = -(eEl/T)f_0, \quad \gamma \ll 1. \tag{22.15}$$

The electric current generated in the gas is determined by the electron mobility

$$b = \frac{\overline{v_z}}{-eE} = \frac{1}{-eEN_e}\int v \cos\theta \,.\, f\, d^3p = -\frac{1}{3eEN_e}\int vf_1\, d^3p, \tag{22.16}$$

where N_e is the electron number density.‡ A simple calculation with f_1 from (22.15) gives for the mobility in a weak field

$$b_0 = 2^{3/2}l/3\pi^{1/2}(mT)^{1/2}. \tag{22.17}$$

This expression satisfies, as it should, the Einstein relation $D = bT$, where D is the diffusion coefficient (11.10).

The significance of the inequality $\gamma \ll 1$ as the criterion of a weak field can be understood from the following simple arguments. It is evident that the influence of the field on the electron distribution will be weak so long as the energy acquired by an electron in its mean free time is small compared with that which it loses to an atom in a collision. The former energy is eEl, and the latter is

$$\delta\epsilon \sim V\delta P \sim Vp \sim \surd(T/M)\surd(Tm),$$

†This is always true at sufficiently low electron temperatures, since for slow particles the cross-section tends to a limit independent of the energy (see *QM*, § 132).

‡Because of the orthogonality of different Legendre polynomials, only the term f_0 out of all those in the expansion (22.7) contributes to the normalization integral, and only $f_1 \cos\theta$ contributes to $\overline{v_z}$.

where P and V are the momentum and speed of the atom; the change δP is of the order of the electron momentum. The criterion follows from a comparison of the two expressions.

In the opposite case of strong fields ($\gamma \gg 1$), we find†

$$\left.\begin{aligned} f_0(p) &= A\exp(-3\epsilon^2/\gamma^2T^2), \\ A &= 3^{3/4}N_e/2^{3/2}\pi\Gamma(\tfrac{3}{4})(m\gamma T)^{3/2}, \end{aligned}\right\} \tag{22.18}$$

$$f_1 = -6\sqrt{(m/M)}\epsilon f_0/T\gamma. \tag{22.19}$$

The mean electron energy is

$$\bar{\epsilon} = \frac{2}{\pi}\sqrt{\left(\frac{2M}{3m}\right)}\Gamma^2(\tfrac{5}{4})eEl = 0.43eL\sqrt{(M/m)}, \tag{22.20}$$

and the electron mobility

$$b = 4\Gamma(\tfrac{5}{4})l^{1/2}/3^{3/4}\pi^{1/2}(mM)^{1/4}(eE)^{1/2}. \tag{22.21}$$

It remains to ascertain the condition for the expansion (22.7) to converge. For this purpose, we note that its successive terms are related in order of magnitude by

$$eEf_{n-1}/mv \sim vf_n/l: \tag{22.22}$$

after substitution of (22.7), multiplication by $P_n(\cos\theta)$, and integration over $d\cos\theta$, the term in f_{n-1} remains on the left-hand side of the transport equation, and the term in f_n in the collision integral. When $\gamma \ll 1$, the mean electron energy $\bar{\epsilon} \sim T$, and we have from (22.22)

$$f_n/f_{n-1} \sim eEl/T \ll (m/M)^{1/2} \ll 1.$$

In strong fields, with $\gamma \gg 1$, the mean energy $\bar{\epsilon} \sim eEl(M/m)^{1/2}$, so that again

$$f_n/f_{n-1} \sim (m/M)^{1/2} \ll 1.$$

Thus the expansion is convergent, since m/M is small.‡

§23. Fluctuations in a weakly ionized non-equilibrium gas

In this section we shall discuss fluctuations of the electron distribution function in a non-equilibrium steady state of a weakly ionized gas which is spatially homogeneous and is in a constant and uniform electric field **E**.

†Formula (22.18) is more simply derived by solving afresh equation (22.11) (with $T = 0$) than by taking the limit in (22.12).

‡Note, however, that the corrections $f_2, f_3, \ldots$ could not be determined by means of (22.1), since this equation is based on the Fokker–Planck approximation, in which quantities of higher order in m/M are neglected.

Only the time correlation of the fluctuations, and not their spatial correlation, will be considered. Then it is appropriate to use, in place of the exact (fluctuating) space-varying distribution function $f(t, \mathbf{r}, \mathbf{p})$, the function averaged over the whole volume of the gas,

$$f(t, \mathbf{p}) = \frac{1}{\mathcal{V}} \int f(t, \mathbf{r}, \mathbf{p})\, d^3x, \tag{23.1}$$

denoted in this section by the same letter f but without the argument $\mathbf{r}$; it fluctuates only with time. The function $\bar{f}(\mathbf{p})$ with respect to which f fluctuates is the distribution (22.8) found in the preceding section.

For the system in question, it is of most interest to find not the fluctuations of the distribution function itself but the related fluctuations of the electric current density $\mathbf{j}$. The correlation functions for these quantities are related by the obvious formula

$$\langle \delta j_\alpha(t) \delta j_\beta(0) \rangle = e^2 \int \langle \delta f(t, \mathbf{p}) \delta f(0, \mathbf{p}') \rangle v_\alpha v'_\beta\, d^3p\, d^3p', \tag{23.2}$$

where of course $\delta \mathbf{j}$ is the current density fluctuation averaged over the gas volume.†

The solution of the problem for a non-equilibrium gas is based on the general method given in §20.‡ The correlation function $\langle \delta f(t, \mathbf{p}) \delta f(0, \mathbf{p}') \rangle$ satisfies (with respect to the variables t and $\mathbf{p}$) the transport equation (22.1), which here corresponds to the equation (20.13) in the general method. A similar equation is satisfied by the function

$$\mathbf{g}(t, \mathbf{p}) = \int \langle \delta f(t, \mathbf{p}) \delta f(0, \mathbf{p}') \rangle \mathbf{v}'\, d^3p', \tag{23.3}$$

and the required current correlation function can be expressed in terms of this:

$$\langle \delta j_\alpha(t) \delta j_\beta(0) \rangle = e^2 \int g_\beta(t, \mathbf{p}) v_\alpha\, d^3p. \tag{23.4}$$

Thus we arrive at the equation

$$\frac{\partial \mathbf{g}}{\partial t} - e\left(\mathbf{E} \cdot \frac{\partial}{\partial \mathbf{p}}\right)\mathbf{g} = \frac{1}{p^2}\frac{\partial}{\partial p}\left[p^2 B\left(\frac{v}{T}\mathbf{g} + \frac{\partial \mathbf{g}}{\partial p}\right)\right] - Nv \int [\mathbf{g}(t, p, \theta) - \mathbf{g}(t, p, \theta')]\, d\sigma, \tag{23.5}$$

with B given by (22.3).

†This averaging corresponds to an experiment in which the fluctuations of the total current in the gas are measured: such a fluctuation is equal to that of the averaged current density in a given direction, multiplied by the cross-section of the sample.

‡The study of this problem by P. J. Price (1959) was the first instance of a calculation of fluctuations in a non-equilibrium system. Here we shall follow the analysis by V. L. Gurevich and R. Katilyus (1965).

The transport equation (22.1) takes account of the collisions of electrons only with molecules, not with other electrons. There is therefore no mechanism here to establish a single-time correlation between electrons with different momenta, and the "initial" condition for the function $\mathbf{g}(t, \mathbf{p})$ is the same as in the equilibrium state. Since we are concerned with the fluctuation of the distribution function averaged over the whole volume of the gas, the constancy of the number of particles (electrons) has to be taken into account.† According to (20.17), we then have

$$\langle \delta f(0, \mathbf{p}) \delta f(0, \mathbf{p}') \rangle = \frac{1}{\mathcal{V}}[\bar{f}(\mathbf{p})\delta(\mathbf{p}-\mathbf{p}') - \frac{1}{N_e}\bar{f}(\mathbf{p})\bar{f}(\mathbf{p}')]$$

(where N_e is the electron density), and hence for the initial function

$$\mathbf{g}(0, \mathbf{p}) = \frac{1}{\mathcal{V}}\bar{f}(\mathbf{p})(\mathbf{v}-\mathbf{V}), \tag{23.6}$$

where $\mathbf{V}$ is the mean electron velocity in the state having the distribution $\bar{f}(\mathbf{p})$. This velocity is, of course, parallel to the field $\mathbf{E}$; we may write it as

$$\mathbf{V} = -eb\mathbf{E}, \tag{23.7}$$

where b is the mobility. The constancy of the total number of electrons signifies that $\int \delta f \, d^3p = 0$, and so

$$\int \mathbf{g}(t, \mathbf{p}) \, d^3p = 0. \tag{23.8}$$

Following the method described in § 19, we take the one-sided Fourier transform of (23.5), multiplying by $e^{i\omega t}$ and integrating with respect to t from 0 to ∞. The term $e^{i\omega t}\partial\mathbf{g}/\partial t$ is integrated by parts with the initial condition (23.6) and the condition $\mathbf{g}(\infty, \mathbf{p}) = 0$. The result is

$$-i\omega\mathbf{g}^{(+)} - e(\mathbf{E} \cdot \partial/\partial\mathbf{p})\mathbf{g}^{(+)} - \frac{1}{p^2}\frac{\partial}{\partial p}\left[\frac{mTp^3}{Ml}\left(\frac{v}{T}\mathbf{g}^{(+)} + \frac{\partial\mathbf{g}^{(+)}}{\partial p}\right)\right]$$
$$+ N_e v\int [\mathbf{g}^{(+)}(\mathbf{p}) - \mathbf{g}^{(+)}(\mathbf{p}')] \, d\sigma = \frac{1}{\mathcal{V}}\bar{f}(\mathbf{p})(\mathbf{v}-\mathbf{V}), \tag{23.9}$$

where

$$\mathbf{g}^{(+)}(\omega, \mathbf{p}) = \int_0^\infty e^{i\omega t}\mathbf{g}(t, \mathbf{p}) \, dt. \tag{23.10}$$

†As we are concerned only with the influence on fluctuations of the departure from equilibrium caused by the presence of the field, we neglect fluctuations of the total number of electrons resulting from ionization and recombination processes. These fluctuations may be strictly absent when all the electrons are formed by impurities with low ionization potential; the total number of electrons is then simply equal to the total number of impurity atoms. We also neglect fluctuations in the concentration of neutral molecules. The relative fluctuation of this concentration is certainly small in comparison with the corresponding quantity for electrons, since the electron density is much less than the molecular concentration.

From (23.8), this equation is to be solved with the added condition

$$\int \mathbf{g}^{+}(\omega, \mathbf{p})\, d^3p = 0. \tag{23.11}$$

If the solution of (23.9) is known, the required spectral expansion of the current correlation function can be found by simple integration: we write

$$(j_\alpha j_\beta)_\omega = \int_{-\infty}^{\infty} dt \int e^{i\omega t} \langle \delta f(t, \mathbf{p}) \delta f(0, \mathbf{p}') \rangle v_\alpha v'_\beta \, d^3p \, d^3p'$$

and proceed exactly as in the derivation of (19.14), obtaining

$$(j_\alpha j_\beta)_\omega = e^2 \int \{ g_\beta^{(+)}(\omega, \mathbf{p}) v_\alpha + g_\alpha^{(+)}(-\omega, \mathbf{p}) v_\beta \}\, d^3p. \tag{23.12}$$

We shall take the specific case of a constant mean free path l. In the equilibrium state, with no electric field, the function $\bar{f}$ is the equilibrium Maxwellian distribution $f_0(p)$. The solution of equation (23.9) is then

$$\mathbf{g}^{(+)} = \frac{\mathbf{p}}{p} \frac{f_0(p)}{\mathscr{V}} \frac{l}{1 - i\omega l/v}, \tag{23.13}$$

as is easily seen, since

$$\int (\mathbf{p} - \mathbf{p}')\, d\sigma = \sigma_t \mathbf{p}. \tag{23.14}$$

If $\omega\tau_p \ll 1$ (where $\tau_p \sim l/v_T$ is the relaxation time as regards the direction of the momentum), we can neglect the term $-i\omega l/v$ in the denominator of (23.13). A calculation of the integral (23.12) then leads to the result

$$(j_\alpha j_\beta)_\omega = (2T\sigma/\mathscr{V})\delta_{\alpha\beta}, \tag{23.15}$$

where $\sigma = e^2 N_e b_0$ is the conductivity of the gàs in a weak field, and b_0 the mobility in a weak field, given by (22.17). The result (23.15) is, of course, in agreement with Nyquist's general formula for equilibrium fluctuations of the current (see *SP* 2, §78). For let us consider a cylindrical volume of gas parallel to the x-axis. Since the current density is already volume-averaged, the total current $J = j_x S$, where S is the cross-sectional area of the cylinder. From (23.15), we then have

$$(J^2)_\omega = 2T\sigma S^2/\mathscr{V} = 2T\sigma S/L = 2T/R, \tag{23.16}$$

where $L = \mathscr{V}/S$ is the length of the sample, and $R = L/\sigma S$ its resistance.†

When $\mathbf{E} \neq 0$, equation (23.9) is solved by successive approximation, in the same way as (22.6), but whereas the latter equation determined a scalar function, (23.9) is

†In comparing with *SP* 2 (78.1), it must be remembered that $\hbar\omega \ll T$ and that by the condition $\omega\tau_p \ll 1$ there is no dispersion of the conductivity, so that $Z = R$.

written for a vector function. The leading terms in the expansion of such a function (depending on the constant vector **E** and the variable vector **p**) may be written as

$$\mathbf{g}^{(+)}(\omega, \mathbf{p}) = h(\omega, p)\mathbf{n} + \mathbf{e}\{g_0(\omega, p) + \mathbf{n}\,.\,\mathbf{e}g_1(\omega, p)\}, \tag{23.17}$$

with $g_1 \ll g_0$; here $\mathbf{n} = \mathbf{p}/p$, $\mathbf{e} = \mathbf{E}/E$. The function $\bar{f}(\mathbf{p})$ is

$$\bar{f}(\mathbf{p}) = f_0(p) + \mathbf{n}\,.\,\mathbf{e}f_1(p), \tag{23.18}$$

with f_0 and $f_1 = eElf_0'/v$ as calculated in §22.

We substitute (23.17)–(23.18) in (23.9) and separate the terms odd and even with respect to **p**. Again assuming $\omega\tau_p \ll 1$, we find, collecting the odd terms,

$$(v/l)\{h\mathbf{n} + g_1\mathbf{e}(\mathbf{n}\,.\,\mathbf{e})\} - \mathbf{e}(\mathbf{e}\,.\,\partial g_0/\partial\mathbf{p})eE = f_0\mathbf{v}/\mathscr{V};$$

terms which are certainly small (in the ratio m/M) in comparison with those given are omitted. Hence

$$h(p) = lf_0(p)/\mathscr{V}, \quad g_1(\omega, p) = (eElm/p)\partial g_0(\omega, p)/\partial p. \tag{23.19}$$

The terms even in **p** have to satisfy equation (23.9) only after averaging over the directions of **p**, in accordance with the fact that (23.17) gives only the leading terms in the expansion of the function sought. A straightforward calculation using the expressions (23.19) leads to the following equation for the function $g_0(\omega, p)$:

$$-i\omega g_0 + \frac{1}{p^2}\frac{\partial}{\partial p}(p^2 S) = \frac{1}{\mathscr{V}}\left\{eEbf_0 + \frac{2eEl}{3p}\frac{\partial}{\partial p}(pf_0)\right\}, \tag{23.20}$$

where

$$S = -\frac{1}{lM}\left(p^2 g_0 + mpT\frac{\partial g_0}{\partial p}\right) - \frac{e^2E^2lm}{3p}\frac{\partial g_0}{\partial p}.$$

This is to be solved with the added condition

$$\int g_0(\omega, p)\, d^3p = 0, \tag{23.21}$$

to which (23.11) reduces on substitution of (23.17).

When the function $\mathbf{g}^{(+)}$ is known, the required current correlation function is found by means of (23.12). Substituting the expansion (23.17) and making a simple transformation with the use of (23.19), we obtain

$$(j_\alpha j_\beta)_\omega = \delta_{\alpha\beta}\frac{2e^2l}{3\mathscr{V}}\int vf_0\, d^3p - E_\alpha E_\beta\frac{2le^3}{3E}\int [g_0(\omega, p) + g_0(-\omega, p)]\frac{d^3p}{p}. \tag{23.22}$$

The term $-i\omega g_0$ in (23.20) becomes important when $\omega \sim mv/Ml$, i.e. when $\omega\tau_\epsilon \sim 1$, where τ_ϵ is the relaxation time as regards the electron energy. Thus the dispersion of the current fluctuations begins at these frequencies.

In the general case, equation (23.20) is very complicated. As an illustration, let us take the case of low frequencies ($\omega\tau_\epsilon \ll 1$) and strong fields satisfying the condition $\gamma \gg 1$, where γ is the parameter (22.13). Because of the latter condition, the function $f_0(p)$ is given by the expression (22.18). The calculation of the integral in the first term in (23.22) gives

$$\delta_{\alpha\beta}\frac{2^{3/2}}{3^{5/4}\Gamma(3/4)}\frac{N_e e^2 l}{\mathcal{V}}\left(\frac{eEl}{m}\right)^{1/2}\left(\frac{M}{m}\right)^{1/4}.$$

In the second term in (23.22), we shall make only an estimate without numerical factors. Equation (23.20) (without the term $-i\omega g_0$) gives

$$g_0 \sim (eEl^2M/\mathcal{V}p^2)f_0.$$

The integral can then be estimated as

$$e^3lE(g_0/p)p^3.$$

The resulting expression for the current correlation function is

$$(j_\alpha j_\beta)_\omega = \frac{N_e e^2 l}{\mathcal{V}}\left(\frac{eEl}{m}\right)^{1/2}\left(\frac{M}{m}\right)^{1/4}\left[0.6\delta_{\alpha\beta} - \beta\frac{E_\alpha E_\beta}{E^2}\right], \tag{23.23}$$

where $\beta \sim 1$ is a numerical constant.

§24. Recombination and ionization

The equilibrium degree of ionization in a partly ionized gas is established by various collisional ionization events and the converse recombination events between colliding charged particles. In the simple case where the gas contains, apart from electrons, only one type of ion, the process of establishment of ionization equilibrium is described by an equation of the form

$$dN_e/dt = \beta - \alpha N_e N_i, \tag{24.1}$$

where β is the number of electrons formed per unit volume and per unit time in collisions of neutral atoms or through photo-ionization of atoms; this number is independent of the electron density N_e and ion density N_i present. The second term gives the decrease in the number of electrons due to recombination with ions; α is called the *recombination coefficient.*

The recombination process is usually very slow in comparison with the other processes of establishment of equilibrium in a plasma. This is because the formation of a neutral atom in an ion–electron collision requires the removal of the energy released (the binding energy of the electron in the atom). The energy may be radiated as a photon in *radiative recombination*; the slowness of the process is then due to the small quantum-electrodynamical emission probability. The energy released may also be transferred to a third particle, a neutral atom; the slowness of the process is then due to the small probability of three-body collisions. The result is that it is often reasonable to consider recombination in conditions where the distribution of all particles may be regarded as Maxwellian.

In equilibrium, the derivative dN_e/dt is zero. It follows that the quantities α and β in (24.1) are related by

$$\beta = \alpha N_{0e} N_{0i}, \tag{24.2}$$

where N_{0e} and N_{0i} are the equilibrium electron and ion densities given by the appropriate thermodynamic formulae; see *SP* 1, § 104.†

The coefficient of radiative recombination is calculated directly from the recombination cross-section σ_{rec} in a collision between an electron and an ion at rest (the speed of the ion is negligible in comparison with that of the electron):

$$\alpha = \langle v_e \sigma_{\text{rec}} \rangle, \tag{24.3}$$

where the averaging is over the Maxwellian distribution of electron speeds v_e; see Problem 1.

Radiative recombination is, however, important only in a sufficiently rarefied gas, when three-body collisions are entirely negligible. In a less rarefied gas, the principal mechanism is recombination involving a third particle, a neutral atom, and it is this mechanism that we shall consider in more detail.

In collisions with atoms, the energy of the electron changes by small amounts. The recombination process therefore begins with the formation of a highly excited atom, and the electron gradually "descends" to lower and lower levels in further collisions of this atom. This type of process may be regarded as a "diffusion in energy" of the captured electron, and so the Fokker–Planck equation may be applied to it (L. P. Pitaevskiĭ, 1962).

Let us consider the distribution function of the captured electrons with respect to their (negative) energies ϵ. The most important "diffusion" is naturally in the energy range $|\epsilon| \sim T$. The temperature here must always be treated as small in comparison with the ionization potential I of the atoms; when $T \sim I$, the gas is almost completely ionized (cf. *SP* 1, § 104).

The Fokker–Planck equation is

$$\partial f/\partial t = -\partial s/\partial \epsilon, \quad s = -B \partial f/\partial \epsilon - Af. \tag{24.4}$$

As usual, the coefficient A can be expressed in terms of B by means of the condition that $s = 0$ when $f = f_0$, where f_0 is the equilibrium distribution. The flux s then becomes

$$s = -Bf_0 \frac{\partial}{\partial \epsilon}(f/f_0). \tag{24.5}$$

The "diffusion coefficient" $B(\epsilon)$ is determined by the general rule as

$$B(\epsilon) = \frac{1}{2} \sum (\Delta\epsilon)^2/\delta t, \tag{24.6}$$

†In radiative recombination, equilibrium of the state presupposes that of the radiation in the plasma.

where $\Delta\epsilon$ is the change in the excitation energy of the atom in a collision with an unexcited atom; the calculation of $B(\epsilon)$ from this formula involves solving the mechanical problem of the collision and then averaging with respect to the velocity of the unexcited atom (see Problem 2).

To find the function $f_0(\epsilon)$, we note that the equilibrium distribution with respect to momenta and coordinates for an electron in the Coulomb field of a charge ze (the ion charge) is given by Boltzmann's formula:

$$f_0(\mathbf{p}, \mathbf{r}) = (2\pi mT)^{-3/2} e^{-\epsilon/T}, \quad \epsilon = p^2/2m - ze^2/r \tag{24.7}$$

(see below concerning its normalization); the motion of the electron with $|\epsilon| \sim T \ll I$ is quasi-classical, and this allows us to use the classical expression for the energy ϵ. The distribution function with respect to ϵ is therefore

$$f_0(\epsilon)\, d\epsilon = (2\pi mT)^{-3/2} e^{|\epsilon|/T} \tau(\epsilon)\, d\epsilon, \tag{24.8}$$

where $\tau(\epsilon)$ is the volume in phase space corresponding to the range $d\epsilon$:

$$\tau(\epsilon) = \int \delta\left(|\epsilon| + \frac{p^2}{2m} - \frac{ze^2}{r}\right) d^3x\, d^3p. \tag{24.9}$$

Replacing $d^3x\, d^3p$ by $4\pi r^2\, dr \,.\, 4\pi p^2\, dp$ and carrying out the integration, we find

$$\tau(\epsilon) = \sqrt{2}\pi^3 (ze^2)^3 m^{3/2}/|\epsilon|^{5/2}. \tag{24.10}$$

To formulate the conditions which determine the appropriate solution of equations (24.4) and (24.5), it is convenient to suppose that the electron density present in the gas is $N_e \gg N_{0e}$; then we can neglect the rate of ionization β in (24.1), so that the decrease of N_e is due only to recombination. Under these conditions, the constant value of the flux s in the stationary solution of (24.4) gives directly the value of the recombination coefficient ($s = \text{constant} = -\alpha$) if $f(\epsilon)$ is suitably normalized: at the highest levels ($|\epsilon| \ll T$) the electrons are in equilibrium with free electrons, and this means that we must have

$$f(\epsilon)/f_0(\epsilon) \to 1 \quad \text{as} \quad |\epsilon| \to 0, \tag{24.11}$$

and the normalization of $f_0(\epsilon)$ must correspond to one free electron per unit volume, as has been ensured in (24.7).

To find the second boundary condition (as $\epsilon \to -\infty$), we note that the distribution at deep levels of the excited atom is not perturbed by the presence of free electrons, and is independent of their number: it is proportional to the equilibrium number N_{0e}, not to the actual number N_e. When $N_e \gg N_{0e}$, this situation is expressed by the boundary condition

$$f(\epsilon)/f_0(\epsilon) \to 0 \quad \text{as} \quad |\epsilon| \to \infty. \tag{24.12}$$

Integrating the equation $s = \text{constant}$ with the boundary condition (24.11), we have

$$f/f_0 = \text{constant} \times \int_0^{|\epsilon|} \frac{d|\epsilon|}{Bf_0} + 1.$$

The constant is $-\alpha$ if it is determined so as to satisfy the condition (24.12). Thus we find as the final result

$$\frac{1}{\alpha} = \int_0^{\infty} \frac{d|\epsilon|}{Bf_0} = \frac{2T^{3/2}}{\pi^{3/2}(ze^2)^3} \int_0^{\infty} \frac{e^{-|\epsilon|/T}|\epsilon|^{5/2}}{B(-|\epsilon|)} d|\epsilon|. \tag{24.13}$$

This formula relates to a process in which the "third body" is an unexcited atom. If the gas is strongly ionized (which is still compatible with the condition $T \ll I$) and sufficiently dense, recombination with a second electron as the third body may become the principal process. The recombination rate then becomes proportional to $N_e^2 N_i$, so that the recombination coefficient itself, defined as before by (24.1), is proportional to N_e. Since the energy relaxation in electron collisions is quick, the method described above for calculating the recombination coefficient is inapplicable in this case.

PROBLEMS

PROBLEM 1. Find the radiative recombination coefficient for capture of an electron to the ground state of a hydrogen atom at temperatures $T \ll I = e^4 m/2\hbar^2$, the ionization potential of this atom.

SOLUTION. The cross-section for recombination of a slow electron with a proton at rest, to the ground level of the hydrogen atom, is

$$\sigma_{\text{rec}} = \frac{2^{10}\pi^2(e^2/\hbar c)a_B{}^2/I^2}{3(2.71\ldots)^4 m^2 c^2 v_e{}^2},$$

where v_e is the speed of the electron, and $a_B = \hbar^2/me^2$ the Bohr radius; see *RQT*, formulae (56.13) and (56.14). The mean value $\langle v_e^{-1}\rangle = (2m/\pi T)^{1/2}$. The result is, in accordance with (24.3),

$$\alpha = \frac{2^{10}\pi^{3/2}}{3(2.71)^4}\left(\frac{e^2}{\hbar c}\right)^3 \frac{a_B{}^3 I}{\hbar}\left(\frac{I}{T}\right)^{1/2} = 35\left(\frac{e^2}{\hbar c}\right)^3 \frac{a_B{}^3 I}{\hbar}\left(\frac{I}{T}\right)^{1/2}.$$

PROBLEM 2. Determine the recombination coefficient given by (24.13), neglecting the influence of the electron binding in the excited atom on its collision with the unexcited atom, and assuming that the transport cross-section for these collisions is independent of the velocity.

SOLUTION. The "diffusion coefficient" $B(\epsilon)$ is calculated as in §22; the result is

$$B(\epsilon) = (N/3m)\langle v_{\text{at}}^2\rangle\langle\sigma_t p^3\rangle, \tag{1}$$

where N is the density of atoms in the gas, m the electron mass and v_{at} the relative speed of the excited and unexcited atoms. The speeds v_{at} have a Maxwellian distribution, with the particle mass represented by the reduced mass $\frac{1}{2}M$ (where M is the mass of the atom); hence $\langle v_{\text{at}}^2\rangle = 6T/M$. Next, p in (1) is the electron momentum in the field of the ion; the averaging of $\sigma_t p^3$ is taken over the region of the electron phase space $\tau(\epsilon)$ corresponding to a given value of $|\epsilon|$. With $\sigma_t = \text{constant}$, we find

$$\begin{aligned}\langle\sigma_t p^3\rangle &= \frac{\sigma_t}{\tau(\epsilon)}\int p^3 \delta\left(|\epsilon| + \frac{p^2}{2m} - \frac{ze^2}{r}\right) d^3x\, d^3p \\ &= \frac{32\sqrt{2}}{3\pi}\sigma_t m|\epsilon|^{3/2}.\end{aligned}$$

Thus

$$B = 64\sqrt{2T}\sigma_t N|\epsilon|^{3/2}/3\pi M,$$

and a calculation from (24.13) then gives finally

$$\alpha = 32\sqrt{(2\pi)}m^{1/2}(ze^2)^3\sigma_t N/3MT^{5/2}. \tag{2}$$

The neglect of the electron binding in the atom is legitimate if the perturbation frequency caused by the atom near the electron ($\sim d/\bar{v}_{at}$, where d is the atomic dimension) is large in comparison with the frequency of rotation of an electron with energy $|\epsilon| \sim T$. This leads to the condition $T \ll (e^2/d)(m/M)^{1/2}$

§25. Ambipolar diffusion

Let us consider the diffusion of charged particles in a weakly ionized gas. As in §22, the degree of ionization is assumed to be so small that collisions between charged particles may be neglected in comparison with those between charged particles and neutral atoms. Even under these conditions, the diffusion of the two types of charged particles (electrons and ions) is not independent, because an electric field arises in the diffusion process (W. Schottky, 1924).

The diffusion equations are the equations of continuity for the electrons (e) and the ions (i):

$$\left.\begin{aligned}\partial N_e/\partial t + \operatorname{div} \mathbf{i}_e = 0,\\ \partial N_i/\partial t + \operatorname{div} \mathbf{i}_i = 0,\end{aligned}\right\} \tag{25.1}$$

the fluxes being expressed in terms of the number densities of particles and their gradients by

$$\left.\begin{aligned}\mathbf{i}_e &= -N_e b_e e\mathbf{E} - D_e \nabla N_e,\\ \mathbf{i}_i &= \quad N_i b_i e\mathbf{E} - D_i \nabla N_i,\end{aligned}\right\} \tag{25.2}$$

where D_e and D_i are the diffusion coefficients and b_e and b_i the mobilities of the electrons and ions.† These are related by the Einstein formulae

$$D_e = Tb_e, \quad D_i = Tb_i, \tag{25.3}$$

which express the condition for the fluxes (25.2) to be zero in equilibrium. Using these relations and expressing the field in terms of its potential by $\mathbf{E} = -\nabla\varphi$, we can rewrite equations (25.1) as

$$\partial N_e/\partial t = D_e \operatorname{div}[\nabla N_e - (eN_e/T)\nabla\varphi], \tag{25.4}$$

$$\partial N_i/\partial t = D_i \operatorname{div}[\nabla N_i + (eN_i/T)\nabla\varphi]. \tag{25.5}$$

To these we must add Poisson's equation for the potential:

$$\Delta\varphi = -4\pi e(N_i - N_e). \tag{25.6}$$

†The ion charge is taken to be $z_i = 1$, as is usually true when the degree of ionization of the gas is small.

The equations (25.4)–(25.6) are considerably simplified if the densities N_e and N_i have almost homogeneous distributions. We can then put $N_e \approx N_i \approx \text{constant} \equiv N_0$ in the coefficients of $\nabla\varphi$ in (25.4) and (25.5), and eliminate φ by means of (25.6). The result is

$$\frac{\partial N_e}{\partial t} = D_e\left[\triangle N_e - \frac{N_e - N_i}{a^2}\right], \tag{25.7}$$

$$\frac{\partial N_i}{\partial t} = D_i\left[\triangle N_i - \frac{N_e - N_i}{a^2}\right], \tag{25.8}$$

where $a^{-2} = 4\pi e^2 N_0/T$, and a is the Debye length for electrons or ions; see §31 below.

Although the electron and ion scattering cross-sections are in general of the same order of magnitude, their diffusion coefficients are quite different, because of the difference in their mean thermal speeds v_T:

$$D_e/D_i \sim v_{Te}/v_{Ti} \sim \sqrt{(M/m)}, \tag{25.9}$$

so that $D_e \gg D_i$. This results in some unusual features of the diffusion process.

Let us consider the variation with time of a slight perturbation of the electron and ion densities, whose characteristic dimensions $L \gg a$. In the initial stage of the process, when the variable parts of the densities are $|\delta N_e| \sim |\delta N_i| \sim |\delta N_e - \delta N_i|$, the first terms on the right of equations (25.7) and (25.8) are small in comparison with the second terms:

$$\triangle N_e \sim \delta N_e/L^2 \ll (\delta N_e - \delta N_i)/a^2. \tag{25.10}$$

Noting also that from (25.9) $|\partial N_i/\partial t| \ll |\partial N_e/\partial t|$, we have

$$\frac{\partial}{\partial t}(\delta N_e - \delta N_i) = -(D_e/a^2)(\delta N_e - \delta N_i),$$

whence

$$\delta N_e - \delta N_i = (\delta N_e - \delta N_i)_0 \exp(-D_e t/a^2). \tag{25.11}$$

From this, we see that in a time $\tau_{e1} \sim a^2/D_e$ the difference $|\delta N_e - \delta N_i|$ becomes small in comparison with δN_e and δN_i themselves, i.e. the gas becomes quasi-neutral.

The next stage of the process consists in the development of the electron distribution to reach the equilibrium form (for a given ion distribution), determined by the condition for the right-hand side of (25.7) to be zero:

$$\delta N_e - \delta N_i = a^2 \triangle N_e \approx a^2 \triangle N_i \sim (a^2/L^2)\delta N_i. \tag{25.12}$$

This stage follows the diffusion equation (25.7), with the characteristic time $\tau_{e2} \sim L^2/D_e$, which is small compared with the characteristic ion diffusion time $\tau_i \sim L^2/D_i$; the ion distribution may therefore still be regarded as unaltered.

The final relaxation of the electron and ion density perturbations takes place according to (25.8), which after the substitution of (25.12) becomes

$$\partial N_i/\partial t = 2D_i \triangle N_i. \tag{25.13}$$

Thus, during a time $\sim \tau_i$, the electrons and ions diffuse together ($\delta N_e \approx \delta N_i$) with a diffusion coefficient twice that of the ions; this process is called *ambipolar diffusion*. Half of the coefficient is due to the intrinsic diffusion of the ions, and half to the electric field resulting from the accelerating electrons.

Lastly, let us note that equation (25.13) has a broader range of applicability than follows from the proof given. Even if the perturbation is not weak, the movement of the electrons quickly gives them a Boltzmann distribution in the field and equalizes the electron and ion densities, i.e. leads to quasi-neutrality. Then

$$N_e = N_i = N_0 e^{e\varphi/T}, \quad e\varphi = T \log(N_i/N_0). \tag{25.14}$$

Substitution of (25.14) in (25.5) again gives (25.13), but without the assumption that the perturbation is small.

§26. Ion mobility in solutions of strong electrolytes

The equations given in §25 are easily generalized to the case where ions of different kinds are present. They are also applicable to the movement of ions in solutions of strong electrolytes.† In the limit of infinite dilution of the solution (i.e. as its concentration tends to zero), the mobility of each kind a of ions tends to a constant limit $b_a^{(0)}$, and the diffusion coefficient correspondingly tends to

$$D_a^{(0)} = Tb_a^{(0)}. \tag{26.1}$$

The present section is concerned with the calculation of the correction terms in the first order (with respect to the small concentration) for the ion mobilities in a weak solution.‡ This also gives the correction terms for the conductivity of the solution. In an electric field $\mathbf{E}$, a force $ez_a\mathbf{E}$ acts on each ion, which thereby acquires a directed velocity $b_a ez_a \mathbf{E}$. The current density in the solution is therefore

$$\mathbf{j} = \mathbf{E}\sum_a ez_a N_a \,.\, b_a ez_a,$$

where N_a is the concentration (number per unit volume) of ions of type a; the conductivity is thus

$$\sigma = e^2 \sum_a N_a z_a^2 b_a. \tag{26.2}$$

†These are substances which dissociate completely into ions when dissolved

‡The theory given below was worked out by P. Debye and E. Hückel (1923) and L. Onsager (1927).

The theory given below is based on the same ideas as that of the thermodynamic properties of plasmas and strong electrolytes, namely that around each ion there is formed an inhomogeneous distribution of charges (an ion cloud), which screens the field of the ion. The corresponding formulae have been derived for a plasma in *SP* 1, §§78 and 79. The formulae for a solution of a strong electrolyte differ only in the presence of a permittivity $\epsilon \neq 1$ of the solvent, and will be given below.

The screening cloud alters the mobility of the ion because of two different effects. First, the movement of the ion in the external electric field changes the charge distribution in the cloud, and this causes an additional field on the ion. Second, the movement of the cloud causes a movement of the liquid and hence a "drift" of the ion. The two corrections are called respectively *relaxation* and *electrophoretic* corrections.

Relaxation correction

Let us first calculate the corrections of the first kind. Since the screening cloud results from the existence of a correlation between the positions of different ions, it is a question of the influence of the external field $\mathbf{E}$ on the correlation functions.

We shall define the pair correlation function w_{ab} so that $N_a w_{ab}(\mathbf{r}_a, \mathbf{r}_b)\, dV_a$ is the number of ions of type a in a volume dV_a around the point $\mathbf{r}_a$, if there is one ion of type b at the point $\mathbf{r}_b$; the types a and b may be the same or different. Evidently

$$w_{ab}(\mathbf{r}_a, \mathbf{r}_b) = w_{ba}(\mathbf{r}_b, \mathbf{r}_a), \tag{26.3}$$

and $w_{ab} \to 1$ as $|\mathbf{r}_a - \mathbf{r}_b| \to \infty$. In equilibrium, the functions w_{ab} depend only on the distances $|\mathbf{r}_a - \mathbf{r}_b|$; in an external field, this is not so.†

The correlation functions, like any distribution functions, satisfy equations in the form of continuity equations in the appropriate space—here, the configuration space of the two particles:

$$\partial w_{ab}/\partial t + \mathrm{div}_a\, \mathbf{j}_a + \mathrm{div}_b\, \mathbf{j}_b = 0, \tag{26.4}$$

where $\mathbf{j}_a$ and $\mathbf{j}_b$ are the probability fluxes for the a and b particles, and the suffixes to div show the variables ($\mathbf{r}_a$ or $\mathbf{r}_b$) with respect to which the differentiation is performed.

The flux $\mathbf{j}_a$ is

$$\mathbf{j}_a = -Tb_a^{(0)} \nabla_a w_{ab} + b_a^{(0)} z_a e w_{ab} (\mathbf{E} - \nabla_a \varphi_b), \tag{26.5}$$

and $\mathbf{j}_b$ is the same with the suffixes a and b interchanged. The first term in (26.5) describes the diffusional motion of the type a ions, which occurs even in the absence of an external field. The second term is the ion flux density due to the forces exerted by the external field $\mathbf{E}$ and by the field $-\nabla_a \varphi_b$ at the point $\mathbf{r}_a$ resulting from the modified cloud, with the condition that a type b ion is at the

†The method of correlation functions as applied to the equilibrium state of a plasma (or an electrolyte) is described in *SP* 1, §79.

point $\mathbf{r}_b$. The potential $\varphi_b = \varphi_b(\mathbf{r}_a, \mathbf{r}_b)$ of the latter field satisfies Poisson's equation:

$$\Delta_a\varphi_b(\mathbf{r}_a, \mathbf{r}_b) = -(4\pi/\epsilon)\Big[\sum_c ez_cN_cw_{cb}(\mathbf{r}_a, \mathbf{r}_b) + ez_b\delta(\mathbf{r}_a - \mathbf{r}_b)\Big]. \quad (26.6)$$

The first term in the square brackets is the mean charge density of ions of all types in the cloud, the second term is the charge density localized (by the condition imposed) at the point $\mathbf{r}_b$. The factor $1/\epsilon$ gives the reduction of the field in the dielectric solvent.

Assuming the solution to be sufficiently dilute, we neglect triple correlations between the positions of the ions. In that approximation, the pair correlation functions w_{ab} are almost unity, and the quantities

$$\omega_{ab} = w_{ab} - 1 \quad (26.7)$$

are small. The potentials φ_a are of the same order of smallness. Neglecting second-order terms, we can rewrite (26.5) as

$$\mathbf{j}_a = b_a^{(0)}[-T\nabla_a\omega_{ab} + ez_a(1+\omega_{ab})\mathbf{E} - ez_a\nabla_a\varphi_b]. \quad (26.8)$$

In equation (26.6), we can simply replace w_{ab} by ω_{ab}, since the solution is electrically neutral on average ($\Sigma ez_cN_c = 0$):

$$\Delta_a\varphi_b(\mathbf{r}_a, \mathbf{r}_b) = -(4\pi/\epsilon)\Big[\sum_c ez_cN_c\omega_{cb}(\mathbf{r}_a, \mathbf{r}_b) + ez_b\delta(\mathbf{r}_a - \mathbf{r}_b)\Big]. \quad (26.9)$$

In a uniform constant field $\mathbf{E}$, the functions w_{ab} are independent of time, and they involve the coordinates of the two points only in the form $\mathbf{r} = \mathbf{r}_a - \mathbf{r}_b$, with $\nabla_a w_{ab} = -\nabla_b w_{ab}$. Substitution of $\mathbf{j}_a$ from (26.8) and $\mathbf{j}_b$ from the analogous expression in (26.4) now gives

$$T(b_a^{(0)} + b_b^{(0)})\Delta\omega_{ab}(\mathbf{r}) + ez_ab_a^{(0)}\Delta\varphi_b(\mathbf{r}) + ez_bb_b^{(0)}\Delta\varphi_a(-\mathbf{r})$$
$$= (z_ab_a^{(0)} - z_bb_b^{(0)})e\mathbf{E}\cdot\nabla\omega_{ab}(\mathbf{r}), \quad (26.10)$$

where all derivatives are taken with respect to $\mathbf{r}$.

Assuming the external field weak, we can solve the problem by successive approximation with respect to $\mathbf{E}$. In the zero-order approximation, when $\mathbf{E} = 0$, the potentials $\varphi_a^{(0)}(\mathbf{r})$ are even functions of $\mathbf{r}$. Since all the functions ω_{ab} and φ_a must tend to zero as $r \to \infty$, we then find from (26.10)

$$T(b_a^{(0)} + b_b^{(0)})\omega_{ab}^{(0)} + e(b_a^{(0)}z_a\varphi_b^{(0)} + b_b^{(0)}z_b\varphi_a^{(0)}) = 0. \quad (26.11)$$

We seek the solution in the form

$$\omega_{ab}^{(0)}(\mathbf{r}) = z_az_b\omega^{(0)}(\mathbf{r}), \quad e\varphi_a^{(0)}(\mathbf{r}) = -Tz_a\omega^{(0)}(\mathbf{r}). \quad (26.12)$$

Then equation (26.11) is satisfied identically, and from (26.9) we obtain an equation for $\omega^{(0)}(\mathbf{r})$:

$$\Delta\omega^{(0)}(\mathbf{r}) - \omega^{(0)}(\mathbf{r})/a^2 = (4\pi e^2/\epsilon T)\delta(\mathbf{r}), \tag{26.13}$$

where

$$a^{-2} = (4\pi e^2/\epsilon T)\sum_c N_c z_c^2. \tag{26.14}$$

The solution of this is

$$\omega^{(0)}(\mathbf{r}) = -\frac{e^2}{\epsilon T}\frac{e^{-r/a}}{r}. \tag{26.15}$$

The quantity a is the Debye screening distance in the electrolyte solution.

In the next approximation, we put

$$\varphi_a = \varphi_a^{(0)} + \varphi_a^{(1)}, \quad \omega_{ab} = \omega_{ab}^{(0)} + \omega_{ab}^{(1)}, \tag{26.16}$$

where the superscript (1) marks small corrections to the zero-order values. Being scalars, all these corrections have the form $\mathbf{E}\cdot\mathbf{r}f(r)$, where the $f(r)$ are functions only of the magnitude r; hence all the $\omega_{ab}^{(1)}$ and $\varphi_a^{(1)}$ are odd functions of $\mathbf{r}$. Since, from (26.3),

$$\omega_{ab}^{(1)}(\mathbf{r}_1, \mathbf{r}_2) \equiv \omega_{ab}^{(1)}(\mathbf{r}) = \omega_{ba}^{(1)}(\mathbf{r}_2, \mathbf{r}_1) \equiv \omega_{ba}^{(1)}(-\mathbf{r}),$$

it follows also that

$$\omega_{ab}^{(1)}(\mathbf{r}) = -\omega_{ba}^{(1)}(\mathbf{r}), \tag{26.17}$$

if we remember that everywhere $\mathbf{r} = \mathbf{r}_a - \mathbf{r}_b$. If the ions a and b are of the same kind, an interchange of suffixes cannot alter the function $\omega_{ab}^{(1)}(\mathbf{r})$, and therefore (26.17) shows that such $\omega_{aa}^{(1)} = 0$. Thus the corrections $\omega_{ab}^{(1)}$ occur only for the correlation functions of pairs of different ions.

To simplify the subsequent calculations, we shall take the case of an electrolyte with ions of only two kinds. Then only one function $\omega_{12}^{(1)}(\mathbf{r}) = -\omega_{21}^{(1)}(\mathbf{r})$ is non-zero, and substitution of (26.16) in Poisson's equation (26.9) gives

$$\Delta\varphi_2^{(1)}(\mathbf{r}) = -\frac{4\pi e}{\epsilon} z_1 N_1 \omega_{12}^{(1)}(\mathbf{r}), \tag{26.18}$$

where $\mathbf{r} = \mathbf{r}_1 - \mathbf{r}_2$. With the condition of electrical neutrality of the solution, and the above-mentioned symmetry properties of the functions, it is easily seen that the potential $\varphi_1^{(1)}(\mathbf{r})$ satisfies a similar equation, and therefore $\varphi_1^{(1)}(\mathbf{r}) = \varphi_2^{(1)}(\mathbf{r})$.

On substituting (26.16) in (26.10), we retain only the term in $\omega_{12}^{(0)}$ on the right, obtaining

$$T(b_1^{(0)} + b_2^{(0)})\Delta\omega_{12}^{(1)}(\mathbf{r}) + e(b_1^{(0)}z_1 - b_2^{(0)}z_2)\Delta\varphi_2^{(1)}(\mathbf{r}) = (b_1^{(0)}z_1 - b_2^{(0)}z_2)ez_1z_2\mathbf{E}\cdot\nabla\omega^{(0)}(\mathbf{r}). \tag{26.19}$$

The equations (26.18) and (26.19) are solved by Fourier analysis. A set of algebraic equations for the Fourier components $\omega^{(1)}_{12\mathbf{k}}$ and $\varphi^{(1)}_{2\mathbf{k}}$ is obtained, differing from (26.18) and (26.19) by the change of operators $\nabla \to i\mathbf{k}$, $\triangle \to -\mathbf{k}^2$. The Fourier component of the function $\omega^{(0)}(\mathbf{r})$ (26.15) on the right of (26.19) is

$$\omega^{(0)}_{\mathbf{k}} = -\frac{e^2}{\epsilon T}\frac{4\pi}{k^2 + 1/a^2}.$$

The final result for the Fourier component of the potential is

$$\varphi^{(1)}_{2\mathbf{k}} = \frac{4\pi e^2 z_1 z_2 q}{\epsilon T a^2}\frac{i\mathbf{k}\,.\,\mathbf{E}}{k^2(k^2 + 1/a^2)(k^2 + q/a^2)}, \tag{26.20}$$

where

$$q = \frac{b^{(0)}_1 z_1 - b^{(0)}_2 z_2}{(z_1 - z_2)(b^{(0)}_1 + b^{(0)}_2)}. \tag{26.21}$$

Since z_1 and z_2 have opposite signs, it is evident that $0 < q < 1$.

The function $\varphi^{(1)}_2(\mathbf{r}_1, \mathbf{r}_2)$ is the additional potential at $\mathbf{r}_1$ when there is an ion 2 at $\mathbf{r}_2$. The corresponding field strength is

$$\mathbf{E}^{(1)}_2(\mathbf{r}) = -\nabla_1\varphi^{(1)}_2(\mathbf{r}_1, \mathbf{r}_2) = -\nabla\varphi^{(1)}_2(\mathbf{r}).$$

Its value for $\mathbf{r}_1 = \mathbf{r}_2$ (i.e. $\mathbf{r} = 0$) gives the required field which acts on the ion 2 itself and thus alters its mobility.

The Fourier component $\mathbf{E}^{(1)}_{2\mathbf{k}} = -i\mathbf{k}\varphi^{(1)}_{2\mathbf{k}}$. Hence

$$\mathbf{E}^{(1)}_2(0) = \left[-\int i\mathbf{k}\varphi^{(1)}_{2\mathbf{k}} e^{i\mathbf{k}\,.\,\mathbf{r}}\frac{d^3k}{(2\pi)^3}\right]_{\mathbf{r}=0} = -\int i\mathbf{k}\varphi^{(1)}_{2\mathbf{k}}\,d^3k/(2\pi)^3.$$

Substitution of (26.20) here leads to the integral

$$\mathbf{I} = \int \frac{\mathbf{k}(\mathbf{k}\,.\,\mathbf{E})}{k^2(k^2 + 1/a^2)(k^2 + q/a^2)}\frac{d^3k}{(2\pi)^3}.$$

The averaging over the directions of $\mathbf{k}$ replaces $\mathbf{k}(\mathbf{k}\,.\,\mathbf{E})$ by $\frac{1}{3}k^2\mathbf{E}$, and the integral with respect to k is then calculated from the residues of the integrand at the poles $k = i/a$ and $k = i\sqrt{q}/a$:

$$\mathbf{I} = \mathbf{E}a/12\pi(1 + \sqrt{q}).$$

Thus the total field acting on the ion 2 is

$$\mathbf{E} + \mathbf{E}^{(1)}_2(0) = \left[1 - \frac{e^2|z_1 z_2|q}{3\epsilon T a(1 + \sqrt{q})}\right]\mathbf{E}. \tag{26.22}$$

A similar result is obtained for the field acting on the ion 1, as is evident from the symmetry of the expression (26.22) with respect to the suffixes 1 and 2. Multiplying the field (26.22) by $b^{(0)}ez$, we obtain the velocity acquired by the ion, and if this is written in the form $bez\mathbf{E}$ it follows that the expression in square brackets gives the ratio $b/b^{(0)}$. Thus the required relaxation correction to the ion mobility is found to be

$$b_{\text{rel}} = -b\frac{e^2|z_1z_2|q}{3\epsilon Ta(1+\sqrt{q})}. \tag{26.23}$$

This effect reduces the mobility.

ELECTROPHORETIC CORRECTION

Let us now go on to calculate the correction due to the movement of the solvent, formulating the problem as follows. We consider a particular ion in the solution, together with the screening cloud around it. This cloud carries an electric charge density $\delta\rho = \Sigma ez_a\delta N_a$, where δN_a is the difference between the density of ions of type a in the cloud and the mean value N_a in the solution. Thus forces with volume density $\mathbf{f} = \mathbf{E}\delta\rho$ act on the liquid carrying this cloud, in the presence of an electric field $\mathbf{E}$. These forces cause the liquid to move, which in turn carries along the central ion in question.

The distribution of ions in the cloud is related to the field potential φ there by Boltzmann's formula:

$$\delta N_a = N_a[e^{-z_ae\varphi/T} - 1] \approx -z_ae\varphi N_a/T.$$

Since the field $\mathbf{E}$ is weak, we may neglect the deformation of the ion cloud in the present problem. In a spherically symmetrical cloud, the potential is

$$\varphi = (ez_b/r)e^{-r/a},$$

where ez_b is the charge on the central ion, and a is determined by (26.14); cf. *SP* 1, §78. The total charge density in the cloud is therefore

$$\delta\rho = -\frac{e^2\varphi}{T}\sum_a N_az_a^2 = -\frac{ez_b}{4\pi a^2}\frac{e^{-r/a}}{r}. \tag{26.24}$$

Since the motion caused by the field $\mathbf{E}$ is slow, the liquid may be regarded as incompressible, and so

$$\operatorname{div}\mathbf{v} = 0. \tag{26.25}$$

For the same reason, the term quadratic in the velocity may be omitted from the Navier–Stokes equation, which then reduces (for a steady motion) to

$$\eta\triangle\mathbf{v} - \nabla P + \mathbf{f} = 0, \tag{26.26}$$

where P is the pressure and η the viscosity of the solvent.

Taking Fourier components in (26.25) and (26.26), we have

$$\mathbf{k}\,.\,\mathbf{v}_{\mathbf{k}} = 0, \quad -\eta k^2 \mathbf{v}_{\mathbf{k}} - i\mathbf{k}P_{\mathbf{k}} + \mathbf{E}\delta\rho_{\mathbf{k}} = 0.$$

Multiplying the second equation scalarly by $i\mathbf{k}$, we find $P_{\mathbf{k}} = -i\mathbf{k}\,.\,\mathbf{E}\delta\rho_{\mathbf{k}}/k^2$, and so

$$\mathbf{v}_{\mathbf{k}} = \frac{\delta\rho_{\mathbf{k}}}{\eta}\frac{k^2\mathbf{E} - \mathbf{k}(\mathbf{k}\,.\,\mathbf{E})}{k^4}. \tag{26.27}$$

The Fourier component of the charge density (26.24) is

$$\delta\rho_{\mathbf{k}} = -ez_b/(a^2k^2 + 1). \tag{26.28}$$

The required velocity of the liquid at the point $r = 0$, where the central ion is situated, is given by the integral

$$\mathbf{v}(0) = \int \mathbf{v}_{\mathbf{k}}\, d^3k/(2\pi)^3.$$

Substitution of $\mathbf{v}_{\mathbf{k}}$ from (26.27) and (26.28), and integration over the direction of $\mathbf{k}$, gives

$$\mathbf{v}(0) = -\mathbf{E}\frac{ez_b}{(2\pi)^3\eta}\frac{8\pi}{3}\int_0^\infty \frac{dk}{k^2a^2 + 1},$$

and finally

$$\mathbf{v}(0) = -(ez_b/6\pi\eta a)\mathbf{E}.$$

This velocity is added to the velocity $ez_b b_b^{(0)}\mathbf{E}$ acquired by the ion through the direct action of the field. It is therefore clear that the required electrophoretic correction to the mobility is

$$b_{\text{el-ph}} = -1/6\pi a\eta, \tag{26.29}$$

the same for ions of all types. The total correction is given by the sum of the two expressions (26.23) and (26.29). Both are negative, and through the factor $1/a$ are proportional to the square root of the concentration.

CHAPTER III

COLLISIONLESS PLASMAS

§ 27. The self-consistent field

A WIDE field of application of transport theory occurs for *plasmas*, by which we shall mean completely ionized gases.† The thermodynamic theory of the equilibrium state of plasmas has been discussed in *SP* 1, §§78–80; *SP* 2, §85. Chapters III–V will be devoted to the transport properties of plasmas. To avoid complications having no fundamental significance, we shall (where necessary) regard the plasma as having only two components: electrons with charge $-e$, and positive ions of one type with charge ze.

As in ordinary gases, the plasma must be sufficiently rarefied if the transport equation is to be applicable to it; the gas must be almost ideal. However, because Coulomb forces decrease only slowly, this condition is more stringent for a plasma than for a gas of neutral particles. At present making no distinction between particles having different charges, we can write the condition for the plasma to be almost ideal as

$$T \gg e^2/\bar{r} \sim e^2 N^{1/3}, \tag{27.1}$$

where T is the temperature of the plasma, N the total number of particles per unit volume and $\bar{r} \sim N^{-1/3}$ the mean distance between them. This condition states that the mean interaction energy of two ions is small in comparison with their mean kinetic energy. It may be differently expressed by using the *Debye length a* of the plasma, defined by

$$1/a^2 = (4\pi/T) \sum_a N_a (z_a e)^2, \tag{27.2}$$

where the summation is over all types of ion; it may be noted (see *SP* 1, § 78) that a determines the distance at which the Coulomb field of a charge in the plasma is screened. With $a \sim (T/4\pi N e^2)^{1/2}$ in (27.1), we have

$$e^2 N^{1/3}/T \sim \bar{r}^2/4\pi a^2 \ll 1: \tag{27.3}$$

in a rarefied plasma, the mean distance between the particles must be small compared with the Debye length, i.e. the ion cloud around a charge must in fact

†The term is due to I. Langmuir (1923), who laid the foundations of the systematic theoretical study of plasmas.

contain many particles. The small ratio (27.3) acts as a "gaseousness parameter" for the plasma.

The plasma will be assumed classical throughout Chapters III–V, except in § 40. This implies the fulfilment of only a very weak condition: the plasma temperature must be high in comparison with the degeneracy temperature of its electron component,

$$T \gg \hbar^2 N^{2/3}/m, \tag{27.4}$$

where m is the electron mass (cf. *SP* 1, § 80).

The transport equation for each type of particle in the plasma (electrons and ions) is

$$\frac{\partial f}{\partial t} + \mathbf{v} \cdot \frac{\partial f}{\partial \mathbf{r}} + \dot{\mathbf{p}} \cdot \frac{\partial f}{\partial \mathbf{p}} = C(f), \tag{27.5}$$

where f is the coordinate and momentum distribution function of the particles concerned, and C is their collision integral (with particles of any kind). The derivative $\dot{\mathbf{p}}$ is determined by the force acting on the particle. This force is in turn expressed in terms of the electric and magnetic fields, due to all the other particles, at the position of the particle considered. The following point now arises, however.

For neutral particles (atoms or molecules), because of the rapid decrease of the interaction forces, there are noticeable changes in their motion, interpretable as collisions, only at small impact parameters, of the order of atomic dimensions. Between such collisions, the particles move as if free; for this reason, $\dot{\mathbf{p}} = 0$ is assumed on the left-hand side of the transport equation for ordinary gases. In a plasma, however, because of the long-range Coulomb forces, a noticeable change in the motion of the particles occurs even at large impact parameters; the Coulomb forces are screened in the plasma only at distances $\sim a$, which by the condition (27.3) are large even in comparison with the distances between particles (see *SP* 1, § 78, and § 31 below, Problem 1). However, not all such cases are to be interpreted as collisions in the transport equation. In transport theory, random collisions are the mechanism causing an approach to the state of equilibrium, with a corresponding increase in the entropy of the system. But collisions at large ($\gtrsim a$) impact parameters cannot act as such a relaxation mechanism. The reason is that the interaction of two charged particles at such distances is actually a collective effect involving many particles. Accordingly, the effective field which can describe this interaction is also generated by a large number of particles, i.e. is macroscopic. The whole process then becomes a macroscopically certain and not a random one; such processes cannot cause the entropy of the system to increase. They must therefore be excluded from the scope of the term "collisions" as applied to the right-hand side of the transport equations.

This distinction corresponds to the representation of the exact microscopic values of the electric field $\mathbf{e}$ and the magnetic field $\mathbf{h}$, acting on a particle in the plasma, as

$$\mathbf{e} = \mathbf{E} + \mathbf{e}', \quad \mathbf{h} = \mathbf{B} + \mathbf{h}', \tag{27.6}$$

where $\mathbf{E}$ and $\mathbf{B}$ are the fields averaged over regions containing many particles and having dimensions large compared with the distances between the particles but small compared with the Debye length. The terms $\mathbf{e}'$ and $\mathbf{h}'$ then describe the random fluctuations of the fields, which cause random changes in the motion of the particles, i.e. collisions.

In (27.6), the precise significance of $\mathbf{E}$ and $\mathbf{B}$ is that of being the mean fields at the position of a given particle. Since the plasma is assumed rarefied, the correlation between simultaneous positions of particles in it may be neglected. Then the position of each given particle is in no way distinctive, and so $\mathbf{E}$ and $\mathbf{B}$ may be regarded as just the fields averaged in the ordinary sense of macroscopic electrodynamics. These fields determine the Lorentz force which is to replace $\dot{\mathbf{p}}$ in (27.5).

In this chapter, the phenomena discussed will be those in which collisions between plasma particles are unimportant. Such a plasma is said to be *collisionless*. The precise conditions for collisions to be negligible depend in general on the specific formulation of the problem, but a necessary condition is usually that the effective collision frequency ν (the reciprocal of the mean free time of a particle) should be small in comparison with the frequency ω of variation of the macroscopic fields $\mathbf{E}$ and $\mathbf{B}$ in the process concerned:

$$\nu \ll \omega. \tag{27.7}$$

Because of this condition, the collision integral in the transport equation is small in comparison with $\partial f/\partial t$. Collisions may be neglected even if the particle mean free path $l \sim \bar{v}/\nu$ is large compared with the distance L over which the field varies (the field "wavelength"). Putting $1/L \sim k$, we can write this condition as

$$\nu \ll k\bar{v}. \tag{27.8}$$

The collision integral is small in comparison with the term $\mathbf{v} \,.\, \nabla f$ on the left-hand side of the transport equation.

When the collision integral is neglected, the transport equations for the electron and ion distribution functions f_e and f_i become†

$$\left.\begin{aligned} \frac{\partial f_e}{\partial t} + \mathbf{v}\cdot\frac{\partial f_e}{\partial \mathbf{r}} - e\left(\mathbf{E} + \frac{\mathbf{v}\times\mathbf{B}}{c}\right)\cdot\frac{\partial f_e}{\partial \mathbf{p}} = 0, \\ \frac{\partial f_i}{\partial t} + \mathbf{v}\cdot\frac{\partial f_i}{\partial \mathbf{r}} + ze\left(\mathbf{E} + \frac{\mathbf{v}\times\mathbf{B}}{c}\right)\cdot\frac{\partial f_i}{\partial \mathbf{p}} = 0. \end{aligned}\right\} \tag{27.9}$$

To these we must add the averaged Maxwell's equations

$$\operatorname{curl}\mathbf{E} = -\frac{1}{c}\frac{\partial \mathbf{B}}{\partial t}, \quad \operatorname{div}\mathbf{B} = 0, \quad \operatorname{curl}\mathbf{B} = \frac{1}{c}\frac{\partial \mathbf{E}}{\partial t} + \frac{4\pi}{c}\mathbf{j}, \quad \operatorname{div}\mathbf{E} = 4\pi\rho, \tag{27.10}$$

†Strictly speaking, in the presence of a magnetic field the phase space of the particle is to be defined as the $(\mathbf{r}, \mathbf{P})$ space, where $\mathbf{P} = \mathbf{p} - e\mathbf{A}(t, \mathbf{r})/c$ is the generalized momentum. But $d^3x\, d^3P = d^3x\, d^3p$, since the addition of $\mathbf{A}$ changes only the zero of momentum at each point in space. We can therefore continue to relate the distribution function to $d^3x\, d^3p$.

where ρ and $\mathbf{j}$ are the mean charge density and current density, which can be expressed in terms of the distribution functions by the obvious formulae

$$\left.\begin{aligned}\rho &= e\int (zf_i - f_e)\, d^3p,\\ \mathbf{j} &= e\int (zf_i - f_e)\mathbf{v}\, d^3p.\end{aligned}\right\} \tag{27.11}$$

Equations (27.9)–(27.11) form a coupled set of equations to determine simultaneously the distribution functions f_e, f_i and the fields $\mathbf{E}$, $\mathbf{B}$; the fields thus determined are said to be *self-consistent*. The self-consistent field was brought into the transport equations by A. A. Vlasov (1937); equations (27.9)–(27.11) are called the *Vlasov equations*.

In accordance with the foregoing discussion, the time variation of the distribution functions in a collisionless plasma with a self-consistent field is not associated with an increase of entropy, and therefore cannot in itself bring about the establishment of statistical equilibrium. This is also seen directly from the form of equations (27.9), in which $\mathbf{E}$ and $\mathbf{B}$ occur formally only as external fields imposed on the plasma.

Each of the transport equations (27.9) has the form

$$df/dt = 0, \tag{27.12}$$

where the total derivative signifies differentiation along the paths of the particles. The general solution of such an equation is an arbitrary function of all integrals of the motion of a particle in the fields $\mathbf{E}$ and $\mathbf{B}$.

§28. Spatial dispersion in plasmas

We can rewrite equations (27.10) in a form more usual in macroscopic electrodynamics, by including the electric induction $\mathbf{D}$ as well as the field $\mathbf{E}$. We define the electric polarization vector $\mathbf{P}$ by the relations

$$\partial\mathbf{P}/\partial t = \mathbf{j}, \quad \operatorname{div}\mathbf{P} = -\rho; \tag{28.1}$$

the compatibility of these is guaranteed by the equation of continuity $\operatorname{div}\mathbf{j} = -\partial\rho/\partial t$ (the definition will be further discussed later in this section). Then equations (27.10) become, with $\mathbf{D} = \mathbf{E} + 4\pi\mathbf{P}$,

$$\left.\begin{aligned}\operatorname{curl}\mathbf{E} &= -(1/c)\partial\mathbf{B}/\partial t, \quad \operatorname{div}\mathbf{B} = 0,\\ \operatorname{curl}\mathbf{B} &= (1/c)\partial\mathbf{D}/\partial t, \quad \operatorname{div}\mathbf{D} = 0.\end{aligned}\right\} \tag{28.2}$$

In weak fields, the relation between the induction $\mathbf{D}$ and the field $\mathbf{E}$ is linear,† but even in ordinary media the relation is not instantaneous: the value of $\mathbf{D}(t, \mathbf{r})$ at a

†The condition for the field to be weak will be formulated in §29.

time t depends in general on the values of $\mathbf{E}(t, \mathbf{r})$ not only at that instant but also at all previous instants (see *ECM*, §58). In a plasma, it is additionally a non-local relation: the value of $\mathbf{D}(t, \mathbf{r})$ at a point $\mathbf{r}$ depends on the values of $\mathbf{E}(t, \mathbf{r})$ not only at that point but in general also throughout the plasma. This is because a "free" (i.e. collisionless) movement of particles in the plasma is governed by the field values all along their trajectories.

The general linear relation between the functions $\mathbf{D}(t, \mathbf{r})$ and $\mathbf{E}(t, \mathbf{r})$ may be written (on the assumption that the unperturbed plasma is in a steady state)

$$D_\alpha(t, \mathbf{r}) = E_\alpha(t, \mathbf{r}) + \int_{-\infty}^{t} \int K_{\alpha\beta}(t - t', \mathbf{r}, \mathbf{r}') E_\beta(t', \mathbf{r}')\, d^3x'\, dt'.$$

For a spatially homogeneous plasma, the kernel $K_{\alpha\beta}$ of the integral operator depends only on the difference $\mathbf{r} - \mathbf{r}'$ of the arguments. Writing $\mathbf{r} - \mathbf{r}' = \boldsymbol{\rho}$, $t - t' = \tau$, we can rewrite the above relation as

$$D_\alpha(t, \mathbf{r}) = E_\alpha(t, \mathbf{r}) + \int_0^\infty \int K_{\alpha\beta}(\tau, \boldsymbol{\rho}) E_\beta(t - \tau, \mathbf{r} - \boldsymbol{\rho})\, d^3\rho\, d\tau. \tag{28.3}$$

In the usual way, an expansion as a Fourier series or integral allows the field to be represented as a set of plane waves in which $\mathbf{E}$ and $\mathbf{D}$ are proportional to $e^{i(\mathbf{k}\cdot\mathbf{r} - \omega t)}$. For such waves, the relation between $\mathbf{D}$ and $\mathbf{E}$ becomes

$$D_\alpha = \epsilon_{\alpha\beta}(\omega, \mathbf{k}) E_\beta, \tag{28.4}$$

where the permittivity tensor is

$$\epsilon_{\alpha\beta}(\omega, \mathbf{k}) = \delta_{\alpha\beta} + \int_0^\infty \int K_{\alpha\beta}(\tau, \boldsymbol{\rho}) e^{i(\omega\tau - \mathbf{k}\cdot\boldsymbol{\rho})}\, d^3\rho\, d\tau. \tag{28.5}$$

From this definition it follows at once that

$$\epsilon_{\alpha\beta}(-\omega, -\mathbf{k}) = \epsilon^*_{\alpha\beta}(\omega, \mathbf{k}). \tag{28.6}$$

Thus the non-localness of the relation between $\mathbf{E}$ and $\mathbf{D}$ has the result that the permittivity of the plasma depends on the wave vector as well as on the frequency. This is referred to as *spatial dispersion*, in the same way as the frequency dependence is called *time dispersion* or *frequency dispersion*.

Returning to equations (28.1) and (28.2), we may recall that in the formulation of Maxwell's equations for variable fields in ordinary media the introduction of the dielectric polarization $\mathbf{P}$ is accompanied by that of the magnetization $\mathbf{M}$, the mean microscopic current being divided into two parts $\partial\mathbf{P}/\partial t$ and $c\,\mathrm{curl}\,\mathbf{M}$; in a plane wave, these become $-i\omega\mathbf{P}$ and $ic\mathbf{k}\times\mathbf{M}$. In the presence of spatial dispersion, however, when all quantities depend on $\mathbf{k}$ in any event, this division is inappropriate.

If the current $\mathbf{j}$ and the charge density ρ are fully included in the definition of the polarization $\mathbf{P}$, as in (28.1), the latter in general depends on both the electric field $\mathbf{E}$

and the magnetic field **B**. The field **B** can be expressed in terms of **E** by means of the first pair of Maxwell's equations (28.2), which contain only these two quantities, i.e. (for a plane wave) from $\mathbf{k}\times\mathbf{E} = \omega\mathbf{B}/c$ and $\mathbf{k}\cdot\mathbf{B} = 0$. The polarization **P** is thereby expressed in terms of **E** alone, as is implied in the definition of $\epsilon_{\alpha\beta}$ by (28.3)–(28.5).

The dependence on the wave vector creates a distinctive direction in the function $\epsilon_{\alpha\beta}(\omega, \mathbf{k})$, namely that of its argument **k**. Hence, when spatial dispersion is present, the permittivity is a tensor even in an isotropic medium. The general form of such a tensor may be written

$$\epsilon_{\alpha\beta}(\omega, \mathbf{k}) = \epsilon_t(\omega, k)(\delta_{\alpha\beta} - k_\alpha k_\beta/k^2) + \epsilon_l(\omega, k)k_\alpha k_\beta/k^2. \tag{28.7}$$

On multiplication by E_β, the first term in (28.7) gives a contribution to the induction **D** that is perpendicular to the wave vector, and the second term a contribution that is parallel to **k**. For fields **E** perpendicular or parallel to **k**, the relation between **D** and **E** reduces respectively to $\mathbf{D} = \epsilon_t\mathbf{E}$ or $\mathbf{D} = \epsilon_l\mathbf{E}$. The scalar functions ϵ_t and ϵ_l are called the *transverse* and *longitudinal* permittivities. They depend on two independent variables: the frequency ω, and the magnitude k of the wave vector. When $\mathbf{k}\to 0$, the distinctive direction disappears, and the tensor $\epsilon_{\alpha\beta}$ must then reduce to the form $\epsilon(\omega)\delta_{\alpha\beta}$, where $\epsilon(\omega)$ is the ordinary scalar permittivity, which takes account only of frequency dispersion. Correspondingly, the limiting values of the functions ϵ_t and ϵ_l are equal:

$$\epsilon_t(\omega, 0) = \epsilon_l(\omega, 0) = \epsilon(\omega). \tag{28.8}$$

According to (28.6), the scalar functions ϵ_l and ϵ_t have the property

$$\epsilon_l(-\omega, k) = \epsilon_l^*(\omega, k), \quad \epsilon_t(-\omega, k) = \epsilon_t^*(\omega, k). \tag{28.9}$$

The spatial dispersion does not affect the properties of ϵ_l and ϵ_t as functions of the complex variable ω. All the results (see *ECM*, §62) for the permittivity $\epsilon(\omega)$ of ordinary media without spatial dispersion remain valid for these functions.

In this chapter, we shall consider only isotropic plasmas. It must be emphasized that this implies not only the absence of an external magnetic field, but also the isotropy of the momentum distribution of the particles (in a plasma unperturbed by a field). Otherwise, further distinctive directions appear, and the tensor structure of $\epsilon_{\alpha\beta}$ is more complicated.

It has already been mentioned that the origin of the spatial dispersion in the plasma is related to the dependence of the "free" movement of the particles on the field values along their trajectories. In practice, of course, the movement of a particle at any point is significantly affected by the field values not along the whole trajectory but only along fairly short sections of it. The order of magnitude of these lengths may be governed by two processes: collisions, which perturb the free movement along the trajectory, and averaging of the oscillating field during the time of flight of the particle. The characteristic distance for the first process is the mean free path $l \sim \bar{v}/\nu$ of the particle, and that for the second process is the distance $\bar{v}/\omega$ which a particle moving with mean speed $\bar{v}$ traverses in one period of the field.

In the expression (28.3), the range of correlation between the values of **D** and **E** at different points in space corresponds to the distances r_{cor} over which the function $K_{\alpha\beta}(\tau, \boldsymbol{\rho})$ decreases significantly. We can therefore say that the order of magnitude of these distances is given by the smaller of l and $\bar{v}/\omega$, taken for the particles (electrons or ions) that have the higher value of it.† If $\nu \ll \omega$, then $\bar{v}/\omega$ is the smaller, and

$$r_{cor} \sim \bar{v}/\omega. \tag{28.10}$$

The spatial dispersion is considerable when $kr_{cor} \gtrsim 1$, and disappears when $kr_{cor} \ll 1$; in the latter case, $e^{-i\mathbf{k}\cdot\boldsymbol{\rho}} \approx 1$ in (28.5), and the integral no longer depends on **k**. With r_{cor} from (28.10), we therefore find that the spatial dispersion is important for waves whose phase velocity ω/k is comparable with or less than the mean speed of the particles in the plasma. In the opposite limiting case, with

$$\omega \gg k\bar{v}, \tag{28.11}$$

the spatial dispersion is not significant.

It is important to note that the values of r_{cor} in plasmas may be large compared with the mean distances ($\sim N^{-1/3}$) between the particles. This is the condition that makes possible the macroscopic description of spatial dispersion in terms of the permittivity even when the dispersion is considerable. It has been mentioned in *ECM*, §83, that in ordinary media the correlation length is represented by the atomic dimensions, and hence the condition for the macroscopic theory to be applicable already requires the inequality $kr_{cor} \ll 1$ to be satisfied (the wavelength must be large in comparison with atomic dimensions); for this reason, the spatial dispersion in such media (as manifested, for example, in the natural optical activity) is never more than a small correction.

§29. The permittivity of a collisionless plasma

In the general case of arbitrary **k**, when spatial dispersion plays an important part, the calculation of the permittivity requires the use of the transport equation. We shall do so on the assumptions that the dielectric polarization of the plasma involves only electrons, and that the movement of the ions is unimportant (this is called an *electron plasma*). The condition for such assumptions to be admissible, and the generalization of the results, will be discussed in §31.

For a weak field, we look for the electron distribution function in the form $f = f_0 + \delta f$, where f_0 is the stationary isotropic homogeneous distribution function unperturbed by the field, and δf the change in it due to the field. Neglecting the second-order terms in the transport equation, we have

$$\frac{\partial \delta f}{\partial t} + \mathbf{v}\cdot\frac{\partial \delta f}{\partial \mathbf{r}} = e\left(\mathbf{E} + \frac{\mathbf{v}\times\mathbf{B}}{c}\right)\cdot\frac{\partial f_0}{\partial \mathbf{p}}.$$

In an isotropic plasma, the distribution function depends only on the magnitude of

†This is correct for ϵ_t. For ϵ_l (if $l \ll \bar{v}/\omega$), because of particle diffusion along the field, $r_{cor} \sim (l\bar{v}/\omega)^{1/2}$.

the momentum. For such a function, the direction of the vector $\partial f_0/\partial \mathbf{p}$ is the same as that of $\mathbf{p} = m\mathbf{v}$, and its scalar product with $\mathbf{v}\times\mathbf{B}$ is zero. In the linear approximation, therefore, the magnetic field does not affect the distribution function. The resulting equation for δf is

$$\frac{\partial \delta f}{\partial t} + \mathbf{v}\cdot\frac{\partial \delta f}{\partial \mathbf{r}} = e\mathbf{E}\cdot\frac{\partial f_0}{\partial \mathbf{p}}. \tag{29.1}$$

The function δf is assumed to be proportional to $\exp[i(\mathbf{k}\,.\,\mathbf{r} - \omega t)]$, like the field $\mathbf{E}$. Then (29.1) gives

$$\delta f = \frac{e\mathbf{E}}{i(\mathbf{k}\,.\,\mathbf{v} - \omega)}\cdot\frac{\partial f_0}{\partial \mathbf{p}}. \tag{29.2}$$

The condition for the field to be small results from the requirement that δf be small in comparison with f_0. The coefficient of $\partial f_0/\partial \mathbf{p}$ in (29.2) is the amplitude of the momentum acquired by the electron in the field $\mathbf{E}$. This must be small in comparison with the mean momentum $m\bar{v}$ determined according to the distribution f_0.

In an unperturbed plasma, the electron charge density is balanced at every point by the ion charges, and the current density is identically zero, since the plasma is isotropic. The charge density and current density in the plasma perturbed by the field are

$$\rho = -e\int \delta f\, d^3p, \quad \mathbf{j} = -e\int \mathbf{v}\delta f\, d^3p. \tag{29.3}$$

These quantities, like δf, are proportional to $\exp[i(\mathbf{k}\,.\,\mathbf{r} - \omega t)]$, and according to (28.1) their relation to the dielectric polarization is given by

$$i\mathbf{k}\,.\,\mathbf{P} = -\rho, \quad -i\omega\mathbf{P} = \mathbf{j}. \tag{29.4}$$

The method of taking the integrals in (29.3) requires more precise specification, however, since the function δf has a pole at

$$\omega = \mathbf{k}\,.\,\mathbf{v}. \tag{29.5}$$

In order to attach a meaning to the integral, we shall consider not a strictly harmonic field ($\propto e^{-i\omega t}$) but one which is applied with infinite slowness from $t = -\infty$ onwards. This description of the field corresponds to adding to its frequency an infinitesimal positive imaginary part, i.e. replacing ω by $\omega + i\delta$ with $\delta \to +0$: we then have $E \propto e^{-i\omega t}e^{t\delta} \to 0$ as $t \to -\infty$, and the unlimited increase of the field caused by the factor $e^{t\delta}$ is unimportant as $t \to \infty$, since the causality principle shows that it cannot affect what is observed at finite times t (whereas with $\delta < 0$ the field would have been large in the past, and this would prevent the use of the approximation linear in the field). Thus the rule for avoiding the poles (29.5) is expressed by

$$\omega \to \omega + i0; \tag{29.6}$$

this was first established by L. D. Landau (1946).

The rule (29.6) can also be arrived at from a different standpoint, by including in the transport equation an infinitesimal collision integral $C(f) = -\nu\delta f$. The addition of such a term on the right of (29.1) is equivalent to the change $\omega \to \omega + i\nu$ in the term $\partial\delta f/\partial t = -i\omega\delta f$; then, as $\nu \to 0$, we again have the rule (29.6).†

The integrations using the rule (29.6) involve integrals of the form

$$\int_{-\infty}^{\infty} \frac{f(z)dz}{z - i\delta}, \quad \delta > 0.$$

In such an integral, the path of integration in the complex z-plane passes below the point $z = i\delta$; as $\delta \to 0$, this is equivalent to an integration along the real axis but passing along an infinitesimal semicircle below the pole $z = 0$. The contribution to the integral from this semicircle is given by half the residue of the integrand, and the result is

$$\int_{-\infty}^{\infty} \frac{f(z)}{z - i0} dz = \mathrm{P} \int_{-\infty}^{\infty} \frac{f(z)}{z} dz + i\pi f(0), \tag{29.7}$$

where P denotes the principal value of the integral. This may also be written symbolically

$$\frac{1}{z - i0} = \mathrm{P}\frac{1}{z} + i\pi\delta(z), \tag{29.8}$$

where P now denotes that the principal values are taken in subsequent integrations.

Let us calculate the longitudinal part of the plasma permittivity, using the first relation (29.4), in which we substitute $\delta\rho$ from (29.3) and (29.2):

$$i\mathbf{k}\,.\,\mathbf{P} = -e^2\mathbf{E}\cdot\int \frac{\partial f_0}{\partial \mathbf{p}} \frac{d^3p}{i(\mathbf{k}\,.\,\mathbf{v} - \omega - i0)}.$$

Let the field $\mathbf{E}$, and therefore $\mathbf{P}$, be parallel to $\mathbf{k}$. Then $4\pi\mathbf{P} = (\epsilon_l - 1)\mathbf{E}$. We thus arrive at the following formula for the longitudinal permittivity of a plasma having any stationary distribution function $f(p)$ (the suffix 0 to f will be omitted):

$$\epsilon_l = 1 - \frac{4\pi e^2}{k^2}\int \mathbf{k}\cdot\frac{\partial f}{\partial \mathbf{p}} \frac{d^3p}{\mathbf{k}\,.\,\mathbf{v} - \omega - i0}. \tag{29.9}$$

We take the x-axis along $\mathbf{k}$. Only f in the integrand in (29.9) depends on p_y and p_z. The formula may therefore be written differently by using the distribution function only with respect to $p_x = mv_x$:

$$f(p_x) = \int f(p)\, dp_y\, dp_z.$$

Then

$$\epsilon_l = 1 - \frac{4\pi e^2}{k}\int_{-\infty}^{\infty} \frac{df(p_x)}{dp_x} \frac{dp_x}{kv_x - \omega - i0}. \tag{29.10}$$

In an isotropic plasma, $f(p_x)$ is an even function.

†In this analysis, there are essentially two limits taken: those of small fields (linearization of the equation) and of $\nu \to 0$. It should be noted that the former limit is taken first. The necessity of this sequence arises from that of fulfilling the condition $\delta f \ll f_0$ in linearization; when $\nu = 0$, the increment δf tends to infinity at $\mathbf{k}\,.\,\mathbf{v} = \omega$.

An important result may be noted immediately. The permittivity of a collisionless plasma is a complex quantity; the imaginary part of the integral (29.10) is determined by (29.7). This important result will be discussed further in § 30; here, we shall consider the analytical properties of the function of the frequency ω that is defined by the integral (29.10). It is known from the general properties of the permittivity that this function can have singularities only in the lower half of the complex ω-plane (see *ECM*, § 62); this is a consequence of the definition (28.5). It is, however, useful to see how it also follows directly from (29.10), and to elucidate the relationship between these singularities and the properties of the distribution function $f(p_x)$.

With a change of notation for the variable of integration, we can write the integral in (29.10) as

$$\int_C \frac{df(z)}{dz} \frac{dz}{z - \omega/k}. \tag{29.11}$$

The integration is taken along the real axis in the complex z-plane, but passing below the point $z = \omega/k$ (Fig. 7a). The integral (29.11) then defines an analytic function throughout the upper half-plane of ω also: for any such ω the pole $z = \omega/k$ is passed beneath, as it should be. In the analytical continuation of this function to the lower half-plane, however, the need to pass beneath the pole demands an appropriate shift of the contour of integration (Fig. 7b). But the function $df(z)/dz$, which is regular for real z, in general has singularities for complex z (at z_0, say), some of which are in the lower half-plane. It is impossible to bring the integration contour C away from the pole $z = \omega/k$ when this pole comes close to one of the singularities z_0, and C is pinched between the two points. Thus the function (29.11) has singularities in the lower half-plane of ω at values of ω/k that coincide with the singularities of $df(z)/dz$.

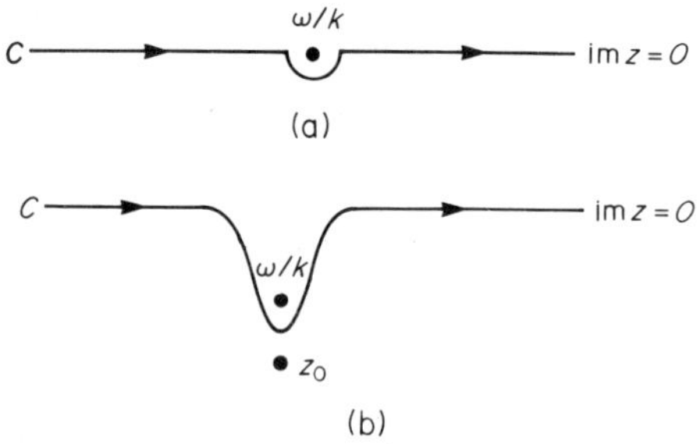

FIG. 7.

§ 30. Landau damping

It has already been noted that the permittivity of a collisionless plasma is a complex quantity $\epsilon_l = \epsilon_l' + i\epsilon_l''$. Separating the imaginary part by means of (29.8), we have

$$\epsilon_l'' = -4\pi^2 e^2 \int \frac{\partial f}{\partial \mathbf{p}} \cdot \frac{\mathbf{k}}{k^2} \delta(\omega - \mathbf{k} \cdot \mathbf{v})\, d^3p, \tag{30.1}$$

or

$$\epsilon_l'' = -\frac{4\pi^2 e^2 m}{k^2}\left[\frac{df(p_x)}{dp_x}\right]_{v_x=\omega/k}. \tag{30.2}$$

The complex permittivity signifies that there is dissipation of the electric field energy in the medium. The formulae for the mean energy Q of a monochromatic electric field dissipated per unit time and volume are as follows. If this field is written in the complex form

$$\mathbf{E} = \mathbf{E}_{\omega k} e^{i(\mathbf{k}\cdot\mathbf{r}-\omega t)},$$

then in the general case of an anisotropic medium†

$$Q = \frac{i\omega}{8\pi}\cdot\tfrac{1}{2}[\epsilon_{\beta\alpha}^*(\omega,\mathbf{k}) - \epsilon_{\alpha\beta}(\omega,\mathbf{k})]E_\alpha^* E_\beta; \tag{30.3}$$

the dissipation is governed by the anti-Hermitian part of the tensor $\epsilon_{\alpha\beta}$. If this tensor is symmetric, the part in question is just the imaginary part:

$$Q = (\omega/8\pi)\epsilon_{\alpha\beta}''(\omega,\mathbf{k})E_\alpha E_\beta^*. \tag{30.4}$$

In the case of a longitudinal field, only the imaginary part of the longitudinal permittivity remains:

$$Q = (\omega/8\pi)\epsilon_l''|\mathbf{E}|^2. \tag{30.5}$$

Substitution of (30.2) then gives

$$Q = -|\mathbf{E}|^2\frac{\pi m e^2\omega}{2k^2}\left[\frac{df(p_x)}{dp_x}\right]_{v_x=\omega/k}. \tag{30.6}$$

Thus dissipation occurs even in a collisionless plasma, a phenomenon predicted by L. D. Landau (1946) and known as *Landau damping*. Being independent of collisions, it is fundamentally different from dissipation in ordinary absorbing media: the collisionless dissipation does not involve an increase of the entropy and is therefore a thermodynamically reversible process, an aspect to which we shall return in §35.

The mechanism of Landau damping is closely connected with spatial dispersion.

†This expression is derived from the general formula

$$Q = \langle\mathbf{E}\cdot\dot{\mathbf{D}}\rangle/4\pi,$$

where the angle brackets denote time averaging; see *ECM*, §61. Here it is assumed that $\mathbf{E}$ and $\mathbf{D}$ are real. If $\mathbf{E}$ is written in the complex form, it has to be replaced in the expression for Q by $\frac{1}{2}(\mathbf{E}+\mathbf{E}^*)$. The corresponding vector $\mathbf{D}$ has the components

$$\tfrac{1}{2}\{\epsilon_{\alpha\beta}(\omega,\mathbf{k})E_\beta + \epsilon_{\alpha\beta}(-\omega,-\mathbf{k})E_\beta^*\},$$

and $\dot{\mathbf{D}}$ has the components

$$\tfrac{1}{2}i\omega\{-\epsilon_{\alpha\beta}(\omega,\mathbf{k})E_\beta + \epsilon_{\alpha\beta}(-\omega,-\mathbf{k})E_\beta^*\}.$$

On averaging the product $\mathbf{E}\cdot\dot{\mathbf{D}}$ and using the property (28.6), we obtain (30.3); cf. the next-but-one footnote.

As is seen from (30.6), the dissipation is due to electrons whose speed in the direction of propagation of the electric wave is equal to the phase velocity of the wave, $v_x = \omega/k$; such electrons are said to be moving in phase with the wave.† The field is stationary with respect to these electrons, and can therefore do work on them that is not zero on averaging over time as it is for other electrons with respect to which the field oscillates. It is instructive to examine this mechanism more closely, with a direct derivation of (30.6) that does not use the transport equation.

Let an electron be moving along the x-axis in a weak electric field

$$E(t, x) = \mathrm{re}\{E_0 e^{i(kx-\omega t)} e^{t\delta}\} \tag{30.7}$$

in that direction; the factor $e^{t\delta}$ describes the slow application of the field beginning at $t = -\infty$. We shall seek the speed $v_x \equiv w$ and the coordinate x of the moving electron in the form

$$w = w_0 + \delta w, \quad x = x_0 + \delta x,$$

where δw and δx are the corrections to the unperturbed motion $x_0 = w_0 t$ with constant speed w_0. The equation of motion of the electron, linearized with respect to small quantities, is

$$\begin{aligned} m\frac{d\delta w}{dt} &= -eE(t, x_0) \\ &= -e\,\mathrm{re}\{E_0 e^{ikt(w_0-\omega/k)} e^{t\delta}\}. \end{aligned}$$

Hence

$$\left.\begin{aligned} \delta w &= -\frac{e}{m}\mathrm{re}\frac{E(t, x_0)}{ik(w_0-\omega/k)+\delta}, \\ \delta x &= -\frac{e}{m}\mathrm{re}\frac{E(t, x_0)}{[ik(w_0-\omega/k)+\delta]^2}. \end{aligned}\right\} \tag{30.8}$$

The mean work done by the field on the electron per unit time is

$$\begin{aligned} q &= -e\langle wE(t, x)\rangle \\ &= -e\langle (w_0+\delta w)E(t, x_0+\delta x)\rangle \\ &\approx -ew_0\left\langle \frac{\partial E}{\partial x_0}\delta x\right\rangle - e\langle \delta w\,.\,E(t, x_0)\rangle, \end{aligned}$$

or, in complex form, ‡

$$q = -\tfrac{1}{2}e\,\mathrm{re}\{w_0\delta x\partial E^*/\partial x_0 + \delta w\,.\,E^*\}.$$

†Note that the difference $\omega - \mathbf{k}\,.\,\mathbf{v}$ is the field frequency in a frame of reference moving with the electron.

‡If two quantities A and B periodic in time are written in complex form ($\propto e^{-i\omega t}$), then

$$\langle \mathrm{re}\,A\,.\,\mathrm{re}\,B\rangle = \tfrac{1}{4}\langle (A+A^*)(B+B^*)\rangle.$$

On averaging, the products AB and A^*B^*, which contain $e^{-2i\omega t}$ and $e^{2i\omega t}$, give zero, leaving

$$\langle \mathrm{re}\,A\,.\,\mathrm{re}\,B\rangle = \tfrac{1}{4}(AB^* + A^*B) = \tfrac{1}{2}\mathrm{re}(AB^*).$$

Substitution of E, δx and δw from (30.7) and (30.8) gives, after a simple reduction,

$$q = \frac{e^2}{2m}|E|^2 \frac{d}{dw_0} \frac{w_0 \delta}{\delta^2 + k^2(w_0 - \omega/k)^2}.$$

It now remains to sum q over electrons with all initial momenta $p_x = m w_0$:

$$Q = \int_{-\infty}^{\infty} q f(p_x)\, dp_x$$

$$= -\tfrac{1}{2} e^2 |E|^2 \int_{-\infty}^{\infty} \frac{w_0 \delta}{\delta^2 + k^2(w_0 - \omega/k)^2} \frac{df}{dp_x}\, dp_x$$

(with integration by parts). The passage to the limit is made by means of the formula

$$\lim_{\delta \to 0} \frac{\delta}{\delta^2 + z^2} = \pi \delta(z) \tag{30.9}$$

and leads directly to (30.6).

In accordance with the reversibility of the collisionless dissipation, the thermodynamic conditions do not require Q to be positive as they do for true dissipation. The expression (30.6) is always positive when the distribution $f(p)$ is isotropic (see Problem). For anisotropic distributions, however, Q may be negative: the electrons then transfer energy to the wave on average, not from it.† Such cases are closely associated with the possible instability of the plasma (see §61); the condition $Q > 0$ (and hence $\epsilon'' > 0$) is thus the result of the stability of the plasma state only.

From the standpoint of the above-mentioned physical picture of Landau damping, the presence of the derivative df/dp_x in (30.6) may be intuitively interpreted as follows. The energy exchange with the field involves particles with speeds v_x close to ω/k; those with $v_x < \omega/k$ gain energy from the wave, while those with $v_x > \omega/k$ lose energy to it. The wave loses energy if the former are rather more numerous than the latter.

PROBLEM

Show that the collisionless dissipation Q is always positive in an isotropic plasma.

SOLUTION. In an isotropic plasma, f is a function only of $p^2 = p_x^2 + p_\perp^2$, where p_x and $\mathbf{p}_\perp$ are the longitudinal and transverse (with respect to $\mathbf{k}$) components of $\mathbf{p}$. We write

$$\frac{df(p_x)}{dp_x} = \frac{d}{dp_x} \int_0^{\infty} f(p_x^2 + p_\perp^2) \pi d(p_\perp^2)$$

$$= 2\pi p_x \int_0^{\infty} f'(p_x^2 + p_\perp^2)\, d(p_\perp^2),$$

and, since $f(p^2) \to 0$ as $p^2 \to \infty$, obtain

$$df(p_x)/dp_x = -2\pi p_x f(p_x^2),$$

so that $df/dp_x < 0$ when $p_x = \omega/k > 0$.

†The intuitive derivation of (30.6) given above does not depend on the isotropy of the distribution, nor does the expression (30.2) (see §32).

§31. Permittivity of a Maxwellian plasma

We can apply formula (29.10) to an electron plasma with an equilibrium (Maxwellian) electron distribution

$$f(p_x) = \frac{N_e}{(2\pi m T_e)^{1/2}} \exp\left(-\frac{p_x^2}{2mT_e}\right), \tag{31.1}$$

where T_e is the temperature of the electron gas; the quantities pertaining to electrons are given the suffix e with a view to the later consideration of the ion component also. The result is

$$\epsilon_l(\omega, k) = 1 + \frac{1}{(ka_e)^2}\left[1 + F\left(\frac{\omega}{\sqrt{2} k v_{Te}}\right)\right], \tag{31.2}$$

where the function $F(x)$ is defined by the integral†

$$\begin{aligned} F(x) &= \frac{x}{\sqrt{\pi}} \int_{-\infty}^{\infty} \frac{e^{-z^2}\, dz}{z - x - i0} \\ &= \frac{x}{\sqrt{\pi}} \mathrm{P} \int_{-\infty}^{\infty} \frac{e^{-z^2}\, dz}{z - x} + i\sqrt{\pi} x e^{-x^2} \end{aligned} \tag{31.3}$$

and the parameters used are

$$v_{Te} = \sqrt{(T_e/m)}, \quad a_e = \sqrt{(T_e/4\pi N_e e^2)}. \tag{31.4}$$

The quantity v_{Te} is a mean thermal speed of the electrons; a_e is the Debye length, determined by their charge, temperature and density.

The limiting expressions for the function $F(x)$ for large and small x are easily found directly from the definition (31.3). For $x \gg 1$, we write

$$\frac{x}{\sqrt{\pi}} \mathrm{P} \int_{-\infty}^{\infty} \frac{e^{-z^2}\, dz}{z - x} = -\frac{1}{\sqrt{\pi}} \int_{-\infty}^{\infty} e^{-z^2}\left(1 + \frac{z}{x} + \frac{z^2}{x^2} + \cdots\right) dz.$$

The integrals of the terms odd in x are zero; the remainder give

$$F(x) + 1 \approx -\frac{1}{2x^2} - \frac{3}{4x^4} + i\sqrt{\pi} x e^{-x^2}, \quad x \gg 1. \tag{31.5}$$

For $x \ll 1$, we first make the change of variable of integration $z = u + x$, and then

†The various representations of $F(x)$, and detailed numerical tables of it, are given by V. N. Faddeeva and N. M. Terent'ev, *Tables of Values of the Function* $w(z) = e^{-z^2}[1 - (2i/\sqrt{\pi}) \int_0^z e^{t^2}\, dt]$ *for Complex Argument*, Pergamon, Oxford, 1961. The function $w(x)$ which they tabulate is related to $F(x)$ by $F(x) = i\sqrt{\pi} x w(x)$. See also M. Abramowitz and I. A. Stegun, *Handbook of Mathematical Functions*, National Bureau of Standards, Washington, 1964; Dover Publishing Company, New York, 1965. Another convenient, and more comprehensive, tabulation is that of $Z(x) = F(x)/x$ by B. D. Fried and S. D. Conte, *The Plasma Dispersion Function*, Academic Press, New York, 1961.

expand in powers of x:

$$\frac{x}{\sqrt{\pi}} \mathrm{P} \int_{-\infty}^{\infty} \frac{e^{-z^2}\,dz}{z-x} = \frac{xe^{-x^2}}{\sqrt{\pi}} \mathrm{P} \int_{-\infty}^{\infty} e^{-u^2-2ux} \frac{du}{u}$$
$$\approx \frac{x}{\sqrt{\pi}} \mathrm{P} \int_{-\infty}^{\infty} e^{-u^2}\left(\frac{1}{u} - 2x\right) du.$$

The principal value of the integral of the first term (odd in u) is zero, and the second term gives

$$F(x) \approx -2x^2 + i\sqrt{\pi}x, \quad x \ll 1. \tag{31.6}$$

These formulae can be used to write down limiting expressions for the permittivity. At high frequencies, we have

$$\epsilon_l = 1 - \frac{\Omega_e^2}{\omega^2}\left(1 + \frac{3k^2 v_{Te}^2}{\omega^2}\right)$$
$$+ i\sqrt{\frac{\pi}{2}} \frac{\omega \Omega_e^2}{(kv_{Te})^3} \exp\left(-\frac{\omega^2}{2k^2 v_{Te}^2}\right) \quad \text{for} \quad \omega/kv_{Te} \gg 1. \tag{31.7}$$

The parameter

$$\Omega_e = v_{Te}/a_e = \sqrt{(4\pi N_e e^2/m)} \tag{31.8}$$

is the *plasma frequency* or *Langmuir frequency* for electrons. In the case $\omega/kv_{Te} \gg 1$, spatial dispersion leads, as it should, to only small corrections in the permittivity, and the imaginary part of ϵ_l is exponentially small, since in a Maxwelliam distribution only an exponentially small fraction of the electrons have speeds $v_x = \omega/k \gg v_{Te}$. The limiting value of the permittivity, which is independent of k, is

$$\epsilon(\omega) = 1 - (\Omega_e/\omega)^2. \tag{31.9}$$

This expression applies to both the longitudinal and the transverse permittivity; see (28.8). It is easily derived by straightforward arguments without the use of the transport equation. As $k \to 0$, the wave field may be regarded as uniform, and the electron equation of motion $m\dot{\mathbf{v}} = -e\mathbf{E}$ then gives $\mathbf{v} = e\mathbf{E}/im\omega$, so that the current density due to the electrons is

$$\mathbf{j} = -(e^2 N_e/im\omega)\mathbf{E}.$$

We also have

$$\mathbf{j} = -i\omega\mathbf{P} = -i\omega \frac{\epsilon(\omega) - 1}{4\pi} \mathbf{E}.$$

A comparison of the two expressions gives formula (31.9).

In the opposite limiting case of low frequencies,

$$\epsilon_l = 1 + \left(\frac{\Omega_e}{kv_{Te}}\right)^2 \left[1 - \left(\frac{\omega}{kv_{Te}}\right)^2 + i\sqrt{\frac{\pi}{2}} \frac{\omega}{kv_{Te}}\right] \quad \text{for} \quad \omega/kv_{Te} \ll 1. \tag{31.10}$$

It should be noted that the spatial dispersion eliminates the pole at $\omega = 0$ of the permittivity in an ordinary conducting medium, and that the imaginary part of the permittivity is relatively small (though not exponentially small) at low frequencies also, in this case because of the smallness of the electron phase volume in which the condition $\mathbf{k}\cdot\mathbf{v} = \omega$ is satisfied.

It has been shown in §29 that the function $\epsilon_l(\omega)$ defined by the integral (29.10) has no singularity in the upper half-plane of ω, and its singularities in the lower half-plane are determined by those of $df(p_x)/dp_x$ as a function of the complex variable p_x. For a Maxwellian distribution, however, the function

$$df(p_x)/dp_x \propto p_x \exp(-p_x^2/2mT)$$

has no singularity at a finite distance anywhere in the complex p_x-plane, i.e. it is an entire function. Hence the permittivity of a Maxwellian collisionless plasma is also an entire function of ω, having no singularity for finite ω.

So far, we have considered only the contribution to the permittivity from the electron component of the plasma. The contribution from the ion component is calculated in exactly the same way, and the two contributions to $\epsilon_l - 1$ are simply added. We thus have the evident generalization of (31.2):

$$\epsilon_l - 1 = \frac{1}{(ka_e)^2}\left[F\left(\frac{\omega}{\sqrt{2}kv_{Te}}\right)+1\right] + \frac{1}{(ka_i)^2}\left[F\left(\frac{\omega}{\sqrt{2}kv_{Ti}}\right)+1\right]. \tag{31.11}$$

The suffixes e and i denote quantities pertaining to electrons and ions;

$$\left.\begin{aligned} v_{Ti} = (T_i/M)^{1/2}, \quad a_i = v_{Ti}/\Omega_i = [T_i/4\pi N_i(ze)^2]^{1/2},\\ \Omega_i^2 = 4\pi N_i(ze)^2/M, \end{aligned}\right\} \tag{31.12}$$

where M and ze are the ion mass and charge. The expression (31.11) pertains to a "two-temperature" plasma, in which each component has the equilibrium distribution but with a different temperature, so that the electrons and ions are not in equilibrium with each other. This occurs naturally, in view of the large difference in mass which impedes the exchange of energy in electron–ion collisions.

The most usual situation is that where $T_i \lesssim T_e$ and $v_{Ti} \ll v_{Te}$. Since we also always have $\Omega_i \ll \Omega_e$, it is easy to deduce that when $\omega \gg kv_{Te} \gg kv_{Ti}$ the contribution of the ions is negligible and so formula (31.7) is valid. In the opposite limiting case, we have

$$\epsilon_l - 1 = \frac{1}{(ka_e)^2} + \frac{1}{(ka_i)^2} + i\sqrt{\frac{\pi}{2}}\frac{\omega}{(ka_i)^2 kv_{Ti}}, \tag{31.13}$$

$$\omega \ll kv_{Ti} \ll kv_{Te}.$$

The case where $kv_{Ti} \ll \omega \ll kv_{Te}$ will be discussed in §32.

All the calculations in §§30 and 31 have been made for the longitudinal part of the permittivity. The calculation of the transverse permittivity is of less interest, because the transverse field usually reduces to ordinary electromagnetic waves for which the frequency and the wave number are related by $\omega/k = c/\sqrt{\epsilon_t}$; then $\omega/k > c \gg v_{Te}$, i.e. $\omega \gg kv_{Te}$, so that the spatial dispersion is small and the permittivity is given by (31.9). For these waves there is also no Landau damping; since the phase velocity of the wave exceeds the speed of light, the plasma contains no particles capable of moving in phase with the wave. (The proof of this statement requires, strictly speaking, a relativistic treatment; see Problem 4.)

PROBLEMS

PROBLEM 1. Find the potential of the electric field due to a small test charge e_1 at rest in the plasma.

SOLUTION. When the plasma polarization is taken into account, the field is determined by the equation $\operatorname{div}\mathbf{D} = 4\pi e_1\delta(\mathbf{r})$. For a constant field, the Fourier components of the induction and the potential are related by $\mathbf{D_k} = \epsilon_l(0,k)\mathbf{E_k} = -i\mathbf{k}\epsilon_l(0,k)\varphi_\mathbf{k}$. Hence we have for $\varphi_\mathbf{k}$ the equation

$$i\mathbf{k}\,.\,\mathbf{D_k} = k^2\epsilon_l(0,k)\varphi_\mathbf{k} = 4\pi e_1.$$

Taking $\epsilon_l(0,k)$ from (31.13), we have

$$\varphi_\mathbf{k} = \frac{4\pi e_1}{k^2+1/a^2}, \quad 1/a^2 = 1/a_e^2 + 1/a_i^2.$$

The corresponding function of the coordinates is

$$\varphi = (e_1/r)e^{-r/a};$$

thus the permittivity (31.13) describes the screening of a static charge in accordance with *SP* 1, §78. The condition for the charge to be small is $e_1 \ll Na^3e$, i.e. e_1 must be small in comparison with the charge on the plasma particles in a volume $\sim a^3$.

PROBLEM 2. Calculate the transverse permittivity of a plasma.

SOLUTION. With the electron polarization $\mathbf{P} = -\mathbf{j}/i\omega$ calculated by using $\mathbf{j}$ from (29.3), we obtain for the permittivity tensor†

$$\epsilon_{\alpha\beta} = \delta_{\alpha\beta} - \frac{4\pi e^2}{\omega}\int \frac{v_\alpha}{\mathbf{k}\,.\,\mathbf{v}-\omega-i0}\frac{\partial f}{\partial p_\beta}d^3p. \tag{1}$$

The transverse part of $\epsilon_{\alpha\beta}$ is separated as

$$\epsilon_t = \tfrac{1}{2}[\epsilon_{\alpha\alpha} - \epsilon_{\alpha\beta}k_\alpha k_\beta/k^2],$$

and is given by the integral

$$\epsilon_t = 1 - \frac{2\pi e^2}{\omega}\int \mathbf{v}_\perp\cdot\frac{\partial f}{\partial \mathbf{p}_\perp}\frac{d^3p}{\mathbf{k}\,.\,\mathbf{v}-\omega-i0}, \tag{2}$$

where $\mathbf{p}_\perp = m\mathbf{v}_\perp$ is the momentum component transverse to $\mathbf{k}$. For a Maxwellian distribution f, integration over $d^2p_\perp$ gives finally

$$\epsilon_t - 1 = \frac{\Omega_e^2}{\omega^2}F\left(\frac{\omega}{\sqrt{2}kv_{Te}}\right) \tag{3}$$

†This expression does not assume that the plasma is isotropic.

with the function F given by (31.3); the ions make an analogous contribution to $\epsilon_l - 1$. In the limiting cases,

$$\epsilon_l - 1 = -\frac{\Omega_e^2}{\omega^2}\left[1+\left(\frac{kv_{Te}}{\omega}\right)^2\right] + i\sqrt{\frac{\pi}{2}}\frac{\Omega_e}{\omega k a_e}\exp\left(-\frac{\omega^2}{2k^2v_{Te}^2}\right) \tag{4}$$

$$(\omega \gg kv_{Te} \gg kv_{Ti}),$$

$$\epsilon_l - 1 = -\frac{1}{(ka_e)^2} - \frac{1}{(ka_i)^2} + i\sqrt{\frac{\pi}{2}}\frac{\Omega_e}{\omega k a_e} \tag{5}$$

$$(\omega \ll kv_{Ti} \ll kv_{Te}).$$

PROBLEM 3. Determine the permittivity of an ultra-relativistic electron plasma, with the temperature $T_e \gg mc^2$ (V. P. Silin, 1960).

SOLUTION. The transport equation retains its form (27.9) even in the relativistic case. Accordingly, such formulae as (29.9) and (2) in Problem 2 remain valid. In the ultra-relativistic case, the electron speed $v \approx c$, the electron energy is cp, and the equilibrium distribution function is

$$f(p) = (N_e c^3/8\pi T_e^3)e^{-cp/T_e}.$$

The longitudinal permittivity is found to be

$$\epsilon_l - 1 = \frac{4\pi e^2 c}{kT_e}\int_{-1}^{1}\int_0^\infty \frac{f(p)\cos\theta \,.\, 2\pi p^2\, dp\, d\cos\theta}{kc\cos\theta - \omega - i0}, \tag{6}$$

where θ is the angle between **k** and **v**. Integration of f over $2\pi p^2\, dp$ gives $\frac{1}{2}N_e$, and then integration over $d\cos\theta$, passing below the pole $\cos\theta = \omega/k$, leads to the result

$$\left.\begin{aligned} \epsilon_l'(\omega, k) - 1 &= \frac{4\pi N_e e^2}{k^2 T_e}\left[1+\frac{\omega}{2kc}\log\left|\frac{\omega - ck}{\omega + ck}\right|\right], \\ \epsilon_l''(\omega, k) &= \pi\omega/2kc \quad \text{when} \quad \omega/k < c, \\ &= 0 \quad \text{when} \quad \omega/k > c. \end{aligned}\right\} \tag{7}$$

Similarly, starting from (2), we find for the transverse permittivity

$$\left.\begin{aligned} \epsilon_t'(\omega, k) - 1 &= \frac{\pi e^2 N_e c}{\omega k T_e}\left[\left(1-\frac{\omega^2}{c^2k^2}\right)\log\left|\frac{\omega - ck}{\omega + ck}\right| - \frac{2\omega}{ck}\right], \\ \epsilon_t''(\omega, k) &= \pi(1-\omega^2/c^2k^2) \quad \text{when} \quad \omega/k < c, \\ &= 0 \quad \text{when} \quad \omega/k > c. \end{aligned}\right\} \tag{8}$$

PROBLEM 4. Find the imaginary part of ϵ_l for a non-relativistic ($T_e \ll mc^2$) electron plasma when $\omega/k \sim c \gg v_{Te}$ (V. P. Silin 1960).

SOLUTION. From (29.9), which is valid for any electron speeds, we find by integration over $d\cos\theta$

$$\epsilon''(\omega, k) = \frac{8\pi^3 e^2\omega}{k^3 T_e}\int_{p_m}^\infty \frac{f(p)p^2}{v}\, dp, \quad p_m = \frac{mc\omega}{\sqrt{(c^2k^2 - \omega^2)}} \tag{9}$$

(the pole $\cos\theta = \omega/kv$ lies on the contour of integration with respect to $\cos\theta$ only if $\omega/kv < 1$; the lower limit of integration over dp therefore corresponds to $v = \omega/k$). The distribution function for $T_e \ll mc^2$, valid for all electron speeds, is

$$f(p) = \frac{N_e}{(2\pi m T_e)^{3/2}}\exp\left(\frac{mc^2}{T_e} - \frac{\epsilon(p)}{T_e}\right),$$

$$\epsilon = c(p^2 + m^2c^2)^{1/2},$$

the value of the normalization integral being governed by the range $\epsilon - mc^2 \approx p^2/2m \sim T_e \ll mc^2$. In the integral (9), with $\omega/k \sim c \gg v_{Te}$, the important range of values of p is that near the lower limit. Putting in the exponential

$$\epsilon(p) \approx \epsilon(p_m) + [d\epsilon/dp]_{p=p_m}(p - p_m)$$
$$= \epsilon(p_m) + (\omega/k)(p - p_m),$$

and in the coefficient of the exponential $p \approx p_m$, $v \approx \omega/k$, and integrating with respect to $p - p_m$ from 0 to ∞, we obtain

$$\epsilon_l'' = \sqrt{\frac{\pi}{2}} \frac{\omega \Omega_e^2}{(kv_{Te})^3} \frac{1}{1-(\omega/kc)^2} \exp\left\{-\frac{mc^2}{T_e}\left[\frac{1}{\sqrt{[1-(\omega/kc)^2]}} - 1\right]\right\}.$$

This gives the form in which ϵ_l'' tends to zero as $\omega/kc \to 1$.

§32. Longitudinal plasma waves

Spatial dispersion makes possible the propagation of longitudinal electric waves in plasmas. The dependence of the frequency on the wave number (the *dispersion relation*) for these waves is given by the equation

$$\epsilon_l(\omega, k) = 0. \tag{32.1}$$

For, when $\epsilon_l = 0$, the longitudinal electric field **E** has $\mathbf{D} = 0$. Putting also $\mathbf{B} = 0$, we satisfy identically the second pair of Maxwell's equations (28.2). There remains from the first pair $\operatorname{curl} \mathbf{E} = 0$, which is satisfied because the field is longitudinal: $\operatorname{curl} \mathbf{E} = i\mathbf{k} \times \mathbf{E} = 0$.

The roots of (32.1) are complex ($\omega = \omega' + i\omega''$). If the imaginary part of the permittivity $\epsilon_l'' > 0$, they lie in the lower half of the complex ω-plane, i.e. $\omega'' < 0$. The quantity $\gamma = -\omega''$ is the damping rate of the wave, since the damping is proportional to $e^{-\gamma t}$. Of course, a propagating wave exists only if $\gamma \ll \omega'$: the damping rate must be much less than the frequency.

Such a root of equation (32.1) is obtained if we assume that

$$\omega \gg kv_{Te} \gg kv_{Ti}. \tag{32.2}$$

Then the oscillations involve only electrons, and the function $\epsilon_l(\omega, k)$ is given by (31.7). The equation $\epsilon_l = 0$ is solved by successive approximation. In the first approximation, omitting all terms which depend on k, we find that†

$$\omega = \Omega_e, \tag{32.3}$$

i.e. the waves have a constant frequency independent of k. These are called *plasma waves* or *Langmuir waves* (I. Langmuir and L. Tonks 1926). They are long waves, in the sense that

$$ka_e \ll 1, \tag{32.4}$$

as follows from (32.2) with $\omega = \Omega_e$.

†The inclusion of ion oscillations would give only a slight shift of this frequency according to $\omega^2 = \Omega_e^2 + \Omega_i^2$.

To determine the k-dependent correction to the real part of the frequency, it is sufficient to put $\omega = \Omega_e$ in the correction term in ϵ'; then

$$\omega = \Omega_e(1+\tfrac{3}{2}k^2a_e^2) \tag{32.5}$$

(A. A. Vlasov, 1938).

The imaginary part of the frequency in this case is

$$\omega'' = -\tfrac{1}{2}\Omega_e\epsilon_l''(\omega, k), \tag{32.6}$$

and is exponentially small with ϵ_l''. To determine it (and also the coefficient of the exponential), we have to substitute in ϵ_l'' the already corrected value (32.5), obtaining

$$\gamma = \sqrt{\frac{\pi}{8}}\frac{\Omega_e}{(ka_e)^3}\exp\left[-\frac{1}{2(ka_e)^2}-\frac{3}{2}\right] \tag{32.7}$$

(L. D. Landau, 1946). Since $ka_e \ll 1$, the damping rate for plasma waves is in fact found to be exponentially small. It increases with decreasing wavelength, and for $ka_e \sim 1$ (when formula (32.7) is no longer valid) it becomes of the same order of magnitude as the frequency, so that the concept of propagating plasma waves ceases to be meaningful.

The above treatment relates, strictly speaking, only to an isotropic plasma, in which the permittivity tensor reduces, by (28.7), to two scalar quantities ϵ_l and ϵ_t. In an anisotropic plasma, i.e. when the distribution function $f(\mathbf{p})$ depends on the direction of $\mathbf{p}$, there are no strictly longitudinal waves. Under certain conditions, however, "almost longitudinal" waves can propagate, in which the field component $\mathbf{E}^{(t)}$ transverse to $\mathbf{k}$ is small compared with the longitudinal component $\mathbf{E}^{(l)}$:

$$E^{(t)} \ll E^{(l)}. \tag{32.8}$$

To ascertain these conditions, we note first of all that, when $\mathbf{E}^{(t)}$ is neglected, the equation div $\mathbf{D} = 0$ gives

$$\mathbf{k}\,.\,\mathbf{D} \approx k_\alpha\epsilon_{\alpha\beta}E_\beta^{(l)} = k_\alpha k_\beta\epsilon_{\alpha\beta}E^{(l)}/k = 0.$$

This determines the dispersion relation for the waves, and may again be written in the form (32.1) if the "longitudinal" permittivity is defined as

$$\epsilon_l = k_\alpha k_\beta\epsilon_{\alpha\beta}/k^2; \tag{32.9}$$

this now depends on the direction of $\mathbf{k}$. However, $\epsilon_l = 0$ does not imply that $\mathbf{D} = 0$; the quantity

$$D_\alpha \approx \epsilon_{\alpha\beta}E_\beta^{(l)} = \epsilon_{\alpha\beta}k_\beta E^{(l)}/k \equiv \epsilon_\alpha E^{(l)}$$

is not zero (whereas in an isotropic plasma $\epsilon_\alpha \equiv 0$ when $\epsilon_l = 0$). Next, from Maxwell's equation curl $\mathbf{B} = (1/c)\partial\mathbf{D}/\partial t$ we estimate the magnetic field in the wave as

$$B \sim (\omega/ck)\epsilon E^{(l)},$$

and then obtain from the equation $\operatorname{curl}\mathbf{E} = -(1/c)\partial\mathbf{B}/\partial t$ an estimate of the transverse electric field:

$$E^{(t)} \sim (\omega/ck)B \sim (\omega/ck)^2\epsilon E^{(l)}. \tag{32.10}$$

Thus the condition (32.8) for the wave to be "almost longitudinal" is satisfied if the wave is "slow" in the sense that

$$\omega/k \ll c/\sqrt{\epsilon}. \tag{32.11}$$

Lastly, it may be noted that (29.10) remains valid for ϵ_l defined by (32.9) in an anisotropic plasma, as is clear from its derivation from the expression

$$\mathbf{k}\,.\,\mathbf{P} = (k_\alpha\epsilon_{\alpha\beta}E_\beta - \mathbf{k}\,.\,\mathbf{E})/4\pi$$

with a longitudinal field $\mathbf{E}$. It is important here that the Lorentz force $e\mathbf{v}\times\mathbf{B}/c$ in the transport equation may be neglected in comparison with $e\mathbf{E}$ (although its product with $\partial f/\partial\mathbf{p}$ is not identically zero for an anisotropic function $f(\mathbf{p})$): the estimate (32.10) gives

$$|\mathbf{v}\times\mathbf{B}|/cE^{(l)} \sim \omega\epsilon\bar{v}/kc^2 \ll 1.$$

This ratio is small both from the condition (32.11) for a "slow" wave and from the inequality $\bar{v} \ll c$.

PROBLEMS

PROBLEM 1. Determine the dispersion relation for transverse oscillations of a plasma.

SOLUTION. For transverse waves, the dispersion relation is given by $\omega^2 = c^2k^2/\epsilon_t$. The high-frequency oscillations ($\omega \gg kv_{Te}$) correspond to ordinary electromagnetic waves. With ϵ_t from (31.9) (see also §31, Problem 2) we find

$$\omega^2 = c^2k^2 + \Omega_e^2.$$

This is valid for any k; there is no Landau damping, as already noted at the end of §31.

For low-frequency oscillations ($\omega \ll kv_{Te}$), the movement of the ions is again unimportant. For long waves ($ka_e \ll 1$), the principal term in the dispersion relation is

$$\omega = -i\sqrt{\frac{2}{\pi}}\frac{k^3c^2v_{Te}}{\Omega_e^2};$$

the purely imaginary value of ω denotes aperiodic damping, so that wave propagation cannot occur.

PROBLEM 2. Find the dispersion relation for plasma waves in an ultra-relativistic electron plasma (V. P. Silin 1960).

SOLUTION. When $\omega \gg ck$, the formula in §31, Problem 3, gives

$$\epsilon_l(\omega, k) = 1 - \frac{\Omega_{e,\mathrm{rel}}^2}{\omega^2}\left(1 + \frac{3k^2c^2}{\omega^2}\right),$$

where

$$\Omega_{e,\mathrm{rel}}^2 = 4\pi e^2N_ec^2/3T_e.$$

Equating to zero this expression for ϵ_l, we obtain the dispersion relation

$$\omega^2 = \Omega_{e,\mathrm{rel}}^2 + \tfrac{3}{5}c^2k^2 \quad (ck \ll \Omega_{e,\mathrm{rel}}).$$

As k increases, this formula becomes invalid, but we still have $\omega > ck$ (and Landau damping is therefore absent). In the limit of large k, the frequency ω tends to ck according to

$$\omega = ck\left[1 + 2\exp\left(-\frac{2k^2c^2}{3\Omega_{e,\mathrm{rel}}^2} - 2\right)\right].$$

PROBLEM 3. The same as Problem 2, but for transverse waves.

SOLUTION. With the expression for $\epsilon_t(\omega, k)$ derived in §31, Problem 3, we find the dispersion relation

$$\omega^2 = \Omega_{e,\mathrm{rel}}^2 + \tfrac{6}{5}c^2k^2 \quad \text{for} \quad \omega \gg ck.$$

The limiting expression for large k is

$$\omega^2 = \tfrac{3}{2}\Omega_{e,\mathrm{rel}}^2 + c^2k^2.$$

Here too, we have $\omega > ck$, and so there is no damping.

§33. Ion-sound waves

As well as the plasma waves associated with electron oscillations, there can also propagate, in a plasma, waves in which both the electron density and the ion density oscillate significantly. This branch of the oscillation spectrum has weak damping (so that the concept of wave propagation has meaning) when the ion gas temperature in the plasma is small in comparison with the electron temperature:

$$T_i \ll T_e. \tag{33.1}$$

It will be confirmed by the result of the calculation that the phase velocity of these waves satisfies the inequalities

$$v_{Ti} \ll \omega/k \ll v_{Te}. \tag{33.2}$$

The smallness of the Landau damping under these conditions is obvious from the start: since the phase velocity is outside the principal ranges of the thermal speeds for both ions and electrons, only a small fraction of the particles can move in phase with the wave and thus take part in energy exchange with it.

The contribution of the electrons to the permittivity under the conditions (33.2) is given by the limiting formula (31.10), and that of the ions by (31.7) with the electron quantities replaced by ion quantities. To the necessary accuracy, we have

$$\epsilon_l = 1 - \frac{\Omega_i^2}{\omega^2} + \frac{1}{(ka_e)^2}\left[1 + i\sqrt{\frac{\pi}{2}}\frac{\omega}{kv_{Te}}\right]. \tag{33.3}$$

Neglecting at first the relatively small imaginary part, we obtain from the equation $\epsilon_l = 0$

$$\omega^2 = \Omega_i^2\frac{k^2a_e^2}{1+k^2a_e^2} = \frac{zT_e}{M}\frac{k^2}{1+k^2a_e^2}; \tag{33.4}$$

in the latter expression we have used the fact that $N_e = zN_i$.

For the longest waves, with the condition $ka_e \ll 1$, the dispersion relation (33.4) reduces to†

$$\omega = k\sqrt{(zT_e/M)}, \quad ka_e \ll 1. \tag{33.5}$$

The frequency is proportional to the wave number, as in ordinary sound waves. Waves having this dispersion relation are called *ion-sound waves*. The phase velocity of these waves is $\omega/k \sim (T_e/M)^{1/2}$, so that the condition (33.2) is in fact satisfied. Taking account of the imaginary part of ϵ_l in the next approximation, we easily find the damping rate:

$$\gamma = \omega\sqrt{(\pi zm/8M)}. \tag{33.6}$$

This damping is due to the electrons. The contribution of the ions to γ is exponentially small, containing the factor $\exp(-zT_e/2T_i)$.

For shorter wavelengths, in the range $1/a_e \ll k \ll 1/a_i$ (which exists by virtue of the postulated inequality (33.1)), we have from (33.4) simply

$$\omega \approx \Omega_i. \tag{33.7}$$

These are ion waves analogous to electron plasma waves. It is easily verified that here again the conditions (33.2) are satisfied, and the damping is slight. As the wavelength decreases further, however, the damping increases, and for $ka_i \gtrsim 1$ the ion contribution to the damping rate becomes comparable with the frequency, so that wave propagation ceases to have any meaning.

Figure 8 shows diagrammatically the spectrum (dispersion relation) for the low-frequency oscillations considered here (lower curve), in comparison with the spectrum of high-frequency electron plasma waves (upper curve). The broken lines mark the regions where the damping becomes large.

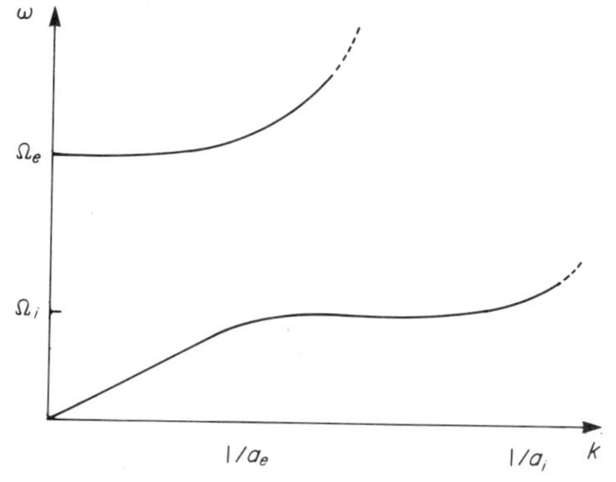

FIG. 8.

†The law (33.5) was discovered by Langmuir and Tonks (1926); the need for the condition (33.1) was noted by G. V. Gordeev (1954).

§34. Relaxation of the initial perturbation

Let us consider the problem of solving the transport equation with a self-consistent field for given initial conditions (L. D. Landau, 1946). We shall take only the case of a purely potential electric field ($\mathbf{E} = -\nabla\varphi$) and zero magnetic field, assuming that only the electron distribution is perturbed (the ion distribution remaining unchanged).

We shall also suppose that the initial perturbation is small: the initial electron distribution function is

$$f(0, \mathbf{r}, \mathbf{p}) = f_0(p) + g(\mathbf{r}, \mathbf{p}), \tag{34.1}$$

where $f_0(p)$ is the equilibrium (Maxwellian) distribution, and $g \ll f_0$. The perturbation remains small at subsequent instants, of course, so that the equations may be linearized; we shall seek the distribution function in the form

$$f(t, \mathbf{r}, \mathbf{p}) = f_0(p) + \delta f(t, \mathbf{r}, \mathbf{p}). \tag{34.2}$$

For the small correction δf and the potential $\varphi(t, \mathbf{r})$ of the self-consistent field (a quantity of the same order of smallness) we find a set of equations comprising the transport equation

$$\frac{\partial \delta f}{\partial t} + \mathbf{v} \cdot \frac{\partial \delta f}{\partial \mathbf{r}} + e\nabla\varphi \cdot \frac{\partial f_0}{\partial \mathbf{p}} = 0 \tag{34.3}$$

and Poisson's equation

$$\Delta\varphi = 4\pi e \int \delta f \, d^3p \tag{34.4}$$

(the equilibrium electron charge is compensated by the ion charge).

Since these equations are linear and do not explicitly involve the coordinates, the required functions δf and φ may be expanded as Fourier integrals with respect to the coordinates, and equations may be written for each Fourier component separately. That is, it suffices to consider solutions having the form

$$\left.\begin{aligned} \delta f(t, \mathbf{r}, \mathbf{p}) &= f_{\mathbf{k}}(t, \mathbf{p}) e^{i\mathbf{k}\cdot\mathbf{r}}, \\ \varphi(t, \mathbf{r}) &= \varphi_{\mathbf{k}}(t) e^{i\mathbf{k}\cdot\mathbf{r}}. \end{aligned}\right\} \tag{34.5}$$

For such solutions, equations (34.3) and (34.4) become

$$\partial f_{\mathbf{k}}/\partial t + i\mathbf{k} \cdot \mathbf{v} f_{\mathbf{k}} + ie\varphi_{\mathbf{k}} \mathbf{k} \cdot \partial f_0/\partial \mathbf{p} = 0, \tag{34.6}$$

$$k^2 \varphi_{\mathbf{k}} = -4\pi e \int f_{\mathbf{k}} \, d^3p. \tag{34.7}$$

To solve these equations, it is sufficient to use a one-sided Fourier transformation, the transform $f_{\omega\mathbf{k}}^{(+)}(\mathbf{p})$ of the function $f_{\mathbf{k}}(t, \mathbf{p})$ being defined as

$$f_{\omega\mathbf{k}}^{(+)}(\mathbf{p}) = \int_0^\infty e^{i\omega t} f_{\mathbf{k}}(t, \mathbf{p}) \, dt. \tag{34.8}$$

The inverse transformation is given by

$$f_{\mathbf{k}}(t, \mathbf{p}) = \int_{-\infty+i\sigma}^{\infty+i\sigma} e^{-i\omega t} f_{\omega\mathbf{k}}^{(+)}(\mathbf{p}) \frac{d\omega}{2\pi}, \tag{34.9}$$

where the integral is taken along a straight line in the complex ω-plane, parallel to and above the real axis ($\sigma > 0$), and also passing above all singularities of $f_{\omega\mathbf{k}}$.†

We multiply both sides of (34.6) by $e^{-i\omega t}$ and integrate with respect to t. Noting that

$$\int_0^\infty \frac{\partial f_{\mathbf{k}}}{\partial t} e^{i\omega t}\, dt = [f_{\mathbf{k}} e^{i\omega t}]_0^\infty - i\omega \int_0^\infty f_{\mathbf{k}} e^{i\omega t}\, dt$$
$$= -g_{\mathbf{k}} - i\omega f_{\omega\mathbf{k}}^{(+)},$$

where $g_{\mathbf{k}}(p) \equiv f_{\mathbf{k}}(0, p)$, and dividing both sides of the equation by $i(\mathbf{k} \cdot \mathbf{v} - \omega)$, we find

$$f_{\omega\mathbf{k}}^{(+)} = \frac{1}{i(\mathbf{k} \cdot \mathbf{v} - \omega)} \left(g_{\mathbf{k}} - ie\varphi_{\omega\mathbf{k}}^{(+)} \mathbf{k} \cdot \frac{\partial f_0}{\partial \mathbf{p}}\right]. \tag{34.10}$$

Similarly, from (34.7),

$$k^2 \varphi_{\omega\mathbf{k}}^{(+)} = -4\pi e \int f_{\omega\mathbf{k}}^{(+)}(p)\, d^3p. \tag{34.11}$$

Substitution of $f_{\omega\mathbf{k}}^{(+)}$ from (34.10) in (34.11) gives an equation for $\varphi_{\omega\mathbf{k}}^{(+)}$ alone, which yields the result

$$\varphi_{\omega\mathbf{k}}^{(+)} = -\frac{4\pi e}{k^2 \epsilon_l(\omega, k)} \int \frac{g_{\mathbf{k}}(\mathbf{p})\, d^3p}{i(\mathbf{k} \cdot \mathbf{v} - \omega)}, \tag{34.12}$$

with the longitudinal permittivity ϵ_l from (29.9). Again using the momentum component $p_x = mv_x$ along $\mathbf{k}$, as in §29, we can rewrite this formula as

$$\varphi_{\omega\mathbf{k}}^{(+)} = -\frac{4\pi e}{k^2 \epsilon_l(\omega, k)} \int_{-\infty}^{\infty} \frac{g_{\mathbf{k}}(p_x)\, dp_x}{i(kv_x - \omega)}, \tag{34.13}$$

where

$$g_{\mathbf{k}}(p_x) = \int g_{\mathbf{k}}(\mathbf{p})\, dp_y\, dp_z.$$

†The transformation (34.8), (34.9) is just the familiar Laplace transformation

$$f_p = \int_0^\infty f(t) e^{-pt} dt, \quad f(t) = \frac{1}{2\pi i} \int_{-i\infty+\sigma}^{i\infty+\sigma} f_p e^{pt}\, dp,$$

in which p has been replaced by $-i\omega$ and the contour of integration correspondingly changed in the expression giving the function $f(t)$ in terms of its transform f_p.

In order to determine the time dependence of the potential by means of the inversion formula

$$\varphi_{\mathbf{k}}(t) = \frac{1}{2\pi}\int_{-\infty+i\sigma}^{\infty+i\sigma} e^{-i\omega t}\varphi_{\omega\mathbf{k}}^{(+)}d\omega, \tag{34.14}$$

it is necessary first to establish the analytical properties of $\varphi_{\omega\mathbf{k}}$ as a function of the complex variable ω.

An expression of the form

$$\varphi_{\omega\mathbf{k}}^{(+)} = \int_0^\infty \varphi_{\mathbf{k}}(t)e^{i\omega t}dt$$

as a function of the complex variable ω is meaningful only in the upper half-plane. The same applies correspondingly to the expression (34.13), where the integration is along a contour (the real p_x-axis) which passes below the pole $p_x = m\omega/k$. We have seen in §29 that the function of the variable ω defined by such an integral, when analytically continued into the lower half-plane, has singularities only at singularities of $g_{\mathbf{k}}(p_x)$. We shall assume that $g_{\mathbf{k}}(p_x)$ as a function of the complex variable p_x is an entire function (i.e. has no singularity at finite p_x); then the integral under consideration also defines an entire function of ω.

It has been noted in §31 that the permittivity ϵ_l of a Maxwellian plasma is again an entire function of ω. Thus the function $\varphi_{\omega\mathbf{k}}$, analytic throughout the ω-plane, is the quotient of two entire functions. Hence it follows that the only singularities (poles) of $\varphi_{\omega\mathbf{k}}$ are the zeros of its denominator, i.e. of $\epsilon_l(\omega, k)$.

These arguments lead to the asymptotic form in which the potential $\varphi_{\mathbf{k}}(t)$ decreases for large values of the time t. In the inversion formula (34.14), the integration is taken along a horizontal line in the ω-plane. However, if $\varphi_{\omega\mathbf{k}}$ is taken to be the analytic function thus defined in the whole plane, we can move the contour of integration into the lower half-plane in such a way as not to cross any of the poles of the function. Let $\omega_k = \omega_k' + i\omega_k''$ be the root of $\epsilon_l(\omega, k) = 0$ that has the smallest (in magnitude) imaginary part, i.e. that lies nearest the real axis. The integration in (34.14) is taken along a contour moved sufficiently far beneath the point $\omega = \omega_k$ and passing round this point (and round other poles lying above it) in the manner shown in Fig. 9. Then only the residue at the pole ω_k is important in the integral (when t is large); the remaining parts of the integral, including that along the horizontal part of the contour, are exponentially small in comparison with that residue, since the integrand contains a factor $e^{-i\omega t}$ which decreases rapidly with increasing $|\mathrm{im}\,\omega|$. Thus the asymptotic law of decrease of the potential is given by

$$\varphi_{\mathbf{k}}(t) \propto e^{-i\omega_k' t}e^{-|\omega_k''|t}, \tag{34.15}$$

i.e. the field perturbation is exponentially damped in the course of time, with damping rate $\gamma_k = |\omega_k''|$.†

†If the initial function $g_{\mathbf{k}}(p_x)$ has a singularity, the competing values of ω include not only the zeros of $\epsilon_l(\omega, k)$ but also the singularities of $\varphi_{\omega\mathbf{k}}$ that result from the singularity of the integral in (34.13). In particular, if $g_{\mathbf{k}}(p_x)$ has a singularity, for instance a break, on the real axis, then $\varphi_{\omega\mathbf{k}}$ will have a singularity at a real value of $\omega = kv_x$. Such a perturbation is not damped at all in a collisionless plasma.

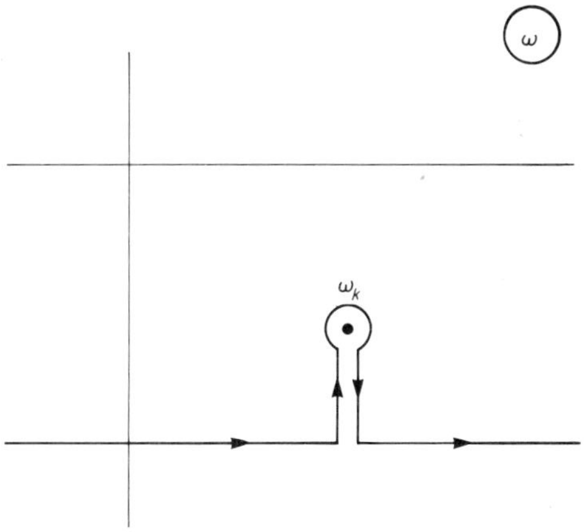

FIG. 9.

For long-wavelength perturbations ($ka_e \ll 1$), the frequency ω'_k and the damping rate γ_k are the same as for plasma waves, and are given by (32.5) and (32.6). The damping rate for such perturbations is exponentially small. In the opposite case of short-wavelength perturbations, when $ka_e \sim 1$, the damping is very strong, and γ_k may even be much larger than ω'_k.†

Lastly, let us consider the properties of the electron distribution function itself. The required function $f_{\mathbf{k}}(t, \mathbf{p})$ is found by substituting (34.10) in the integral (34.9). Besides the poles in the lower half-plane that arise from $\varphi_{\omega\mathbf{k}}$, the integrand has a pole at the point $\omega = \mathbf{k}\cdot\mathbf{v}$ on the real axis. This pole determines the asymptotic behaviour of the integral for large t. The residue there gives

$$f_{\mathbf{k}}(t, \mathbf{p}) \propto e^{-i\mathbf{k}\cdot\mathbf{v}t}. \tag{34.16}$$

Thus the perturbation of the distribution function is not damped in the course of time, but the distribution becomes a more and more rapidly oscillating function of the velocity, the period of oscillation being $\sim 1/kt$. Hence the density perturbation, i.e. the integral $\int f_{\mathbf{k}}\, d^3p$, is damped, like the potential $\varphi_{\mathbf{k}}$.‡

The variation of the distribution function in accordance with (34.16) pertains to times when the field may be regarded as damped; this formula corresponds simply to the free dispersal of the particles, each with its constant velocity. A function of the form

$$f(t, \mathbf{r}, \mathbf{p}) = g(\mathbf{p})\, e^{i(\mathbf{k}\cdot\mathbf{r} - \mathbf{k}\cdot\mathbf{v}t)} \tag{34.17}$$

†The question may be raised of the source of the large damping when the "phase velocity" ω'_k/k lies outside the main range of thermal speeds. In fact, however, the ratio ω'/k cannot be called the phase velocity when $\gamma > \omega'$. If we again expand a function of the form $e^{-i\omega' t}e^{-\gamma t}$ as a Fourier integral, it will contain components with all frequencies from 0 to γ, and accordingly with "phase velocities" from 0 to $\sim\gamma/k$.

‡We may anticipate, however, by noting that the oscillatory nature of the distribution function for large t causes a large increase in the effective number of Coulomb collisions and thus accelerates the ultimate damping of the perturbation; see §41, Problem.

is the solution of the free-particle transport equation

$$\frac{\partial f}{\partial t}+\mathbf{v}\cdot\frac{\partial f}{\partial \mathbf{r}}=0 \tag{34.18}$$

with a given initial ($t=0$) velocity distribution and a periodic ($\propto e^{i\mathbf{k}\cdot\mathbf{r}}$) coordinate distribution.

§35. Plasma echoes

The thermodynamic reversibility of Landau damping manifests itself in some unusual non-linear phenomena called *plasma echoes*. They result from the undamped oscillations of the distribution function (34.16) which remain after the collisionless relaxation of the density (and field) perturbations in the plasma. They are essentially kinematic in origin, and unconnected with the existence of a self-consistent electric field in the plasma. This will first of all be illustrated by the example of a gas of uncharged particles without collisions.

Let a perturbation in the gas be specified at the initial instant, such that the distribution function, remaining Maxwellian with respect to speeds at each point in space, varies periodically in the x-direction:

$$\delta f = A_1 \cos k_1 x \,.\, f_0(p) \quad \text{at} \quad t=0; \tag{35.1}$$

in this section, $p=mv$ denotes the x-component of the momentum, and the distribution function is assumed to be already integrated with respect to p_y and p_z. The perturbation of the gas density, i.e. the integral $\int \delta f \,.\, dp$, varies in the same manner in the x-direction at $t=0$. Subsequently, the perturbation of the distribution function varies according to

$$\delta f = A_1 \cos k_1(x-vt)\,.\, f_0(p), \tag{35.2}$$

which corresponds to a free movement of each particle in the x-direction with its own speed v. The density perturbation, however, is damped (in a time $\sim 1/k_1 v_T$) because the integral $\int \delta f\, dp$ is made small by the speed-oscillatory factor $\cos k_1(x-vt)$ in the integrand. The asymptotic form of the damping at times $t \gg 1/k_1 v_T$ is given by

$$\delta N = \int \delta f\, dp \propto \exp(-\tfrac{1}{2}k_1^2 v_T^2 t^2), \tag{35.3}$$

the integral being estimated by the saddle-point method.

Now let the distribution function be again modulated at a time $t=\tau \gg 1/k_1 v_T$, with amplitude A and a new wave number $k_2 > k_1$. The resulting density perturbation is in turn damped (in a time $\sim 1/k_2 v_T$), but reappears at a time

$$\tau' = k_2\tau/(k_2-k_1), \tag{35.4}$$

since the second modulation creates in the distribution function for $t = \tau$ a second-order term of the form

$$\delta f^{(2)} = A_1 A_2 \cos(k_1 x - k_1 v\tau) \cos k_2 x \,.\, f_0(p), \tag{35.5}$$

whose further development at $t > \tau$ changes it into

$$\begin{aligned}\delta f^{(2)} &= A_1 A_2 f_0(p) \cos[k_1 x - k_1 vt] \cos[k_2 x - k_2 v(t-\tau)] \\ &= \tfrac{1}{2} A_1 A_2 f_0(p) \{\cos[(k_2 - k_1)x - (k_2 - k_1)vt + k_2 v\tau] \\ &\quad + \cos[(k_2 + k_1)x - (k_2 + k_1)vt + k_2 v\tau]\}.\end{aligned}$$

We now see that at $t = \tau'$ the oscillatory dependence of the first term on v disappears, so that this term makes a finite contribution to the perturbation of the gas density with wave number $k_2 - k_1$. The resulting *echo* is then damped in a time $\sim 1/v_T(k_2 - k_1)$, and the final stage of this damping follows a law similar to (35.3).

Let us now consider this effect in an electron plasma (R. W. Gould, T. M. O'Neil and J. H. Malmberg 1967). The mechanism is as before, but the particular law of damping is altered by the influence of the self-consistent field.

We shall suppose that the perturbations are created by pulses from an external potential φ^{ex}, due to "extrinsic" charges, applied to the plasma at times $t = 0$ and τ:

$$\varphi^{ex} = \varphi_1 \delta(t) \cos k_1 x + \varphi_2 \delta(t - \tau) \cos k_2 x; \tag{35.6}$$

here it is assumed that $k_2 > k_1$ and $\tau \gg 1/k_1 v_T$, $1/\gamma(k_1)$, with $\gamma(k)$ the Landau damping rate.

The perturbation of the distribution function ($f = f_0 + \delta f$) satisfies the collisionless transport equation, which with the second-order term is

$$\frac{\partial \delta f}{\partial t} + v \frac{\partial \delta f}{\partial x} + e \frac{\partial \varphi}{\partial x} \frac{df_0}{dp} = -e \frac{\partial \varphi}{\partial x} \frac{\partial \delta f}{\partial p}. \tag{35.7}$$

The potential φ of the field created in the plasma, including the "extrinsic" part φ^{ex}, satisfies the equation

$$\triangle(\varphi - \varphi^{ex}) = 4\pi e \int \delta f \, dp. \tag{35.8}$$

We shall seek the solution of these equations as Fourier integrals:

$$\delta f = \int f_{\omega' k'} e^{i(k'x - \omega' t)} \frac{d\omega' dk'}{(2\pi)^2},$$

$$\varphi = \int \varphi_{\omega'' k''} e^{i(k''x - \omega'' t)} \frac{d\omega'' dk''}{(2\pi)^2}.$$

Substituting these expressions, multiplying the equations by $e^{-i(kx - \omega t)}$, and integrating over $dx\, dt$, we obtain

$$(kv - \omega) f_{\omega k} + ek\varphi_{\omega k}\, df_0/dp = -e \int (k - k')\varphi_{\omega - \omega', k - k'} \frac{df_{\omega' k'}}{dp} \frac{d\omega' dk'}{(2\pi)^2}, \tag{35.9}$$

$$-k^2 \varphi_{\omega k} = 4\pi e \int f_{\omega k}\, dp - k^2 \varphi^{ex}_{\omega k}, \tag{35.10}$$

where

$$\varphi^{ex}_{\omega k} = \pi\varphi_1[\delta(k+k_1)+\delta(k-k_1)] + \pi\varphi_2[\delta(k+k_2)+\delta(k-k_2)]e^{i\omega\tau}.$$

In the linear approximation, i.e. when the right-hand side of (35.9) is neglected, the solution of these equations is

$$f^{(1)}_{\omega k} = -e\frac{df_0}{dp}\frac{k}{kv-\omega}\varphi^{(1)}_{\omega k}, \quad \varphi^{(1)}_{\omega k} = \varphi^{ex}_{\omega k}/\epsilon_l(\omega, k), \tag{35.11}$$

where ϵ_l is the permittivity (29.10). This solution corresponds to perturbations damped from $t = 0$ and $t = \tau$, with respective damping rates $\gamma(k_1)$ and $\gamma(k_2)$.

In the second approximation, we have to substitute (35.11) on the right of (35.9), obtaining for the second-order terms in the perturbations of the distribution function and the potential the equations

$$(kv-\omega)f^{(2)}_{\omega k} + ek\varphi^{(2)}_{\omega k}\, df_0/dp = dI_{\omega k}/dp, \tag{35.12}$$

$$k^2\varphi^{(2)}_{\omega k} = -4\pi e\int f^{(2)}_{\omega k}\, dp, \tag{35.13}$$

where

$$I_{\omega k} = -e\int (k-k')\varphi^{(1)}_{\omega-\omega', k-k'}f^{(1)}_{\omega' k'}\, d\omega'\, dk'/(2\pi)^2. \tag{35.14}$$

The effect under consideration, namely the echo with wave number $k_2 - k_1$, will be contained in the terms on the right of (35.12) which involve $\delta[k \pm (k_2-k_1)]$. Let us collect together the terms of this kind in $I_{\omega k}$. At $t = \tau$, the perturbation $\varphi^{(1)}$ due to the pulse φ_1 applied at $t = 0$ is already damped. It is therefore evident that, on substituting (35.11) in (35.14), we have to take into account only the φ_2 term in $\varphi^{(1)}_{\omega k}$; the relevant terms, of the form

$$I_{\omega k} = I_\omega(k_1, k_2)\delta(k-k_2+k_1) + I_\omega(-k_1, -k_2)\delta(k+k_2-k_1), \tag{35.15}$$

are then obtained from the terms in $f^{(1)}_{\omega k}$ that contain φ_1. After carrying out the integration over dk' in (35.14), we find as the result

$$I_\omega(k_1, k_2) = \tfrac{1}{4}e^2\varphi_1\varphi_2 k_1 k_2\frac{df_0}{dp}\int_{-\infty}^{\infty}\frac{e^{i(\omega-\omega')\tau}\, d\omega'}{(k_1 v+\omega')\epsilon_l(\omega', k_1)\epsilon_l(\omega-\omega', k_2)}, \tag{35.16}$$

where the integration variable ω' is as usual to be understood as $\omega' + i0$.†

The integral (35.16) can be calculated on the basis of the assumption that τ is large ($\tau \gg 1/kv_T, 1/\gamma$). To do this we move the contour of integration into the lower

†In the calculation, it must be remembered that ϵ_l depends only on $|\mathbf{k}|$, and so, in the notation of this section (where $k \equiv k_x$), we have $\epsilon_l(\omega, -k) = \epsilon_l(\omega, k)$.

half of the complex ω'-plane; the contour must still pass above the poles and thus becomes "looped round" them. These are at the zeros of the functions ϵ_l and at the point $\omega' = -k_1 v - i0$. The former have a non-zero negative imaginary part $-\gamma(k_1)$ or $-\gamma(k_2)$, and their contributions to the integral (the residues at the poles) are damped as $e^{-\gamma\tau}$ with increasing τ. An undamped contribution comes only from the real pole $\omega' = -k_1 v - i0$. We thus have

$$I_\omega(k_1, k_2) = -e^2 \cdot \tfrac{1}{2} i\pi \frac{df_0}{dp} \frac{\varphi_1 \varphi_2 k_1 k_2 e^{i(\omega + k_1 v)\tau}}{\epsilon_l(-k_1 v, k_1)\epsilon_l(\omega + k_1 v, k_2)}. \tag{35.17}$$

Returning to the equations (35.12) and (35.13), and substituting $f^{(2)}_{\omega k}$ from the former in the latter, we find

$$\varphi^{(2)}_{\omega k} = -\frac{4\pi e}{k^2 \epsilon_l(\omega, k)} \int_{-\infty}^{\infty} \frac{dI_{\omega k}}{dp} \frac{dp}{kv - \omega - i0}. \tag{35.18}$$

In calculating the derivative $dI_{\omega k}/dp$, only the exponential factor in (35.17) is to be differentiated, since $k_1 v_T \tau \gg 1$.

Now collecting the expressions (35.15)–(35.18) and applying an inverse Fourier transformation, we obtain the required echo potential with wave number $k_3 = k_2 - k_1$:

$$\varphi^{(2)}(t, x) = \mathrm{re}\{A(t) e^{ik_3 x}\}. \tag{35.19}$$

The amplitude $A(t)$ will be written in the asymptotic limit as $t - \tau \to \infty$. In that limit, the integral with respect to ω is determined by the residue of the integrand at the pole $\omega = k_3 v - i0$ only. The final result is

$$A(t) = -i\pi e^3 \varphi_1 \varphi_2 \tau \frac{k_1^2 k_2}{k_3^2} \int_{-\infty}^{\infty} \frac{df_0}{dp} \frac{e^{-ivk_3(t-\tau')} dp}{\epsilon_l(k_3 v, k_3)\epsilon_l(-k_1 v, k_1)\epsilon_l(k_2 v, k_2)}, \tag{35.20}$$

where $\tau' = k_2 \tau / k_3$.

This expression, the echo amplitude, is greatest at $t = \tau'$, and its maximum value is proportional to τ, the time interval between the two pulses. On either side of the maximum, $A(t)$ decreases, but in different ways. In the limit as $t - \tau' \to \infty$, the integral (35.20) is determined by the residue of the integrand at the pole with the negative imaginary part of least magnitude, which is at $\epsilon_l(k_3 v, k_3) = 0$, the imaginary part being im $v = -\gamma(k_3)/k_3$.† On the other side of the maximum, with $t - \tau' \to -\infty$, the integral is determined by the residue at the pole at $\epsilon_l(-k_1 v, k_1) = 0$, for which im $v = \gamma(k_1)/k_1$; the contour of integration must then be moved into the upper half of the complex v-plane. The result is

$$\left.\begin{aligned} A(t) &\propto \exp[-\gamma(k_3)(t - \tau')] \quad \text{as} \quad t - \tau' \to \infty, \\ A(t) &\propto \exp[-(k_3/k_1)\gamma(k_1)(\tau' - t)] \quad \text{as} \quad t - \tau' \to -\infty. \end{aligned}\right\} \tag{35.21}$$

†It is assumed that all wave numbers $k \ll 1/a_e$. Then $\gamma(k)$ is exponentially small, and decreases with increasing k. Since $k_3 < k_2$, the pole at $\epsilon_l(k_2 v, k_2) = 0$ then certainly lies further from the real v-axis than that at $\epsilon_l(k_3 v, k_3) = 0$.

Thus the echo amplitude, before reaching its maximum, increases with a growth rate $k_3\gamma(k_1)/k_1$, and beyond the maximum decreases at a rate $\gamma(k_3)$. Figure 10 illustrates this behaviour. The first two curves show the variation of the potential in the two pulses applied at $t = 0$ and $t = \tau$; the third curve shows the form of the echo. The corresponding growth and decay rates are marked beside the curves.

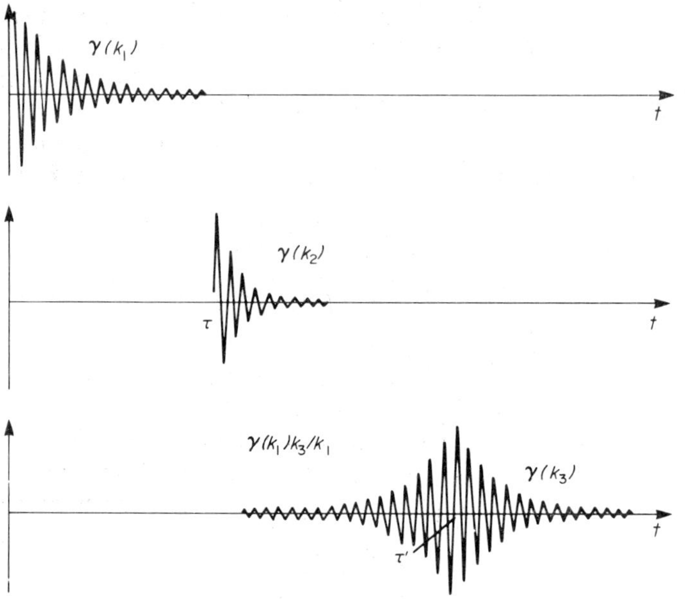

FIG. 10.

These calculations have neglected collisions. Hence the condition for the quantitative formula (35.20) to be valid is that at a given time t the oscillations of the distribution function have not yet been damped out by collisions. By means of the results to be obtained in §41, Problem, we may formulate this condition as

$$\nu(v_T)(kv_T)^2 t^3 \ll 1,$$

where $\nu(v_T)$ is the mean frequency of Coulomb collisions for an electron.

§36. **Adiabatic electron capture**

Let us consider the distribution of plasma electrons in a slowly applied potential electric field. Let L be the order of magnitude of the extent of the field and τ the characteristic time of variation of the field. We shall suppose that

$$\tau \gg L/\bar{v}_e, \tag{36.1}$$

and also that τ is small compared with the electron mean free time, so that the plasma is again collisionless.

By virtue of the condition (36.1), the field may be regarded as stationary during the passage of an electron through it. To the same accuracy, the electron distribution function in the field is stationary. It has been noted at the end of §27 that the solution of the collisionless transport equation depends only on the integrals of the motion of the particle; for a stationary distribution, these can only be the ones that do not depend explicitly on the time.

We shall take only the one-dimensional case, in which the field potential φ depends on only one coordinate, x. Since the motion in the y- and z-directions is then unimportant, we shall consider the distribution function only f with respect to the momentum p_x (and the coordinate x).

In the one-dimensional case, the equation of motion has two integrals, of which only one is not dependent explicitly on the time (in a steady field), namely the electron energy

$$\epsilon = p_x^2/2m + U(x), \tag{36.2}$$

where $U(x) = -e\varphi(x)$.† Hence the stationary distribution function will depend on p_x and x only in the combination (36.2):

$$f = f[\epsilon(x, p_x)]. \tag{36.3}$$

The form of the function $f(\epsilon)$ depends on the boundary conditions.

Let the field $U(x)$ form a potential barrier (Fig. 11a). The function $f(\epsilon)$ is then determined by the distribution of the electrons reaching the barrier from infinity. For instance, if the electrons far from the barrier in either direction have an equilibrium distribution, uniform in space, with temperature T_e, then the Boltzmann distribution occurs throughout space:

$$f = \frac{N_0}{(2\pi m T_e)^{1/2}} \exp(-\epsilon/T_e). \tag{36.4}$$

The density of the electron gas is everywhere distributed according to the formula

$$N_e(x) = N_0 \exp[-U(x)/T_e], \tag{36.5}$$

where N_0 is the density far from the barrier.

Next let the field be a potential well (Fig. 11b). Then the distribution of electrons with positive energy ϵ is again determined by that of particles coming from infinity; with an equilibrium distribution at infinity, the electrons with $\epsilon > 0$ have a Boltzmann distribution throughout space. But in this case, as well as such particles, there are some with $\epsilon < 0$, which execute a finite motion within the potential well; they are "trapped". At infinity there are no particles with $\epsilon < 0$, and so the previous arguments in which the energy was regarded as a strictly conserved quantity are inadequate to give the distribution of trapped particles. We must take into account also the variation of the energy in a field that is not strictly stationary, and in consequence this distribution is found to depend in general on previous events, namely the way in which the field has been applied (A. V. Gurevich, 1967).

†The second integral of the motion may be, for instance, the initial (at some specified instant) value x_0 of the particle coordinate, expressed as a function of time and the variable coordinate along the path, $x_0(t, x)$.

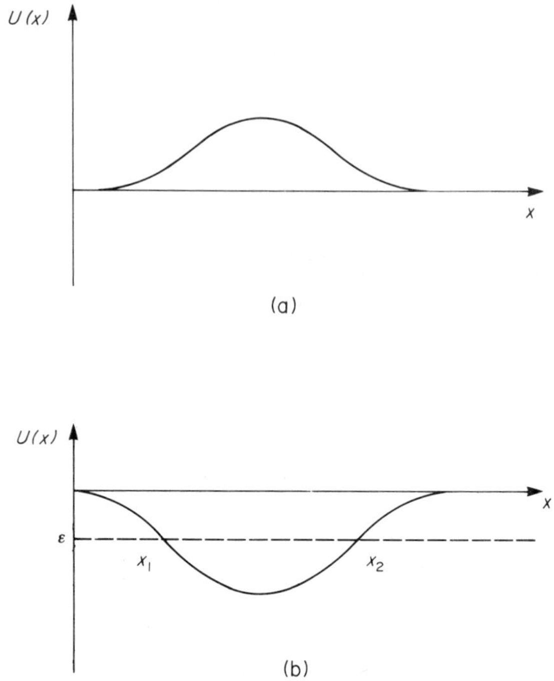

FIG. 11.

From the condition (36.1), the field varies only slightly during the period of the finite motion of the trapped particles. In such cases there is a conserved *adiabatic invariant*, the integral

$$I(t, \epsilon) = \frac{1}{2\pi} \cdot 2 \int_{x_1}^{x_2} [2m(\epsilon - U(t, x))]^{1/2}\, dx, \tag{36.6}$$

taken between the limits of the motion (for given ϵ and t). This quantity acts here as an integral of the motion, in terms of which the trapped particle distribution function is to be expressed:

$$f_{tr} = f_{tr}(I(t, \epsilon)), \tag{36.7}$$

the energy ϵ in turn being assumed expressed in terms of x and p_x by (36.2). The form of the function (36.7) is determined by the fact that the distribution function is a continuous function of ϵ when the field is applied slowly. Hence, for the limiting value of the trapped particle energy, the function $f_{tr}(I)$ must be the distribution function for particles executing an infinite motion above the well.

The case of the potential well as shown in Fig. 11b is, however, particularly simple because the limiting energy has the constant value zero if the field is applied gradually. It then follows from the boundary condition stated that f_{tr} reduces to a constant:

$$f_{tr} = f(0), \tag{36.8}$$

where $f(\epsilon)$ is the distribution function for particles above the well. We shall now determine the spatial distribution of electrons in this case if $f(\epsilon)$ is the Boltzmann function (36.4).

Adding the numbers of electrons with $\epsilon > 0$ and $\epsilon < 0$ gives

$$N_e = 2\int_{p_1}^{\infty} f(\epsilon)\, dp_x + 2\int_0^{p_1} f(0)\, dp_x,$$

$$p_1 = (2m|U|)^{1/2};$$

the factors 2 take account of particles with $p_x > 0$ and $p_x < 0$. Substituting $f(\epsilon)$ from (36.4), we have

$$N_e(t, x) = N_0\{e^{|U|/T_e}[1 - \Phi(\sqrt{(|U|/T_e)})] + 2\sqrt{(|U|/\pi T_e)}\}, \tag{36.9}$$

where

$$\Phi(\xi) = \frac{2}{\sqrt{\pi}}\int_0^{\xi} e^{-u^2}\, du. \tag{36.10}$$

When $\xi \ll 1$, expanding the integrand in (36.10) in powers of u, we find

$$\Phi(\xi) \approx \frac{2}{\sqrt{\pi}}(\xi - \tfrac{1}{3}\xi^3).$$

The distribution of electrons trapped in a shallow well ($|U| \ll T_e$) is therefore

$$N_e = N_0\left[1 + \frac{|U|}{T_e} - \frac{4}{3\sqrt{\pi}}\left(\frac{|U|}{T_e}\right)^{3/2}\right]. \tag{36.11}$$

The first correction term is the same as would result from the Boltzmann formula (36.5), but the next correction is already different from the Boltzmann form.

When $\xi \gg 1$, the difference $1 - \Phi(\xi)$ is exponentially small, $\propto \exp(-\xi^2)$. For a deep well ($|U| \gg T_e$), therefore, only the second term in the braces in (36.9) is important, and

$$N_e(t, x) = 2N_0(|U|/\pi T_e)^{1/2}. \tag{36.12}$$

As $|U|$ increases, the density increases much more slowly than would follow from the Boltzmann formula.

§37. **Quasi-neutral plasmas**

The equations of plasma dynamics allow a considerable simplification for a group of phenomena in which the characteristic scales of length and time satisfy the following conditions. (1) The characteristic dimension L of inhomogeneities in the plasma is assumed large compared with the electron Debye length:

$$a_e/L \ll 1. \tag{37.1}$$

(2) The rate of the process is assumed to be governed by the motion of the ions, so that the characteristic scale of speed is given by v_{Ti}, which is small in comparison with the electron speeds. (3) The motion of the ions causes a slow change in the electric potential that is adiabatically followed by the electron distribution.

Let δN_e and δN_i be the changes in the electron and ion densities in the perturbed plasma. These changes give rise to a mean uncompensated charge density $\delta\rho = e(z\delta N_i - \delta N_e)$ in the plasma. The potential of the electric field due to the charges is determined by Poisson's equation,

$$\Delta\varphi = -4\pi e(z\delta N_i - \delta N_e). \tag{37.2}$$

In order of magnitude, $\Delta\varphi \sim \varphi/L^2$. Hence

$$\left|\frac{z\delta N_i - \delta N_e}{\delta N_e}\right| \sim \frac{1}{4\pi e L^2}\left|\frac{\varphi}{\delta N_e}\right|. \tag{37.3}$$

If the field is weak ($e\varphi \ll T_e$), the change in the electron density is

$$\delta N_e \sim e\varphi N_e/T_e$$

(cf. (36.11)), and then

$$\left|\frac{z\delta N_i - \delta N_e}{\delta N_e}\right| \sim \frac{a_e^2}{L^2} \ll 1. \tag{37.4}$$

This inequality remains valid for a strong perturbation, with $e\varphi \sim T_e$: then $\delta N_e \sim N_e$, and (37.4) again follows from (37.3).

Thus the uncompensated charge density due to the perturbation is small compared with the perturbations of the electron and ion charge densities separately; such a plasma is said to be *quasi-neutral*. For the range of phenomena under consideration, this property enables the potential distribution in the plasma to be determined from just the "equation of quasi-neutrality",

$$N_e = zN_i, \tag{37.5}$$

together with the transport equation for ions and the equation giving the "adiabatic" distribution of electrons.†

At the initial instant, of course, (if we are considering a problem with initial conditions) the electron densities may be specified arbitrarily and need not satisfy the inequality (37.4). The resulting strong electric field, however, creates a movement of

†It must be emphasized that this result itself applies to plasmas with or without collisions. Note also that, since the derivation of the inequality (37.4) does not involve the assumption of a weak field, the quasi-neutrality property occurs even when the electromagnetic properties of the plasma cannot be described by means of the permittivity (i.e. by assuming a linear relation between **D** and **E**).

the electrons, which rapidly (in the characteristic "electronic" times) restores quasi-neutrality; this process in the diffusion case has been analysed in §25.

The passage from the electrodynamic equation (37.2) to the condition (37.5) signifies not only a considerable simplification of the equations of plasma dynamics but also a fundamental change in their dimensionality structure: the potential φ appears in the transport equation and in the electron distribution only as a product with the charge e, and the charge does not occur at all in the condition (37.5), in contrast to equation (37.2). Hence, by the change

$$e\varphi \to \psi, \tag{37.6}$$

the charge e is completely eliminated from the equations, and the length parameter (the Debye length a_e) disappears with it.

Since the equations contain no length parameter, self-similar motions of the plasma are possible. These occur when length parameters are also absent from the initial and boundary conditions, in which case all functions can depend on the coordinates and time only in the combination $\mathbf{r}/t$. For instance, let the plasma initially be confined to the half-space $x < 0$. At time $t = 0$, the constraint is removed, and the plasma begins to expand into a vacuum. The electrons begin to move first, so that the electron density forms a transition layer near the boundary, with characteristic width $\sim a_e$. After a time $t_1 \gg a_e/v_{Te}$, the electron motion is damped and the electron density follows the potential adiabatically in accordance with the Boltzmann formula. The variation of all quantities is then governed by the motion of the ions. Consequently, in a time $t_2 \gg a_e/v_{Ti} \gg a_e/v_{Te}$, the boundary is spread over distances large compared with a_e. The plasma becomes quasi-neutral, and the motion is self-similar.†

We can write down the dynamical equations for a quasi-neutral plasma in an expanded form, taking the particular case in which the electron density everywhere has a Boltzmann distribution:

$$N_e = N_0 e^{\psi/T_e}; \tag{37.7}$$

as shown in §36, this distribution is not affected by a slowly varying field, if the field has no potential wells. From (37.7) and (37.5), the potential can be expressed directly in terms of the ion distribution function:

$$\begin{aligned} \psi &= T_e \log(zN_i/N_0) \\ &= T_e \log\left[(z/N_0) \int f_i \, d^3p\right]. \end{aligned} \tag{37.8}$$

Substituting this expression in the transport equation for ions (with self-consistent field $\mathbf{E} = -\nabla\varphi$), we obtain

$$\frac{\partial f_i}{\partial t} + \mathbf{v} \cdot \frac{\partial f_i}{\partial \mathbf{r}} - zT_e \frac{\partial f_i}{\partial \mathbf{p}} \cdot \frac{\partial}{\partial \mathbf{r}} \log \int f_i \, d^3p = 0. \tag{37.9}$$

†For further details see A. V. Gurevich, L. V. Pariĭskaya and L. P. Pitaevskiĭ, *Soviet Physics JETP* **22**, 449, 1966.

Although this equation is non-linear, its solutions are independent of the mean plasma density: if $f_i(t, \mathbf{r})$ is a solution, so is Cf_i, with any constant factor C.

§38. Fluid theory for a two-temperature plasma

The theoretical treatment is particularly simple for a two-temperature plasma with

$$T_e \gg T_i. \tag{38.1}$$

We have already seen in §33 that in this case undamped ion-sound waves can propagate in the plasma, with speed $\sim(T_e/M)^{1/2}$. This speed is characteristic of the propagation of perturbations in plasmas. Since it is large in comparison with the thermal speeds of the ions, by (38.1), the thermal spread of the ion speeds can be neglected in most problems of plasma motion. The motion of the ion component of the plasma is then described in the one-fluid approximation by a velocity $\mathbf{v} \equiv \mathbf{v}_i$, which is a specified function of position in space (and of the time), and satisfies the equation

$$M\, d\mathbf{v}/dt = ez\mathbf{E},$$

or

$$\partial\mathbf{v}/\partial t + (\mathbf{v}\,.\,\nabla)\mathbf{v} = (ez/M)\mathbf{E}. \tag{38.2}$$

This is to be combined with the equation of continuity

$$\partial N_i/\partial t + \operatorname{div}(N_i\mathbf{v}) = 0 \tag{38.3}$$

and Poisson's equation, which determines the electric field potential φ (and hence the field $\mathbf{E} = -\nabla\varphi$):

$$\Delta\varphi = -4\pi e(zN_i - N_e). \tag{38.4}$$

The electron distribution follows adiabatically the field distribution in plasma motions with speeds $v \lesssim (T_e/M)^{1/2} \ll v_{Te}$. As we have seen in §36, the specific expression for the electron density N_e then depends considerably on the nature of the field. For a field without potential wells, it is given simply by the Boltzmann formula (37.7), so that (38.4) becomes

$$\Delta\varphi = -4\pi eN_0(zN_i/N_0 - e^{e\varphi/T_e}). \tag{38.5}$$

Equations (38.2), (38.3) and (38.5) form a complete set of equations for the functions $\mathbf{v}$, N and φ. They can be further simplified for a quasi-neutral plasma: in this case, (37.8) gives

$$e\varphi = T_e \log(zN_i/N_0), \quad e\mathbf{E} = -(T_e/N_i)\nabla N_i, \tag{38.6}$$

and (38.2) may be rewritten as

$$\frac{\partial \mathbf{v}}{\partial t}+(\mathbf{v}\,.\,\nabla)\mathbf{v}=-\frac{zT_e}{M}\frac{\nabla N_i}{N_i}. \tag{38.7}$$

The equations (38.3) and (38.7) are formally identical with the mechanical equations for an isothermal ideal gas having particle mass M and temperature zT_e. The speed of sound in such a gas is $(zT_e/M)^{1/2}$, in accordance with the expression (33.5) for the speed of ion-sound waves. In this approximation, there is no dispersion of the waves.

The above analogy with fluid mechanics needs a considerable reservation. The equations of fluid mechanics have by no means always solutions continuous throughout space. The absence of a continuous solution in ordinary fluid mechanics signifies the formation of shock waves, i.e. surfaces on which the physical quantities are discontinuous. In collisionless fluid mechanics, there are no shock waves, since these are essentially due to energy dissipation, which in this case does not occur. The absence of continuous solutions then implies that the assumption of a quasi-neutral plasma is violated in some region of space. In such regions (conventionally called *collisionless shock waves*), the dependence of the physical quantities on the coordinates and the time is oscillatory, and the characteristic wavelength of these oscillations is determined not only by the characteristic dimensions of the problem but also by an intrinsic property of the plasma, namely its Debye radius (R. Z. Sagdeev 1964).†

Let us now return to the more general equations (38.2)–(38.4), which do not assume a quasi-neutral plasma. An important property of these equations is that they have one-dimensional solutions, in which all quantities depend on the variables t and x only in the combination $\xi = x - ut$ with constant u. Such solutions describe waves propagating with speed u and without change of profile. If we change to a frame of reference moving with speed u relative to the original frame, the plasma motion becomes stationary. The most interesting solutions of this type are those periodic in space and those which decrease at infinity in both directions; the latter are known as *solitary waves* or *solitons* (A. A. Vedenov, E. P. Velikhov and R. Z. Sagdeev 1961).

If differentiation with respect to ξ is denoted by a prime, we have from (38.2) and (38.3)

$$(v-u)v' = -(e/M)\varphi', \quad (N_i v)' - uN_i' = 0;$$

for simplicity, we take $z = 1$.

Integrating these equations with the boundary conditions $\varphi = 0$, $v = 0$, $N_i = N_0$ as $\xi \to \infty$, we find

$$(e/M)\varphi = \tfrac{1}{2}u^2 - \tfrac{1}{2}(u-v)^2, \tag{38.8}$$

$$N_i = N_0 u/(u-v) = N_0 u/[u^2 - 2e\varphi/M]^{1/2}. \tag{38.9}$$

†Such structures have been given for several particular cases by A. V. Gurevich and L. P. Pitaevskiĭ, *Soviet Physics JETP* **38**, 291, 1974.

Equation (38.4) gives $\varphi'' = -4\pi e(N_i - N_e)$ or, after multiplication by $2\varphi'$ and integration,

$$\varphi'^2 = -8\pi e \int_0^{\varphi} [N_i(\varphi) - N_e(\varphi)] d\varphi. \tag{38.10}$$

The function $N_i(\varphi)$ is taken from (38.9), and $N_e(\varphi)$ is given by the formulae in §36. In the wave considered, we everywhere have $\varphi > 0$, as is seen from (38.8). The potential energy of an electron in such a field is $U = -e\varphi < 0$, i.e. the field forms a potential well for electrons.

Equation (38.10) reduces to quadratures the problem of determining the wave profile $\varphi(\xi)$. The speed u is directly related to the wave amplitude, i.e. the maximum value of $\varphi(\xi)$, which we denote by φ_m. When $\varphi = \varphi_m$, we must have $\varphi' = 0$. Equating to zero the integral on the right of (38.10) (and carrying out the integration of the first term), we obtain

$$\frac{Mu^2}{e}\left[1 - \left(1 - \frac{2e}{Mu^2}\varphi_m\right)^{1/2}\right] = \frac{1}{N_0}\int_0^{\varphi_m} N_e(\varphi)\, d\varphi, \tag{38.11}$$

which in principle determines u as a function of φ_m. Here it is evident that we must have

$$2e\varphi_m/Mu^2 < 1. \tag{38.12}$$

This condition in general places an upper limit on the possible values of the wave amplitude φ_m, and therefore of the speed u.

For collisions to be entirely negligible, it is necessary that the field frequency ω be large in comparison with the characteristic collision frequencies of both electrons (ν_e) and ions (ν_i). Since $\nu_e \sim (M/m)^{1/2}\nu_i \gg \nu_i$ (see §43), a situation can occur in which $\nu_e \gg \omega \gg \nu_i$. The collisions then again have no effect on the motion of the ions, but the electrons may be regarded as having a Boltzmann distribution even when there are potential wells present.

PROBLEM

Determine the profile and speed of a weak solitary wave ($e\varphi_m/T_e \ll 1$) in a plasma of electrons distributed according to (36.11) (A. V. Gurevich, 1967).

SOLUTION. In (36.11), all terms must be retained, since the formation of the solitary wave is due to the final non-linear term there. A calculation with (38.11) gives

$$u^2 = \frac{T_e}{M}\left[1 + \frac{16}{15}\left(\frac{e\varphi_m}{\pi T_e}\right)^{1/2}\right].$$

The wave profile is found by integration of (38.10):

$$\varphi = \varphi_m \cosh^{-4}\left[\frac{x}{\sqrt{15}a_e}\left(\frac{e\varphi_m}{\pi T_e}\right)^{1/4}\right].$$

§39. Solitons in a weakly dispersing medium

The existence (in a medium without dissipation) of non-linear waves with a stationary profile is closely related to the presence of dispersion. In a non-dispersing medium, the wave cannot be stationary if non-linearity is taken into account; the rate of propagation of various points on the profile is dependent on the amplitude at those points, which causes a distortion of the profile. For example, in the dynamics of an ideal compressible fluid, non-linear effects cause a gradual increase in the slope of the forward edge of the wave; see *FM*, §94. The dispersion, for its part, causes a gradual smoothing-out of the profile, and the two effects may cancel out, leaving a stationary profile.

In this section, the phenomena concerned will be investigated in a general form for a fairly wide class of cases of wave propagation in a non-dissipative weakly dispersive medium, including weak non-linearity.

Let u_0 be the rate of wave propagation in the linear approximation, when dispersion is neglected. In this approximation, for a one-dimensional wave propagating in one direction parallel to the x-axis, all quantities depend on x and t only in the combination $\xi = x - u_0 t$. This property can be expressed in differential form as

$$\partial b/\partial t + u_0 \partial b/\partial x = 0,$$

where b denotes any of the quantities oscillating in the wave.

A constant speed u_0 corresponds to a wave dispersion relation $\omega = u_0 k$. In a dispersive medium, this relation is just the first term in the expansion of the function $\omega(k)$ in powers of the small quantity k. Including the next term, we have†

$$\omega = u_0 k - \beta k^3, \tag{39.1}$$

where β is a constant that may in principle be either positive or negative.

The differential equation which describes (in the linear approximation) the propagation, in one direction, of a wave in a medium with such a dispersion is

$$\frac{\partial b}{\partial t} + u_0 \frac{\partial b}{\partial x} + \beta \frac{\partial^3 b}{\partial x^3} = 0,$$

since this gives (39.1) for a wave in which $b \propto \exp(-i\omega t + ikx)$.

Lastly, the inclusion of non-linearity causes terms of higher order in b to appear in the equation. These terms must certainly satisfy the condition of vanishing for a constant b (independent of x), corresponding simply to a homogeneous medium. Considering only the term containing the derivative of lowest order (k is small), we write the equation of propagation of a slightly non-linear wave as

$$\frac{\partial b}{\partial t} + u_0 \frac{\partial b}{\partial x} + \beta \frac{\partial^3 b}{\partial x^3} + \alpha b \frac{\partial b}{\partial x} = 0, \tag{39.2}$$

where α is a constant parameter which again may in principle have either sign.‡

†The fact that $\omega(k)$ can be expanded in odd powers of k follows from a consideration of quantities that must be real. The initial set of physical equations of motion for the medium contains only real quantities and parameters. The imaginary unit i occurs only through the substitution in these equations of a solution proportional to $\exp(-i\omega t + ikx)$. The dispersion relation resulting from this substitution therefore determines $i\omega$ as a function of ik with real coefficients; the expansion of such a function must contain only odd powers of ik. In the general case of a dissipative medium, $\omega(k)$ is complex: $\omega = \omega' + i\omega''$, and the statement made then refers to the expansion of the real part $\omega'(k)$ of the frequency. The expansion of $\omega''(k)$ will, for the same reasons, contain only even powers of k.

‡It must be emphasized, however, to avoid misunderstanding, that this form of slight non-linearity is not at all universal. For example, the slight non-linearity in wave propagation in plasmas that results from the last term in the electron distribution (36.11) (used in §38, Problem) would correspond to a term $\propto \sqrt{b}\, \partial b/\partial x$ in an equation such as (39.2).

To simplify this equation, we replace x with a new variable ξ and b with a new unknown function a, defined by

$$\xi = x - u_0 t, \quad a = \alpha b. \tag{39.3}$$

This gives

$$\frac{\partial a}{\partial t} + a\frac{\partial a}{\partial \xi} + \beta\frac{\partial^3 a}{\partial \xi^3} = 0, \tag{39.4}$$

the *Korteweg–de Vries equation.*† We shall take first of all the particular case where $\beta > 0$.

We shall be interested in solutions which describe waves with a stationary profile. In such solutions, the function $a(t, \xi)$ depends only on the difference $\xi - v_0 t$ with some constant v_0:

$$a = a(\xi - v_0 t), \tag{39.5}$$

and the wave propagation speed is

$$u = u_0 + v_0. \tag{39.6}$$

Substituting (39.5) in (39.4) and denoting differentiation with respect to ξ by a prime, we obtain the equation

$$\beta\alpha''' + aa' - v_0 a' = 0. \tag{39.7}$$

This is invariant under the change

$$a \to a + V, \quad v_0 \to v_0 + V \tag{39.8}$$

with any constant V.

The first integral of equation (39.7) is

$$\beta a'' + \tfrac{1}{2}a^2 - v_0 a = \tfrac{1}{2}c_1.$$

Multiplication by $2a'$ and another integration gives

$$\beta a'^2 = -\tfrac{1}{3}a^3 + v_0 a^2 + c_1 a + c_2. \tag{39.9}$$

Instead of the three constants v_0, c_1, c_2 it is convenient to use the three roots of the cubic on the right of (39.9). If these are denoted by a_1, a_2, a_3, then

$$\beta a'^2 = -\tfrac{1}{3}(a - a_1)(a - a_2)(a - a_3). \tag{39.10}$$

The constant v_0 is related to the new constants by

$$v_0 = \tfrac{1}{3}(a_1 + a_2 + a_3). \tag{39.11}$$

†It was derived by D. J. Korteweg and G. de Vries (1895) for waves on the surface of shallow water.

We shall be concerned only with solutions of (39.10) such that $|a(\xi)|$ is bounded; an unlimited increase of $|a|$ would contradict the assumption of slight non-linearity. It is easy to see that this condition is not satisfied if the roots a_1, a_2, a_3 are not all real. For, let a_1 and a_2 $(=a_1^*)$ be complex; then the right-hand side of (39.10) becomes $\frac{1}{3}|a-a_1|^2(a_3-a)$ and there is nothing to prevent a from tending to $-\infty$.

Thus the constants a_1, a_2, a_3 must be real; let them be ordered so that $a_1 > a_2 > a_3$. Since the expression on the right of (39.10) must be positive, the function $a(\xi)$ can vary only in the range $a_1 \geqslant a \geqslant a_2$. We can put $a_3 = 0$ without loss of generality; this may always be achieved by a transformation of the type (39.8). With this choice, we rewrite (39.10) as

$$\beta a'^2 = \tfrac{1}{3}(a_1 - a)(a - a_2)a. \tag{39.12}$$

The nature of the solution depends on whether a_2 is zero. If $a_2 = 0$, $a_1 > 0$, integration gives

$$a(\xi) = a_1 \cosh^{-2}(\tfrac{1}{2}\xi\sqrt{(a_1/3\beta)}); \tag{39.13}$$

the zero of ξ is taken at the maximum of the function. (Here and henceforward, to simplify the notation, we write the wave profile as a function of $\xi = x$ at some given instant $t = 0$.) This solution describes a solitary wave, or soliton: as $\xi \to \pm\infty$, the function $a(\xi)$ vanishes, together with its derivatives. The constant a_1 gives the soliton amplitude, and the width decreases as $a_1^{-1/2}$ with increasing amplitude. According to (39.11), $v_0 = \frac{1}{3}a_1$, and the speed of the soliton is therefore

$$u = u_0 + \tfrac{1}{3}a_1. \tag{39.14}$$

This exceeds u_0, and increases with the amplitude.

The non-linearity, it will be remembered, is assumed slight for processes described by the Korteweg–de Vries equation. The condition has an obvious meaning: for example, if a is the change in the density of the medium, this change must be small in comparison with the unperturbed density. At the same time, the degree of non-linearity of these processes is expressed by a further dimensionless parameter $L(a_1/\beta)^{1/2}$, where L is a characteristic length and a_1 the amplitude of the perturbation. This parameter defines the relative importance of non-linearity and dispersion, and may be either small (if dispersion predominates) or large (if non-linearity predominates). For a soliton, whose width $L \sim (\beta/a_1)^{1/2}$, the parameter is of the order of unity.

Let us now take the case where $a_2 \neq 0$. The solution of (39.12) then describes a wave of infinite extent, periodic in space. Integration of the equation gives

$$\begin{aligned}\xi &= \int_a^{a_1} \frac{\sqrt{(3\beta)}\, da}{[a(a_1 - a)(a - a_2)]^{1/2}} \\ &= (12\beta/a_1)^{1/2} F(s, \varphi),\end{aligned} \tag{39.15}$$

where $F(s, \varphi)$ is an elliptic integral of the first kind:

$$F(s, \varphi) = \int_0^{\varphi} \frac{d\varphi}{(1 - s^2 \sin^2 \varphi)^{1/2}}, \tag{39.16}$$

with†

$$\sin\varphi = \sqrt{\frac{a_1 - a}{a_1 - a_2}}, \quad s = \sqrt{\left(1 - \frac{a_2}{a_1}\right)}; \tag{39.17}$$

the zero of ξ is taken at one of the maxima of the function $a(\xi)$.

Inverting the formula (39.15) by means of the Jacobi elliptic function, we have

$$a = a_1 \,\mathrm{dn}^2(\sqrt{(a_1/12\beta)}\xi, s). \tag{39.18}$$

This is a periodic function, whose period (wavelength) in the coordinate x is

$$\lambda = 4\sqrt{(3\beta/a_1)}F(\tfrac{1}{2}\pi, s) = 4\sqrt{(3\beta/a_1)}K(s), \tag{39.19}$$

where $K(s)$ is a complete elliptic integral of the first kind. The mean value over a period of (39.18) is

$$\bar{a} = \frac{1}{\lambda}\int_0^\lambda a(\xi)d\xi = a_1 E(s)/K(s), \tag{39.20}$$

where $E(s)$ is a complete elliptic integral of the second kind. It is natural to consider a periodic wave in which the mean value of the oscillating quantity is zero. This may always be achieved by means of the transformation (39.8), subtracting the quantity (39.20) from the function (39.18). The wave propagation speed is then

$$u = u_0 + [\tfrac{1}{3}(a_1 + a_2) - a_1 E(s)/K(s)]. \tag{39.21}$$

Small oscillation amplitudes $a_1 - a_2$ correspond to parameter values $s \ll 1$. Using the approximation

$$\mathrm{dn}(z, s) \approx 1 - \tfrac{1}{4}s^2 + \tfrac{1}{4}s^2 \cos 2z, \quad s \ll 1,$$

we find that the solution (39.18) becomes in this case, as it should, the harmonic wave

$$a = \tfrac{1}{2}(a_1 + a_2) + \tfrac{1}{2}(a_1 - a_2)\cos kx, \quad k = \sqrt{(a_1/3\beta)}.$$

The speed (39.21) becomes $u = u_0 - \frac{1}{3}a_1 = u_0 - \beta k^2$ in accordance with (39.1).

The opposite limiting case of large amplitudes (in the wave model under consideration) corresponds to $a_2 \to 0$ and $s \to 1$. With the limiting expression

$$K(s) \approx \tfrac{1}{2}\log[16/(1 - s^2)], \quad s^2 \to 1,$$

we find that in this limit the wavelength increases logarithmically:

$$\lambda = \sqrt{(12\beta/a_1)}\log(16a_1/a_2). \tag{39.22}$$

†The parameter of the elliptic integral is denoted by s instead of the usual k, in order to avoid confusion with the wave number.

Thus the successive antinodes of the wave move further apart. The wave profile near each antinode is obtained from (39.18) by taking the limit of $\operatorname{dn} z$ for $s = 1$, which is valid for finite z: $\operatorname{dn} z = 1/\cosh z$. The result is again (39.13). In the limit $s \to 1$, therefore, a periodic wave separates into a sequence of widely spaced solitons.

So far we have assumed that $\beta > 0$. The case where $\beta < 0$ does not need special consideration, since a change in the sign of β in (39.4) is equivalent to the change $\xi \to -\xi$, $a \to -a$. Since this converts the argument $\xi - v_0 t$ in (39.5) into $-\xi - v_0 t$, the wave propagation speed becomes $u = u_0 - v_0$. For instance, the above results for the soliton are altered only in that the function $a(\xi)$ becomes negative, and the speed of the soliton $u < u_0$.

The Korteweg–de Vries equation has certain specific properties which lead to a number of general theorems. These are based on the formal relation between the equation and the eigenvalue problem for an equation of the Schrödinger type (C. S. Gardner, J. M. Greene, M. D. Kruskal and R. M. Miura, 1967).

Let us consider the equation

$$\frac{\partial^2 \psi}{\partial \xi^2} + \left[\frac{1}{6\beta} a(t, \xi) + \epsilon\right]\psi = 0, \tag{39.23}$$

again taking the particular case $\beta > 0$. This has the form of Schrödinger's equation with $-a(t, \xi)$ as the potential energy depending on t as a parameter. Let $a(t, \xi)$ be positive in some range of ξ, and tend to zero as $\xi \to \pm\infty$. Then the equation (39.23) has eigenvalues ϵ corresponding to a finite motion in the potential well $-a(t, \xi)$; since a depends on t, these eigenvalues will in general also depend on t. We shall show that the eigenvalues ϵ do *not* depend on t if $a(\xi, t)$ satisfies the Korteweg–de Vries equation (39.4).

Expressing a from (39.23) as

$$a = -6\beta(\psi''/\psi + \epsilon),$$

and substituting in (39.4), we find by a direct calculation

$$\psi^2 d\epsilon/dt = (\psi' A - \psi A')', \tag{39.24}$$

where

$$A(t, \xi) = 6\beta\left(\frac{1}{\beta}\frac{\partial \psi}{\partial t} - \frac{3}{\psi}\psi'\psi'' + \psi''' - \tfrac{1}{6}\epsilon\psi'\right); \tag{39.25}$$

here it is important that the right-hand side of (39.24) is expressed in terms of the derivative with respect to ξ of an expression that vanishes as $\xi \to \pm\infty$ (the eigenfunctions of the discrete spectrum of (39.23) are zero at infinity). Integration of (39.24) with respect to ξ over the whole range from $-\infty$ to ∞ therefore gives

$$\frac{d\epsilon}{dt}\int_{-\infty}^{\infty} \psi^2 \, d\xi = 0$$

and, since this normalization integral of ψ is not zero, it follows that $d\epsilon/dt = 0$.

We shall now show that equation (39.23) has only one discrete eigenvalue for a stationary "potential" $a(\xi)$ in the form (39.13), corresponding to one soliton. With this potential, (39.23) becomes

$$\psi'' + \left(\frac{U_0}{\cosh^2 \alpha\xi} + \epsilon\right)\psi = 0, \tag{39.26}$$

where

$$U_0 = a_1/6\beta, \quad \alpha = (a_1/12\beta)^{1/2}. \tag{39.27}$$

The discrete eigenvalues of (39.26) are

$$\epsilon_n = -\alpha^2(s-n)^2, \quad s = \tfrac{1}{2}[-1+\sqrt{(1+4U_0/\alpha^2)}], \quad n = 0, 1, 2, \ldots,$$

where we must have $n < s$; see *QM*, §23, Problem 4. With the parameter values (39.27), $s = 1$, so that there is only one eigenvalue,

$$\epsilon = -a_1/12\beta. \tag{39.28}$$

If, however, $a(t, \xi)$ represents an assembly of solitons at large intervals (so that there is no "interaction" between them), the eigenvalue spectrum of equation (39.23) consists of the "levels" (39.28) in each potential well, each level being determined by the amplitude a_1 of the corresponding soliton.

Since the soliton propagation speed increases with the amplitude, a soliton of large amplitude will always eventually overtake one of smaller amplitude. An arbitrary initial assembly of solitons at large intervals, therefore, will ultimately, after a series of mutual "collisions", become an assembly of solitons arranged in order of increasing amplitude (all perturbations described by the Korteweg–de Vries equation propagate in the same direction).

The above results lead immediately to an interesting conclusion: the initial and final assemblies of solitons have the same total number and amplitudes, differing only in their order. This follows directly from the fact that each isolated soliton corresponds to one eigenvalue ϵ, and the eigenvalues do not depend on the time.

Any positive ($a > 0$) initial perturbation occupying a finite region of space and evolving in accordance with the Korteweg–de Vries equation ultimately breaks up into an assembly of isolated solitons whose amplitudes are independent of time. These amplitudes and the number of solitons can in principle be found by determining the spectrum of discrete eigenvalues of equation (39.23) with the initial distribution $a(0, \xi)$ as the potential. If the initial perturbation contains also regions with $a < 0$, however, its evolution will create a wave packet which gradually spreads out and does not break up into solitons.

To avoid misunderstanding, however, we should specify more precisely what is meant by the initial perturbation in the Korteweg–de Vries equation. An actual perturbation formed in the medium at some instant evolves (in a way described by the complete wave equation, of the second order in the time) and in general separates into two perturbations propagating in the positive and negative x-directions. The "initial" perturbation for the Korteweg–de Vries equation is to be taken as one of these two, immediately after the separation.

PROBLEM

Determine the coefficients α and β in equation (39.2) for ion-sound waves in a plasma with $T_i \ll T_e$.

SOLUTION. The dispersion coefficient β is found from (33.4) by expanding in terms of the small quantity ka_e:

$$\beta = a_e^2 u_0,$$

where $u_0 = (zT_e/M)^{1/2}$.

In determining the non-linearity coefficient α, we may neglect the dispersion, i.e. consider the limiting case $k \to 0$. In this limit, the plasma may always be regarded as quasi-neutral, and accordingly be described by the mechanical equations of an isothermal ideal gas (38.3), (38.7). Putting $N_i = N_0 + \delta N$, we can write these equations as far as the second-order terms in the small quantities δN and v. In these second-order terms, we can put $v = u_0 \delta N/N_0$, as in the linear approximation for a wave propagating in the positive x-direction (u_0 being the speed of the waves in the linear approximation). The equations then become

$$\frac{\partial \delta N}{\partial t} + N_0 \frac{\partial v}{\partial x} = -\frac{\partial}{\partial x}(v\delta N) = -\frac{2u_0}{N_0}\delta N \frac{\partial \delta N}{\partial x},$$

$$\frac{\partial v}{\partial t} + \frac{u_0^2}{N_0}\frac{\partial \delta N}{\partial x} = \frac{u_0^2}{N_0^2}\delta n \frac{\partial \delta N}{\partial x} - v\frac{\partial v}{\partial x} = 0.$$

Differentiating the first equation with respect to t, the second with respect to x, and eliminating $\partial^2 v/\partial t \partial x$, we find

$$\left(\frac{\partial}{\partial t} - u_0\frac{\partial}{\partial x}\right)\left(\frac{\partial}{\partial t} + u_0\frac{\partial}{\partial x}\right)\delta N = -\frac{2u_0}{N_0}\frac{\partial}{\partial t}\left(\delta N \frac{\partial \delta N}{\partial x}\right).$$

With the same accuracy, we replace the derivative $\partial/\partial t$ on the right, and in the difference $\partial/\partial t - u_0\partial/\partial x$ on the left, by $-u_0\partial/\partial x$. Lastly, cancelling the differentiation $\partial/\partial x$ on both sides and comparing the result with (39.2), we have

$$\alpha = u_0/N_0.$$

§40. Permittivity of a degenerate collisionless plasma

In calculating the permittivity of a collisionless plasma in §§29 and 31, we entirely neglected all quantum effects. The results thus obtained are restricted, first of all, as to temperature by the condition that degeneracy be absent; for electrons, this condition is

$$T \gg \epsilon_F \sim \hbar^2 N_e^{2/3}/m, \tag{40.1}$$

where $\epsilon_F = p_F^2/2m$ and p_F is the limiting momentum of the Fermi distribution at $T = 0$, which is related to the number density of electrons by $p_F^3/3\pi^2\hbar^3 = N_e$.

Moreover, the possibility of applying the classical Boltzmann equation to a plasma in an external field involves the imposition of certain conditions on the field frequency ω and wave vector $\mathbf{k}$. The characteristic distances ($\sim 1/k$) over which the field varies must be large compared with the electron de Broglie wavelength $\hbar/\bar{p}$, and the corresponding uncertainty ($\sim \hbar k$) in the momentum must be small compared with the width ($\sim T/\bar{v}$) of the region over which the electron thermal distribution extends. For a non-degenerate plasma, $\bar{p} \sim T/\bar{v} \sim (mT)^{1/2}$, so that the two conditions coincide. For a degenerate plasma, $\bar{p} \sim p_F$, $v \sim v_F = p_F/m$, but

$T/\bar{v} \lesssim \bar{p}$ since $T \lesssim \epsilon_F$. Thus in either case it is sufficient if

$$\hbar k\bar{v} \ll T. \tag{40.2}$$

Lastly, the frequency must satisfy the condition

$$\hbar\omega \ll \epsilon_F: \tag{40.3}$$

the field energy quantum must be small in comparison with the mean electron energy (but this condition is usually unimportant).

Let us now consider the dielectric properties of the plasma without imposing the conditions (40.1)–(40.3) on the electron component; the ion component may remain non-degenerate. We shall calculate the electronic part of the permittivity. The condition for the interaction of the plasma particles to be negligible will again be assumed satisfied:

$$e^2 N_e^{1/3} \ll \bar{\epsilon}; \tag{40.4}$$

when $\bar{\epsilon} \sim \epsilon_F$, this becomes $N_e^{1/3} \gg me^2/\hbar^2$, or $e^2/\hbar v_F \ll 1$ (cf. *SP* 1, § 80; *SP* 2, § 85).

The abandonment of the condition (40.2) means that the quantum-mechanical equation for the density matrix must be applied from the start. Since the interaction between electrons is neglected, we can immediately write a closed equation for the one-particle density matrix $\rho_{\sigma_1\sigma_2}(t, \mathbf{r}_1, \mathbf{r}_2)$ (where σ_1 and σ_2 are the spin indices). We shall assume that the electron distribution is independent of the spin, i.e. the dependence of the density matrix on the spin indices is separated as a factor $\delta_{\sigma_1\sigma_2}$, which will be omitted. The spin-independent density matrix $\rho(t, \mathbf{r}_1, \mathbf{r}_2)$ satisfies the equation

$$i\hbar\partial\rho/\partial t = (\hat{H}_1 - \hat{H}_2^*)\rho, \tag{40.5}$$

where $\hat{H}$ is the electron Hamiltonian in the external field, and the suffixes 1 and 2 indicate the variable ($\mathbf{r}_1$ or $\mathbf{r}_2$) on which the operator acts (see *QM*, § 14). This equation replaces the classical Liouville's theorem $df/dt = 0$ for the classical one-particle distribution function.

As in § 29, we shall calculate the longitudinal permittivity. Accordingly, we consider an electric field with a scalar potential $\varphi(t, \mathbf{r})$, and the electron Hamiltonian thus becomes

$$\hat{H} = -(\hbar^2/2m)\triangle - e\varphi(t, \mathbf{r}). \tag{40.6}$$

Assuming the field to be weak, we put

$$\rho = \rho_0(\mathbf{r}_1 - \mathbf{r}_2) + \delta\rho(t, \mathbf{r}_1, \mathbf{r}_2), \tag{40.7}$$

where ρ_0 is the density matrix of the unperturbed stationary and homogeneous (but not necessarily equilibrium) state of the gas; the homogeneity implies that ρ_0 depends only on the difference $\mathbf{R} = \mathbf{r}_1 - \mathbf{r}_2$. The density matrix $\rho_0(\mathbf{R})$ is related to the (unperturbed) electron momentum distribution function $n_0(\mathbf{p})$ by

$$n_0(\mathbf{p}) = \mathcal{N}_e \int \rho_0(\mathbf{R}) e^{-i\mathbf{p}\cdot\mathbf{R}/\hbar}\, d^3x, \qquad (40.8)$$

where $\mathcal{N}_e$ is the total number of electrons; see *SP* 2 (7.20). The function $n(\mathbf{p})$ is defined as the occupation numbers of quantum states of electrons with definite values of the momentum and the spin component. The number of states in an element d^3p of momentum space and with either value of the spin component is $2\, d^3p/(2\pi\hbar)^3$. Hence $n(\mathbf{p})$ is related to the distribution function $f(\mathbf{p})$ used previously by

$$f(\mathbf{p}) = 2n(\mathbf{p})/(2\pi\hbar)^3. \qquad (40.9)$$

Substituting (40.7) in (40.8) and omitting terms of the second order of smallness, we obtain a linear equation for the small correction to the density matrix:

$$\left[i\hbar\frac{\partial}{\partial t} + \frac{\hbar^2}{2m}(\triangle_1 - \triangle_2)\right]\delta\rho(t, \mathbf{r}_1, \mathbf{r}_2) = -e[\varphi(t, \mathbf{r}_1) - \varphi(t, \mathbf{r}_2)]\rho_0(\mathbf{r}_1 - \mathbf{r}_2). \qquad (40.10)$$

Let†

$$\varphi(t, \mathbf{r}) = \varphi_{\omega\mathbf{k}} e^{i(\mathbf{k}\cdot\mathbf{r} - \omega t)}. \qquad (40.11)$$

Then the dependence of the solution of (40.10) on the sum $\mathbf{r}_1 + \mathbf{r}_2$ (and on the time) can be separated by putting

$$\delta\rho = \exp[i\mathbf{k}\cdot\tfrac{1}{2}(\mathbf{r}_1 + \mathbf{r}_2) - i\omega t]g_{\omega\mathbf{k}}(\mathbf{r}_1 - \mathbf{r}_2). \qquad (40.12)$$

Substituting this expression in (40.10), we obtain an equation for $g_{\omega k}(\mathbf{R})$:

$$\left[\hbar\omega + \frac{\hbar^2}{2m}(\nabla + \tfrac{1}{2}i\mathbf{k})^2 - \frac{\hbar^2}{2m}(\nabla - \tfrac{1}{2}i\mathbf{k})^2\right]g_{\omega\mathbf{k}}(\mathbf{R}) = -e\varphi_{\omega\mathbf{k}}(e^{i\mathbf{k}\cdot\mathbf{R}^2} - e^{i\mathbf{k}\cdot\mathbf{R}^2})\rho_0(\mathbf{R}).$$

We can now apply a Fourier expansion with respect to $\mathbf{R}$. Multiplying both sides by $\exp(-i\mathbf{p}\cdot\mathbf{R}/\hbar)$ and integrating over d^3x, and using (40.8), we obtain

$$[\hbar\omega - \epsilon(\mathbf{p} + \tfrac{1}{2}\hbar\mathbf{k}) + \epsilon(\mathbf{p} - \tfrac{1}{2}\hbar\mathbf{k})]g_{\omega\mathbf{k}}(\mathbf{p}) = -(e\varphi_{\omega\mathbf{k}}/\mathcal{N}_e)[n_0(\mathbf{p} - \tfrac{1}{2}\hbar\mathbf{k}) - n_0(\mathbf{p} + \tfrac{1}{2}\hbar\mathbf{k})],$$

where $\epsilon(\mathbf{p}) = \mathbf{p}^2/2m$, or equivalently

$$g_{\omega\mathbf{k}}(\mathbf{p}) = (e\varphi_{\omega\mathbf{k}}/\hbar\mathcal{N}_e)[n_0(\mathbf{p} + \tfrac{1}{2}\hbar\mathbf{k}) - n_0(\mathbf{p} - \tfrac{1}{2}\hbar\mathbf{k})]/(\omega - \mathbf{k}\cdot\mathbf{v}). \qquad (40.13)$$

†The Hamiltonian (40.6) must be Hermitian, and so the function φ in it (and consequently in (40.10) also) is real. But, having written down equation (40.10), which is linear, we can solve it for each of the complex monochromatic field components separately.

The value of the density matrix at $\mathbf{r}_1 = \mathbf{r}_2 \equiv \mathbf{r}$ determines the number density of particles in the system: $N = 2\mathcal{N}\rho(t, \mathbf{r}, \mathbf{r})$; see *SP* 2 (7.19). Hence the change in the electron density due to the field is

$$\delta N_e = 2\mathcal{N}_e\delta\rho(t, \mathbf{r}, \mathbf{r}) = 2\mathcal{N}_e e^{i(\mathbf{k}\cdot\mathbf{r}-\omega t)} g_{\omega\mathbf{k}}(\mathbf{R} = 0),$$

or, expressing $g_{\omega\mathbf{k}}(\mathbf{R} = 0)$ in Fourier components,

$$\delta N_e = 2\mathcal{N}_e e^{i(\mathbf{k}\cdot\mathbf{r}-\omega t)} \int g_{\omega\mathbf{k}}(\mathbf{p})\, d^3p/(2\pi\hbar)^3. \tag{40.14}$$

The corresponding change in the charge density is $-e\delta N_e$.

The permittivity is now calculated as in §29: starting from the relation between the charge density and the dielectric polarization vector ($-e\delta N_e = -\operatorname{div}\mathbf{P} = -i\mathbf{k}\cdot\mathbf{P}$), we write

$$\begin{aligned} e\delta N_e &= i(\epsilon_l - 1)\mathbf{E}\,.\,\mathbf{k}/4\pi \\ &= k^2(\epsilon_l - 1)\varphi_{\omega\mathbf{k}}/4\pi. \end{aligned}$$

We thus arrive at the following formula for the electronic part of the longitudinal permittivity of a plasma having an electron distribution function $n(\mathbf{p})$ (the suffix 0 being now omitted);

$$\epsilon_l(\omega, k) - 1 = -\frac{4\pi e^2}{\hbar k^2} \int \frac{n(\mathbf{p} + \frac{1}{2}\hbar\mathbf{k}) - n(\mathbf{p} - \frac{1}{2}\hbar\mathbf{k})}{\mathbf{k}\,.\,\mathbf{v} - \omega - i0} \frac{2d^3p}{(2\pi\hbar)^3} \tag{40.15}$$

(Yu. L. Klimontovich and V. P. Silin, 1952); the pole in the integral is, as usual, avoided in the manner specified by the Landau rule.

In the quasi-classical case, when the conditions (40.2) and (40.3) are satisfied, the functions $n(\mathbf{p} \pm \frac{1}{2}\hbar\mathbf{k})$ can be expanded in powers of $\mathbf{k}$. Then

$$n(\mathbf{p} + \tfrac{1}{2}\hbar\mathbf{k}) - n(\mathbf{p} - \tfrac{1}{2}\hbar\mathbf{k}) \approx \hbar\mathbf{k}\,.\,\partial n(\mathbf{p})/\partial\mathbf{p},$$

and (40.15) becomes the previous formula (29.9) when the relation (40.9) is taken into account. It must be emphasized, however, that the distribution $n(\mathbf{p})$ in (40.15) may relate to a degenerate plasma.

Let us apply this formula to a completely degenerate electron plasma at $T = 0$, with $n(\mathbf{p}) = 1$ for $p < p_F$ and $n(\mathbf{p}) = 0$ for $p > p_F$. Changing the variable of integration in each term in (40.15) by $\mathbf{p} \pm \frac{1}{2}\hbar\mathbf{k} \to \mathbf{p}$, we obtain

$$\epsilon_l - 1 = \frac{4\pi e^2}{\hbar k^2} \int_{p<p_F} \left\{ \frac{1}{\omega_+ - \mathbf{k}\,.\,\mathbf{v} + i0} - \frac{1}{\omega_- - \mathbf{k}\,.\,\mathbf{v} + i0} \right\} \frac{2d^3p}{(2\pi\hbar)^3},$$

where $\omega_\pm = \omega \pm \hbar k^2/2m$. An elementary though fairly laborious integration leads to the result

$$\epsilon_l(\omega, k) - 1 = \frac{3\Omega_e^2}{2k^2 v_F^2}\{1 - g(\omega_+) + g(\omega_-)\}, \tag{40.16}$$

$$g(\omega) = \frac{m(\omega^2 - k^2 v_F^2)}{2\hbar k^3 v_F} \log\frac{\omega + kv_F}{\omega - kv_F};$$

the logarithm is to be taken as $\log|u| - i\pi$ if its argument $u < 0$. The "plasma frequency" Ω_e is again defined as $\Omega_e = (4\pi N_e e^2/m)^{1/2}$.

In the quasi-classical limit $\hbar k \ll p_F$, $\hbar\omega \ll \epsilon_F$,† formula (40.16) leads to a simple expression that does not involve $\hbar$:

$$\epsilon_l - 1 = \frac{3\Omega_e^2}{k^2 v_F^2}\left[1 - \frac{\omega}{2kv_F}\log\frac{\omega + kv_F}{\omega - kv_F}\right] + \begin{cases} 0 & \text{for } |\omega| > kv_F, \\ i\,.\,3\pi\Omega_e^2\omega/2(kv_F)^3 & \text{for } |\omega| < kv_F. \end{cases} \tag{40.17}$$

The static case is of particular interest. When $\omega = 0$, the expression (40.16) as a function of k has a singularity at the point where $\hbar k$ is equal to the diameter of the Fermi sphere,

$$\hbar k = 2p_F; \tag{40.18}$$

at this point, the argument of one of the logarithms is zero. Near it,

$$\epsilon_l(0, k) - 1 = (e^2/2\pi\hbar\epsilon_F)[1 - \xi\log(1/|\xi|)], \tag{40.19}$$

$$\xi = (\hbar k - 2p_F)/2p_F, \quad |\xi| \ll 1.$$

We shall show that the presence of this singularity (called the *Kohn singularity*) leads to a change in the nature of the screening of the field of charges in the plasma, which ceases to be exponential.‡

We write (40.19) in the form

$$\epsilon_l(0, k) = \beta - \alpha\xi\log(1/|\xi|), \tag{40.20}$$

where $\alpha = e^2/2\pi\hbar v_F$, and the constant β may include the non-singular contribution from the non-degenerate ionic component of the plasma.

The Fourier component of the field due to a small point charge e_1 at rest in the plasma is expressed in terms of the permittivity by

$$\varphi_{\mathbf{k}} = 4\pi e_1/k^2\epsilon_l(0, k); \tag{40.21}$$

see §31, Problem 1. For the potential $\varphi(r)$ as a function of the distance from the charge e_1, we have

$$\begin{aligned}\varphi(r) &= \int \varphi_{\mathbf{k}} e^{i\mathbf{k}\cdot\mathbf{r}}\, d^3k/(2\pi)^3 \\ &= \frac{1}{2\pi^2 r}\,\mathrm{im}\int_0^\infty \varphi_{\mathbf{k}} e^{ikr} k\, dk.\end{aligned} \tag{40.22}$$

†At $T = 0$, these conditions are sufficient, since the limiting value of ϵ_l as $\hbar k v_F/\epsilon_F \to 0$ and $T \to 0$ is independent of the order of passage to the limits. The relation between $\hbar k v_F$ and T is therefore unimportant.

‡The physical consequences of the singularity which results when the condition (40.18) is satisfied were noted by W. Kohn (1959).

As $k \to 0$, $\varphi(k)$ tends to a constant and has no singularity. Hence the asymptotic form of the integral in (40.22) as $r \to \infty$ is determined by the singularity of this function at $\hbar k = 2p_F$. Near that point,

$$\varphi_{\mathbf{k}} = \frac{e_1 \pi \hbar^2}{\beta p_F^2}\left[1 + \frac{\alpha}{\beta}\,\xi \log\frac{1}{|\xi|}\right].$$

The contribution from this region to the asymptotic value of the integral is

$$\varphi(r) \approx (2e_1\alpha/\pi\beta^2 r)\,\mathrm{im}(e^{2ip_F r/\hbar} J),$$

$$J = \int_{-\infty}^{\infty} \xi \log\frac{1}{|\xi|}\, e^{2ip_F r\xi/\hbar}\, d\xi;$$

because of the rapid convergence (see below), the integration with respect to ξ can be taken from $-\infty$ to ∞.

To calculate the integral J, we divide it into two parts, from $-\infty$ to 0 and from 0 to ∞, and in each part we turn the contour of integration in the complex ξ-plane until it lies along the positive imaginary axis. Then, putting $\xi = iy$, we obtain

$$J = \int_0^{\infty} e^{-2p_F r y/\hbar}\left[\log\frac{1}{-iy} - \log\frac{1}{iy}\right] y\, dy.$$

The difference in the brackets is just $i\pi$, so that $J = i\pi(\hbar/2p_F r)^2$. The final result is

$$\varphi(r) \approx \frac{e_1 \alpha \hbar^2}{2\beta^2 p_F^2}\,\frac{\cos 2p_F r/\hbar}{r^3}. \tag{40.23}$$

Thus the potential of the screened field far from the charge oscillates with an amplitude that decreases according to a power law. This result, derived for a degenerate plasma with $T = 0$, remains valid for low temperatures at distances $r \ll \hbar v_F/T$.

PROBLEM

Determine the spectrum of electron oscillations of a degenerate plasma at $T = 0$ in the quasi-classical range of values of k.

SOLUTION. The function $\omega(k)$ is given by $\epsilon_l(\omega, k) = 0$, with ϵ_l from (40.17). For small k ($kv_F \ll \Omega_e$), it is found that $kv_F/\omega \ll 1$: expanding $\epsilon_l(\omega, k)$ in powers of this ratio, we find

$$\omega = \Omega_e\left[1 + \frac{3}{10}\left(\frac{kv_F}{\Omega_e}\right)^2\right] \tag{1}$$

(A. A. Vlasov 1938).† This part of the spectrum corresponds to ordinary plasma oscillations; cf. (32.5).

For large k ($kv_F \gg \Omega_e$, but still $\hbar k \ll p_F$), we find that $\omega \approx kv_F$. Solving the equation $\epsilon_l = 0$ by successive approximation, we find

$$\omega = kv_F[1 + 2\exp(-2k^2 v_F^2/3\Omega_e^2 - 2)] \tag{2}$$

†The condition $\hbar\Omega_e \ll \epsilon_F$ for the frequency Ω_e to be quasi-classical in a degenerate plasma is the same as the condition (40.4) for an ideal plasma.

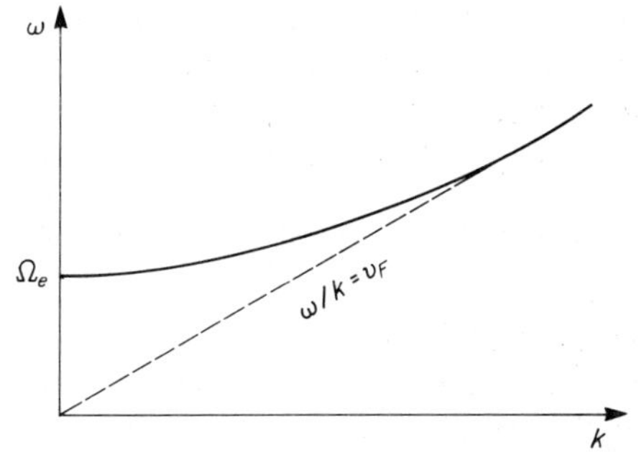

FIG. 12.

(I. I. Gol'dman, 1947). This part of the spectrum is analogous to zero sound in an uncharged Fermi gas; cf. *SP* 2 (4.16).

Figure 12 shows diagrammatically the form of the spectrum. We have everywhere $\omega/k > v_F$; since $T = 0$, there are no particles with $v > v_F$, and the Landau damping is exactly zero.

CHAPTER IV

COLLISIONS IN PLASMAS

§41. The Landau Collision Integral

THE study of the properties of plasmas, with collisions between particles taken into account, must start with the derivation of the transport equation for the electron and ion distribution functions.

The specific feature of this case is the slow decrease of the Coulomb interaction forces between charged particles. If the Boltzmann collision integral is used as it stands, the result is that the integrals diverge for large distances between the colliding particles. This means that distant encounters are important. At large distances, the particles are deflected with only a slight change in their momenta. The collision integral can consequently be put in a form similar to that which it has in the Fokker–Planck equation. In contrast to this, however, the collision integral is no longer linear in the required distribution functions. But the relative smallness of the changes of momentum in collisions certainly means that the process described by the collision integral may be treated as diffusion in momentum space. Accordingly, it may be written as

$$C(f) = -\operatorname{div}_{\mathbf{p}} \mathbf{s} \equiv -\partial s_\alpha / \partial p_\alpha,$$

where $\mathbf{s}$ is the particle flux in momentum space. The problem is to express this flux in terms of the distribution functions.

We write as

$$w f(\mathbf{p}) f'(\mathbf{p}')\, d^3q\, d^3p'$$

the number of collisions occurring per unit time between a particle with momentum $\mathbf{p}$ and particles with momentum $\mathbf{p}'$ in the range d^3p'; in a collision, $\mathbf{p}$ and $\mathbf{p}'$ become respectively $\mathbf{p}+\mathbf{q}$ and $\mathbf{p}'-\mathbf{q}$. The conservation of momentum in collisions has been taken into account here. The arguments t and $\mathbf{r}$ of the distribution functions are omitted, for brevity. The particles $\mathbf{p}$ and $\mathbf{p}'$ may be of the same or different types (electrons, ions). We shall take w to be expressed as a function of half the sum of the momenta of each particle before and after the collision, and of the momentum transfer $\mathbf{q}$:

$$w(\mathbf{p}+\tfrac{1}{2}\mathbf{q}, \mathbf{p}'-\tfrac{1}{2}\mathbf{q}; \mathbf{q});$$

it depends, of course, also on the types of particle involved. By virtue of the principle of detailed balancing (2.8), the function w is symmetrical with regard to the interchange of the initial and final particles:

$$w(\mathbf{p}+\tfrac{1}{2}\mathbf{q}, \mathbf{p}'-\tfrac{1}{2}\mathbf{q}; \mathbf{q}) = w(\mathbf{p}+\tfrac{1}{2}\mathbf{q}, \mathbf{p}'-\tfrac{1}{2}\mathbf{q}; -\mathbf{q}). \tag{41.1}$$

The function w contains a delta-function factor which expresses the conservation of energy in collisions (the conservation of momentum has already been taken into account).

Let us consider a unit area at some point $\mathbf{p}$ in momentum space (of particles of a given kind), perpendicular to the p_α-axis. By definition, the flux component s_α is the excess of particles (of that kind) crossing this area from left to right per unit time over those crossing it from right to left. The movement in momentum space is due to collisions. If a particle receives in a collision an α-component of momentum equal to q_α (>0), the result of such collisions will be for those particles to cross the area from left to right whose values of this component before the collision were from $p_\alpha - q_\alpha$ to p_α. Hence the total number of particles crossing the area from left to right is

$$\sum \int_{q_\alpha>0} d^3q \int d^3p' \int_{p_\alpha-q_\alpha}^{p_\alpha} w(\mathbf{p}+\tfrac{1}{2}\mathbf{q}, \mathbf{p}'-\tfrac{1}{2}\mathbf{q}; \mathbf{q}) f(\mathbf{p}) f'(\mathbf{p}')\, dp_\alpha.$$

The summation is taken over all types of particle to which the primed quantities refer (including, of course, the given type to which the unprimed quantities refer). Similarly, the number of particles crossing that area from right to left may be written

$$\sum \int_{q_\alpha>0} d^3q \int d^3p' \int_{p_\alpha-q_\alpha}^{p_\alpha} w(\mathbf{p}+\tfrac{1}{2}\mathbf{q}, \mathbf{p}'-\tfrac{1}{2}\mathbf{q}; -\mathbf{q}) f(\mathbf{p}+\mathbf{q}) f'(\mathbf{p}'-\mathbf{q})\, dp_\alpha.$$

Because of (41.1), the functions w in the two integrals are the same. The difference of these integrals therefore has in the integrand the difference

$$f(\mathbf{p})f'(\mathbf{p}') - f(\mathbf{p}+\mathbf{q})f'(\mathbf{p}'-\mathbf{q}').$$

We now use the fact that the momentum transfer $\mathbf{q}$ is small (more precisely, that the values of $\mathbf{q}$ important in the integrals are small in comparison with $\mathbf{p}$ and $\mathbf{p}'$). Expanding the above difference in powers of $\mathbf{q}$, we find as far as the first-order terms

$$\left[-\frac{\partial f(\mathbf{p})}{\partial p_\beta} f'(\mathbf{p}') + f(\mathbf{p}) \frac{\partial f'(\mathbf{p}')}{\partial p'_\beta}\right] q_\beta.$$

We can then, to the same accuracy, put in the integrands

$$w(\mathbf{p}+\tfrac{1}{2}\mathbf{q}, \mathbf{p}'-\tfrac{1}{2}\mathbf{q}; \mathbf{q}) \approx w(\mathbf{p}, \mathbf{p}'; \mathbf{q}).$$

The integration over dp_α, which covers the short range from $p_\alpha - q_\alpha$ to p_α, may be replaced by a simple multiplication by the length q_α of this range. The result is

$$s_\alpha = \sum \int_{q_\alpha>0} d^3q \int w(\mathbf{p}, \mathbf{p}'; \mathbf{q}) \left[f(\mathbf{p}) \frac{\partial f'(\mathbf{p}')}{\partial p'_\beta} - f'(\mathbf{p}') \frac{\partial f(\mathbf{p})}{\partial p_\beta}\right] q_\alpha q_\beta\, d^3p'. \tag{41.2}$$

Because of (41.1), $w(\mathbf{p}, \mathbf{p}'; \mathbf{q})$ is an even function of $\mathbf{q}$, and therefore so is the whole integrand in (41.2). This enables us to replace the integral over the half-space $q_\alpha > 0$ by half the integral over the whole of $\mathbf{q}$-space.

In rewriting (41.2) we also express w in terms of the collision cross-section by

$$w\,d^3q = |\mathbf{v} - \mathbf{v}'|\,d\sigma.$$

As already explained in connection with the form (3.9) for the collision integral, we can then suppose that the number of independent integrations has already been decreased by using the law of conservation of energy. Thus the momentum flux in the momentum space of particles of each type is

$$s_\alpha = \sum \int \left[f(\mathbf{p}) \frac{\partial f'(\mathbf{p}')}{\partial p'_\beta} - f'(\mathbf{p}') \frac{\partial f(\mathbf{p})}{\partial p_\beta} \right] B_{\alpha\beta}\,d^3p', \tag{41.3}$$

where

$$B_{\alpha\beta} = \tfrac{1}{2} \int q_\alpha q_\beta |\mathbf{v} - \mathbf{v}'|\,d\sigma. \tag{41.4}$$

It remains to calculate the quantities $B_{\alpha\beta}$ for collisions of particles with Coulomb interaction.

For a deviation through a small angle, the change $\mathbf{q}$ in the momentum of the colliding particles is perpendicular to their relative velocity $\mathbf{v} - \mathbf{v}'$. The tensor $B_{\alpha\beta}$ also is therefore transverse to this vector:

$$B_{\alpha\beta}(v_\beta - v'_\beta) = 0. \tag{41.5}$$

It may be noted at once that this automatically ensures the vanishing of the fluxes (41.3) for an equilibrium distribution of all the particles. With Maxwellian distributions f and f' (having the same temperature T), the integrand in (41.3) becomes

$$(ff'/T)(v'_\beta - v_\beta)B_{\alpha\beta} = 0.$$

The vector $\mathbf{v} - \mathbf{v}'$ is also the only vector on which the tensor $B_{\alpha\beta}$ can depend. Such a tensor, transverse to $\mathbf{v} - \mathbf{v}'$, must have the form

$$B_{\alpha\beta} = \tfrac{1}{2} B \left[\delta_{\alpha\beta} - \frac{(v_\alpha - v'_\alpha)(v_\beta - v'_\beta)}{(\mathbf{v} - \mathbf{v}')^2} \right],$$

where the scalar

$$B = B_{\alpha\alpha} = \tfrac{1}{2} \int q^2 |\mathbf{v} - \mathbf{v}'|\,d\sigma.$$

Let χ be the angle of deviation of the relative velocity (in the centre-of-mass system of the two particles). For small values of this angle, the momentum change has magnitude $q \approx \mu|\mathbf{v}-\mathbf{v}'|\chi$, where μ is the reduced mass of the particles. Hence

$$B = \tfrac{1}{2}\mu^2|\mathbf{v}-\mathbf{v}'|^3 \int \chi^2\, d\sigma = \mu^2|\mathbf{v}-\mathbf{v}'|^3\sigma_t,$$

where

$$\sigma_t = \int (1-\cos\chi)\, d\sigma \approx \tfrac{1}{2}\int \chi^2\, d\sigma$$

is the transport cross-section. The differential cross-section for small-angle scattering in a Coulomb field is given by the Rutherford formula

$$d\sigma \approx \frac{4(ee')^2\, do}{\mu^2(\mathbf{v}-\mathbf{v}')^4\chi^4} \approx \frac{8\pi(ee')^2}{\mu^2(\mathbf{v}-\mathbf{v}')^4}\frac{d\chi}{\chi^3}, \tag{41.6}$$

where e and e' are the charges on the colliding particles. Hence the transport cross-section is

$$\sigma_t = \frac{4\pi(ee')^2}{\mu^2(\mathbf{v}-\mathbf{v}')^4}L, \quad L = \int d\chi/\chi. \tag{41.7}$$

The value of $B_{\alpha\beta}$ is consequently

$$B_{\alpha\beta} = \frac{2\pi(ee')^2}{|\mathbf{v}-\mathbf{v}'|}L\left[\delta_{\alpha\beta} - \frac{(v_\alpha - v'_\alpha)(v_\beta - v'_\beta)}{(\mathbf{v}-\mathbf{v}')^2}\right]. \tag{41.8}$$

The integral L is logarithmically divergent. The divergence at the lower limit has a physical cause—the slowness of the decrease of the Coulomb forces, which leads to a high probability of small-angle scattering. In reality, however, in an electrically neutral plasma the Coulomb field of a particle at sufficiently large distances is screened by other charges; let $\chi_{\min}$ denote the order of magnitude of the smallest angles for which the scattering can still be regarded as Coulomb scattering. The divergence at the upper limit arises simply because all formulae have been written on the assumption of small angles, and cease to be valid when $\chi \sim 1$. Since a logarithm of a large argument is fairly insensitive to small changes in that argument, we can take the limits of integration from estimates of their orders of magnitude, writing

$$L = \log(1/\chi_{\min}). \tag{41.9}$$

This quantity is called the *Coulomb logarithm*. It must be stressed at once that such a method of determining it restricts the whole discussion to what is called *logarithmic accuracy*, which neglects quantities small compared not only with the large quantity $1/\chi_{\min}$ but also with its logarithm.

A practical estimate of $\chi_{\min}$ depends on whether the scattering of particles is to be described in classical or quantum-mechanical terms; the expression (41.8) itself is valid in either case, since purely Coulomb scattering is described by the Rutherford formula in both classical and quantum mechanics.†

†In the quantum case, the exchange effect has to be taken into account in the scattering of like particles (electrons). This effect does not, however, alter the limiting form (41.6) of the cross-section at small angles.

The Coulomb field of a particle in a plasma is screened at distances of the order of the Debye length a. In the classical case, $\chi_{\min}$ is defined as the scattering angle for a passage at impact parameter $\sim a$. The corresponding momentum change $q \sim |ee'|/a\bar{v}_{\text{rel}}$, the product of the force $\sim |ee'|/a^2$ and the transit time $\sim a/\bar{v}_{\text{rel}}$.† Dividing q by the momentum $\sim \mu\bar{v}_{\text{rel}}$, we have $\chi_{\min} \sim |ee'|/a\mu\bar{v}_{\text{rel}}^2$. The condition for classical scattering is $|ee'|/\hbar\bar{v}_{\text{rel}} \gg 1$; see *QM*, § 127. Thus we have

$$L = \log(a\mu\bar{v}_{\text{rel}}^2/|ee'|) \quad \text{with} \quad |ee'|/\hbar\bar{v}_{\text{rel}} \gg 1. \tag{41.10}$$

In the opposite limiting case $|ee'|/\hbar\bar{v}_{\text{rel}} \ll 1$, the scattering must be treated quantum-mechanically, using the Born approximation. The scattering cross-section in this case is expressed in terms of the Fourier component of the scattering potential with wave vector $\mathbf{q}/\hbar$. The contribution to this component from the screening charge "cloud" (with dimensions $\sim a$) becomes small when $qa/\hbar \gtrsim 1$; this is then the condition for purely Coulomb scattering. The angle $\chi_{\min}$ is therefore found from the condition

$$q_{\min}a/\hbar \sim \mu\bar{v}\chi_{\min}a/\hbar \sim 1.$$

In this case, then,

$$L = \log(\mu a\bar{v}_{\text{rel}}/\hbar) \quad \text{with} \quad |ee'|/\hbar\bar{v}_{\text{rel}} \ll 1. \tag{41.11}$$

When $|ee'| \sim \hbar\bar{v}_{\text{rel}}$, the expressions (41.10) and (41.11) coincide, of course.

Let us now write the final expression for the fluxes in momentum space by substituting (41.8) in (41.3):

$$s_\alpha = \sum 2\pi(ee')^2 L \int \left(f\frac{\partial f'}{\partial p'_\beta} - f'\frac{\partial f}{\partial p_\beta}\right)\frac{(\mathbf{v}-\mathbf{v}')^2\delta_{\alpha\beta} - (v_\alpha - v'_\alpha)(v_\beta - v'_\beta)}{|\mathbf{v}-\mathbf{v}'|^3}\, d^3p'. \tag{41.12}$$

The corresponding transport equations are

$$\frac{\partial f}{\partial t} + \mathbf{v}\cdot\frac{\partial f}{\partial \mathbf{r}} + e(\mathbf{E} + \mathbf{v}\times\mathbf{B}/c)\cdot\frac{\partial f}{\partial \mathbf{p}} = -\operatorname{div}_{\mathbf{p}}\mathbf{s}, \tag{41.13}$$

where e is the charge on the particles to which f relates, i.e. $-e$ for electrons and ze for ions. The collision integral in the logarithmic approximation for a gas with Coulomb interaction between the particles was determined by L. D. Landau (1936).

The validity of the Landau collision integral depends on the fulfilment of certain conditions. The characteristic lengths $1/k$ over which the distribution function varies significantly must be large compared with the screening radius a, and the characteristic times $1/\omega$ must be large compared with $a/\bar{v}_{\text{rel}}$; in the logarithmic approximation, however, it is sufficient in practice if these conditions are satisfied

†Here, and in all analogous places below, $\bar{v}_{\text{rel}}$ is the mean relative velocity $|\mathbf{v}-\mathbf{v}'|$ of the two particles. If they are of the same type it is the mean value $\bar{v}$; if they are of different types, it is the larger of $\bar{v}$ and $\bar{v}'$.

in the weak form

$$ka < 1, \quad \omega < \bar{v}_{\text{rel}}/a,$$

with $<$ in place of $\ll$.

PROBLEM

It has been shown in §34 that, after the perturbations of the electron density with wave vector **k** have been removed by Landau damping, the perturbations of the distribution function continue to oscillate as $e^{-i\mathbf{k}.\mathbf{v}t}$ (34.16). Find the damping of these oscillations by Coulomb collisions at times $t \gg 1/k\bar{v}$.

SOLUTION. We seek the distribution function in the form

$$f = f_0 + \delta f, \quad \delta f = a(t, \mathbf{v})e^{-i\mathbf{k}.\mathbf{v}t + i\mathbf{k}.\mathbf{r}}, \tag{1}$$

where δf is the perturbation of the equilibrium distribution f_0, and a is a slowly varying function of the velocity (changing appreciably only over ranges $\sim \bar{v} \gg 1/kt$). On substituting (1) in (41.12), we need retain in the integrand only the term

$$-f_0(\mathbf{p}')\partial\delta f(\mathbf{p})/\partial\mathbf{p} \approx (i/m)\mathbf{k}t\delta f(\mathbf{p})f_0(\mathbf{p}');$$

the remaining terms make only a small contribution, either because the integral is made small by the rapidly oscillating factor $\exp(-i\mathbf{k}.\mathbf{v}'t)$ or because they do not contain the factor $kt \gg 1/\bar{v}$. For the latter reason, too, only the exponential factor need be differentiated in the calculation of $\text{div}_{\mathbf{p}}\mathbf{s}$. The transport equation then gives

$$\partial a/\partial t = -k_\alpha k_\beta b_{\alpha\beta}t^2 a,$$

where in order of magnitude the coefficients $b_{\alpha\beta} \sim \overline{v^2}\nu$, with ν the collision frequency. Hence

$$a(t, \mathbf{v}) = a_0(\mathbf{v}) \exp\{-\tfrac{1}{3}k_\alpha k_\beta b_{\alpha\beta}t^3\}, \tag{2}$$

and so the damping time of the oscillations is

$$\tau_d \sim \nu^{-1/3}(k\bar{v})^{-2/3}.$$

Since the whole theory of Landau damping is valid only if $k\bar{v} \gg \nu$, we have $\tau_d \ll 1/\nu$. The result (2) is correct if the exponent in it is small compared with the exponent kvt in (1); for this to be so, we must have $t \ll (\nu k\bar{v})^{-1/2}$. In this time, the oscillations are damped by a factor $\exp[-\sqrt{(k\bar{v}/\nu)}]$.

§42. Energy transfer between electrons and ions

The large difference between the electron mass m and the ion mass M impedes the transfer of energy between electrons and ions: when a heavy and a light particle collide, the energy of each is almost unchanged. The establishment of equilibrium among the electrons alone and the ions alone therefore takes place considerably more quickly than that between electrons and ions. This may easily give rise to a situation where the electron and ion components of the plasma each have a Maxwellian distribution but with different temperatures T_e and T_i, of which T_e is usually the greater.

The difference between the electron and ion temperatures causes an energy transfer between the two components of the plasma, which may be determined as follows (L. D. Landau 1936).

We shall temporarily denote by a prime the quantities pertaining to electrons, leaving unmarked those pertaining to ions. The change in the ion energy per unit volume and unit time is given by the integral

$$dE/dt = \int \epsilon C(f)\, d^3p = -\int \epsilon\, \text{div}_{\mathbf{p}}\mathbf{s}\, d^3p,$$

or, integrating by parts,

$$dE/dt = \int \mathbf{s}\,.\,(\partial\epsilon/\partial\mathbf{p})\, d^3p = \int \mathbf{s}\,.\,\mathbf{v}\, d^3p; \tag{42.1}$$

the integral over an infinite surface in momentum space is, as usual, zero.

In the sums (41.3) which determine the electron and ion fluxes in momentum space, the only terms remaining are those which correspond to electron–ion collisions; the electron–electron and ion–ion terms are zero for Maxwellian distributions. Substituting in these remaining terms the Maxwellian distributions with temperatures T' and T, we obtain for the ion flux

$$s_\alpha = \int ff'\left(\frac{v_\beta}{T} - \frac{v'_\beta}{T'}\right)B_{\alpha\beta}\, d^3p'.$$

From (41.5), $B_{\alpha\beta}v'_\beta = B_{\alpha\beta}v_\beta$; making this change and substituting the flux in (42.1), we find

$$\frac{dE}{dt} = \left(\frac{1}{T} - \frac{1}{T'}\right)\int\int ff'v_\alpha v_\beta B_{\alpha\beta}\, d^3p\, d^3p'. \tag{42.2}$$

Since the electrons have small masses, their mean speeds are large compared with those of the ions. We can therefore put $v'_\alpha - v_\alpha \approx v'_\alpha$ in $B_{\alpha\beta}$. Then the quantities $B_{\alpha\beta}$ no longer depend on the v_α, and in (42.2) we can carry out the integration over d^3p:

$$\int fv_\alpha v_\beta\, d^3p = \tfrac{1}{3}\delta_{\alpha\beta}N\overline{v^2} = \delta_{\alpha\beta}NT/M.$$

Thus

$$\frac{dE}{dt} = \frac{NT}{M}\left(\frac{1}{T} - \frac{1}{T'}\right)\int f'B\, d^3p'. \tag{42.3}$$

Finally, substituting here from (41.8) $B = 4\pi e^4z^2L/v'$ (where ze is the ion charge), and noting that for a Maxwellian distribution

$$\int f'\, d^3p'/v' = N'\surd(2m/\pi T'),$$

we obtain

$$\frac{dE}{dt} = \frac{4NN'z^2e^4\surd(2\pi m)L}{MT'^{3/2}}(T' - T). \tag{42.4}$$

The same expression, with the opposite sign, gives the decrease in the energy of the electron component of the plasma, $-dE'/dt$. Expressing the electron energy per unit volume in terms of the electron temperature by $E' = 3N'T'/2$ and resuming the use of the suffixes e and i for electron and ion quantities, we can write the following final expression for the rate of change of the electron temperature:

$$\frac{dT_e}{dt} = -\frac{T_e - T_i}{\tau_{ei}^{\epsilon}}, \quad \tau_{ei}^{\epsilon} = \frac{T_e^{3/2} M}{8N_i z^2 e^4 L_e (2\pi m)^{1/2}}. \tag{42.5}$$

The Coulomb logarithm here is

$$L_e = \begin{cases} \log(aT_e/ze^2) & \text{for } ze^2/\hbar v_{Te} \gg 1, \\ \log[\sqrt{(mT_e)}a/\hbar] & \text{for } ze^2/\hbar v_{Te} \ll 1. \end{cases} \tag{42.6}$$

The quantity τ_{ei}^{ϵ} is the relaxation time for the establishment of electron–ion equilibrium.

§43. Mean Free Path of Plasma Particles

We have seen from the derivation in §41 that the transport cross-section σ_t (41.7) serves as a collision parameter in the transport equation. It must therefore also occur in the definition of the mean free path.

For electron–electron (ee) and electron–ion (ei) collisions, the reduced mass $\mu \sim m$, and, since the speeds of the electrons are much greater than those of the ions, we have

$$\mu(\mathbf{v}_e - \mathbf{v}_i)^2 \sim m v_{Te}^2 \sim T_e.$$

Hence follows the estimate of the electron mean free path

$$l_e \sim T_e^2/4\pi e^4 N L_e, \tag{43.1}$$

with L_e from (42.6). The factors z are omitted from the estimates, it being assumed that $z_i \sim 1$. The electron mean free time τ_e (or its reciprocal, the collision frequency ν_e) is

$$\tau_e \sim 1/\nu_e \sim l_e/v_{Te} \sim T_e^{3/2} m^{1/2}/4\pi e^4 N L_e. \tag{43.2}$$

Note that

$$l_e/a_e \sim (1/L_e)(T_e/N^{1/3}e^2)^{3/2},$$

and $l_e \gg a_e$ by virtue of the condition (27.1) for the plasma to be rarefied. Accordingly, the collision frequency is small in comparison with the electron plasma frequency:

$$\nu_e \ll v_{Te}/a_e = \Omega_e. \tag{43.3}$$

Similarly, the ion mean free path with regard to ion–ion collisions is

$$l_i \sim T_i^2/4\pi e^4 N L_i, \quad L_i = \log(aT_i/e^2), \tag{43.4}$$

where L_i is the Coulomb logarithm with ion quantities in place of electron ones. The corresponding mean free time is

$$\tau_{ii} \sim 1/\nu_{ii} \sim T_i^{3/2} M^{1/2}/4\pi e^4 N L_i. \tag{43.5}$$

The quantity τ_e determines the order of magnitude of the relaxation time for the establishment of local thermal equilibrium of the electron component of the plasma, and τ_{ii} the corresponding time for the ion component. Since the frequencies ν_{ee} and ν_{ei} for *ee* and *ei* collisions are of the same order, τ_e is not the relaxation time for the establishment of equilibrium between the electrons and the ions; it describes only the rate of momentum transfer from the electrons to the ions, not the rate of energy exchange between them. The relaxation time for electron–ion equilibrium is given by the quantity τ_{ei}^{ϵ} determined in §42. A comparison of these various times shows that

$$\tau_{ee} : \tau_{ii} : \tau_{ei}^{\epsilon} \sim 1 : (M/m)^{1/2} : (M/m). \tag{43.6}$$

The mean free path can be used to estimate the transport coefficients of the plasma. To estimate the electrical conductivity σ, we use the familiar elementary "kinetic theory of gases" formula. The particles (carriers) with charge e and mass m, in free motion during a time τ, acquire from the electric field E an "ordered" speed $V \sim \tau eE/m$. The electric current density created by this motion is $j \sim eNV$. The conductivity, which is the proportionality factor between j and E, is therefore

$$\sigma \sim e^2 N\tau/m \sim e^2 Nl/mv_T, \tag{43.7}$$

where l, m and v_T are to be taken as the quantities pertaining to the lighter particles, i.e. the electrons. An estimate using this formula gives

$$\sigma \sim T_e^{3/2}/e^2 m^{1/2} L_e. \tag{43.8}$$

The thermal conductivity is estimated similarly with the elementary formula (7.10). Electrons play the main part, and we have $\kappa \sim N_e l_e v_{Te} c_e$ (where $c_e \sim 1$ is the electronic specific heat), whence

$$\kappa \sim T_e^{5/2}/e^4 m^{1/2} L_e. \tag{43.9}$$

The viscosity of the plasma, unlike the electrical and thermal conductivities, is due mainly to the movement of the ions, since most of the momentum of the plasma is concentrated in the ion component. Moreover, the momentum of an ion is not much changed by collisions with electrons, and for this reason it is sufficient to consider *ii* collisions only. From (8.11), the viscosity is estimated as $\eta \sim N_i M l_i v_{Ti}$, whence

$$\eta \sim M^{1/2} T_i^{5/2}/e^4 L_i. \tag{43.10}$$

The calculation of the coefficients for the expressions (43.8)–(43.10) requires the solution of the linearized transport equation with the Landau collision integral, which can be done only by approximate numerical methods. As an example, in a hydrogen plasma ($z = 1$) the coefficients in the expressions for σ, κ and η are respectively 0.6, 0.9 and 0.4.

§44. Lorentzian plasmas

In calculating the electron contribution to the transport coefficients in a plasma, it is in general necessary to take account of both *ei* and *ee* collisions. However, if the ion charge is sufficiently large, *ei* collisions may have a predominant effect: the *ee* collision cross-section is proportional to $(e^2)^2$ and the frequency ν_{ee} of such collisions is proportional also to the electron density N_e; similarly, the frequency of *ei* collisions is proportional to $(ze^2)^2N_i = e^4 z_i N_e$, so that $\nu_{ei} \gg \nu_{ee}$ if $z \gg 1$. A plasma in which *ee* collisions may be neglected in comparison with *ei* collisions is called a *Lorentzian plasma*. Although this is not a very realistic case, it is interesting both methodologically and for possible application to other systems.†

Since the ion speeds are small compared with the electron speeds, they may be neglected in the first approximation, i.e. the ions may be regarded as at rest with a given distribution. In the problem of the behaviour of the plasma in an external electric field there is a preferred direction, that of the field **E**. If the electron distribution function differs only slightly from the equilibrium form, $f = f_0(p) + \delta f$, the small correction δf is linear in the field, i.e. has the form $\delta f = \mathbf{p} \cdot \mathbf{E} g(p)$. Under these conditions, the electron–ion collision integral has the same form as that of the collision integral in § 11 for the problem of diffusion of a light gas in a heavy gas:

$$C(f) = -\nu_{ei}(v)\delta f, \tag{44.1}$$

with the speed-dependent *effective collision frequency*

$$\nu_{ei}(v) = N_i v \sigma_t^{(ei)}, \tag{44.2}$$

and $\sigma_t^{(ei)}$ the transport cross-section for scattering of electrons by ions. With this quantity given by (41.7), and $zN_i = N_e$, we find

$$\nu_{ei}(v) = 4\pi z e^4 N_e L / m^2 v^3. \tag{44.3}$$

In the rest of this section we shall write $\nu(v)$ simply, omitting the suffixes *ei*.

Let us now calculate the permittivity of a Lorentzian plasma in a spatially uniform (wave vector $\mathbf{k} = 0$) but periodically varying ($\propto e^{-i\omega t}$) electric field. The correction δf to the equilibrium distribution depends on the time in the same manner, and the transport equation for it is

$$-i\omega\delta f - e\mathbf{E} \,.\, \partial f_0/\partial \mathbf{p} + \nu(v)\delta f = 0. \tag{44.4}$$

†For example, a weakly ionized gas, where *ei* collisions are replaced by collisions between electrons and neutral atoms.

Noting also that $\partial f_0/\partial \mathbf{p} = -\mathbf{v}f_0/T$, we therefore have

$$\delta f = -(e/T)\mathbf{E}\cdot\mathbf{v}f_0/[\nu(v) - i\omega]. \tag{44.5}$$

The permittivity is found by means of the relation (29.4): $-i\omega\mathbf{P} = \mathbf{j}$, or

$$-i\omega(\epsilon - 1)\mathbf{E}/4\pi = -e\int \mathbf{v}\delta f\, d^3p. \tag{44.6}$$

Substitution of (44.5) and averaging over the directions of $\mathbf{v}$ (with $\langle v_\alpha v_\beta\rangle = \frac{1}{3}\delta_{\alpha\beta}v^2$), we obtain

$$\epsilon(\omega) = 1 - \frac{4\pi e^2}{3\omega T}\int\frac{v^2 f_0\, d^3p}{\omega + i\nu(v)}. \tag{44.7}$$

In the limit $\omega \gg \nu$,† this gives

$$\epsilon(\omega) = 1 - \frac{4\pi e^2 N_e}{m\omega^2} + i\frac{4\pi e^2 N_e}{3\omega^3 T}\langle v^2\nu(v)\rangle, \tag{44.8}$$

where the averaging is over the Maxwellian electron distribution. Calculating the mean value for $\nu(v)$ as in (44.3), we obtain

$$\epsilon(\omega) = 1 - \frac{\Omega_e^2}{\omega^2} + i\frac{4\sqrt{(2\pi)}}{3}\frac{ze^4 L N_e}{T^{3/2}m^{1/2}}\frac{\Omega_e^2}{\omega^3} \quad (\omega \gg \nu). \tag{44.9}$$

However, the range of validity of this formula also has an upper limit given by the general condition (41.4) for the logarithmic approximation in the collision integral to be applicable, $\omega \ll v_{Te}/a_e = \Omega_e$ (the frequency must be small in comparison with the electron plasma frequency).‡

Formula (44.9) has a special significance, since it is valid for any (not only large) values of z. When $\omega \gg \nu$, the collisions cause only small corrections, and so the *ei* and *ee* collisions may be treated independently. In the absence of ions, a uniform electric field would cause only a shift of the electrons as a whole, and collisions in such a system could not cause dissipation (represented by the imaginary part ϵ'' of the permittivity); under such conditions, the dissipation is due only to the *ei* collisions taken into account in (44.9).

In the opposite limiting case, when $\omega \ll \nu$, the permittivity is

$$\epsilon = i\,.\,4\pi\sigma/\omega, \quad \sigma = (e^2N_e/3T)\langle v^2/\nu(v)\rangle. \tag{44.10}$$

The quantity σ in this limiting expression is the static conductivity of the plasma; see *ECM*, § 58. The calculation with $\nu(v)$ from (44.3) gives

$$\sigma = \frac{4\sqrt{2}}{\pi^{3/2}}\frac{T^{3/2}}{ze^2 L m^{1/2}}. \tag{44.11}$$

†The symbol ν (without argument) means the value of $\nu(v)$ for $v = v_T$. In the present case, $\nu = 4\pi z e^4 N_e L/m^{1/2}T_e^{3/2}$.

‡The calculation of ϵ'' for $\omega \gg \Omega_e$ is discussed in § 48.

The same result could, of course, also be reached by a direct calculation of the electric current density

$$\mathbf{j} = -e\int \mathbf{v}\delta f\, d^3p,$$

with δf from (44.5) (with $\omega = 0$).

We shall also calculate the other transport coefficients for a Lorentzian plasma, which are related to its behaviour in a constant ($\omega = 0$) electric field and a temperature gradient. Let us first recall the definition of these coefficients (see *ECM*, § 25).

The conditions of thermal equilibrium require, as we know, not only the constancy of temperature but also that of $\mu + U$ throughout the medium, where μ is the chemical potential of the particles and U their energy in the external field. In the present case, we are concerned with equilibrium with respect to the electrons, so that μ is to be taken as their chemical potential, and $U = -e\varphi$, where φ is the electric field potential. Accordingly, the electric current $\mathbf{j}$ and the dissipative energy flux $\mathbf{q}'$ are simultaneously zero only if $T = \text{constant}$ and $\mu - e\varphi = \text{constant}$, i.e. $\nabla T = 0$, $\nabla\mu + e\mathbf{E} = 0$. The expressions for $\mathbf{j}$ and $\mathbf{q}'$ are written as follows, satisfying the condition stated:

$$\mathbf{E} + \frac{1}{e}\nabla\mu = \frac{1}{\sigma}\mathbf{j} + \alpha\nabla T, \tag{44.12}$$

$$\mathbf{q}' = \mathbf{q} - (\varphi - \mu/e)\mathbf{j} = \alpha T\mathbf{j} - \kappa\nabla T. \tag{44.13}$$

Here σ is the electrical conductivity of the medium, κ the thermal conductivity, α the thermoelectric coefficient; the relation between the coefficients of ∇T in (44.12) and $\mathbf{j}$ in (44.13) follows from Onsager's principle. The quantity $(\varphi - \mu/e)\mathbf{j}$ which is subtracted from the total energy flux is the convective energy flux.†

To calculate the transport coefficients, we start from the transport equation

$$-e\mathbf{E}\cdot\frac{\partial f_0}{\partial \mathbf{p}} + \mathbf{v}\cdot\frac{\partial f_0}{\partial \mathbf{r}} = -\nu(v)\delta f. \tag{44.14}$$

Substituting the equilibrium distribution in the form‡

$$f_0 = \exp[(\mu - \epsilon)/T], \tag{44.15}$$

we find

$$\delta f = -\frac{f_0}{T\nu(v)}(e\mathbf{E} + \nabla\mu)\,.\,\mathbf{v} + f_0\frac{\mu - \epsilon}{T^2\nu(v)}\mathbf{v}\,.\,\nabla T. \tag{44.16}$$

The thermoelectric coefficient is calculated from the coefficient in the equation

†The relations (44.12) and (44.13) are written with a different notation in *ECM*, § 25, where φ and $\mathbf{E}$ stood for $\varphi - \mu/e$ and $\mathbf{E} + \nabla\mu/e$. That definition is permissible in the phenomenological approach, but not appropriate in the kinetic theory, where $-e\mathbf{E}$ must be the force acting on the electron.

‡The use of the same letter ϵ to denote both the permittivity and the electron energy $\frac{1}{2}mv^2$ is unlikely to cause any misunderstanding.

$\mathbf{j} = -\alpha\sigma\nabla T$ with $\mathbf{E} + \nabla\mu/e = 0$. We write

$$\mathbf{j} = -e\int \mathbf{v}\delta f\, d^3p$$
$$= -\frac{e}{T^2}\int f_0 \frac{\mu - \epsilon}{\nu(v)} \mathbf{v}(\mathbf{v}\,.\,\nabla T)\, d^3p,$$

and find after the averaging over directions

$$\alpha = \frac{N_e e}{3\sigma T^2}\left\langle \frac{v^2(\mu - \epsilon)}{\nu(v)} \right\rangle = \frac{1}{eT}\left\{\mu - \frac{\langle v^2\epsilon/\nu(v)\rangle}{\langle v^2/\nu(v)\rangle}\right\}. \tag{44.17}$$

Calculation with $\nu(v)$ from (44.3) gives†

$$\alpha = \frac{1}{e}\left(\frac{\mu}{T} - 4\right). \tag{44.18}$$

To calculate the thermal conductivity we note that for $\mathbf{j} = 0$ we must have $\mathbf{E} + \nabla\mu/e = \alpha\nabla T$. Substituting this value in (44.16), with α from (44.18), gives

$$\delta f = \frac{f}{T\nu(v)}\left(4 - \frac{\epsilon}{T}\right)\mathbf{v}\,.\,\nabla T.$$

With this function, the calculation of the energy flux

$$\mathbf{q} = \int \mathbf{v}\epsilon\delta f\, d^3p$$

leads to

$$\kappa = \frac{N_e}{3T^2}\left\langle \frac{v^2\epsilon(4T - \epsilon)}{\nu(v)} \right\rangle \tag{44.19}$$

and finally

$$\kappa = \frac{16\sqrt{2}}{\pi^{3/2}} \frac{T^{5/2}}{ze^4Lm^{1/2}}. \tag{44.20}$$

PROBLEM

Find the collisional part of the damping of electron plasma waves.

SOLUTION. If the imaginary part of the permittivity is small, the contributions to it from Landau damping and from collisions are additive. Then ϵ is given by (44.9), and equating it to zero gives $\omega = \Omega_e - i\gamma$, where the damping ratio is

$$\gamma = \frac{\nu_{ei}}{3\sqrt{(2\pi)}} = \frac{2\sqrt{(2\pi)}}{3}\frac{ze^4LN_e}{m^{1/2}T_e^{3/2}}.$$

†In classical statistical physics, the chemical potential contains a term of the form ζT with an indefinite constant ζ (corresponding to the indefinite additive constant in the entropy). This in turn gives an indefinite constant ζ/e in α. The indefiniteness does not, however, influence any observable effects, since such terms $(\zeta/e)\nabla T$ cancel on the two sides of (44.12). If f_0 is written in the form (44.15), this fixes the choice of the constant ζ: $\mu = T\log[N_e/(2\pi mT)^{3/2}]$.

The ratio

$$\gamma/\Omega_e = (\sqrt{2}zL/3)(e^2N_e^{1/3}/T_e)^{3/2} \ll 1$$

by virtue of the condition for the plasma to be rarefied, and this justifies the use of (44.9).

§45. Runaway electrons

The rapid decrease of the Coulomb cross-section with increasing speed of the colliding particles has the result, as we shall see, that in an electric field, however weak, the distribution function of sufficiently fast electrons in a plasma is highly distorted.

An electron moving at the thermal speed v in an electric field E acquires during its mean free time a directed speed

$$V \sim eEl/mv \sim eE/mvN_e\sigma_t(v) \sim v^3mE/4\pi e^3LN_e,$$

with the cross-section σ_t from (41.7). For $v \sim v_c$, where

$$v_c = (4\pi N_e e^3L/mE)^{1/2}, \tag{45.1}$$

we have $V \sim v$, and for $v > v_c$ the mean free path and time are governed by the speed V. The momentum acquired by the electron in the mean free time is then

$$eEl/V \sim eE/VN_e\sigma_t(V) \sim V^3m^2E/4\pi e^3LN_e \sim mV(V/v_c)^2.$$

The momentum transferred by the electron at the end of its free path is $\sim mV$. Hence we see that electrons with sufficiently high speeds will be accelerated without limit; these are called *runaway electrons*. If $v_c \gg (T_e/m)^{1/2}$, the phenomenon will be observed only in the tail of the Maxwellian distribution; the electric field must then satisfy the condition

$$E \ll E_c = 4\pi e^3LN_e/T_e. \tag{45.2}$$

Under these circumstances the problem of runaway electrons may be solved as a stationary one. The majority of the electrons, whose distribution is Maxwellian, act as a large reservoir from which a small steady flux moves towards high energies.†

It is evident, from the fact that the runaway electrons result from directed acceleration by the electric field, that they move chiefly at small angles θ to the direction of the field. If, however, we seek to calculate only the flux of runaway electrons, their distribution function need not be determined completely; it is sufficient to find the energy distribution $\bar{f}$ averaged over angles.

The transport equation for the electron momentum distribution in the electric field is

$$\frac{\partial f}{\partial t} - e\mathbf{E}\cdot\frac{\partial f}{\partial \mathbf{p}} + \operatorname{div}_{\mathbf{p}}\mathbf{s} = 0, \tag{45.3}$$

†The phenomenon of runaway electrons was pointed out by H. Dreicer (1958); the quantitative theory given here is due to A. V. Gurevich (1960).

where $\mathbf{s}$ is the collisional flux in momentum space. In spherical polar coordinates p, θ, φ in momentum space, with the polar axis along the force $-e\mathbf{E}$, we have

$$-e\mathbf{E}\,.\,\partial f/\partial \mathbf{p} = eE[\cos\theta\,\partial f/\partial p - (\sin\theta/p)\partial f/\partial\theta]$$
$$= eE\left[\frac{\cos\theta}{p^2}\frac{\partial}{\partial p}(p^2 f) - \frac{1}{p\sin\theta}\frac{\partial}{\partial\theta}(f\sin^2\theta)\right].$$

The divergence of the flux is

$$\mathrm{div}_{\mathbf{p}}\mathbf{s} = \frac{1}{p^2}\frac{\partial}{\partial p}(p^2 s_p) + \frac{1}{p\sin\theta}\frac{\partial}{\partial\theta}(s_\theta\sin\theta).$$

We average (45.3) over angles, i.e. multiply by $2\pi\sin\theta\, d\theta/4\pi$ and integrate. All terms containing $\partial/\partial\theta$ disappear, and the factor $\cos\theta$ may in the first approximation be replaced by unity. We thus obtain for the averaged function $\bar{f}$ the equation

$$\frac{\partial\bar{f}}{\partial t} + \frac{eE}{p^2}\frac{\partial}{\partial p}(p^2\bar{f}) + \frac{1}{p^2}\frac{\partial}{\partial p}(p^2\overline{s_p}) = 0. \tag{45.4}$$

This contains only the radial component of the flux in momentum space, which is related to the energy transfer in collisions. The contribution to it from *ei* collisions is evidently small in comparison with that from *ee* collisions.

Since the runaway electrons are only a very small fraction of the total, in calculating the flux s_p we need consider only their collisions with the main mass of Maxwellian electrons (not with one another), whose speeds are much less than those of the runaway electrons. Under these conditions there is no need to calculate s_p afresh; we can write by direct analogy with (22.5)

$$s_p = -T_e\nu_{ee}(v)m\left[\frac{\partial f}{\partial p} + \frac{p}{mT_e}f\right], \tag{45.5}$$

where $\nu_{ee}(v) = 4\pi e^4 N_e L/m^2v^3$ is the frequency of Coulomb collisions between fast and slow electrons (cf. (44.3)).† Since the expression (45.5) refers to electrons with speeds $v \sim v_c$, we have for the Coulomb logarithm

$$L = \log(mv_c^2 a/e^2). \tag{45.6}$$

The quantity

$$\overline{S_p} = \overline{s_p} + eE\bar{f} \tag{45.7}$$

is seen from the form of (45.4) to be the total radial flux (from collisions and from the action of the field) in momentum space. According to the foregoing discussion, the distribution of runaway electrons may be sought in a stationary form, i.e. the time

†In deriving (22.5), we used only the smallness of the energy transfer in collisions, and that of the target particle speed relative to the incident electron speed. To change to the present case, we need only replace M in (22.5) by m and take as the mean free path l that for *ee* collisions.

derivative in the transport equation (45.4) may be neglected. Then

$$4\pi p^2\overline{S_p} = \text{constant} \equiv n_{\text{run}}. \tag{45.8}$$

This equation, with $\overline{s_p}$ given by (45.5), is a differential equation determining the distribution function $\bar{f}$. The constant n_{run} gives the required total number of runaway electrons per unit time and volume.

We use the dimensionless variable u and constant b defined by

$$u = p/p_c, \quad b = E/E_c, \quad p_c = (mT_e/b)^{1/2}. \tag{45.9}$$

Then equation (45.8) becomes

$$-\frac{b}{u}\frac{d\bar{f}}{du} - (1-u^2)\bar{f} = C, \tag{45.10}$$

where the constant C differs from n_{run} by a constant factor. Since we assume the field $E \ll E_c$, the parameter $b \ll 1$; in the present problem, this is the small parameter which gives the order of the approximation.†

The solution of (45.10) is

$$\bar{f} = F - CF\int_0^u (u/F)\, du, \tag{45.11}$$

where

$$F = \frac{N_e}{(2\pi m T_e)^{3/2}}\exp\left\{\frac{1}{2b}(\tfrac{1}{2}u^4 - u^2)\right\} \tag{45.12}$$

is the solution of the homogeneous equation. The normalization factor in F is determined from the condition that as $u \to 0$ the function $\bar{f}$ should become the Maxwellian distribution

$$f_0 = \frac{N_e}{(2\pi m T_e)^{3/2}}\exp(-u^2/2b).$$

As $u \to \infty$, the function F increases without limit, whereas $\bar{f}(u)$ must remain finite. Hence we have the condition $\bar{f}/F \to 0$ as $u \to \infty$, from which the constant C is found:‡

$$C = \frac{N_e}{(2\pi m T_e)^{3/2}}\left[\int_0^\infty \exp\left\{-\frac{1}{2b}(\tfrac{1}{2}u^4 - u^2)\right\} u\, du\right]^{-1}. \tag{45.13}$$

The integral is calculated by the saddle-point method, the exponent being expanded near its maximum at $u = 1$. This yields the following dependence of the number of runaway electrons per unit time and volume on the field E:

$$n_{\text{run}} \sim N_e \nu_{ee}(v_{Te}) \exp(-E_c/4E). \tag{45.14}$$

†In particular, analysis of the angular part of the transport equation shows that the directions of motion of the runaway electrons lie in the range of angles $\theta \sim b^{1/4}$.

‡The formulation of the boundary conditions here is analogous to that in § 24.

The coefficient of the exponential here is only dimensionally correct; a more accurate calculation would go beyond the approximation being used and call for a more accurate solution of the transport equation from the start.

§46. Convergent collision integrals

The transport equation with the Landau collision integral allows problems of plasma physics to be solved only with logarithmic accuracy: the large argument of the Coulomb logarithm is not fully determined. This uncertainty is due to the divergence of the integrals at large and small scattering angles. As already mentioned, the divergence at large angles has no fundamental significance: it results only from the expansion in powers of the momentum transfer $\mathbf{q}$, and does not occur in the Boltzmann collision integral itself. The divergence at small angles is due to neglecting the screening effect of the plasma on the scattering of particles by one another in it. To calculate the collision integral to a higher than logarithmic accuracy, we must consistently take account of screening throughout, and not only when determining the range of integration in the Coulomb logarithm.

It has been noted in § 41 that the conditions for the collision integral with screened interaction between charged particles to be applicable require that the distribution functions should not vary greatly in times $\sim a/\bar{v}_{rel}$ and over distances $\sim a$. These same conditions enable us to treat the screening of the charges macroscopically as the result of dielectric polarization of the plasma.

We shall consider the problem in two limiting cases: (1) when the Born approximation of quantum mechanics can be applied to particle collisions, (2) when the collision process is quasi-classical.

The Born case

The first case occurs when

$$|ee'|/\hbar\bar{v}_{rel} \ll 1. \tag{46.1}$$

The influence of the dielectric medium on particle scattering is most clearly expressed in the language of the diagram technique. In the Born approximation, the scattering of two particles is described (in the non-relativistic case) by the diagram†

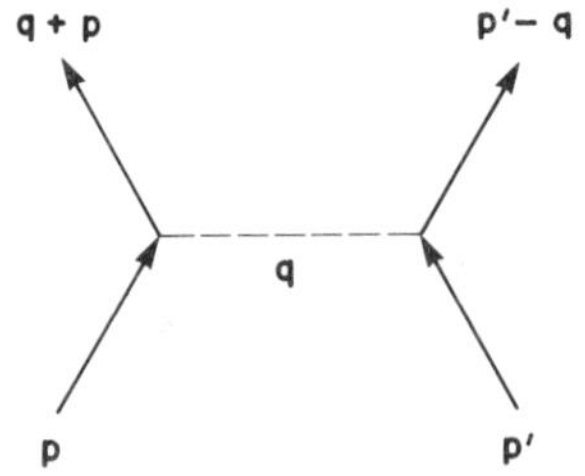

(46.2)

†As in §41, the unprimed and primed quantities refer to the two colliding particles (which may be of the same or different types).

in which the broken line corresponds to the function $4\pi/\mathbf{q}^2$, the Fourier component of the Coulomb potential of a unit charge ($\mathbf{q}$ being the momentum transferred in scattering). The only effect of the medium is to replace this function by the potential component in the medium, $4\pi/q_\alpha q_\beta \epsilon_{\alpha\beta}$, where $\epsilon_{\alpha\beta}(\omega, \mathbf{q}/\hbar)$ is the permittivity tensor of the medium, and $\hbar\omega$ is the energy transferred; cf. *SP* 2, § 85. The scattering amplitude correspondingly contains an extra factor $q^2/q_\alpha q_\beta \epsilon_{\alpha\beta}$, and the cross-section contains the squared modulus of this. Thus

$$d\sigma = d\sigma_{\mathrm{Ru}} q^4/|\epsilon_{\alpha\beta} q_\alpha q_\beta|^2. \tag{46.3}$$

For simplicity, we shall henceforward assume the plasma to be isotropic. Then the tensor $\epsilon_{\alpha\beta}$ reduces to two scalars ϵ_t and ϵ_l, and the product $\epsilon_{\alpha\beta} q_\alpha q_\beta = \epsilon_l q^2$ involves only one of these; we shall omit the suffix l, and denote by ϵ the longitudinal permittivity.

Thus the scattering cross-section becomes

$$d\sigma = d\sigma_{\mathrm{Ru}}/|\epsilon(\omega, q/\hbar)|^2, \tag{46.4}$$

where $d\sigma_{\mathrm{Ru}}$ is the ordinary Rutherford cross-section for scattering in a vacuum.† Note also that the energy transferred in a collision is related to the momentum transfer by

$$\hbar\omega = \mathbf{q} \cdot \mathbf{V}, \tag{46.5}$$

where $\mathbf{V}$ is the velocity of the centre of mass of the colliding particles.‡ The magnitude of the vector $\mathbf{q}$ is related to the scattering angle χ in the centre-of-mass system by the usual formula

$$q = 2\mu|\mathbf{v} - \mathbf{v}'| \sin \tfrac{1}{2}\chi, \tag{46.6}$$

where $\mu = mm'/(m + m')$.

The collision integral which automatically gives a correct treatment of large and small scattering angles, and is free from divergence, is obtained by substituting (46.4) in the usual Boltzmann integral (cf. (3.9)):

$$C(f) = \sum \int \{f(\mathbf{p}+\mathbf{q})f'(\mathbf{p}'-\mathbf{q}) - f(\mathbf{p})f'(\mathbf{p}')\} \frac{|\mathbf{v}-\mathbf{v}'|\, d\sigma_{\mathrm{Ru}}}{|\epsilon(\omega, q/\hbar)|^2} d^3p'; \tag{46.7}$$

the summation is over all types of particle to which the primed quantities refer.

The transport equation with the collision integral (46.7) is very complicated, not only because the integrand cannot be expanded in powers of $\mathbf{q}$, but because the permittivity of the plasma is itself defined in terms of the distribution functions sought. An important simplification is attainable only in the case of a slight departure from equilibrium, when the transport equation can be linearized. Then the permittivity is to be calculated with the equilibrium distribution functions, and so is independent of the corrections sought.

†In the scattering of identical particles (through angles that are not small), $d\sigma_{\mathrm{Ru}}$ is to be taken as the Coulomb scattering cross-section with exchange effects (see *QM*, § 137).

‡This is easily shown by expressing the velocities $\mathbf{v}$ and $\mathbf{v}'$ of the particles in terms of $\mathbf{V}$ and the velocity $\mathbf{v} - \mathbf{v}'$ of their relative motion, and using the fact that V and $|\mathbf{v} - \mathbf{v}'|$ are unchanged in scattering.

THE QUASI-CLASSICAL CASE

Let us now go to the opposite limiting case, where

$$|ee'|/\hbar \bar{v}_{\text{rel}} \gg 1 \tag{46.8}$$

and the quasi-classical approximation is applicable to the scattering of particles. In this case, we cannot take into account the effect of the medium on the scattering in the same manner for small and large scattering angles (as was possible in the Born case); it is necessary to consider these two ranges separately and then join the results at intermediate angles.

The field of a charge e moving with velocity $\mathbf{v}$ in a dielectric medium is given by the equation

$$\operatorname{div} \mathbf{D} = 4\pi e \delta(\mathbf{r} - \mathbf{v}t).$$

In Fourier components, this gives for the field potential†

$$\varphi_{\mathbf{k}} = \frac{4\pi e}{k^2 \epsilon(\mathbf{k} \cdot \mathbf{v}, k)} e^{-i\mathbf{k} \cdot \mathbf{v}t}. \tag{46.9}$$

For small scattering angles, the change in the momentum of the particle is given (see *Mechanics*, §20) by the classical formula

$$\mathbf{q} = -\int (\partial U/\partial \mathbf{r})\, dt, \tag{46.10}$$

where U is the interaction energy of the two particles, and the integration is taken along the straight path $\mathbf{r} = \boldsymbol{\rho} + \mathbf{v}'t$ ($\boldsymbol{\rho}$ being the impact parameter vector).‡ Expressing the energy $U = e'\varphi$ as a Fourier integral:

$$U = 4\pi ee' \int \frac{e^{i(\mathbf{k} \cdot \mathbf{r} - \omega t)}}{k^2 \epsilon(\omega, k)} \frac{d^3k}{(2\pi)^3}, \tag{46.11}$$

with $\omega = \mathbf{k} \cdot \mathbf{v}$, and substituting in (46.10), we obtain

$$\mathbf{q} = -4\pi i ee' \int \frac{d^3k}{(2\pi)^3} \left\{ \frac{\mathbf{k} e^{i\mathbf{k} \cdot \boldsymbol{\rho}}}{k^2 \epsilon(\omega, k)} \int_{-\infty}^{\infty} e^{-i\mathbf{k} \cdot (\mathbf{v} - \mathbf{v}')t}\, dt \right\}.$$

The inner integral gives $2\pi\delta(k_\parallel)/|\mathbf{v} - \mathbf{v}'|$, where $k_\parallel$ is the component of the vector $\mathbf{k}$ in the

†The derivation of (46.9) assumes a linear relation between $\mathbf{D}$ and $\mathbf{E}$, and therefore a sufficiently weak field. This condition is certainly satisfied (in a slightly non-ideal gas) at distances $r \gtrsim a$, from which arises the divergence that is to be eliminated by the use of formula (46.9). These distances correspond to values of $k \lesssim 1/a$, for which the permittivity is considerably different from unity.

‡It does not matter whether $\mathbf{q}$ is calculated as the change in momentum of each of the colliding particles, or as the change in momentum of their relative motion.

direction of $\mathbf{v}-\mathbf{v}'$. Then, eliminating the delta function by integration over $dk_{\parallel}$, we find

$$\mathbf{q} = -\frac{4\pi iee'}{|\mathbf{v}-\mathbf{v}'|}\int \frac{\mathbf{k}_{\perp}e^{i\mathbf{k}_{\perp}\cdot\boldsymbol{\rho}}}{k_{\perp}^{2}\epsilon(\omega, k_{\perp})}\frac{d^2k_{\perp}}{(2\pi)^2}, \tag{46.12}$$

where $\mathbf{k}_{\perp}$, like $\boldsymbol{\rho}$, is a two-dimensional vector in the plane perpendicular to $\mathbf{v}-\mathbf{v}'$. The frequency is

$$\omega = \mathbf{k}_{\perp}\,.\,\mathbf{v} = \mathbf{k}_{\perp}\,.\,\mathbf{V}. \tag{46.13}$$

In the rest of this section we shall omit the suffix $\perp$ and denote by $\mathbf{k}$ this two-dimensional vector.

We now calculate from (46.12) the quantities

$$B_{\alpha\beta} = \tfrac{1}{2}\int q_{\alpha}q_{\beta}|\mathbf{v}-\mathbf{v}'|\,d^2\rho, \tag{46.14}$$

which appear in the collision integral when it is expanded in powers of the small quantity $\mathbf{q}$; the cross-section $d\sigma$ in (41.4) is here written as the impact area $d^2\rho$. Writing the product of two integrals (46.12) as a double integral over $d^2k\,d^2k'$, we carry out the integration over $d^2\rho$ by means of the formula

$$\int e^{i\boldsymbol{\rho}\,.\,(\mathbf{k}+\mathbf{k}')}\,d^2\rho = (2\pi)^2\delta(\mathbf{k}+\mathbf{k}').$$

The integration over d^2k' then simply removes the delta function, leaving

$$B_{\alpha\beta} = \frac{2e^2e'^2}{|\mathbf{v}-\mathbf{v}'|}\int \frac{k_{\alpha}k_{\beta}\,d^2k}{k^4|\epsilon(\mathbf{k}\,.\,\mathbf{V}, k)|^2}, \tag{46.15}$$

where we have also used the fact that $\epsilon(-\omega, k) = \epsilon^*(\omega, k)$ by (28.9). These integrals converge for small k, since $|\epsilon|^{-2}\to 0$ as $\omega, k\to 0$.†

Equation (46.15) involves the permittivity at the non-zero frequency $\omega = \mathbf{k}\,.\,\mathbf{V}$; it is therefore sometimes said to take account of *dynamic screening*. The integrand in (46.15) depends on the direction of $\mathbf{V}$ through the argument $\mathbf{k}\cdot\mathbf{V}$ of the function ϵ. This dependence disappears when the integral is calculated in the logarithmic approximation, in which the integration is limited to the range from $k\sim 1/a$ to $k\sim \mu\overline{v^2}/|ee'|$. The most important values of k in the integral are those far from both these limits; in that range, $|\epsilon|^2 = 1$, and the integral reduces to $\int k_{\alpha}k_{\beta}\,d^2k/k^4$. Averaging the integrand over all directions of $\mathbf{k}$ in the plane perpendicular to $\mathbf{v}-\mathbf{v}'$, we return to the previous expression (41.8) with $L = \int dk/k$.

To eliminate the divergence at large momentum transfers we must, as already mentioned, join the collision integral expanded in powers of $\mathbf{q}$ to the unexpanded integral (J. Hubbard 1961, O. Aono 1962).

†The elimination of the divergence due to Coulomb field screening in the Landau collision integral is due to R. Balescu (1960) and A. Lenard (1960). The completely convergent expression (46.7) was given by A. A. Rukhadze and V. P. Silin (1961).

Let us consider the difference

$$C_{cl}(f) - C_B(f), \tag{46.16}$$

where C_{cl} is the required convergent collision integral, and C_B is given by (46.7), which in the Born case is the correct collision integral, but here has only an auxiliary role.

We divide the range of variation of the scattering angle into two parts: (I) $\chi < \chi_1$, (II) $\chi > \chi_1$, where χ_1 is chosen so that

$$|ee'|/\mu a\bar{v}_{rel}^2 \ll \chi_1 \ll 1. \tag{46.17}$$

In classical scattering through small angles in a Coulomb field, the scattering angle χ is related to the impact parameter ρ by

$$\rho = 2|ee'|/\mu(\mathbf{v}-\mathbf{v}')^2\chi.$$

Hence the value $\chi = \chi_1$ corresponds, with the condition (46.17), to $\rho = \rho_1 \ll a$, so that the screening at this distance is unimportant and the scattering may in fact be regarded as purely Coulomb scattering. The same applies to the whole of the range $\rho < \rho_1$, i.e. $\chi > \chi_1$. The scattering cross-section in this range consequently has the Rutherford form, and the corresponding contribution to the collision integral is

$$C_{cl}^{II}(f) = \sum \int_{\chi>\chi_1} [f(\mathbf{p}+\mathbf{q})f'(\mathbf{p}'-\mathbf{q}) - f(\mathbf{p})f'(\mathbf{p}')]|\mathbf{v}-\mathbf{v}'|\, d\sigma_{Ru}.$$

The contribution from the range $\chi > \chi_1$ to the integral (46.7) is exactly similar: in that range, $q > q_1$, and by the condition (46.8)

$$q_1/\hbar \sim \mu\bar{v}_{rel}\chi_1/\hbar \gg |ee'|/\hbar\bar{v}_{rel}a \gg 1/a,$$

so that we may put $|\epsilon|^2 = 1$ in (46.7). Thus a contribution to the difference (46.16) arises only from the range $\chi < \chi_1$ ($\rho > \rho_1$), which remains to be considered.

Throughout this range, the momentum transfer is small, and so the collision integral may be expanded in powers of $\mathbf{q}$. The quantities $B_{\alpha\beta}$ which appear in the expanded C_{cl} are calculated as the integrals (46.14) with $\mathbf{q}$ from (46.12). The contribution to these integrals from the range $\rho > \rho_1$ is

$$(B_{\alpha\beta})_{cl}^{I} = \frac{(ee')^2}{2\pi^2|\mathbf{v}-\mathbf{v}'|} F_{\alpha\beta},$$

$$F_{\alpha\beta} = \int_{\rho_1}^{\infty} d^2\rho \left(\int_0^{\infty} \frac{ik_\alpha e^{i\mathbf{k}\cdot\boldsymbol{\rho}}}{k^2\epsilon}\, d^2k \int_0^{\infty} \frac{ik_\beta e^{i\mathbf{k}\cdot\boldsymbol{\rho}}}{k^2\epsilon}\, d^2k \right), \tag{46.18}$$

where the limits in the double integrals (over $d^2\rho$ and d^2k) are conventionally shown by the limits for ρ and k. We can rewrite the $F_{\alpha\beta}$ identically as

$$F_{\alpha\beta} = \int_0^\infty d^2\rho \left(\int_0^{q_1/\hbar} \ldots d^2k\right)_\alpha \left(\int_0^{q_1/\hbar} \ldots d^2k\right)_\beta$$
$$- \int_0^{\rho_1} d^2\rho \left(\int_0^{q_1/\hbar} \ldots d^2k\right)_\alpha \left(\int_0^{q_1/\hbar} \ldots d^2k\right)_\beta$$
$$+ \int_{\rho_1}^\infty d^2\rho \left(\int_0^\infty \ldots d^2k\right)_\alpha \left(\int_{q_1/\hbar}^\infty \ldots d^2k\right)_\beta$$
$$+ \int_{\rho_1}^\infty d^2\rho \left(\int_{q_1/\hbar}^\infty \ldots d^2k\right)_\alpha \left(\int_0^{q_1/\hbar} \ldots d^2k\right)_\beta. \qquad (46.19)$$

The first term here, when transformed as in the derivation of (46.15), gives a contribution to (46.18)

$$\frac{2(ee')^2}{|\mathbf{v}-\mathbf{v}'|} \int_0^{q_1/\hbar} \frac{k_\alpha k_\beta}{k^4|\epsilon|^2} d^2k.$$

This expression is the same as would be obtained by an expansion of the integral (46.7) taken over the range $\chi < \chi_1$;† it therefore makes no contribution to the difference (46.16).

To transform the remaining terms in (46.19), we note that we may put $\epsilon = 1$ in their integrands: the integrals then remain convergent, and their values are determined by the range $k \sim q_1/\hbar$, in which $ka \gg 1$ and therefore $|\epsilon| \approx 1$. It is also important that, by virtue of the condition (46.8),

$$q_1\rho_1/\hbar = 2|ee'|/\hbar v_{\text{rel}} \gg 1; \qquad (46.20)$$

we need therefore retain only the terms that remain finite as $q_1\rho_1/\hbar \to 0$. In this limit, the third and fourth terms in (46.19) vanish. Thus there remains only

$$(B_{\alpha\beta})^{\text{I}}_{\text{cl}} - (B_{\alpha\beta})^{\text{I}}_B = -\frac{(ee')^2}{2\pi^2|\mathbf{v}-\mathbf{v}'|} \int_0^{\rho_1} d^2\rho \left(\int_0^{q_1/\hbar} ik_\alpha e^{i\mathbf{k}\cdot\boldsymbol{\rho}} \frac{d^2k}{k^2} \int_0^{q_1/\hbar} ik_\beta e^{i\mathbf{k}\cdot\boldsymbol{\rho}} \frac{d^2k}{k^2}\right), \qquad (46.21)$$

where the suffixes cl and B signify that the values of $B_{\alpha\beta}$ relate to the expansions of the integrals C_{cl} and C_B respectively.

Each of the two integrals over d^2k is parallel to the vector $\boldsymbol{\rho}$; after integration over these directions (in the plane perpendicular to $\mathbf{v}-\mathbf{v}'$), we obtain for the difference (46.21) the expression (41.8) with the opposite sign and

$$L = \int_0^{\rho_1} \rho\, d\rho \left[\frac{i}{2\pi} \int_0^{q_1/\hbar} \int_0^{2\pi} \cos\varphi \,.\, e^{ik\rho\cos\varphi}\, d\varphi\, dk\right]^2.$$

†The Rutherford cross-section for scattering through small angles, expressed in terms of $\mathbf{q}$, is

$$d\sigma_{\text{Ru}} = \frac{4(ee')^2}{q^4|\mathbf{v}-\mathbf{v}'|^2} d^2q,$$

if we use the formulae $q \approx \mu|\mathbf{v}-\mathbf{v}'|\chi$, $do \approx d^2q/\mu^2(\mathbf{v}-\mathbf{v}')^2$.

Using the familiar integral representation of the Bessel functions, and the equation $J_0'(x) = -J_1(x)$, we can rewrite this integral as

$$L = \int_0^{\rho_1} \rho\, d\rho \left[\int_0^{q_1/\hbar} J_1(k\rho)\, dk \right]^2$$

$$= \int_0^{\rho_1 q_1/\hbar} [J_0(x) - 1]^2 \frac{dx}{x},$$

or, integrating by parts,

$$L = \log(q_1\rho_1/\hbar) + 2 \int_0^{\infty} J_1(x)[J_0(x) - 1] \log x\, dx.$$

Here we have used the fact that the parameter $\rho_1 q_1/\hbar$ (which does not involve the auxiliary quantity χ_1) is large; accordingly, the upper limit is replaced by infinity in the remaining integral, and in the first term we put $J_0(q_1\rho_1/\hbar) \approx 0$. With the values

$$\int_0^{\infty} J_1(x) \log x\, dx = -C + \log 2,$$

$$\int_0^{\infty} J_0(x) J_1(x) \log x\, dx = \tfrac{1}{2}(\log 2 - C),$$

where $C = 0.577\ldots$ is Euler's constant ($\gamma = e^C = 1.78\ldots$), and with (46.20), we have finally

$$L = \log \frac{\gamma |ee'|}{\hbar |\mathbf{v} - \mathbf{v}'|}. \tag{46.22}$$

The total result of these calculations is that in the quasi-classical case the collision integral without divergence may be expressed as

$$C_{cl}(f) = C_B(f) - C_L(f), \tag{46.23}$$

where C_B is given by (46.7), and C_L is the Landau collision integral with the Coulomb logarithm (46.22). It must be emphasized that, in the latter, $|\mathbf{v} - \mathbf{v}'|$ is the exact variable, not the mean value $\bar{v}_{rel}$.

Because of the approximations made in the derivation, this result is of course valid only with "improved logarithmic" accuracy: the transport equation with the collision integral (46.23) enables us to improve the accuracy of the calculations only as regards determining the exact coefficient in the argument of the large logarithm. To this accuracy, the quantity $\hbar$ naturally drops out of all the results; in (46.23) it acts only as an auxiliary parameter.

PROBLEMS

PROBLEM 1. In the Born case with improved logarithmic accuracy, calculate the imaginary part of the permittivity of a singly charged ($z = 1$) equilibrium ($T_i = T_e$) plasma for frequencies $\omega \gg \nu_{ei}$.

SOLUTION. In calculating ϵ'' when $\omega \gg \nu$, we need consider only *ei* collisions, as explained in connection with the derivation of (44.8). Since the collision integral (46.7) differs from the usual Boltzmann integral only by the factor $|\epsilon|^{-2}$ in front of $d\sigma_{\mathrm{Ru}}$, the required ϵ'' can be calculated by means of the same formula (44.8):

$$\epsilon''(\omega) = \frac{4\pi e^2 N_i N_e}{3T\omega^3}\langle v_e^3 \langle \sigma_t \rangle_i \rangle_e, \tag{1}$$

where $\langle \ldots \rangle_e$ and $\langle \ldots \rangle_i$ denote averaging over the equilibrium distribution of electron velocities $\mathbf{v}_e$ and ion velocities $\mathbf{v}_i$ respectively. The only difference from the calculations in §44 is that σ_t is now defined as

$$\sigma_t = \int (1-\cos\chi)|\epsilon(\mathbf{q}\,.\,\mathbf{v}_i/\hbar,\, q/\hbar)|^{-2}\, d\sigma_{\mathrm{Ru}}, \tag{2}$$

and that σ_t has to be averaged over the ion velocities (which of course are not negligible here); in the argument $\omega = \mathbf{q}\,.\,\mathbf{V}/\hbar$ of the function ϵ, the velocity of the centre of mass of the electron and the ion has been replaced, as an approximation, by the velocity of the ion. The Rutherford cross-section is written as

$$d\sigma_{\mathrm{Ru}} = \frac{(ze^2)^2 m^2}{4p_e^4}\cdot\frac{2\pi \sin\chi\, d\chi}{\sin^4 \frac{1}{2}\chi} = \frac{8\pi(ze^2)^2 m^2}{p_e^2 q^3}\, dq, \tag{3}$$

where

$$q = 2p_e \sin\tfrac{1}{2}\chi, \quad 1-\cos\chi = q^2/2p_e^2, \quad 0 \le q \le 2p_e,$$

and $\mathbf{p}_e = m\mathbf{v}_e$ is the electron momentum.

The function $\epsilon(\omega, q/\hbar) - 1$ is determined by (31.11), and consists of electronic and ionic parts. Since its argument in (2) $\hbar\omega = \mathbf{q}\,.\,\mathbf{v}_i \ll qv_e$, the electronic part may be taken with $\omega = 0$. Then

$$\epsilon(\mathbf{q}\,.\,\mathbf{v}_i/\hbar,\, q/\hbar) - 1 = (\hbar^2/q^2 a_e^2)\{2 + F(v_{iq}/\sqrt{2}v_{Ti})\}, \tag{4}$$

where v_{iq} is the component of $\mathbf{v}_i$ along q, and we have used the fact that $a_i = a_e$ if $z = 1$.

Substitution of (3) and (4) in (2) and an obvious change of variables gives

$$\langle \sigma_t \rangle_i = \frac{2\sqrt{\pi} e^4}{p_e^4}\int_0^{4p_e^2 a_e^2/\hbar^2}\int_0^\infty \frac{\zeta e^{-\xi^2}\, d\xi\, d\zeta}{[\zeta + 2 + F'(\xi)]^2 + [F''(\xi)]^2},$$

where $F = F' + iF''$. The integration over $d\zeta$ is elementary; in substituting the limits, we must note that $\hbar^2/p_e^2 a_e^2 \ll 1$, and reject all terms $\sim \hbar^2/p_e^2 a_e^2$ and those of higher orders. The result is

$$\langle \sigma_t \rangle_i = (4\pi e^4/m^2 v_e^4)[\log(2mv_e a_e/\hbar + A)], \tag{5}$$

where

$$A = \frac{1}{\sqrt{\pi}}\int_0^\infty e^{-\xi^2}\left\{\frac{2+F'}{F''}\left[\tan^{-1}\frac{2+F'}{F''} - \tfrac{1}{2}\pi\right] - \tfrac{1}{2}\log[(2+F')^2 + F''^2]\right\} d\xi.$$

Here we have used the fact that F' and F'' are respectively even and odd functions. A numerical calculation gives $A = -0.69$.

The averaging in (1) is carried out by means of the formulae

$$\langle v^{-1} \rangle = (2m/\pi T)^{1/2},$$

$$\left\langle \frac{\log v}{v} \right\rangle = \left(\frac{m}{2\pi T}\right)^{1/2}\left[\log\frac{2T}{m} - C\right],$$

C being Euler's constant. The final result is

$$\epsilon'' = \frac{4\sqrt{(2\pi)}}{3}\frac{e^4 N_e}{T^{3/2} m^{1/2}}\frac{\Omega_e^2}{\omega^3} L_B, \quad L_B = \log\frac{\alpha_B (mT)^{1/2} a_e}{\hbar},$$

$$\log \alpha_B = \tfrac{3}{2}\log 2 - \tfrac{1}{2}C + A = \log 1.06 \tag{6}$$

(V. I. Perel' and G. M. Éliashberg 1961).

PROBLEM 2. The same as Problem 1, but in the quasi-classical case.

SOLUTION. From (46.23) the expression for σ_t in the quasi-classical case is found by subtracting $\log(\gamma e^2/\hbar v)$ from the logarithm in (5):

$$\langle\sigma_t\rangle_i = \frac{2\pi e^4}{m^2 v_e^4}\left[\log\frac{2mv_e^2 a_e}{\gamma e^2} + A\right]. \tag{7}$$

For ϵ'', we get formula (6), with the logarithm L_B replaced by

$$L_{cl} = \log(Ta_e\alpha_{cl}/e^2), \quad \log\alpha_{cl} = 2\log 2 - 2C + A = \log 0.63. \tag{8}$$

PROBLEM 3. Determine with improved logarithmic accuracy the rate of energy transfer from electrons to ions in a singly charged ($z = 1$) plasma, assuming the temperature difference between the electrons and the ions to be small ($\delta T = T_e - T_i \ll T_e$).†

SOLUTION. Since the ratio m/M is small (and therefore so is the energy transfer per event), it is clear from the start that the equation for the electron distribution function reduces to one of the Fokker–Planck type. It is (see §21)

$$\frac{\partial f_e}{\partial t} = \frac{1}{p_e^2}\frac{\partial}{\partial p_e}\left\{p_e^2 B(p_e)\left[\frac{\partial f_e}{\partial p_e} + \frac{v_e}{T_i}f_e\right]\right\}.$$

We multiply by $p_e^2/2m$ and integrate over $4\pi_e^2\, dp_e$. After integrating by parts, we find as the rate of change of the electron energy

$$\frac{dE_e}{dt} = -\int Bv_e\left[\frac{\partial f_e}{\partial p_e} + \frac{v_e}{T_i}f_e\right]d^3p_e.$$

Assuming the electron distribution function Maxwellian, and the temperature difference small, we have

$$\frac{dE_e}{dt} = \left(\frac{1}{T_e} - \frac{1}{T_i}\right)\int Bv_e^2 f_e\, d^3p_e$$
$$\approx -(\delta T/T^2)N_e\langle Bv_e^2\rangle_e. \tag{9}$$

The coefficient B, as in (21.11), is expreseed in terms of the mean square of the change in the electron momentum in a collision with an ion:

$$B = \frac{\Sigma(\Delta p_e)^2}{2\delta t} = \tfrac{1}{2}N_i v_e \int \langle(\Delta p_e)^2\rangle_i\, d\sigma. \tag{10}$$

The value of Δp_e is given by (46.5):

$$\Delta p_e \approx -\mathbf{V}\,.\,\mathbf{q}/v_e \approx -\mathbf{v}_i\,.\,\mathbf{q}/v_e \equiv v_{iq}q/v_e.$$

Substituting in (10) and thence in (9), and using the relation between q and the scattering angle χ from Problem 1, we get

$$dE_e/dt = -(\delta T/T^2)Nm^2\langle v_e^3\langle v_{iq}^2\sigma_t\rangle_i\rangle_e \quad (N_i = N_e \equiv N). \tag{11}$$

Formula (11) is exactly analogous to (1) in Problem 1, and the subsequent calculations are therefore practically identical. In the Born case,

$$\langle v_{iq}^2\sigma_t\rangle_i = \frac{4\pi e^4 T}{m^2 M v_e^4}\left[\log\frac{2mv_e a_e}{\hbar} + A_1\right],$$

where A_1 is an integral which differs from A in Problem 1 by having an additional factor $2\xi^2$ in the integrand; a numerical calculation gives $A_1 = -0.52$. The averaging over electron velocities is as in Problem 1. The final result is

$$\frac{dE_e}{dt} = -\frac{4\sqrt{(2\pi m)}N^2 e^2}{MT^{3/2}}L_B\,\delta T, \quad L_B = \log(\beta_B(mT)^{1/2}a_e/\hbar), \tag{12}$$

†This problem was discussed by R. R. Ramazashvili, A. A. Rukhadze and V. P. Silin (1962).

where

$$\log \beta_B = \tfrac{3}{2}\log 2 - \tfrac{1}{2}C + A_1 = \log 1.26.$$

Similarly, in the quasi-classical case, we obtain an expression in the same form (12) but with L_B replaced by

$$L_{cl} = \log(Ta_e\beta_{cl}/e^2), \quad \log \beta_{cl} = 2\log 2 - 2C + A_1 = \log 0.75. \tag{13}$$

Formulae (12) and (13) refine the results of §42 by determining (for the case of a small temperature difference) the numerical factor in the argument of the logarithm in (42.6).

§47. Interaction via plasma waves

In some cases, the inclusion of dynamic screening of the Coulomb interaction of particles in a plasma not only refines the argument of the Coulomb logarithm, but also leads to qualitatively new effects. To investigate these, we can put the collision integral in a form that gives the exact contribution from small-angle scattering, but only with logarithmic accuracy the contribution from large-angle scattering.

In the quasi-classical case, the large scattering angles ($\chi \sim 1$) arise from small impact parameters:

$$\rho \lesssim |ee'|/\mu\bar{v}_{\text{rel}}^2.$$

The required collision integral has the Landau form, with the $B_{\alpha\beta}$ from (46.15):

$$B_{\alpha\beta} = \frac{2(ee')^2}{|\mathbf{v}-\mathbf{v}'|}\int \frac{k_\alpha k_\beta\, d^2k}{k^4|\epsilon(\mathbf{k}\,.\,\mathbf{V}, k)|^2}, \tag{47.1}$$

where the integration is taken over the range up to

$$k_{\max} \sim \mu\bar{v}_{\text{rel}}^2/|ee'|. \tag{47.2}$$

In the opposite (Born) case, the required form of the collision integral is found by expanding the integrand in (46.7) in powers of $\mathbf{q}$. The result is again a Landau integral, with the $B_{\alpha\beta}$ given by the same formula (47.1), except that

$$k_{\max} \sim \mu\bar{v}_{\text{rel}}/\hbar; \tag{47.3}$$

the value of k is $q/\hbar$ for a momentum transfer $q \sim \mu\bar{v}_{\text{rel}}$. Let us mention once more that the physical significance of the cut-off at large k is the same in the classical and Born cases: it takes place at scattering angles $\chi \sim 1$. The different relations between k and χ in the two cases lead, however, to different expressions for $k_{\max}$.

The Landau collision integral with the $B_{\alpha\beta}$ from (47.1) is called the *Balescu–Lenard integral*.† We can rewrite (47.1) in a form more convenient for the subsequent analysis:

$$B_{\alpha\beta} = 2(ee')^2\int_{-\infty}^{\infty}\int_{k\le k_{\max}} \delta(\omega-\mathbf{k}\,.\,\mathbf{v})\delta(\omega-\mathbf{k}\,.\,\mathbf{v}')\frac{k_\alpha k_\beta\, d^3k\, d\omega}{k^4|\epsilon(\omega, k)|^2}, \tag{47.4}$$

†The formal derivation of this integral will be given at the end of §51.

where the integration is now over three-dimensional (instead of two-dimensional) vectors $\mathbf{k}$. The two delta functions in the integrand ensure the equality $\mathbf{k}\cdot\mathbf{v} = \mathbf{k}\cdot\mathbf{v}'$, i.e. that $\mathbf{k}$ is transverse to $\mathbf{v}-\mathbf{v}'$. The integration over ω replaces the argument ω in $\epsilon(\omega, k)$ by the necessary value $\omega = \mathbf{k}\cdot\mathbf{v} = \mathbf{k}\cdot\mathbf{v}' = \mathbf{k}\cdot\mathbf{V}$.

Note that the factor $|\epsilon(\omega, k)|^{-2}$ in the integrand in (47.4) becomes infinite for values of $\omega = \mathbf{k}\cdot\mathbf{V}$ and $\mathbf{k}$ such that $\epsilon(\omega, k) = 0$, i.e. for values corresponding to the dispersion relation for longitudinal plasma waves. These values of $\mathbf{k}$ may make a large contribution to the collision integral. This contribution may be physically described as the result of interaction between particles by their emission and absorption of plasma waves. The effect will be considerable, however, only if the plasma contains sufficiently many particles whose speeds are comparable with or greater than the phase velocity $v_{\text{ph}} = \omega/k$ of the waves, since only such particles can satisfy the necessary relation $\omega = \mathbf{k}\cdot\mathbf{V}$.

Let us consider a plasma in which the electrons and ions have different temperatures T_e and T_i. When $T_e \approx T_i$, only electron plasma waves with phase velocity $v_{\text{ph}} \gg v_{Te}$ can propagate in the plasma without appreciable damping; the number of electrons that can "exchange" waves in this case is therefore exponentially small.

If $T_e \gg T_i$, however, ion-sound waves also are able to propagate in the plasma; their phase velocity satisfies the inequalities

$$v_{Ti} \ll \omega/k \ll v_{Te}. \tag{47.5}$$

These waves can make an important contribution to the collision integral between electrons (V. P. Silin 1962).

Let $B^{(\text{pl})}_{\alpha\beta}$ denote the part due to this effect in the electron–electron quantities $B^{(ee)}_{\alpha\beta}$. It arises from the range of integration in (47.4) that lies near the root of $\epsilon(\omega, k) = 0$ corresponding to the dispersion relation for ion-sound waves. This root $\omega(k)$ has a small imaginary part (the damping ratio of the wave); when ω has real values in the range of integration, the real part of the function $\epsilon = \epsilon' + i\epsilon''$ passes through zero, while the imaginary part remains small. Using formula (30.9), we write the factor $|\epsilon|^{-2}$ in the integrand in (47.4) as

$$\frac{1}{|\epsilon|^2} = \frac{1}{\epsilon'^2 + \epsilon''^2} = \frac{\pi}{|\epsilon''|}\delta(\epsilon').$$

For the electron–electron collision integral, the velocities $\mathbf{v}$ and $\mathbf{v}'$ in (47.4) relate to electrons, and because of the inequality $\omega \ll kv_{Te}$ the terms ω may be omitted from the arguments of the two delta functions. Thus the relevant part of $B^{(ee)}_{\alpha\beta}$ is

$$B^{(\text{pl})}_{\alpha\beta} = 2\pi e^4 \int_{-\infty}^{\infty}\int \delta(\mathbf{k}\cdot\mathbf{v})\delta(\mathbf{k}\cdot\mathbf{v}')\delta(\epsilon')\frac{k_\alpha k_\beta\, d^3k\, d\omega}{k^4|\epsilon''(\omega, k)|}, \tag{47.6}$$

the integration over d^3k being over the range (47.5) for a given ω.

We can transform the integral over d^3k to new variables

$$\kappa = \mathbf{k}\cdot\mathbf{n}, \quad k_1 = \mathbf{k}\cdot\mathbf{v}, \quad k_2 = \mathbf{k}\cdot\mathbf{v}',$$

where $\mathbf{n}$ is a unit vector along $\mathbf{v}\times\mathbf{v}'$. Direct calculation of the Jacobian of the transformation shows that d^3k is replaced by $d\kappa\, dk_1\, dk_2/|\mathbf{v}\times\mathbf{v}'|$. The integration over $dk_1\, dk_2$ removes the delta functions (which make $k_1 = k_2 = 0$), and then $\mathbf{k} = \kappa\mathbf{n}$. The variable κ takes both positive and negative values; with integration over positive values only, we write

$$B_{\alpha\beta}^{(\mathrm{pl})} = \frac{2\pi e^4 n_\alpha n_\beta}{|\mathbf{v}\times\mathbf{v}'|} 2 \int\int_{-\infty}^{\infty} \frac{\delta[\epsilon'(\omega,\kappa)]}{\kappa^2|\epsilon''(\omega,\kappa)|}\, d\omega\, d\kappa. \tag{47.7}$$

The permittivity of a two-temperature plasma in the ion-sound wave region (47.5) is given by†

$$\left.\begin{aligned} \epsilon' &= 1 - \Omega_i^2/\omega^2 + 1/k^2a_e^2, \\ \epsilon'' &= \sqrt{\frac{\pi}{2}}\frac{\omega}{k^3}\left\{\frac{\Omega_e^2}{v_{Te}^3} + \frac{\Omega_i^2}{v_{Ti}^3}\exp(-\omega^2/2k^2v_{Ti}^2)\right\}. \end{aligned}\right\} \tag{47.8}$$

The main contribution to the integral over $d\kappa$ in (47.7) comes (as will be confirmed by the subsequent calculation) from the range $a_e\kappa \gg 1$; the last term in $\epsilon'(\omega,\kappa)$ is therefore negligible. Since

$$\delta(1 - \Omega_i^2/\omega^2) = \tfrac{1}{2}\Omega_i[\delta(\omega - \Omega_i) + \delta(\omega + \Omega_i)],$$

the integration over $d\omega$ in (47.7) gives

$$B_{\alpha\beta}^{(\mathrm{pl})} = n_\alpha n_\beta \frac{4\pi e^4\Omega_i}{|\mathbf{v}\times\mathbf{v}'|}\int \frac{d\kappa}{\kappa^2\epsilon''(\Omega_i,\kappa)},$$

or, substituting the expression for ϵ'' and using the variable $\xi = \kappa^2 a_i^2$,

$$B_{\alpha\beta}^{(\mathrm{pl})} = n_\alpha n_\beta \frac{2\sqrt{(2\pi)}e^4 v_{Te}a_e^2}{|\mathbf{v}\times\mathbf{v}'|a_i^2}\int \frac{d\xi}{1 + \exp(-1/2\xi + \frac{1}{2}L_1)}, \tag{47.9}$$

where

$$L_1 = \log(\Omega_i^4 v_{Te}^6/\Omega_e^4 v_{Ti}^6) = \log(z^2MT_e^3/mT_i^3). \tag{47.10}$$

Because of the conditions (47.5), the integration in (47.9) must be taken over the range $(\Omega_i a_i/\Omega_e a_e)^2 \ll \xi \ll 1$. Since the integral converges for small ξ, the lower limit may be taken as zero.

As $L_1 \to \infty$, the integral in (47.9) tends to zero; assuming that L_1 is fairly large, we shall calculate it in the logarithmic approximation, i.e. take only the first term in the expansion in powers of $1/L_1$. The main contribution to the integral comes from the range where the exponential term in the denominator is negligible. For this, we must have $-1/2\xi + \frac{1}{2}L_1 > 1$, i.e. the integral is to be taken from 0 to $1/(L_1 - 1) \approx 1/L_1$,

†See (33.3). The ionic contribution to ϵ'' is also included in (47.8). Although it is exponentially small in the region (47.5), it determines the range of integration in (47.9) below.

which gives simply $1/L_1$.† The final result is therefore

$$B^{(\mathrm{pl})}_{\alpha\beta} = n_\alpha n_\beta \frac{2\sqrt{(2\pi)}e^4 z v_{Te} T_e}{|\mathbf{v}\times\mathbf{v}'|T_i L_1}. \tag{47.11}$$

The total value of the $B^{(ee)}_{\alpha\beta}$ in the electron–electron collision integral is found by adding (47.11) to the ordinary Coulomb expression (41.8), with the Debye length

$$a = (a_e^{-2} + a_i^{-2})^{-1/2} \approx a_i$$

in the argument of the Coulomb logarithm L. The contribution (47.11) of the plasma waves becomes predominant when

$$zT_e/T_i L L_1 \gg 1. \tag{47.12}$$

§48. Plasma absorption in the high-frequency limit

The frequency range in which the formula (44.9) is valid for the imaginary part of the plasma permittivity is limited by the inequalities $\Omega_e \gg \omega \gg \nu_{ei}$. The first inequality is the general condition for the collision integral with screened Coulomb interaction to be applicable. Let us now consider the limit opposite to this, for which

$$\omega \gg \Omega_e. \tag{48.1}$$

We can note immediately that here the real part ϵ' of the permittivity is certainly close to unity, and the imaginary part ϵ'' is small.

Dissipation of the energy of the variable external field is caused by *ei* collisions, whose duration is of the order of, or less than, the period of the field. This means that, for $\omega \gg \Omega_e$, collisions will be important which occur at distances $\sim v_{Te}/\omega \ll v_{Te}/\Omega_e = a_e$. At such distances the Coulomb field of the ions is not screened, and the collisions are therefore purely two-particle ones, not multi-particle as they essentially are when the interaction is screened. Under these conditions, the individual field energy absorption events become processes inverse to bremsstrahlung in pair collisions of charged particles. This enables us to use the principle of detailed balancing to express ϵ'' in terms of the bremsstrahlung cross-section (V. L. Ginzburg 1949).

The dissipation Q of the electromagnetic field energy per unit volume of the medium and per unit time is expressed in terms of ϵ'' by (30.5). In order to relate this quantity to the bremsstrahlung cross-section, we assume that the field is created by a monochromatic plane wave in which the energy density is

$$\mathscr{E} = (\overline{E^2} + \overline{H^2})/8\pi = |\mathbf{E}|^2/8\pi;$$

†In the range that is important in the integral, $\xi \sim 1/L_1$, i.e. $\kappa \sim 1/a_i L_1^{1/2}$. Then $\kappa a_e \sim a_e/a_i L_1^{1/2} \sim (T_e/T_i L_1)^{1/2} \gg 1$, in accordance with the assumption made above.

in the latter expression, it is assumed that **E** is expressed as a complex quantity (cf. the third footnote to §30). Since the permittivity is nearly unity, we put here $\epsilon = 1$. Then formula (30.5) may be written

$$Q = \omega\epsilon''\mathscr{E}. \tag{48.2}$$

On the other hand, the dissipation is equal to the difference between the energy Q_{abs} absorbed in electron–ion collisions and the energy radiated in these collisions. That is, the energy Q_{st} of the stimulated (not the spontaneous) emission, which generates photons coherent with the original field and in that sense indistinguishable from it.

The cross-section for spontaneous emission of a photon, i.e. the ordinary bremsstrahlung, may be written as

$$d\sigma_{sp} = \frac{w(\mathbf{p}', \mathbf{p})}{v}\delta(\epsilon - \epsilon' - \hbar\omega)\frac{d^3k}{(2\pi)^3}d^3p', \tag{48.3}$$

where **k** is the photon wave vector, **p** and **p**′ the initial and final momenta of the electron. The product $N_i v\, d\sigma_{sp}$, where N_i is the number density of ions, is the probability per unit time for the electron to undergo emission of a photon; the function $w(\mathbf{p}', \mathbf{p})$ depends also on the polarization of the photon emitted. Integrating over the directions of **p**′ and **k**, and summing over the polarizations of the photon, we obtain the frequency-differential bremsstrahlung cross-section $d\sigma_\omega$; the delta function in (48.3) is removed by integration with respect to $\epsilon' = p'^2/2m$. Thus

$$d\sigma_\omega = (4m^2v'/\pi vc^3)\bar{w}\omega^2\, d\omega,$$

where $\bar{w}(\mathbf{p}, \mathbf{p}')$ is the value of $w(\mathbf{p}, \mathbf{p}')$ averaged over the directions of **p** and **p**′; this value is independent of the polarization of the photon, and hence the summation over the latter amounts to multiplication by 2. With the "effective emission" κ_ω defined by

$$\hbar\omega\, d\sigma_\omega = \kappa_\omega\, d\omega,$$

we can then write

$$\bar{w} = (\pi vc^3/4m^2v'\hbar\omega^3)\kappa_\omega. \tag{48.4}$$

The stimulated emission cross-section differs from (48.3) only by the factor $N_{\mathbf{ke}}$, the number of photons in the quantum state with wave vector **k**, and polarization **e** parallel to **E** (see *RQT*, §44). Hence the total energy of the stimulated emission is

$$Q_{st} = N_i \sum_{\mathbf{e}} \int N_{\mathbf{ke}}\hbar\omega w(\mathbf{p}', \mathbf{p})f(\mathbf{p})\delta(\epsilon - \epsilon' - \hbar\omega)\frac{d^3k}{(2\pi)^3}d^3p\, d^3p',$$

where $f(\mathbf{p})$ is the electron distribution function. We shall take this function to be Maxwellian, depending only on the magnitude p. Averaging over the directions of **p** and $\mathbf{p}'$, and noting that from the monochromaticity of the field

$$\sum_{\mathbf{e}} \int N_{\mathbf{k}\mathbf{e}} \frac{d^3k}{(2\pi)^3} = \mathscr{E}/\hbar\omega,$$

we can write

$$Q_{st} = N_i \mathscr{E} \int \bar{w} f(p) \delta(\epsilon - \epsilon' - \hbar\omega)\, d^3p\, d^3p'. \tag{48.5}$$

The energy absorbed in the inverse transitions with change of electron momentum $\mathbf{p}' \to \mathbf{p}$ (inelastic electron scattering in the electromagnetic field) is calculated similarly. According to the principle of detailed balancing, the probability functions w which determine the cross-sections for the direct and reverse processes are equal. We therefore get an expression for Q_{abs} that differs from (48.5) only in that the distribution function $f(p)$ is replaced by $f(p')$. The dissipation $Q = Q_{abs} - Q_{st}$; a comparison with (48.2) shows that

$$\epsilon'' = (N_i/\omega) \int \bar{w}[f(p') - f(p)] \delta(\epsilon - \epsilon' - \hbar\omega)\, d^3p\, d^3p'. \tag{48.6}$$

We shall consider only frequencies such that

$$\hbar\omega \ll T. \tag{48.7}$$

Then the difference $p' - p$ is small, and we can put

$$f(p') - f(p) = -(df/d\epsilon)\hbar\omega = (\hbar\omega/T) f(p),$$

and in the remaining factors $p = p'$. Substituting this in (48.6) and expressing $\bar{w}$ in terms of κ_ω by (48.4), we finally obtain the following expression for the imaginary part of the permittivity:

$$\epsilon''(\omega) = N_i N_e (\pi^2 c^3 / T\omega^3) \langle v \kappa_\omega \rangle, \tag{48.8}$$

where the angle brackets denote averaging over the Maxwellian distribution of electrons.

Let us apply this formula in two limiting cases: the quasi-classical case and the Born case. In the first of these, i.e. when

$$ze^2/\hbar v \gg 1, \tag{48.9}$$

the frequency range $\omega \gg \Omega_e$ can be reduced further to

$$mv_{Te}^3/ze^2 \gg \omega \gg \Omega_e. \tag{48.10}$$

The quantity on the left is the reciprocal of the electron time of flight at a distance from the ion such that the scattering angle is of the order of unity. It is easily seen that (48.7) necessarily follows from the conditions (48.9) and (48.10). In the quasi-classical case the effective emission at the frequencies (48.10) in a collision between an electron and an ion at rest is given by

$$\kappa_\omega = (16z^2e^6/3v^2c^3m^2)\log(2mv^3/\gamma\omega ze^2), \tag{48.11}$$

where $\gamma = e^C = 1.78\ldots$, and C is Euler's constant; see *Fields* (70.21). Substituting in (48.8) and carrying out the averaging we obtain†

$$\epsilon'' = \frac{4\sqrt{(2\pi)}}{3}\frac{ze^4N_e}{T^{3/2}m^{1/2}}\frac{\Omega_e^2}{\omega^3}\log\frac{2^{5/2}T^{3/2}}{\gamma^{5/2}\omega ze^2m^{1/2}}. \tag{48.12}$$

In the Born case, i.e. when $ze^2/\hbar v \ll 1$, the effective emission at frequencies $\hbar\omega \ll T$ is given by‡

$$\kappa_\omega = (16z^2e^6/3v^2c^3m^2)\log(2mv^2/\hbar\omega). \tag{48.13}$$

A calculation with (48.8) gives

$$\epsilon'' = \frac{4\sqrt{(2\pi)}}{3}\frac{ze^4N_e}{m^{1/2}T^{3/2}}\frac{\Omega_e^2}{\omega^3}\log\frac{4T}{\gamma\hbar\omega}, \tag{48.14}$$

which differs from (44.9) only in the argument of the logarithm.

§49. Quasi-linear theory of Landau damping

The theory of plasma oscillations in §§29–32 is based on solving the transport equation in the linear approximation of perturbation theory. The condition for its validity is that the correction δf (29.2) to the distribution function be small in comparison with the unperturbed function f_0:

$$\frac{e\mathbf{E}}{|\mathbf{k}\cdot\mathbf{v}-\omega|}\cdot\frac{\partial f_0}{\partial \mathbf{p}} \ll f_0. \tag{49.1}$$

For only slightly damped plasma oscillations with frequency $\approx\Omega_e$ and wave number $k \ll \Omega_e/v_{Te}$, it is thus necessary that

$$(eE/\Omega_e)\partial f_0/\partial p \ll f_0.$$

For a Maxwellian plasma, this condition (with both sides squared) can be written

$$E^2/4\pi \ll N_eT_e. \tag{49.2}$$

†Using the result that

$$\int_0^\infty e^{-x}\log x\,dx = -C.$$

‡See *RQT* (92.16). In going from this formula to (48.14), we have also used the fact that when $\hbar\omega \ll T = mv_{Te}^2$ the electron loses only a small part of its energy by radiation.

This form has a simple physical significance: the wave-field energy density must be much less than the kinetic energy density of the plasma electrons.

The condition (49.2) ensures that the correction δf is small for most of the electrons. However, even if it is satisfied, there are a relatively small number of *resonant particles* for which (49.1) may not be satisfied, which are moving almost in phase with the wave ($\mathbf{k}\cdot\mathbf{v}\approx\omega$) and thus take part in Landau damping; their distribution function may be considerably altered by even a weak field. This change is a non-linear effect, and its nature therefore depends greatly on the spectrum (in ω and $\mathbf{k}$) of the wave field, since it is only in the linear approximation that the various Fourier components of the field act independently on the particles.

Here we shall consider electromagnetic perturbations in a plasma that are an assembly of plasma waves with wave vectors taking a continuous range of values in some interval $\Delta\mathbf{k}$.

If the initial perturbation includes a wide range of wave numbers $k\sim\Omega_e/v_{Te}$, the Landau damping extends to a large number of electrons that are in the same conditions as regards the effect of the field on them. In consequence, the distortion of the distribution function is relatively small at all speeds; the linear theory, with the condition (49.2), is therefore valid throughout the development of the perturbation.

On the other hand, if the perturbation contains wave vectors in only a narrow range $\Delta\mathbf{k}$ near a value $k_0\ll\Omega_e/v_{Te}$, then the resonance range of electron velocities,

$$|\Delta\mathbf{v}|\sim\Delta(\Omega_e/k)\sim(v_0/k_0)|\Delta\mathbf{k}|,\quad \mathbf{v}_0=(\Omega_e/k_0)\mathbf{k}_0/k_0, \tag{49.3}$$

is also small and lies near $v_0\gg v_{Te}$. Thus only a comparatively small number of electrons participate in Landau damping, and the electron distribution function may be greatly changed.

The quantitative theory of this phenomenon will be given here for the case where the perturbation is an almost monochromatic wave whose amplitude and phase are modulated in space according to some statistical law. The spectrum of $\mathbf{k}$ values for the initial perturbation is narrow:

$$|\Delta\mathbf{k}|/k_0\ll 1, \tag{49.4}$$

but at the same time

$$|\Delta\mathbf{k}|/k\gg(1/v_0)(e|\varphi_0|/m)^{1/2}, \tag{49.5}$$

where φ_0 is the order of magnitude of the wave electric field potential amplitude (the significance of this condition will be explained below). By (49.2) (where $E\sim k\varphi_0$) the expression on the right of the inequality (49.5) is small, $e|\varphi_0|/mv_0^2\ll 1$. We shall also assume the field to be uniform on average throughout the plasma; this means that $\mathbf{E}^2$ averaged over the statistical distribution of wave phases and amplitudes is independent of the coordinates. Such an averaging is equivalent to one over regions of space with dimensions $\Delta x\gg 1/|\Delta\mathbf{k}|$.

The field $\mathbf{E}$ at the initial instant is expressed as a Fourier integral:

$$\mathbf{E}=\int\mathbf{E}_{\mathbf{k}}e^{i\mathbf{k}\cdot\mathbf{r}}\,d^3k/(2\pi)^3, \tag{49.6}$$

where $\mathbf{E}_{-\mathbf{k}} = \mathbf{E}_{\mathbf{k}}^*$ since $\mathbf{E}$ is real. The hypothesis (49.4) as to the nature of the initial perturbation signifies that the integration in (49.6) is in practice taken only over the neighbourhoods of the points $\mathbf{k} = \pm\mathbf{k}_0$. The condition of spatial uniformity of the perturbation is easily formulated by writing the quadratic tensor $E_\alpha E_\beta$ as a double integral:

$$E_\alpha E_\beta = \int\int E_{\mathbf{k}\alpha} E_{\mathbf{k}'\beta} e^{i(\mathbf{k}+\mathbf{k}')\cdot\mathbf{r}} \frac{d^3k\, d^3k'}{(2\pi)^6}.$$

After averaging over the statistical distribution, this expression should be independent of $\mathbf{r}$.† For this to be so, the mean value $\langle E_{\mathbf{k}\alpha} E_{\mathbf{k}'\beta}\rangle$ must contain the delta function $\delta(\mathbf{k}+\mathbf{k}')$. Since the plasma waves are longitudinal, we then write

$$\langle E_{\mathbf{k}\alpha} E_{\mathbf{k}'\beta}\rangle = (2\pi)^3 \frac{k_\alpha k_\beta}{k^2} (\mathbf{E}^2)_{\mathbf{k}}\, \delta(\mathbf{k}+\mathbf{k}'). \tag{49.7}$$

This relation is to be regarded as a definition of the quantities symbolically denoted by $(\mathbf{E}^2)_{\mathbf{k}}$. These are real quantities. The expression (49.7) is zero except when $\mathbf{k} = -\mathbf{k}'$, and is symmetrical with respect to the interchange of $\mathbf{k}$ and $\mathbf{k}'$. Hence $(\mathbf{E}^2)_{\mathbf{k}} = (\mathbf{E}^2)_{-\mathbf{k}}$; and a change in the sign of $\mathbf{k}$ is equivalent to taking the complex conjugate. The mean square $\langle \mathbf{E}^2\rangle$ is expressed in terms of these quantities by

$$\langle \mathbf{E}^2\rangle = \int (\mathbf{E}^2)_{\mathbf{k}}\, d^3k/(2\pi)^3. \tag{49.8}$$

The integration in (49.6), and therefore in (49.8), is (as already mentioned) taken over the neighbourhoods of the points $\mathbf{k}_0$ and $-\mathbf{k}_0$. It is, however, more convenient to eliminate $-\mathbf{k}_0$ by putting (49.6) in the form

$$\mathbf{E} = \int_{\mathbf{k}\approx\mathbf{k}_0} \mathbf{E}_{\mathbf{k}} e^{i\mathbf{k}\cdot\mathbf{r}} \frac{d^3k}{(2\pi)^3} + \text{c.c.}, \tag{49.9}$$

where the integration is taken only over the neighbourhood of the point $\mathbf{k} = \mathbf{k}_0$, and c.c. denotes the complex conjugate. Accordingly, (49.8) is written as

$$\langle \mathbf{E}^2\rangle = 2\int_{\mathbf{k}\approx\mathbf{k}_0} (\mathbf{E}^2)_{\mathbf{k}}\, d^3k/(2\pi)^3, \tag{49.10}$$

and the relations (49.7) as

$$\left.\begin{aligned} \langle E_{\mathbf{k}\alpha} E^*_{\mathbf{k}'\beta}\rangle &= (2\pi)^3 (\mathbf{E}^2)_{\mathbf{k}} \frac{k_\alpha k_\beta}{k^2}\, \delta(\mathbf{k}-\mathbf{k}'), \\ \langle E_{\mathbf{k}\alpha} E_{\mathbf{k}'\beta}\rangle &= 0. \end{aligned}\right\} \tag{49.11}$$

†For perturbations of the type under consideration, the integrals $\mathbf{E}_{\mathbf{k}} = \int \mathbf{E}(\mathbf{r}) e^{-i\mathbf{k}\cdot\mathbf{r}}\, d^3x$ in fact diverge, since $\mathbf{E}(\mathbf{r})$ is not zero at infinity. This, however, is not important in the formal derivations, which involve mean squares that are certainly finite.

The time variation of the perturbation (49.9) is represented by

$$\mathbf{E} = \int_{\mathbf{k}\approx\mathbf{k}_0} e^{i(\mathbf{k}\cdot\mathbf{r}-\omega t)}\mathbf{E}_\mathbf{k}(t)\frac{d^3k}{(2\pi)^3} + \text{c.c.}, \tag{49.12}$$

where $\omega(\mathbf{k}) \approx \Omega_e$ is the plasma wave frequency, and the coefficients $\mathbf{E}_\mathbf{k}(t)$ vary slowly on account of Landau damping. The electron distribution function is expressed similarly:

$$f = f_0(t, \mathbf{p}) + \left\{\int_{\mathbf{k}\approx\mathbf{k}_0} f_\mathbf{k}(t, \mathbf{p})e^{i(\mathbf{k}\cdot\mathbf{r}-\omega t)}\frac{d^3k}{(2\pi)^3} + \text{c.c.}\right\}. \tag{49.13}$$

The expression in the braces is the "random" part of the variation of the distribution function, oscillating rapidly in space and time; it vanishes on statistical averaging of the waves. The term $f_0(t, \mathbf{p})$ is the slowly varying averaged distribution.†

Our object is to derive a set of equations to determine the time variation of averaged characteristics of the state of the plasma, namely the functions $(\mathbf{E}^2)_\mathbf{k}$ and $f_0(t, \mathbf{p})$. For such a set of equations to be closed, these characteristics must embrace all electrons participating in the non-linear effects concerned. In turn, the velocity range (49.3) corresponding to the spread of wave vectors $\Delta\mathbf{k}$ must then always widely overlap the amplitude of the electron velocity oscillations due to the field of the waves in resonance with the electrons. This is the condition expressed by the inequality (49.5); $(e|\varphi_0|/m)^{1/2}$ is the order of magnitude of the amplitude in question. For, in coordinates moving with the phase velocity of the wave, the wave field is static and consists of a sequence of potential humps with height $|\varphi_0|$. In these coordinates, a resonant electron oscillates between two humps, and its speed varies in the range between $\pm(2e|\varphi_0|/m)^{1/2}$.

One of the equations relating $(\mathbf{E}^2)_\mathbf{k}$ and f_0 expresses the Landau damping of each Fourier component of the field:

$$d(\mathbf{E}^2)_\mathbf{k}/dt = -2\gamma_\mathbf{k}(\mathbf{E}^2)_\mathbf{k}, \tag{49.14}$$

where

$$\gamma_\mathbf{k} = 2\pi^2e^2\Omega_e\int\frac{\partial f_0}{\partial\mathbf{p}}\cdot\frac{\mathbf{k}}{k^2}\,\delta(\omega - \mathbf{k}\cdot\mathbf{v})\,d^3p \tag{49.15}$$

is, from (32.6) and (30.1), the wave amplitude damping ratio; the factor 2 on the right of (49.14) appears because $(\mathbf{E}^2)_\mathbf{k}$ is quadratic.

The second equation is derived from the transport equation for a collisionless plasma:

$$\frac{\partial f}{\partial t} + \mathbf{v}\cdot\frac{\partial f}{\partial\mathbf{r}} - e\mathbf{E}\cdot\frac{\partial f}{\partial\mathbf{p}} = 0. \tag{49.16}$$

Let us first apply this in the linear approximation to an individual Fourier component of the perturbation. In the last term in the equation, which already contains

†Not to be confused with the Maxwellian equilibrium distribution.

the small quantity $\mathbf{E}_\mathbf{k} e^{i(\mathbf{k}\cdot\mathbf{r}-\omega t)}$, we put $f \approx f_0$. In the first term, we neglect the slow variation of $f_\mathbf{k}$ with t. Then we get for $f_\mathbf{k}$ the usual expression

$$f_\mathbf{k} = \frac{ie\mathbf{E}_\mathbf{k}}{\omega - \mathbf{k}\cdot\mathbf{v}} \cdot \frac{\partial f_0}{\partial \mathbf{p}}, \tag{49.17}$$

where, in the subsequent integrations, ω is to be taken to have the customary meaning $\omega + i0$.

Next, we substitute in (49.16) the complete expressions (49.12) and (49.13) for $\mathbf{E}$ and f, with $f_\mathbf{k}$ from (49.17), and average over the statistical distribution of waves by means of (49.11). All terms linear in the perturbation drop out; the quadratic terms determine the derivative $\partial f_0/\partial t$ as

$$\frac{\partial f_0}{\partial t} = e^2 \frac{\partial}{\partial p_\alpha}\left\{\frac{\partial f_0}{\partial p_\beta}\int_{\mathbf{k}\approx\mathbf{k}_0} \frac{k_\alpha k_\beta}{k^2}(\mathbf{E}^2)_\mathbf{k}\left[\frac{i}{\omega - \mathbf{k}\cdot\mathbf{v} + i0} - \frac{i}{\omega - \mathbf{k}\cdot\mathbf{v} - i0}\right]\frac{d^3k}{(2\pi)^3}\right\}$$

Replacing the difference in the square brackets by $2\pi\delta(\omega - \mathbf{k}\cdot\mathbf{v})$ according to (29.8), we have finally

$$\frac{\partial f_0}{\partial t} = \frac{\partial}{\partial p_\alpha}\left(D_{\alpha\beta}^{(nl)}\frac{\partial f_0}{\partial p_\beta}\right), \tag{49.18}$$

where

$$D_{\alpha\beta}^{(nl)}(\mathbf{p}) = 2\pi e^2 \int_{\mathbf{k}\approx\mathbf{k}_0} (\mathbf{E}^2)_\mathbf{k} \frac{k_\alpha k_\beta}{k^2}\, \delta(\omega - \mathbf{k}\cdot\mathbf{v}) \frac{d^3k}{(2\pi)^3}. \tag{49.19}$$

Equations (49.14) and (49.18) constitute the required complete set. The theory of plasma waves based on them is called the *quasi-linear theory*.†

Equation (49.18) has the form of a diffusion equation in velocity space, with $D_{\alpha\beta}^{(nl)}$ as the diffusion coefficient tensor; the superscript nl indicates that this "diffusion" is due to non-linearity effects. The coefficients as functions of the electron velocity are zero except in a range $\Delta\mathbf{v}$ near $\mathbf{v}_0$, which is related to the spread $\Delta\mathbf{k}$ by (49.3). In this velocity range, diffusion occurs and there is a corresponding distortion of the distribution function (which remains Maxwellian for the bulk of the electrons). The nature of the distortion is evident from the general properties of diffusion processes: diffusion causes a smoothing, or in this case a plateau of width $\sim \Delta v$ on the tail of the function $f_0(p)$ at $v \approx v_0 \gg v_{Te}$, as shown diagrammatically in Fig. 13. With this type of distortion, the principal change is in the derivative $\partial f_0/\partial p$, while f_0 itself remains close to the Maxwellian value.

Let us estimate the relaxation time τ_{nl} for this process. Since the equalization is to take place over a range $\Delta p = m\Delta v$, we have

$$\tau_{nl} \sim m^2(\Delta v)^2/D^{(nl)}. \tag{49.20}$$

To estimate the diffusion coefficient, we note that from (49.10) $(\mathbf{E}^2)_\mathbf{k}(\Delta k/2\pi)^3 \sim \langle\mathbf{E}^2\rangle$. The presence of the delta function in (49.19) is equivalent, in order of magnitude, to

†It was developed by A. A. Vedenov, E. P. Velikhov and R. Z. Sagdeev (1961). Equations (49.14) and (49.18) were independently derived by Yu. A. Romanov and G. F. Filippov (1961) and by W. E. Drummond and D. Pines (1961).

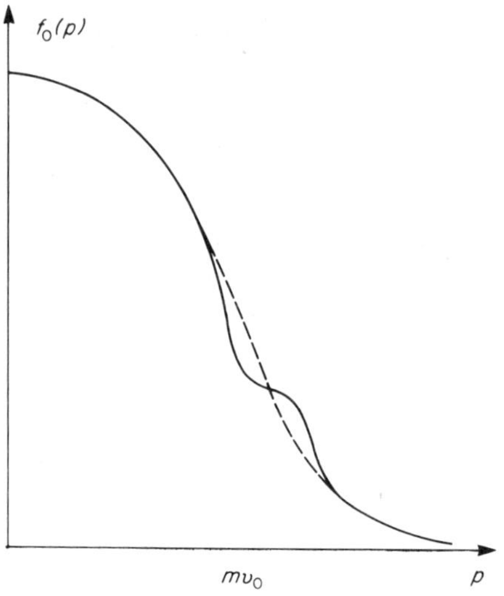

FIG. 13.

multiplying the integral by $1/v_0\Delta k$. Thus

$$D^{(nl)} \sim e^2\langle \mathbf{E}^2\rangle/v_0\Delta k \sim e^2\langle \mathbf{E}^2\rangle/k_0\Delta v. \tag{49.21}$$

Lastly, expressing $\langle \mathbf{E}^2\rangle$ in terms of the amplitude φ_0 of the potential oscillations $(\sim k^2|\varphi_0|^2)$, and substituting (49.21) in (49.20), we find†

$$\tau_{nl} \sim (\Delta v)^3/k_0(e|\varphi_0|/m)^2. \tag{49.22}$$

In the above discussion, it is assumed of course that τ_{nl} is much less than the Landau damping time $1/\gamma$; otherwise, the waves are damped before non-linearity effects can appear. The applicability of (49.14), however, presupposes that $1/\gamma$ is much less than the electron mean free time: $1/\gamma \ll 1/\nu_e$, where ν_e is the mean collision frequency. The latter condition does not guarantee the legitimacy of neglecting collisions in the phenomenon considered, i.e. of using the transport equation in the form (49.16): what is significant as regards the competition with non-linear effects is not the total collisional relaxation time but only the time for collisional relaxation in the range Δv, which we denote by τ_{coll}.

For relaxation in the range Δv near $v_0 \gg v_{Te}$, which contains only a relatively small fraction of all the electrons, the position is similar to that in the problem of runaway electrons. The process is a diffusion in momentum space, the diffusion coefficient

$$D^{(\text{coll})} = m^2\nu_{ee}(v)v_{Te}^2 = 4\pi e^4 L N_e T_e/mv^3 \approx m^2\nu_{ee}(v_{Te})v_{Te}^5/v^3, \tag{49.23}$$

i.e. the coefficient of $\partial f/\partial p$ in the flux (45.5) in momentum space.

†When $\Delta v \sim (e|\varphi_0|/m)^{1/2}$ and the theory given here is, strictly speaking, inapplicable (the sign $\gg$ in (49.5) becoming $\sim$), this estimate gives $\tau_{nl} \sim k_0^{-1}(m/e|\varphi_0|)^{1/2}$. This result was to be expected when the spread Δv of resonant speeds coincides with the speed amplitude of the electrons oscillating in the wave field: τ_{nl} has the same order of magnitude as the period of these oscillations.

The required collisional relaxation time in the range Δv differs from (49.20) in that $D^{(nl)}$ is replaced by $D^{(\text{coll})}$:

$$\tau_{\text{coll}} \sim m^2(\Delta v)^2/D^{(\text{coll})} \sim \frac{(\Delta v)^2}{v_{Te}^2 \nu_{ee}(v_{Te})}\left(\frac{v_0}{v_{Te}}\right)^3. \tag{49.24}$$

When

$$\tau_{nl} \gg \tau_{\text{coll}}, \tag{49.25}$$

i.e. $D^{(nl)} \ll D^{(\text{coll})}$, the non-linear effects play no part: the collisions are able to maintain the Maxwellian distribution near $\mathbf{v}_0$ despite the perturbation from the wave field, and accordingly the Landau damping ratio is given by the usual expression corresponding to the Maxwellian value of the derivative $\partial f_0/\partial \mathbf{p}$ in the neighbourhood of $\mathbf{v}_0$. Thus the inequality (49.25) is the condition for the strictly linear theory of Landau damping to be applicable. The quasi-linear theory given here is valid with the much weaker condition (49.2). The condition (49.25) may be written

$$E^2/4\pi \ll N_e T_e[\sqrt{(4\pi)}L\eta^{3/2}(v_{Te}/v_0)^3 \Delta v/v_0], \tag{49.26}$$

where $\eta = e^2 N^{1/3}/T$ is the gaseousness parameter. The smallness of the factor in the brackets shows that the condition (49.2) is weak in comparison with (49.25).†

In the opposite limiting case where $\tau_{nl} \ll \tau_{\text{coll}}$, the non-linear effects cause a great decrease of the derivative $\partial f_0/\partial p$ in the range concerned, roughly in the ratio $D^{(\text{coll})}/D^{(nl)}$. The Landau damping ratio is correspondingly reduced.

§50. The transport equation for a relativistic plasma

If the velocities of the particles (electrons) in a plasma are not small compared with that of light, the transport equation has to take account of relativistic effects (S. T. Belyaev and G. I. Budker 1956).

We shall first show that the distribution function in phase space, $f(t, \mathbf{r}, \mathbf{p})$, is a relativistic invariant. This is shown by noting that the spatial density and the flux of particles, i.e. the integrals

$$N = \int f\, d^3p, \quad \mathbf{i} = \int \mathbf{v} f\, d^3p,$$

must form the 4-vector $i^k = (cN, \mathbf{i})$; cf. *Fields*, § 28.‡ Since in relativistic mechanics the velocity of a particle with momentum $\mathbf{p}$ and energy ϵ is $\mathbf{v} = \mathbf{p}c^2/\epsilon$, we can write this 4-vector as

$$i^k = c^2 \int (p^k f/\epsilon)\, d^3p, \tag{50.1}$$

†It has already been mentioned that a strictly linear theory is inapplicable if $\tau_{nl} \gg 1/\gamma$. This may be rewritten as

$$E^2/4\pi \ll N_e T_e[(\Delta v/v_{Te})^2(\Delta v/v_0)\gamma/\Omega_e],$$

which may prove to be a weaker condition than (49.26).

‡In this section, the Latin letters k and l denote four-dimensional vector indices. The scalar product of two 4-vectors a and b is denoted by $(ab) \equiv a_k b^k$.

where $p^k = (\epsilon/c, \mathbf{p})$ is the 4-momentum. The expression d^3p/ϵ is a 4-scalar (see *Fields*, § 10). It is therefore clear, since the integral (50.1) is a 4-vector, that f is a 4-scalar.†

Going on now to derive the transport equation, we note that the calculations in § 41 remain valid in the relativistic case as far as the expression (41.3), (41.4) for the flux in momentum space. We need only recalculate the quantities

$$B_{\alpha\beta} = \tfrac{1}{2}\int q_\alpha q_\beta v_{\mathrm{rel}}\, d\sigma. \tag{50.2}$$

The quantity $\mathbf{v}_{\mathrm{rel}}$ is, as before, the relative velocity of the two particles. In relativistic mechanics, however, it is defined as the velocity of one particle in the rest frame of the other, and does not in general reduce to the difference $\mathbf{v}-\mathbf{v}'$ (see *Fields*, § 12).

Let us first ascertain the transformation properties of these quantities. The product

$$v_{\mathrm{rel}}\, d\sigma \,.\, ff'\, d^3p\, d^3p'\, d^3x\, dt$$

is the number of scattering events in the volume d^3x and the time dt, between two particles with momenta in given ranges d^3p and d^3p'; this number is, by definition, invariant. Writing it in the form

$$\epsilon\epsilon' v_{\mathrm{rel}}\, d\sigma \,.\, f \,.\, f' \,.\, (d^3p/\epsilon) \,.\, (d^3p'/\epsilon') \,.\, d^3x\, dt$$

and noting that the last five factors (between points) are invariant, we conclude that the first factor $\epsilon\epsilon' v_{\mathrm{rel}}\, d\sigma$ is also invariant. Hence it follows in turn that the integrals

$$W^{kl} = \tfrac{1}{2}\epsilon\epsilon' \int q^k q^l v_{\mathrm{rel}}\, d\sigma \tag{50.3}$$

form a symmetric 4-tensor. The quantities (50.2) are related to the space components of this 4-tensor by

$$B_{\alpha\beta} = W^{\alpha\beta}/\epsilon\epsilon'. \tag{50.4}$$

We first calculate the 4-tensor (50.3) in a frame of reference where one particle, say e, is at rest. The relativistic cross-section for Rutherford scattering of particles e' by particles e at rest (before the collision) with small scattering angles χ is‡

$$d\sigma = \frac{4(ee')^2\epsilon'^2}{p'^4\chi^4} 2\pi\chi\, d\chi. \tag{50.5}$$

†The distribution function with respect to momenta only, i.e. $f(t, \mathbf{p}) = \int f(t, \mathbf{r}, \mathbf{p})\, d^3x$, is not a 4-scalar, however; such a function is discussed in *Fields*, § 10.

‡This expression applies to the scattering of electrons by either electrons or ions. In the first case it follows from *RQT* (81.7); in the second case, from the cross-section for scattering by a fixed Coulomb centre, *RQT* (80.7).

A calculation similar to the derivation of (41.8) gives the following expression for the space components of the tensor (50.3):

$$W^{\alpha\beta} = 2\pi(ee')^2L(v'^2\delta_{\alpha\beta} - v'_\alpha v'_\beta)mc^2\epsilon'/v'^3. \tag{50.6}$$

The remaining components are to be taken as zero:

$$W^{00} = W^{0\alpha} = 0, \tag{50.7}$$

since the change in the particle energy in the collision (q^0) in this frame of reference is of the second order with respect to the small scattering angle, and so $W^{0\alpha}$ and W^{00} would be of the third or fourth order, whereas the whole calculation of the collision integral is accurate only as far as second-order quantities.

From (50.6) and (50.7),

$$W_k^{\ k} = -W_\alpha^{\ \alpha} = -4\pi(ee')^2Lmc^2\epsilon'/v'.$$

This 4-scalar may be written in an invariant form by noting that in the rest frame of the particle e we have

$$(uu') = \epsilon'/m'c^2, \quad \frac{[(uu')^2 - 1]^{1/2}}{uu'} = v'/c,$$

where $u^k = p^k/mc$, $u'^k = p^k/m'c$ are the 4-velocities of the two particles. Hence

$$W_k^{\ k} = -4\pi(ee')^2Lmm'c^4\frac{(uu')^2}{c[(uu')^2 - 1]^{1/2}}. \tag{50.8}$$

From (50.6) and (50.7) we also find that

$$W^{kl}u_l = W^{kl}u'_l = 0, \tag{50.9}$$

and, since these equations are relativistically invariant in form, they are valid in any frame of reference.

The expression for the 4-tensor W^{kl} valid in any frame of reference must evidently be symmetrical in the two particles. The general form of such a 4-tensor depending only on the 4-vectors u^k and u'^k is

$$W^{kl} = \alpha g^{kl} + \beta(u^ku^l + u'^ku'^l) + \delta(u^ku'^l + u'^ku^l),$$

where α, β and δ are scalars. Determining these from the conditions (50.8) and (50.9), we obtain

$$W^{kl} = 2\pi(ee')^2L\frac{mm'c^4(uu')^2}{c[(uu')^2 - 1]^{3/2}}\times$$

$$\times\{-[(uu')^2 - 1]g^{kl} - (u^ku^l + u'^ku'^l) + (uu')(u^ku'^l + u'^ku^l)\}. \tag{50.10}$$

Lastly, taking the space part of this 4-tensor in an arbitrary frame of reference, we have the following final expression for the quantities $B_{\alpha\beta}$ in the collision integral:

$$B_{\alpha\beta} = 2\pi(ee')^2 L \frac{\gamma\gamma'(1-\mathbf{v}\cdot\mathbf{v}'/c^2)^2}{c[\gamma^2\gamma'^2(1-\mathbf{v}\cdot\mathbf{v}'/c^2)^2-1]^{3/2}} \times$$
$$\times\left\{\left[\gamma^2\gamma'^2\left(1-\frac{\mathbf{v}\cdot\mathbf{v}'}{c^2}\right)^2-1\right]\delta_{\alpha\beta}-\frac{\gamma^2}{c^2}v_\alpha v_\beta-\frac{\gamma'^2}{c^2}v'_\alpha v'_\beta\right.$$
$$\left.+\frac{\gamma^2\gamma'^2}{c^2}\left(1-\frac{\mathbf{v}\cdot\mathbf{v}'}{c^2}\right)(v_\alpha v'_\beta+v'_\alpha v_\beta)\right\}, \tag{50.11}$$

where

$$\gamma = \epsilon/mc^2 = (1-v^2/c^2)^{-1/2},$$
$$\gamma' = \epsilon'/m'c^2 = (1-v'^2/c^2)^{-1/2},$$

are the Lorentz factors for the two particles. Despite its more complex form in comparison with the non-relativistic case, the three-dimensional tensor (50.11) again satisfies the relations

$$B_{\alpha\beta}v_\beta = B_{\alpha\beta}v'_\beta. \tag{50.12}$$

To estimate the Coulomb logarithm, we note that in the relativistic case the Born situation occurs: $ze^2/\hbar v \sim ze^2/\hbar c \ll 1$. Hence, for *ee* and *ei* collisions,

$$L \approx \log(pa/\hbar) \approx \log(T_e a/\hbar c). \tag{50.13}$$

For *ii* collisions, T_e must be replaced by T_i (if the ions too are relativistic), or else the ordinary non-relativistic expressions should be used.

The transport equation with the Coulomb collision integral is valid so long as Rutherford scattering is the principal cause of variation of the electron momentum and energy. The competing process here is bremsstrahlung (and also the Compton effect, if the plasma contains an appreciable number of photons). The Rutherford scattering (transport) cross-section is in order of magnitude

$$\sigma_{\text{Ru}} \sim z^2(e^2/mc^2)^2(mc^2/\epsilon)^2 L \sim z^2(e^2/mc^2)^2(mc^2/T_e)^2 L. \tag{50.14}$$

The cross-section for bremsstrahlung emission of a photon with energy $\hbar\omega \sim T_e$ is

$$\sigma_{\text{br}} \sim (z^2/137)(e^2/mc^2)^2 \log(T_e/mc^2); \tag{50.15}$$

cf. *RQT* (93.17). These cross-sections are comparable if

$$T_e/mc^2 \sim (137L/\log 137L)^{1/2}.$$

PROBLEMS

PROBLEM 1. Find the rate of energy transfer from electrons with temperature $T_e \gg mc^2$ to ions with temperature $T_i \ll Mc^2$.

SOLUTION. The calculations in §42 remain valid as far as (42.3). We take the $B_{\alpha\beta}^{(ei)}$ from (50.4) and (50.6) with $v' \approx c$:

$$B_{\alpha\beta}^{(ei)} = 4\pi e^4 z^2 L/c.$$

The result is

$$\frac{dE_i}{dt} = -\frac{dE_e}{dt} = \left(1 - \frac{T_i}{T_e}\right)\frac{4\pi z^2 e^4 N_i N_e L}{Mc}.$$

Expressing the energy of ultra-relativistic electrons in terms of their temperature by $E_e = 3T_e N_e$ (see *SP* 1, §44, Problem), we obtain

$$\frac{dT_e}{dt} = -(T_e - T_i)\frac{4\pi z^2 e^4 N_i L}{3McT_e}.$$

PROBLEM 2. Find the electrical conductivity of a relativistic Lorentzian plasma.

SOLUTION. When we neglect *ee* collisions and go to the limit $M \to \infty$, the process of solution in the relativistic case is the same as for the non-relativistic problem in §44. The correction to the distribution function in a constant ($\omega = 0$) electric field is again

$$\delta f = -[e\mathbf{E} \cdot \mathbf{v}/T_e \nu_{ei}(p)]f_0$$

(cf. (44.5)), the only difference being that the collision frequency is now determined by the relativistic Rutherford scattering cross-section:

$$\nu_{ei}(p) = N_i v \sigma_t, \quad \sigma_t \approx \int \tfrac{1}{2}\chi^2 d\sigma = 4\pi z^2 e^4 L/v^2 p^2.$$

Calculating the current as the integral $-e\int \mathbf{v}\delta f\, d^3p$, we find as the conductivity

$$\sigma = \langle v^3 p^2 \rangle / 12\pi z e^2 T_e L.$$

In the ultra-relativistic case, $v \approx c$, $\langle p^2 \rangle = 12(T_e/c)^2$, and so

$$\sigma = cT_e/\pi z e^2 L.$$

§51. Fluctuations in plasmas

The theory of fluctuations in plasmas is in principle constructed in the same way as for an ordinary gas (§§19 and 20). The different-time correlation functions such as

$$\langle \delta f_a(t_1, \mathbf{r}_1, \mathbf{p}_1)\delta f_b(t_2, \mathbf{r}_2, \mathbf{p}_2)\rangle, \quad \langle \delta\varphi(t_1, \mathbf{r}_1)\delta f_a(t_2, \mathbf{r}_2, \mathbf{p}_2)\rangle,$$

where φ is the electric field potential and a, b distinguish the types of particle, satisfy (when $t = t_1 - t_2 > 0$) the same equations, namely the linearized transport equation and the linearized Poisson's equation, as the distribution functions $\bar{f}_a$ and the potential $\bar{\varphi}$. To solve these equations, the corresponding single-time correlation functions are needed as an initial condition. But, in contrast to an equilibrium gas of neutral particles, there is in a plasma a single-time correlation between the positions of different particles due to their Coulomb interaction and extending to a large distance ($\sim a$). In the equilibrium case, this correlation is described by the density correlation functions calculated in *SP* 1, §79. In non-equilibrium cases, the determination of the single-time correlation functions is a difficult problem.

This difficulty may, however, be overcome in a general manner for the case of a collisionless plasma. For such a plasma, the problem of fluctuations in a stationary non-equilibrium state has a particularly natural formulation, since in the absence of an external field any distribution functions $\bar{f}_a(\mathbf{p})$ depending only on the particle momenta are a stationary solution of the transport equation. The correlation function of fluctuations relative to such a distribution will depend, as in the equilibrium case, on the coordinates of the two points and on the two times only through the differences $\mathbf{r} = \mathbf{r}_1 - \mathbf{r}_2$, $t = t_1 - t_2$. If the plasma is collisionless, the times t considered are much less than $1/\nu$, where ν is the effective collision frequency. The method given below is applicable in just these conditions; the plasma is throughout treated as collisionless. The method is based on a direct averaging of the products of the exact fluctuating distribution functions $f_a(t, \mathbf{r}, \mathbf{p})$.†

These functions satisfy the equations

$$\frac{df_a}{dt} = \frac{\partial f_a}{\partial t} + \mathbf{v} \cdot \frac{\partial f_a}{\partial \mathbf{r}} - e_a \frac{\partial \varphi}{\partial \mathbf{r}} \cdot \frac{\partial f_a}{\partial \mathbf{p}} = 0, \tag{51.1}$$

where φ is the exact electric field potential, which satisfies the equation

$$\Delta \varphi = -4\pi \sum_a e_a \int f_a \, d^3p. \tag{51.2}$$

Equations (51.1) are the analogue of Liouville's theorem. In these exact equations, collisions have not yet been neglected. The exact distribution functions

$$f_a(t, \mathbf{r}, \mathbf{p}) = \sum \delta[\mathbf{r} - \mathbf{r}_a(t)]\delta[\mathbf{p} - \mathbf{p}_a(t)] \tag{51.3}$$

(with summation over all particles of type a) take account of particle motion along paths $\mathbf{r} = \mathbf{r}_a(t)$ that are exact solutions of the equations of motion for interacting particles. Equations (51.1) are easily verified by direct differentiation of the expressions (51.3), using the equations of motion of particles in a self-consistent field.

Equations (51.1) and (51.2) themselves are not very useful; to apply the distribution functions in the form (51.3) would mean following the motion of each particle separately. However, if they are averaged over physically infinitesimal volumes,‡ the ordinary transport equations are obtained. Putting $f_a = \bar{f}_a + \delta f_a$, $\varphi = \bar{\varphi} + \delta\varphi$, and averaging the equations (without any approximation), we obtain

$$\frac{\partial \bar{f}_a}{\partial t} + \mathbf{v} \cdot \frac{\partial \bar{f}_a}{\partial \mathbf{r}} - e_a \frac{\partial \bar{\varphi}}{\partial \mathbf{r}} \cdot \frac{\partial \bar{f}_a}{\partial \mathbf{p}} = e_a \left\langle \frac{\partial \delta\varphi}{\partial \mathbf{r}} \cdot \frac{\partial \delta f_a}{\partial \mathbf{p}} \right\rangle, \tag{51.4}$$

$$\Delta \bar{\varphi} = -4\pi \sum_a e_a \int \bar{f}_a \, d^3p. \tag{51.5}$$

The right-hand side of (51.4) is the collision integral.§

†It is due to N. Rostoker (1961) and to Yu. L. Klimontovich and V. P. Silin (1962).

‡Or, equivalently, over the initial conditions of the exact mechanical problem, corresponding to a specified macroscopic state.

§We shall return to this expression at the end of the section, and meanwhile note only that it corresponds to the right-hand side of (16.7) in the case where the particles have a Coulomb interaction.

Subtracting (51.4) and (51.5) from the exact equations (51.1) and (51.2), we obtain equations for the fluctuating parts of the distribution functions and the potential. The terms in the transport equation that are quadratic in $\delta\varphi$ and δf_a describe the influence of collisions on the fluctuations. Neglecting these terms and considering the case of spatial homogeneity, i.e. putting

$$\bar{f}_a = \bar{f}_a(\mathbf{p}), \quad \bar{\varphi} = 0, \tag{51.6}$$

we obtain the equations

$$\frac{\partial \delta f_a}{\partial t} + \mathbf{v} \cdot \frac{\partial \delta f_a}{\partial \mathbf{r}} - e_a \frac{\partial \delta\varphi}{\partial \mathbf{r}} \cdot \frac{\partial \bar{f}_a}{\partial \mathbf{p}} = 0, \tag{51.7}$$

$$\triangle\, d\varphi = -4\pi \sum_a e_a \int \delta f_a\, d^3p. \tag{51.8}$$

These equations enable us to express the functions $\delta f_a(t, \mathbf{r}, \mathbf{p})$ at any instant t in terms of their values at some initial instant $t = 0$, and hence to express the correlation function

$$\langle \delta f_a(t_1, \mathbf{r}_1, \mathbf{p}_1)\delta f_b(t_2, \mathbf{r}_2, \mathbf{p}_2)\rangle \tag{51.9}$$

in terms of its value for $t_1 = t_2 = 0$. This initial value of the correlation function, which we denote by $g_{ab}(\mathbf{r}_1 - \mathbf{r}_2, \mathbf{p}_1, \mathbf{p}_2)$, is a largely (see below) arbitrary function. It must be emphasized immediately that this is not the single-time correlation function which (together with the complete different-time correlation function) we are trying to find. The central point which ensures the effectiveness of the method under discussion is that with an arbitrary choice of the function g the correlation function (51.9) thus calculated reduces in the course of time (when t_1 and t_2 are of the order of the Landau damping time) to a function only of the difference $t = t_1 - t_2$, independent of the choice of g. The problem is thereby solved: this limiting function is the required different-time correlation function, and its value for $t_1 - t_2 = 0$ is the single-time correlation function.

To carry out the above programme, we use the components of a Fourier expansion with respect to the coordinates and a one-sided Fourier expansion with respect to the time:

$$\delta f^{(+)}_{a\omega\mathbf{k}}(\mathbf{p}) = \int d^3x \int_0^\infty dt\, e^{-i(\mathbf{k}\cdot\mathbf{r} - \omega t)}\delta f_a(t, \mathbf{r}, \mathbf{p}), \tag{51.10}$$

and similarly for $\varphi^{(+)}_{\omega\mathbf{k}}$. Multiplying equations (51.7) and (51.8) by $e^{-i(\mathbf{k}\cdot\mathbf{r} - \omega t)}$ and integrating over dt from 0 to ∞ and over d^3x, we obtain

$$i(\mathbf{k}\,.\,\mathbf{v} - \omega)\delta f^{(+)}_{a\omega\mathbf{k}} - ie_a\mathbf{k}\,.\,(\partial\bar{f}_a/\partial\mathbf{p})\delta\varphi^{(+)}_{\omega\mathbf{k}} = \delta f_{a\mathbf{k}}(0, \mathbf{p}),$$

$$-k^2\delta\varphi^{(+)}_{\omega\mathbf{k}} = 4\pi \sum_a e_a \int \delta f^{(+)}_{a\omega\mathbf{k}}\, d^3p. \tag{51.11}$$

Similar equations have already been encountered several times (cf. (34.10), (34.11)); they lead to the result

$$\delta\varphi^{(+)}_{\omega\mathbf{k}} = -\frac{4\pi}{k^2\epsilon_l(\omega, k)}\sum_a e_a \int \frac{\delta f_{a\mathbf{k}}(0, \mathbf{p})}{i(\mathbf{k}\cdot\mathbf{v}-\omega)}\, d^3p, \tag{51.12}$$

where ϵ_l is the permittivity of a plasma with distribution $\bar{f}(p)$.† Multiplication of two such expressions, followed by statistical averaging, gives

$$\langle\delta\varphi^{(+)}_{\omega\mathbf{k}}\delta\varphi^{(+)}_{\omega'\mathbf{k}'}\rangle = \frac{16\pi^2}{k^4\epsilon_l(\omega, k)\epsilon_l(\omega', k')}\times$$
$$\times\sum_{a,b} e_a e_b \int \frac{\langle\delta f_{a\mathbf{k}}(0, \mathbf{p})\delta f_{b\mathbf{k}'}(0, \mathbf{p}')\rangle}{i(\mathbf{k}\cdot\mathbf{v}-\omega)i(\mathbf{k}'\cdot\mathbf{v}'-\omega')}\, d^3p\, d^3p'. \tag{51.13}$$

The mean value in the numerator of the integrand is related to the Fourier component $g_{ab\mathbf{k}}(\mathbf{p}_1, \mathbf{p}_2)$ of the "initial" correlation function $g_{ab}(\mathbf{r}_1-\mathbf{r}_2, \mathbf{p}_1, \mathbf{p}_2)$ by

$$\langle\delta f_{a\mathbf{k}}(0, \mathbf{p})\delta f_{b\mathbf{k}'}(0, \mathbf{p}')\rangle = (2\pi)^3\delta(\mathbf{k}+\mathbf{k}')g_{ab\mathbf{k}}(\mathbf{p}_1, \mathbf{p}_2);$$

cf. (19.13). Like any single-time correlation function, the initial correlation function must contain a delta-function term expressing the cases where there is only one particle in coinciding elements of phase space:

$$\delta_{ab}\bar{f}(p)\delta(\mathbf{r}_1-\mathbf{r}_2)\delta(\mathbf{p}_1-\mathbf{p}_2);$$

see (19.6). The Fourier transform of this term is $\delta_{ab}\bar{f}(p)\delta(\mathbf{p}_1-\mathbf{p}_2)$. Thus we must put in (51.13)

$$\langle\delta f_{a\mathbf{k}}(0, \mathbf{p})\delta f_{b\mathbf{k}'}(0, \mathbf{p}')\rangle$$
$$= (2\pi)^3\delta(\mathbf{k}+\mathbf{k}')[\delta_{ab}f_a(p)\delta(\mathbf{p}-\mathbf{p}')+\mu_{\mathbf{k}}(\mathbf{p}, \mathbf{p}')], \tag{51.14}$$

where $\mu_{\mathbf{k}}(\mathbf{p}, \mathbf{p}')$ is an arbitrary smooth (non-singular for real $\mathbf{p}$ and $\mathbf{p}'$) function, the Fourier transform of some function $\mu(\mathbf{r}_1-\mathbf{r}_2, \mathbf{p}_1, \mathbf{p}_2)$ which tends to zero as $|\mathbf{r}_1-\mathbf{r}_2|\to\infty$.

On substitution in (51.13), the term containing this arbitrary function in (51.14) gives

$$\frac{4(2\pi)^5\delta(\mathbf{k}+\mathbf{k}')}{k^4\epsilon_l(\omega, k)\epsilon_l(\omega', k')}\sum_{a,b}\int\frac{\mu_k(\mathbf{p}, \mathbf{p}')\, d^3p\, d^3p'}{i(\mathbf{k}\cdot\mathbf{v}-\omega)i(\mathbf{k}\cdot\mathbf{v}'-\omega')}. \tag{51.15}$$

We shall show that this expression corresponds to a function in the time representation that decreases rapidly with increasing t or t'.

The change from the Laplace transform $\langle\delta\varphi^{(+)}_{\omega\mathbf{k}}\delta\varphi^{(+)}_{\omega'\mathbf{k}'}\rangle$ (see the first footnote to

†We shall assume, merely to simplify the subsequent formulae, that the function $\bar{f}(p)$ is istropic, so that the corresponding permittivity tensor $\epsilon_{\alpha\beta}$ reduces to scalars ϵ_l and ϵ_t.

§34) to a function of the times t_2 and $t_1 = t_2 + t$ is made by means of the formula

$$\langle\delta\varphi_{\mathbf{k}}(t_1)\delta\varphi_{\mathbf{k}'}(t_2)\rangle = \int e^{-i\omega t_1 - i\omega' t_2}\langle\delta\varphi^{(+)}_{\omega\mathbf{k}}\delta\varphi^{(+)}_{\omega'\mathbf{k}'}\rangle\, d\omega\, d\omega'/(2\pi)^2, \tag{51.16}$$

where the integration is taken along contours in the complex ω and ω' planes which pass above all singularities of the integrand. We are interested in the asymptotic form of (51.16) as $t_1, t_2 \to \infty$. To find this, we must lower the contours of integration until they "catch" on the singularities; for instance, a singularity at $\omega = \omega_c$ gives an asymptotic time dependence $\exp(-i\omega_c t)$ of the integral with respect to ω. It is easy to see that (51.15) has singularities only in the lower half-planes of ω and ω' (and not on the real axes), and therefore the asymptotic form of the integral (51.16), with (51.15) as $\langle\delta\varphi^{(+)}_{\omega\mathbf{k}}\delta\varphi^{(+)}_{\omega'\mathbf{k}'}\rangle$, contains only damped terms.

Let us consider, for example, the integral with respect to ω. The factor $1/\epsilon_l(\omega, k)$ in (51.15) has poles at the zeros of $\epsilon_l(\omega, k)$, which are all in the lower half of the ω-plane.† The integral over d^3p in (51.15) has a similar property: it has the form

$$\int \frac{\psi(z)\, dz}{z - \omega/k - i0}$$

with $z \equiv v_x$ the component of $\mathbf{v}$ along $\mathbf{k}$, and the factor $\psi(z)$ can (according to the assumed properties of $\mu_{\mathbf{k}}(\mathbf{p}, \mathbf{p}')$) have singularities only for complex z; an integral of this form has already been discussed at the end of §29, and shown to have poles only in the lower half of the ω-plane.

Thus the undamped part of the correlation function, which we wish to determine, arises only from the contribution of the first term in (51.14) to the integral (51.13):

$$\langle\delta\varphi^{(+)}_{\omega\mathbf{k}}\delta\varphi^{(+)}_{\omega'\mathbf{k}'}\rangle = -\frac{4(2\pi)^5\delta(\mathbf{k}+\mathbf{k}')}{k^4\epsilon_l(\omega, k)\epsilon_l(\omega', k)}\sum_a e_a^2 \int \frac{\bar{f}_a(p)\, d^3p}{(\omega - \mathbf{k}\cdot\mathbf{v} + i0)(\omega' + \mathbf{k}\cdot\mathbf{v} + i0)}. \tag{51.17}$$

The integrand is transformed by putting

$$\frac{1}{(\omega - \mathbf{k}\cdot\mathbf{v} + i0)(\omega' + \mathbf{k}\cdot\mathbf{v} + i0)} = \frac{1}{\omega + \omega' + i0}\left[\frac{1}{\omega - \mathbf{k}\cdot\mathbf{v} + i0} + \frac{1}{\omega' + \mathbf{k}\cdot\mathbf{v} + i0}\right].$$

On the further integration with respect to ω' in (51.16), a contribution not damped as $t \to \infty$ comes from the residue at the pole $\omega' = -\omega - i0$, which is avoided by the contour of integration in the way shown in Fig. 14. In this sense, the factor $1/(\omega + \omega')$ is to be interpreted as $-2\pi i\delta(\omega + \omega')$. The significance of the factors $1/(\omega \pm \mathbf{k}\cdot\mathbf{v})$ in the subsequent integration with respect to ω is given by (29.8), according to which

$$\frac{1}{\omega - \mathbf{k}\cdot\mathbf{v} + i0} - \frac{1}{\omega - \mathbf{k}\cdot\mathbf{v} - i0} = -2\pi i\delta(\omega - \mathbf{k}\cdot\mathbf{v});$$

†It is assumed that the distribution $\bar{f}(p)$ corresponds to a stable state of the plasma, so that the plasma waves are damped. It is evident that only in this case is there any meaning in the problem of stationary fluctuations.

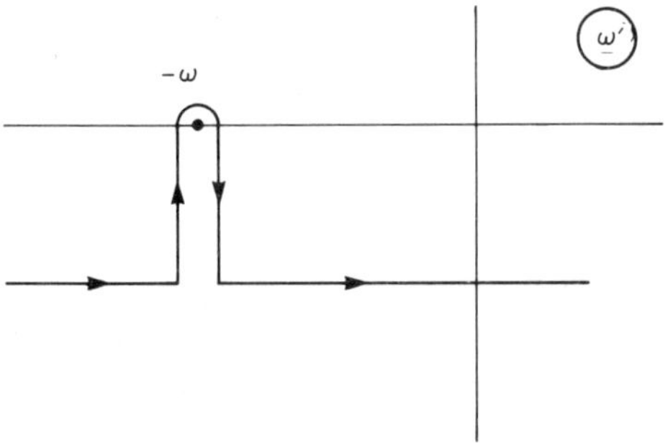

FIG. 14.

this notation implies that the integrations with respect to ω and ω' are taken along the real axis.

Thus, to calculate the correlation function in the asymptotic limit of long times t, we must make in the integral (51.17) the substitution

$$[(\omega - \mathbf{k}.\mathbf{v} + i0)(\omega' + \mathbf{k}.\mathbf{v} + i0)]^{-1} \to -(2\pi)^2 \delta(\omega + \omega')\delta(\omega - \mathbf{k}.\mathbf{v}). \qquad (51.18)$$

The result is†

$$\langle \delta\varphi^{(+)}_{\omega\mathbf{k}} \delta\varphi^{(+)}_{\omega'\mathbf{k}'} \rangle = (2\pi)^4 \delta(\omega + \omega')\delta(\mathbf{k} + \mathbf{k}')(\delta\varphi^2)_{\omega\mathbf{k}}, \qquad (51.19)$$

where

$$(\delta\varphi^2)_{\omega\mathbf{k}} = \frac{32\pi^3}{k^4 |\epsilon_l(\omega, k)|^2} \sum_a e_a^2 \int \bar{f}_a(p)\delta(\omega - \mathbf{k}.\mathbf{v})\, d^3p. \qquad (51.20)$$

It is seen from the definition (51.19) (cf. (19.13)) that the $(\delta\varphi^2)_{\omega\mathbf{k}}$ are the required Fourier transform of the correlation function—the frequency correlation function. Thus formula (51.20) gives the solution of the problem stated for fluctuations of the potential.

The other correlation functions are determined similarly. For example, expressing $\delta f^{(+)}_{a\omega'\mathbf{k}'}$ in terms of $\delta\varphi^{(+)}_{\omega'\mathbf{k}'}$ from (51.11), multiplying by $\delta\varphi^{(+)}_{\omega\mathbf{k}}$ from (51.12), and averaging, we obtain the correlation function for the potential and the distribution function:‡

$$(\delta\varphi\, \delta f_a)_{\omega\mathbf{k}} = \frac{e_a \mathbf{k}}{\mathbf{k}.\mathbf{v} - \omega + i0} \cdot \frac{\partial \bar{f}_a}{\partial \mathbf{p}} (\delta\varphi^2)_{\omega\mathbf{k}} + \frac{8\pi^2 e_a}{k^2 \epsilon_l(\omega, k)} \bar{f}_a(p)\delta(\omega - \mathbf{k}.\mathbf{v}). \qquad (51.21)$$

The order of $\delta\varphi$ and δf_a in $(\delta\varphi\delta f_a)_{\omega\mathbf{k}}$ is significant: by definition (cf. *SP* 1 (122.11)), (51.21) is the Fourier transform of the space–time correlation function

†To avoid misunderstanding, it may be mentioned that this is not the complete expression, but only the part with the singularity in $\omega + \omega'$, which governs the asymptotic form of the correlation function. In the complete expression, not all terms contain $\delta(\omega + \omega')$, since the corresponding function of t_1 and t_2 depends on the difference $t = t_1 - t_2$ only asymptotically for large t_1 and t_2.

‡Note that the avoidance rule for the contour is opposite in the first term ($\omega - i0$ instead of $\omega + i0$). This is because, for $\omega = -\omega'$, $\mathbf{k} = -\mathbf{k}'$, we have $(\mathbf{k}'.\mathbf{v} - \omega' - i0)^{-1} = -(\mathbf{k}.\mathbf{v} - \omega + i0)^{-1}$.

$\langle\delta\varphi(t, \mathbf{r})\delta f_a(0,0)\rangle$. If, however, the correlation function is defined as $\langle\delta f_a(t, \mathbf{r})\delta\varphi(0,0)\rangle$, then we have

$$(\delta f_a \delta\varphi)_{\omega\mathbf{k}} = (\delta\varphi\delta f_a)_{-\omega,-\mathbf{k}} = (\delta\varphi\delta f_a)^*_{\omega\mathbf{k}}; \tag{51.22}$$

cf. *SP* 1 (122.13).

Lastly, the frequency correlation function of the distribution function is

$$\begin{aligned}(\delta f_a\, \delta f_b)_{\omega\mathbf{k}} = {} & 2\pi\delta_{ab}\delta(\mathbf{p}_1-\mathbf{p}_2)\bar{f}_a(p_1)\delta(\omega-\mathbf{k}\cdot\mathbf{v}_1)\\ & + \frac{e_a e_b(\delta\varphi^2)_{\omega\mathbf{k}}}{(\omega-\mathbf{k}\cdot\mathbf{v}_1+i0)(\omega-\mathbf{k}\cdot\mathbf{v}_2-i0)}\left(\mathbf{k}\cdot\frac{\partial\bar{f}_a}{\partial\mathbf{p}_1}\right)\left(\mathbf{k}\cdot\frac{\partial\bar{f}_b}{\partial\mathbf{p}_2}\right)\\ & - \frac{8\pi^2 e_a e_b}{k^2}\left\{\left(\mathbf{k}\cdot\frac{\partial\bar{f}_a}{\partial\mathbf{p}_1}\right)\bar{f}_b\,\frac{\delta(\omega-\mathbf{k}\cdot\mathbf{v}_2)}{\epsilon_l(\omega,k)(\omega-\mathbf{k}\cdot\mathbf{v}_1+i0)}\right.\\ & \left. + \left(\mathbf{k}\cdot\frac{\partial\bar{f}_b}{\partial\mathbf{p}_2}\right)\bar{f}_a\,\frac{\delta(\omega-\mathbf{k}\cdot\mathbf{v}_2)}{\epsilon_l^*(\omega,k)(\omega-\mathbf{k}\cdot\mathbf{v}_2-i0)}\right\}.\end{aligned} \tag{51.23}$$

This is the Fourier transform of the correlation function

$$\langle\delta f_a(t, \mathbf{r}, \mathbf{p}_1)\delta f_b(0, 0, \mathbf{p}_2)\rangle.$$

If the $\bar{f}_a$ in (51.20)–(51.23) are taken to be the Maxwellian functions f_{0a}, we obtain the correlation functions of fluctuations in an equilibrium collisionless plasma.

Let us consider, for example, fluctuations of the potential. For a Maxwellian plasma, the imaginary part of the longitudinal permittivity may be expressed as

$$\epsilon_l''(\omega, k) = \frac{4\pi^2\omega}{k^2 T}\sum_a e_a^2 \int f_{0a}(p)\delta(\omega-\mathbf{k}\cdot\mathbf{v})\, d^3p \tag{51.24}$$

(cf. (30.1)); the generalization to particles of several types is obvious. Substituting this expression in (51.20), we obtain

$$(\delta\varphi^2)_{\omega\mathbf{k}} = 8\pi T\epsilon_l''(\omega, k)/\omega k^2|\epsilon_l(\omega, k)|^2. \tag{51.25}$$

The correlation function of the longitudinal electric field is

$$(E_\alpha E_\beta)_{\omega\mathbf{k}} = k_\alpha k_\beta(\delta\varphi^2)_{\omega\mathbf{k}}. \tag{51.26}$$

This result could, of course, also be derived from the general macroscopic theory of equilibrium electromagnetic fluctuations given in *SP* 2, §§75–77.† According to that theory, the frequency correlation function of the electric field is expressed in terms of the retarded Green's function by a formula which in the classical limit

†The self-consistent field in a plasma is a macroscopic quantity, and the macroscopic theory of fluctuations is therefore applicable to it. The distribution function, however, is not a macroscopic quantity, and its fluctuations always require kinetic treatment.

($\hbar\omega \ll T$) becomes

$$(E_\alpha E_\beta)_{\omega\mathbf{k}} = -(2\omega T/\hbar c^2)\,\mathrm{im}\,D^R_{\alpha\beta}(\omega, \mathbf{k}); \tag{51.27}$$

see *SP* 2 (76.3), (77.2). In a medium with spatial dispersion, the Green's function is†

$$D^R_{\alpha\beta}(\omega, \mathbf{k}) = \frac{4\pi\hbar}{\omega^2\epsilon_t/c^2 - k^2}\left(\delta_{\alpha\beta} - \frac{k_\alpha k_\beta}{k^2}\right) + \frac{4\pi\hbar c^2}{\omega^2\epsilon_l}\frac{k_\alpha k_\beta}{k^2}. \tag{51.28}$$

Substitution of the longitudinal part of this function (the second term) in (51.27) gives (51.25) and (51.26).

Lastly, let us return to (51.4) and show that the expression

$$e_a\frac{\partial}{\partial\mathbf{p}}\cdot\left\langle\frac{\partial\delta\varphi(t, \mathbf{r})}{\partial\mathbf{r}}\,\delta f_a(t, \mathbf{r}, \mathbf{p})\right\rangle \tag{51.29}$$

on the right-hand side is in fact the same as the familiar expression for the collision integral in a plasma. The quantity (51.29) is obtained from the correlation function $\langle\delta\varphi(t, \mathbf{r})\delta f_a(0, 0)\rangle$ by differentiation with respect to $\mathbf{r}$ followed by putting $\mathbf{r} = 0$. Thus we find

$$\left\langle\frac{\partial\delta\varphi}{\partial\mathbf{r}}\,\delta f_a\right\rangle = \int i\mathbf{k}(\delta\varphi\,\delta f_a)_{\omega\mathbf{k}}\,d\omega\,d^3k/(2\pi)^4 = -\int \mathbf{k}\,\mathrm{im}\,(\delta\varphi\,\delta f_a)_{\omega\mathbf{k}}\,d\omega\,d^3k/(2\pi)^4, \tag{51.30}$$

the latter expression being derived by means of (51.22). From (51.21), with (51.20) and (51.24), we have

$$\begin{aligned}\mathrm{im}\,(\delta\varphi\,\delta f_a)_{\omega\mathbf{k}} &= \left\{-\pi\mathbf{k}\cdot\frac{\partial\bar{f}_a}{\partial\mathbf{p}_a}(\delta\varphi^2)_{\omega\mathbf{k}} - \frac{8\pi^2\epsilon''}{k^2|\epsilon|^2}\bar{f}_a\right\}e_a\delta(\omega - \mathbf{k}\,.\,\mathbf{v}_a)\\ &= -\frac{32\pi e_a}{k^4|\epsilon_l|^2}\sum_b e_b^2\int\mathbf{k}\cdot\left\{\frac{\partial\bar{f}_a}{\partial\mathbf{p}_a}\bar{f}_b - \bar{f}_a\frac{\partial\bar{f}_b}{\partial\mathbf{p}_b}\right\}\times\\ &\qquad\times\delta(\omega - \mathbf{k}\,.\,\mathbf{v}_b)\,d^3p_b\,.\,\delta(\omega - \mathbf{k}\,.\,\mathbf{v}_a).\end{aligned}$$

Substitution of this in (51.30) easily converts (51.29) to the Balescu–Lenard collision integral (§ 47).

In this proof, it may appear strange that, to calculate the collision integral, it was sufficient to consider fluctuations in a collisionless plasma. This occurs because the important Fourier components of the electric field in collisions in plasmas are those with $k \gtrsim 1/a \gg 1/l$, so that collisions may be neglected. The situation here is exactly similar to that in the derivation of the Boltzmann transport equation (§ 16): equation (16.10) signifies the neglect of the influence of collisions on the pair correlation function.

†This is obtained from *SP* 2 (75.20), by dividing that expression into transverse and longitudinal parts, and replacing ϵ in the two parts by $\epsilon_t(\omega, k)$ and $\epsilon_l(\omega, k)$ respectively.

CHAPTER V

PLASMAS IN MAGNETIC FIELDS

§52. Permittivity of a collisionless cold plasma

THIS chapter deals with the properties of plasmas in an external magnetic field, which are said to be *magnetoactive*. By forcing the charged particles to move in helical paths along the lines of force, the magnetic field exerts a profound influence on the behaviour of the plasma. In particular, it affects the dielectric properties.

Let us first recall some general properties of the permittivity tensor in the presence of a magnetic field with induction **B** (see *ECM*, §82). As when the field is absent, equation (28.6) is valid:

$$\epsilon_{\alpha\beta}(-\omega, -\mathbf{k}; \mathbf{B}) = \epsilon^*_{\alpha\beta}(\omega, \mathbf{k}; \mathbf{B}). \tag{52.1}$$

According to Onsager's principle, this tensor is symmetrical when the signs of the field and the wave vector are simultaneously changed:

$$\epsilon_{\alpha\beta}(\omega, \mathbf{k}; \mathbf{B}) = \epsilon_{\beta\alpha}(\omega, -\mathbf{k}; -\mathbf{B}). \tag{52.2}$$

If the medium is invariant under spatial inversion (as is an equilibrium plasma), the $\epsilon_{\alpha\beta}$ are even functions of **k**, and (52.2) becomes

$$\epsilon_{\alpha\beta}(\omega, \mathbf{k}; \mathbf{B}) = \epsilon_{\beta\alpha}(\omega, \mathbf{k}; -\mathbf{B}). \tag{52.2a}$$

It must be emphasized, however, that this property occurs only in a medium in thermodynamic equilibrium, unlike (52.1), which follows from the definition of $\epsilon_{\alpha\beta}$.

In the general case, the tensor $\epsilon_{\alpha\beta}$ may be divided into an Hermitian part $\frac{1}{2}(\epsilon_{\alpha\beta} + \epsilon^*_{\beta\alpha})$ and an anti-Hermitian part $\frac{1}{2}(\epsilon_{\alpha\beta} - \epsilon^*_{\beta\alpha})$. The latter determines the dissipation of the field energy in the medium; cf. (30.3).

We shall begin the study of magnetoactive plasmas with the simple case of a "cold" collisionless plasma. The temperature of such a plasma is assumed to be so low that the thermal motion of the particles is negligible; the conditions for this will be formulated below. In this approximation, there is no spatial dispersion, and the permittivity depends only on the frequency of the electric field. There is also no dissipation, and the tensor $\epsilon_{\alpha\beta}$ is therefore Hermitian:

$$\epsilon_{\alpha\beta}(\omega; \mathbf{B}) = \epsilon^*_{\beta\alpha}(\omega; \mathbf{B}). \tag{52.3}$$

From this and (52.1), it follows that

$$\epsilon_{\alpha\beta}(\omega; \mathbf{B}) = \epsilon_{\beta\alpha}(-\omega; \mathbf{B}). \tag{52.4}$$

Separating the Hermitian tensor into real and imaginary parts, $\epsilon_{\alpha\beta} = \epsilon'_{\alpha\beta} + i\epsilon''_{\alpha\beta}$, we have from (52.2) and (52.3)

$$\left.\begin{aligned}\epsilon'_{\alpha\beta}(\omega;\mathbf{B}) &= \epsilon'_{\beta\alpha}(\omega;\mathbf{B}) = \epsilon'_{\alpha\beta}(\omega;-\mathbf{B}),\\ \epsilon''_{\alpha\beta}(\omega;\mathbf{B}) &= -\epsilon''_{\beta\alpha}(\omega;\mathbf{B}) = -\epsilon''_{\alpha\beta}(\omega;-\mathbf{B}).\end{aligned}\right\} \tag{52.5}$$

Thus, in a non-dissipative medium, the $\epsilon'_{\alpha\beta}$ are even functions of the field, and the $\epsilon''_{\alpha\beta}$ are odd functions.

We shall suppose that the anisotropy of the plasma is due only to the presence of a constant and uniform magnetic field, whose induction within the plasma is denoted by $\mathbf{B}_0$. In such a case, the general linear relation between the induction and strength of a weak monochromatic electric field is

$$\mathbf{D} = \epsilon_\perp \mathbf{E} + (\epsilon_\parallel - \epsilon_\perp)\mathbf{b}(\mathbf{b}\,.\,\mathbf{E}) + ig\mathbf{E}\times\mathbf{b}, \tag{52.6}$$

where $\mathbf{b} = \mathbf{B}_0/B_0$; $\epsilon_\perp$, $\epsilon_\parallel$ and g are functions of ω and B_0. This relation is written in tensor form as $D_\alpha = \epsilon_{\alpha\beta}E_\beta$, where

$$\epsilon_{\alpha\beta} = \epsilon_\perp\delta_{\alpha\beta} + (\epsilon_\parallel - \epsilon_\perp)b_\alpha b_\beta + ige_{\alpha\beta\gamma}b_\gamma. \tag{52.7}$$

If the z-axis is taken along $\mathbf{B}_0$, the components of this tensor are

$$\left.\begin{aligned}\epsilon_{xx} &= \epsilon_{yy} = \epsilon_\perp, & \epsilon_{zz} &= \epsilon_\parallel,\\ \epsilon_{xy} &= -\epsilon_{yx} = ig, & \epsilon_{xz} &= \epsilon_{yz} = 0.\end{aligned}\right\} \tag{52.8}$$

From the condition for the tensor (52.7) to be Hermitian, it follows that $\epsilon_\perp$, $\epsilon_\parallel$ and g are real, and from (52.4) it follows that $\epsilon_\perp$ and $\epsilon_\parallel$ are even functions of the frequency, g an odd function. The expression (52.7) necessarily satisfies Onsager's principle.

In weak fields the tensor $\epsilon_{\alpha\beta}$ must be expandable in integral powers of the vector $\mathbf{B}_0$. Hence, as $\mathbf{B}_0 \to 0$, the coefficient $\epsilon_\perp$ tends to a finite limit, the permittivity in the absence of the magnetic field. The difference $\epsilon_\perp - \epsilon_\parallel \propto B_0^2$, and $g \propto B_0$.

The calculation of $\epsilon_{\alpha\beta}$ in this approximation can be made directly from the equations of motion of particles in a variable field $\mathbf{E}$ and a constant field $\mathbf{B}_0$, as in the derivation of (31.9). For example, for electrons

$$m\,d\mathbf{v}/dt = -e\mathbf{E} - e\mathbf{v}\times\mathbf{B}_0/c. \tag{52.9}$$

The velocity $\mathbf{v}$ varies with time in the same way ($\propto e^{-i\omega t}$) as the field $\mathbf{E}$. Neglecting the spatial variation of $\mathbf{E}$ in the region of motion of the particle, we have from (52.9)

$$i\omega\mathbf{v} = e\mathbf{E}/m + e\mathbf{v}\times\mathbf{B}_0/mc.$$

The solution of this algebraic vector equation contains terms parallel to $\mathbf{E}$, $\mathbf{b}$ and $\mathbf{E}\times\mathbf{b}$; if the coefficients in these terms are apppropriately chosen, we obtain

$$\mathbf{v} = -\frac{ie\omega}{m(\omega^2 - \omega_{Be}^2)}\left\{\mathbf{E} - \frac{\omega_{Be}^2}{\omega^2}\mathbf{b}(\mathbf{E}\,.\,\mathbf{b}) - \frac{i\omega_{Be}}{\omega}\mathbf{E}\times\mathbf{b}\right\}, \tag{52.10}$$

where $\omega_{Be} = eB_0/mc$. The polarization **P** due to the motion of the electrons, and therefore the induction **D**, are related to the electron velocity by (29.4):

$$-i\omega\mathbf{P} = -i\omega\frac{\mathbf{D}-\mathbf{E}}{4\pi} = \mathbf{j} = -eN_e\mathbf{v}.$$

The ionic contribution to the polarization is calculated in the same way, and the two contributions are additive. The result is

$$\left.\begin{aligned} \epsilon_\perp &= 1 - \frac{\Omega_e^2}{\omega^2-\omega_{Be}^2} - \frac{\Omega_i^2}{\omega^2-\omega_{Bi}^2}.\\ \epsilon_\parallel &= 1 - \frac{\Omega_e^2+\Omega_i^2}{\omega^2},\\ g &= \frac{\omega_{Be}\Omega_e^2}{\omega(\omega^2-\omega_{Be}^2)} - \frac{\omega_{Bi}\Omega_i^2}{\omega(\omega^2-\omega_{Bi}^2)}. \end{aligned}\right\} \tag{52.11}$$

Here

$$\omega_{Be} = eB_0/mc, \quad \omega_{Bi} = zeB_0/Mc \tag{52.12}$$

are the electron and ion *Larmor frequencies*;† the values of these parameters are an important characteristic of a magnetoactive plasma (they are the frequencies of rotation of charged particles in circular orbits in the magnetic field).

The ratios

$$\omega_{Bi}/\omega_{Be} = zm/M, \Omega_i/\Omega_e = (zm/M)^{1/2} \tag{52.13}$$

are small quantities. The ratio of the frequencies Ω_e and ω_{Be}, or Ω_i and ω_{Bi}, which depend on entirely different parameters (the plasma density, and the field B_0), may vary over a very wide range.

The ion contribution to the permittivity of a magnetoactive plasma may, despite the large mass of the ions, be comparable with or even greater than the electron contribution at sufficiently low frequencies ω. As $\omega \to 0$, the two terms in g cancel, and $g \to 0$, as is easily seen by noting that

$$\Omega_e^2/\omega_{Be} = \Omega_i^2/\omega_{Bi} \tag{52.14}$$

because of the electrical neutrality of the plasma ($N_e = zN_i$). The two terms in g remain of the same order of magnitude when $\omega \sim \omega_{Bi}$, and the ion part of g is negligible when $\omega \gg \omega_{Bi}$. In the transverse permittivity $\epsilon_\perp$, the two terms are comparable only in the range

$$\omega \sim \omega_{Bi}(M/m)^{1/2} \sim (\omega_{Bi}\omega_{Be})^{1/2}.$$

†Also called *cyclotron frequencies* or *gyro-frequencies*.

The ion contribution is negligible here only if

$$\omega \gg (\omega_{Bi}\omega_{Be})^{1/2}. \tag{52.15}$$

Lastly, in the longitudinal permittivity $\epsilon_{\parallel}$ (which contains the sum of Ω_e^2 and Ω_i^2) the ion part is always negligible. Incidentally, $\epsilon_{\parallel}$ is independent of $\mathbf{B}_0$ because the field $\mathbf{E}$ has been assumed uniform: in crossed uniform fields, the magnetic field does not affect the motion of particles parallel to $\mathbf{B}_0$.

Let us finally consider the conditions for the above formulae to be applicable. In applying equation (52.9) to the motion of particles, we have neglected the spatial variation of $\mathbf{E}$ in the region where the particles are. The size of this region in the direction of the constant field $\mathbf{B}_0$ is given by the distance v_T/ω traversed by a particle moving with the mean thermal speed v_T during the period of the variable field. In the directions perpendicular to $\mathbf{B}_0$, the size of the region when $\omega < \omega_B$ is determined by

$$r_B \sim v_T/\omega_B, \tag{52.16}$$

the radius of the circular orbits of particles moving with speed v_T in the magnetic field $\mathbf{B}_0$, called the *Larmor radius* of the particles. The approximation described above requires that these distances be small in comparison with those over which the field $\mathbf{E}$ varies in the relevant directions:

$$v_T|k_z|/\omega \ll 1, \quad v_T k_\perp/\omega_B \ll 1, \tag{52.17}$$

where $k_z \equiv k_\parallel$, and $\mathbf{k}_\perp$, are the components of the wave vector along and across the field $\mathbf{B}_0$. These inequalities must be satisfied for each type of particle in the plasma.

We shall see below that the frequency ω must also not be too close to ω_{Be}, ω_{Bi} or a multiple of these (the conditions (53.17)). Near such frequencies, the spatial dispersion has to be taken into account even if the conditions (52.17) are satisfied. As we shall see in §55, this eliminates the poles of the expressions (52.11) at $\omega^2 = \omega_{Be}^2, \omega_{Bi}^2$.

§53. The distribution function in a magnetic field

The permittivity tensor in a collisionless magnetoactive plasma, with allowance for spatial dispersion, is calculated from the electron and ion distribution functions which are determined by the transport equation.

All formulae will be written for the particular case of electrons. The transport equations for a collisionless plasma have been given in §27. For electrons, the equation is

$$\frac{\partial f}{\partial t} + \mathbf{v}\cdot\frac{\partial f}{\partial \mathbf{r}} - e(\mathbf{E} + \mathbf{v}\times\mathbf{B}/c)\cdot\frac{\partial f}{\partial \mathbf{p}} = 0. \tag{53.1}$$

Let the plasma be in a constant and uniform magnetic field $\mathbf{B}_0$ of any strength, and a weak variable electromagnetic field in which

$$\mathbf{E}, \mathbf{B}' \propto e^{i(\mathbf{k}\,.\,\mathbf{r}-\omega t)}. \tag{53.2}$$

From Maxwell's equations,

$$\omega \mathbf{B}'/c = \mathbf{k} \times \mathbf{E}. \tag{53.3}$$

We substitute in (53.1) $\mathbf{B} = \mathbf{B}_0 + \mathbf{B}'$, and express the distribution function as $f = f_0 + \delta f$, where f_0 is the stationary uniform distribution in the absence of the variable field; the small correction δf depends on t and $\mathbf{r}$ in the same way (53.2) as the fields $\mathbf{E}$ and $\mathbf{B}'$ to which it is proportional. Separating the zero-order and first-order terms (relative to the weak field) in the equation, we obtain†

$$\frac{\partial f}{\partial \mathbf{p}} \cdot \mathbf{v} \times \mathbf{B}_0 = 0, \tag{53.4}$$

$$i(\mathbf{k}\,.\,\mathbf{v} - \omega)\delta f - \frac{e}{c}\mathbf{v} \times \mathbf{B}_0 \cdot \frac{\partial \delta f}{\partial \mathbf{p}} = e\frac{\partial f_0}{\partial \mathbf{p}} \cdot \left\{\mathbf{E} + \frac{1}{\omega}\mathbf{v} \times (\mathbf{k} \times \mathbf{E})\right\}. \tag{53.5}$$

Let v_z and k_z denote the components of the vectors $\mathbf{v}$ and $\mathbf{k}$ along the field $\mathbf{B}_0$; $\mathbf{v}_\perp$ and $\mathbf{k}_\perp$, the components in the plane perpendicular to $\mathbf{B}_0$; φ, the angle between $\mathbf{v}_\perp$ and the plane of $\mathbf{k}_\perp$ and $\mathbf{B}_0$, measured in the direction of rotation of a corkscrew driven along $\mathbf{B}_0$. The variables v_z, $v_\perp$ and φ are cylindrical polar coordinates in $\mathbf{v}$-space. In these variables, (53.5) becomes

$$i(k_z v_z + k_\perp v_\perp \cos\varphi - \omega)\delta f + \omega_{Be}\partial \delta f/\partial \varphi = e\left\{\mathbf{E} + \frac{1}{\omega}\mathbf{v} \times (\mathbf{k} \times \mathbf{E})\right\} \cdot \partial f_0/\partial \mathbf{p}. \tag{53.6}$$

From (53.4) it follows that $\partial f_0/\partial \varphi = 0$, i.e. f_0 can be any function of p_z and $p_\perp$ only:

$$f_0 = f_0(p_z, p_\perp), \tag{53.7}$$

a result that is obvious for a collisionless plasma, since p_z and $p_\perp$ are the variables not affected by the magnetic field.

To simplify the formulae, we use the notation

$$\alpha = (k_z v_z - \omega)/\omega_{Be}, \quad \beta = k_\perp v_\perp/\omega_{Be}, \tag{53.8}$$

$$Q(v_z, v_\perp, \varphi) = \frac{e}{\omega_{Be}}\frac{\partial f_0}{\partial \mathbf{p}} \cdot \left\{\mathbf{E} + \frac{1}{\omega}\mathbf{v} \times (\mathbf{k} \times \mathbf{E})\right\}. \tag{53.9}$$

If f_0 depends only on the electron energy $\epsilon = p^2/2m$, the derivative $\partial f_0/\partial \mathbf{p} = \mathbf{v}\, df_0/d\epsilon$, and its product with the second term in the braces is zero, so that

$$Q = \frac{e}{\omega_{Be}}\frac{df_0}{d\epsilon}\mathbf{v}\,.\,\mathbf{E}. \tag{53.10}$$

†In a cold plasma, there was no need to take account of the Lorentz force exerted by the weak field $\mathbf{B}'$, since it is of the second order of smallness when the intrinsic movement of the particles (in the absence of the field) is neglected.

With this notation, (53.6) becomes

$$\partial\delta f/\partial\varphi + i(\alpha + \beta\cos\varphi)\delta f = Q(\varphi); \tag{53.11}$$

the arguments v_z and $v_\perp$ of Q are omitted. The solution is

$$\delta f = e^{-i(\alpha\varphi+\beta\sin\varphi)}\int_C^{\varphi} e^{i(\alpha\varphi'+\beta\sin\varphi')}Q(\varphi')\,d\varphi',$$

or, with the change of variable of integration $\varphi' = \varphi - \tau$,

$$\delta f = e^{-i\beta\sin\varphi}\int_0^{C-\varphi} e^{i\beta\sin(\varphi-\tau)-i\alpha\tau}Q(\varphi-\tau)\,d\tau.$$

The constant C is determined by the condition that the function δf be periodic in φ with the period 2π. Since the integrand and the coefficient of the integral are periodic in φ, this condition is satisfied if the limits of integration are independent of φ, for which we must take $C = \infty$ or $C = -\infty$. The choice between these two possibilities is decided by the Landau contour rule (29.6): the integration is to be taken with $\omega \to \omega + i0$, i.e. $\alpha \to \alpha - i0$. Such an integral is convergent only with $C = \infty$†. The final result is

$$\begin{aligned}\delta f &= e^{-i\beta\sin\varphi}\int_0^{\infty} e^{i\beta\sin(\varphi-\tau)-i\alpha\tau}Q(\varphi-\tau)\,d\tau \\ &= \int_0^{\infty}\exp\{-i\alpha\tau - 2i\beta\cos(\varphi-\tfrac{1}{2}\tau)\sin\tfrac{1}{2}\tau\}Q(\varphi-\tau)\,d\tau.\end{aligned} \tag{53.12}$$

In the limit $B_0 \to 0$, this should become (29.2). To take the limit, we note that for $\alpha \gg 1$ the important range in the integral is $\tau \ll 1$. Then $\sin(\varphi-\tau) \approx \sin\varphi - \tau\cos\varphi$, and the integral becomes

$$\begin{aligned}\delta f &= Q(\varphi)\int_0^{\infty} e^{-i(\alpha+\beta)\tau}\,d\tau \\ &= Q(\varphi)\int_0^{\infty}\exp\left\{-i\tau\frac{\mathbf{k}\cdot\mathbf{v}-\omega}{\omega_{Be}}\right\}d\tau.\end{aligned}$$

Taking the integral with $\omega \to \omega + i0$, we obtain

$$\delta f = Q\omega_{Be}/i(\mathbf{k}\cdot\mathbf{v}-\omega), \tag{53.13}$$

the required result.

If the field frequency is equal to the Larmor frequency ω_{Be} or a multiple of it, we have *simple* or *multiple cyclotron resonance* (of the electrons). To study the dielectric properties of the plasma near such resonances, a different method of

†This conclusion depends on the sign of ω in the exponent. For ions, the charge $-e$ is replaced by ze, so that $\omega_{Be} \to -\omega_{Bi}$. Then $\alpha \to \alpha + i0$ when $\omega \to \omega + i0$, and C would have to be taken as $-\infty$.

solving equation (53.11) is convenient, based on expanding the function sought in a Fourier series with respect to the variable φ.

With the substitution

$$\delta f = e^{-i\beta \sin\varphi} g \tag{53.14}$$

in (53.11), we obtain for g the equation

$$\partial g/\partial\varphi + i\alpha g = e^{i\beta \sin\varphi} Q(v_z, v_\perp, \varphi).$$

The solution is sought as a Fourier series

$$g = \sum_{s=-\infty}^{\infty} e^{is\varphi} g_s(v_z, v_\perp) \tag{53.15}$$

and the coefficients g_s are found to be

$$\left.\begin{aligned} g_s &= Q_s/i(\alpha + s), \\ Q_s(v_z, v_\perp) &= \frac{1}{2\pi}\int_0^{2\pi} e^{i(\beta \sin\tau - s\tau)} Q(v_z, v_\perp, \tau)\, d\tau. \end{aligned}\right\} \tag{53.16}$$

The expansion (53.15) automatically makes δf a periodic function of φ.

First of all, the expression of δf as the series (53.14), (53.15) allows an immediate formulation of the conditions for spatial dispersion to be negligible. The wave number appears in the terms of the series through the parameters

$$\beta = k_\perp v_\perp/\omega_B, \quad \alpha + s = -(\omega - s\omega_B - k_z v_z)/\omega_B.$$

The permittivity of the plasma is determined by the distribution function at speeds $v \sim v_T$. The wave number does not appear in this function if

$$k_\perp v_\perp \ll \omega_B, \quad |\omega - s\omega_B| \gg |k_z| v_T. \tag{53.17}$$

The first inequality (53.17), and the second one with $s = 0$, are the same as the conditions (52.17). We see that, as well as these conditions, the frequency ω must not be too close to any of the cyclotron resonances.

In the neighbourhood of these resonances, the distribution function can be represented by a single term of the Fourier series under certain conditions, namely

$$|k_z| v_T \ll \omega_B, \quad |\omega - n\omega_B| \ll \omega_B, \tag{53.18}$$

where n is any of the numbers $0, \pm 1, \pm 2 \ldots$ It is easily seen that the nth term in the expansion (53.15) is then large compared with the others:

$$g_n \sim Q_n\omega_B/(|k_z v_T| + |\omega - n\omega_B|) \gg Q_n,$$

whereas $g_s \lesssim Q_s$ for $s \neq n$ (since $|s\omega_B - \omega| \gtrsim \omega_B$). Retaining this one term, we find for the electron distribution function

$$\left.\begin{aligned}\delta f &= Q_n \frac{\omega_{Be}\exp[i\{n\varphi - (k_\perp v_\perp/\omega_{Be})\sin\varphi\}]}{i[k_z v_z - (\omega - n\omega_{Be})]},\\ Q_n &= \frac{1}{2\pi}\int_0^{2\pi} \exp[-i\{n\tau - (k_\perp v_\perp/\omega_{Be})\sin\tau\}]Q(v_z, v_\perp, \tau)\,d\tau.\end{aligned}\right\} \tag{53.19}$$

The dependence of the distribution function on the angle φ is given explicitly by this formula. In particular, when $n = 0$ and $k_\perp \to 0$, the distribution is independent of φ. The source of this property is evident from the condition $\omega \ll \omega_{Be}$, i.e. (53.18) with $n = 0$: the Larmor rotation frequency is much greater than the field variation frequency, and this "averages" the distribution frequency over the angle of rotation.†

§54. **Permittivity of a magnetoactive Maxwellian plasma**

The electron contribution to the permittivity tensor is calculated from the distribution function by means of the formula

$$P_\alpha = \frac{\epsilon_{\alpha\beta} - \delta_{\alpha\beta}}{4\pi} E_\beta = \frac{e}{i\omega}\int v_\alpha\,\delta f\,d^3p \tag{54.1}$$

and the ion contribution is similar with $-ze$ instead of e. For a Maxwellian plasma, the integration over d^3p in this expression can be carried out explicitly.

The function δf is given by the integral (53.12), and from the definition (53.10)

$$Q = -(e\mathbf{E}\,.\,\mathbf{v}/\omega_{Be}T)f_0. \tag{54.2}$$

We can rewrite the integral more compactly by using instead of $\mathbf{k} = (k_z, \mathbf{k}_\perp)$ and $\mathbf{E} = (E_z, \mathbf{E}_\perp)$ the vectors

$$\mathbf{K} = (k_z\tau, 2\tilde{\mathbf{k}}_\perp \sin\tfrac{1}{2}\tau), \quad \tilde{\mathbf{E}} = (E_z, \tilde{\mathbf{E}}_\perp), \tag{54.3}$$

where $\tilde{\mathbf{k}}_\perp$ is $\mathbf{k}_\perp$ turned through $\frac{1}{2}\tau$ in the plane perpendicular to $\mathbf{B}_0$, and $\tilde{\mathbf{E}}_\perp$ is $\mathbf{E}_\perp$ turned through τ. Then δf becomes

$$\delta f = -\frac{e}{T\omega_{Be}}\int_0^\infty \exp\left\{\frac{i}{\omega_{Be}}(\omega\tau - \mathbf{K}\,.\,\mathbf{v})\right\} f_0(p)\tilde{\mathbf{E}}\cdot\mathbf{v}\,d\tau,$$

where $f_0(p)$ is the Maxwellian distribution function.

†The relevant arguments are given more fully for an analogous situation in §1.

This expression is substituted in (54.1), and the variable of integration $\mathbf{p} = m\mathbf{v}$ is changed according to

$$\mathbf{v} = \mathbf{u} - i\mathbf{K}T/m\omega_{Be}.$$

The integration over d^3u is elementary, and the result is

$$\mathbf{P} = \frac{ie^2N_e}{m\omega\omega_{Be}}\int_0^\infty \left(\tilde{\mathbf{E}} - \frac{T}{m\omega_{Be}^2}(\tilde{\mathbf{E}}\,.\,\mathbf{K})\mathbf{K}\right)\exp\left[-\frac{i\omega\tau}{\omega_{Be}} - \frac{\mathbf{K}^2T}{2m\omega_{Be}^2}\right]d\tau. \tag{54.4}$$

According to the definition (54.3),

$$\mathbf{K}^2 = k_z^2\tau^2 + 4k_\perp^2\sin^2\tfrac{1}{2}\tau.$$

By writing (54.4) in components, we find the components of the tensor $\epsilon_{\alpha\beta}$. The coordinate axes are chosen as follows: z along $\mathbf{B}_0$, x along $\mathbf{k}_\perp$, and y along $\mathbf{B}_0\times\mathbf{k}_\perp$ (Fig. 15). A simple calculation gives

$$\epsilon_{\alpha\beta} - \delta_{\alpha\beta} = \frac{i\Omega_e^2}{\omega\omega_{Be}}\int_0^\infty \kappa_{\alpha\beta}\exp\left\{i\tau\frac{\omega + i0}{\omega_{Be}} - \tfrac{1}{2}k_z^2r_{Be}^2\tau^2 - 2k_\perp^2r_{Be}^2\sin^2\tfrac{1}{2}\tau\right\}d\tau, \tag{54.5}$$

where

$$\left.\begin{aligned}
\kappa_{xx} &= \cos\tau - (k_\perp r_{Be})^2\sin^2\tau,\\
\kappa_{yy} &= \cos\tau + 4(k_\perp r_{Be})^2\sin^4\tfrac{1}{2}\tau,\\
\kappa_{zz} &= 1 - (k_z r_{Be})^2\tau^2,\\
\kappa_{xy} &= -\kappa_{yx} = -\sin\tau + 2(k_\perp r_{Be})^2\sin\tau\sin^2\tfrac{1}{2}\tau,\\
\kappa_{xz} &= \kappa_{zx} = -k_zk_\perp r_{Be}^2\tau\sin\tau,\\
\kappa_{yz} &= -\kappa_{zy} = -2k_zk_\perp r_{Be}^2\tau\sin^2\tfrac{1}{2}\tau;
\end{aligned}\right\} \tag{54.6}$$

$r_{Be} = v_{Te}/\omega_{Be}$ is the electron Larmor radius.

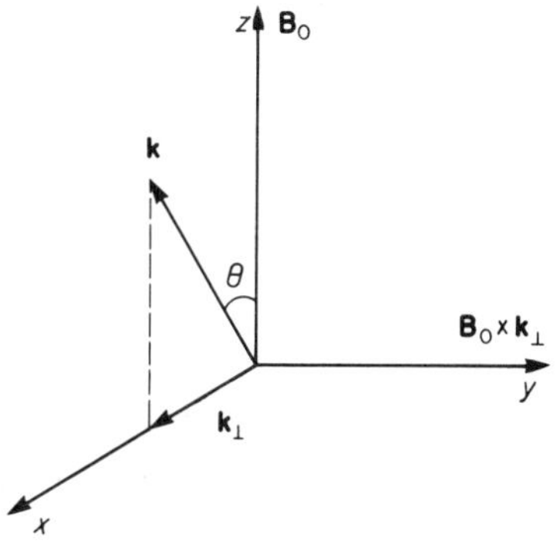

FIG. 15.

The equations

$$\epsilon_{xy} = -\epsilon_{yx}, \quad \epsilon_{xz} = \epsilon_{zx}, \quad \epsilon_{yz} = -\epsilon_{zy}, \tag{54.7}$$

are evident as follows: with fixed coordinates, Onsager's principle gives $\epsilon_{\alpha\beta}(\mathbf{B}_0) = \epsilon_{\beta\alpha}(-\mathbf{B}_0)$, and with the above choice of axes tied to the directions of $\mathbf{B}_0$ and $\mathbf{k}_\perp$ the y- and z-axes are reversed when $\mathbf{B}_0 \to -\mathbf{B}_0$. With these axes, therefore,

$$\left.\begin{aligned} \epsilon_{xy}(\mathbf{B}_0) &= -\epsilon_{yx}(-\mathbf{B}_0), \\ \epsilon_{xz}(\mathbf{B}_0) &= -\epsilon_{zx}(-\mathbf{B}_0), \\ \epsilon_{yz}(\mathbf{B}_0) &= \epsilon_{zy}(-\mathbf{B}_0). \end{aligned}\right\} \tag{54.8}$$

Now $\mathbf{B}_0$ (the direction of the z-axis) is a pseudovector; $\mathbf{k}_\perp$ and $\mathbf{B}_0 \times \mathbf{k}_\perp$ (the directions of the x- and the y-axis) are true vectors. Hence, because of the requirement of invariance under inversion of the coordinates, the components ϵ_{xz} and ϵ_{yz}, which contain one suffix z, must be odd functions of $\mathbf{B}_0$, and all other components must be even functions. Hence equations (54.8) imply (54.7).

Because of the relations (54.7), the Hermitian and anti-Hermitian parts of the various components $\epsilon_{\alpha\beta} = \epsilon'_{\alpha\beta} + i\epsilon''_{\alpha\beta}$ are differently expressed in terms of their real and imaginary parts. The division into Hermitian and anti-Hermitian parts is expressed by the sum

$$(\epsilon_{\alpha\beta}) = \begin{pmatrix} \epsilon'_{xx} & i\epsilon''_{xy} & \epsilon'_{xz} \\ -i\epsilon''_{xy} & \epsilon'_{yy} & i\epsilon''_{yz} \\ \epsilon'_{xz} & -i\epsilon''_{yz} & \epsilon'_{zz} \end{pmatrix} + \begin{pmatrix} i\epsilon''_{x} & \epsilon'_{xy} & i\epsilon''_{xz} \\ -\epsilon'_{xy} & i\epsilon''_{yy} & \epsilon'_{yz} \\ i\epsilon''_{xz} & -\epsilon'_{yz} & i\epsilon''_{zz} \end{pmatrix}. \tag{54.9}$$

Although all the calculations have been made for the electron part of the permittivity, exactly similar formulae are valid for the ion contribution. The change to the latter is made by putting Ω_e, $v_{Te} \to \Omega_i$, v_{Ti} and $\omega_{Be} \to -\omega_{Bi}$, and simultaneously changing the upper limit of the integral in (54.5) to $-\infty$; see the third footnote to § 53. Then, with the change of the variable of integration $\tau \to -\tau$, we return to the previous expressions (54.5), (54.6) with Ω_i, v_{Ti}, ω_{Bi} instead of Ω_e, v_{Te}, ω_{Be}, and a change in sign of κ_{xy} and κ_{yz}. Thus the rule for going from the electron contribution to the ion contribution to the permittivity is to replace the electron parameters by ion parameters and at the same time to change the sign of the components ϵ_{xy} and ϵ_{yz}.

§ 55. **Landau damping in magnetoactive plasmas**

When the thermal movement of plasma particles is taken into account, the tensor $\epsilon_{\alpha\beta}$ acquires an anti-Hermitian part. In a collisionless plasma, since there is no true dissipation of energy, this part of the tensor is due to Landau damping.

We have seen in § 30 that the mechanism of Landau damping depends on the transfer of electromagnetic field energy to particles moving in phase with the wave: the damping involves particles for which $\omega = \mathbf{k}\cdot\mathbf{v}$, i.e. the component of the velocity **v** in the direction of **k** is equal to the phase velocity ω/k of the wave. In a magnetoactive plasma, this condition is somewhat altered: the components of the particle velocity and the wave phase velocity along the constant field $\mathbf{B}_0$ must be equal:

$$v_z k_z = \omega. \tag{55.1}$$

This is because the motion of the particle transversely to $\mathbf{B}_0$ is in circles and cannot involve any systematic transfer of energy from the field to the particle: if the particle moves in phase with the wave and gains energy from it in one part of the circle, a similar amount of energy will be transferred from the particle to the field in the opposite part of the circle.

In a magnetoactive plasma, however, there is a further mechanism of collisionless dissipation, due to the Larmor rotation of the particles. In coordinates moving with the particle at speed v_z along the field $\mathbf{B}_0$, the particle moves in a circular orbit with frequency ω_B. Such a particle is electrodynamically an oscillator radiating at frequency ω_B (synchrotron radiation). When an oscillator is placed in a variable external field, it absorbs at this frequency. The electromagnetic wave frequency in coordinates moving relative to the plasma is modified by the Doppler effect and is $\omega' = \omega - k_z v_z$. The particles concerned in the absorption are therefore those for which

$$\omega - k_z v_z = \omega_B.$$

If $\mathbf{k}_\perp = 0$, the wave field is uniform in the directions transverse to $\mathbf{B}_0$, i.e. the stimulating force on the oscillator is independent of the latter's coordinates. Under these conditions, the oscillator absorbs only at its frequency ω_B. If, however, $\mathbf{k}_\perp \neq 0$, the stimulating force depends on the coordinates of the oscillator, and so there is absorption at multiple frequencies also, i.e. when

$$\omega - k_z v_z = n\omega_B, \tag{55.2}$$

where n is any positive or negative integer. This mechanism of dissipation is called *Landau cyclotron damping*; the cyclotron resonance is simple ($n = \pm 1$) or multiple, according to the value of n.

Thus there can be considerable damping in frequency ranges for which

$$|\omega - n\omega_B| \lesssim |k_z| v_T, \quad n = 0, \pm 1, \pm 2, \ldots, \tag{55.3}$$

the value $n = 0$ corresponding to the condition (55.1). These resonance absorption lines exist at the electron and ion frequencies ω_{Be} and ω_{Bi}.

Mathematically, the conditions (55.1) and (55.2) correspond to the poles, at these points, of the various terms in the Fourier-series expansion (53.14)–(53.16) of the distribution function. The anti-Hermitian parts of the tensor $\epsilon_{\alpha\beta}$ arise from the residues when the poles in the integral (54.1) are avoided by Landau's rule. The passage to the limit $B_0 \to 0$ is mathematically peculiar. In a magnetic field, the pole values v_z (for a given k_z) form a discrete sequence determined by equation (55.2). As the field decreases, the poles come closer together, and in the limit $B_0 = 0$ the pole values v_z depend not on a discrete number n but on the continuous parameter $\mathbf{k}_\perp . \mathbf{v}_\perp$, in accordance with the condition

$$\omega = \mathbf{k} . \mathbf{v} = k_z v_z + \mathbf{k}_\perp . \mathbf{v}_\perp,$$

as shown in going from (53.12) to (53.13).

Let us calculate, as an example, the permittivity tensor in the region of a simple ($n = 1$) electron cyclotron resonance. We shall also suppose that

$$|k_z| v_{Te}/\omega_{Be} \ll 1, \quad k_\perp v_{Te}/\omega_{Be} \ll 1. \tag{55.4}$$

Then it is sufficient to use for the distribution function just one term of the Fourier series—the expression (53.19), corresponding to the given value of n. Because of the second condition (55.4), this function can be expanded in powers of $k_\perp$. When $n = 1$, only the zero-order term need be taken in such an expansion, since cyclotron absorption at the frequency ω_{Be} does not require the external field to be non-uniform in the xy-plane.

Thus we write the distribution function as

$$\delta f = Q_1 \frac{\omega_{Be} e^{i\varphi}}{i[k_z v_z - (\omega - \omega_{Be})]}, \tag{55.5}$$

with

$$Q_1 = -\frac{e f_0}{2\pi T \omega_{Be}} \int_0^{2\pi} \mathbf{E} . \mathbf{v} e^{-i\tau}\, d\tau.$$

Writing

$$\mathbf{E} . \mathbf{v} = E_x v_\perp \cos\tau + E_y v_\perp \sin\tau + E_z v_z,$$

and carrying out the integration, we obtain

$$Q_1 = -\frac{e v_\perp}{2T\omega_{Be}} f_0 (E_x - iE_y). \tag{55.6}$$

With this distribution function, the polarization vector (54.1) has only x- and y-components. After the integration over $v_\perp\, dv_\perp\, d\varphi$, they are

$$P_x = -iP_y = (E_x - iE_y)\frac{e^2 N_e}{2\omega m k_z}\left(\frac{m}{2\pi T}\right)^{1/2} \int_{-\infty}^{\infty} \exp\left(-\frac{m v_z^2}{2T}\right) \frac{dv_z}{v_z - (\omega - \omega_{Be})/k_z - i0 \operatorname{sgn} k_z}.$$

An integral of this form can be expressed in terms of the function F defined by (31.3). We obtain finally as the components of the permittivity tensor†

$$\left.\begin{aligned} \epsilon_{xx} - 1 = \epsilon_{yy} - 1 = i\epsilon_{xy} &= \frac{\Omega_e^2}{2\omega(\omega - \omega_{Be})} F\left(\frac{\omega - \omega_{Be}}{\sqrt{2} v_{Te}|k_z|}\right), \\ \epsilon_{zz} - 1 = \epsilon_{xz} = \epsilon_{yz} &= 0. \end{aligned}\right\} \tag{55.7}$$

†The pole $v_z = (\omega - \omega_{Be})/k_z$ is passed below or above, according to the sign of k_z; this is the reason for the presence of the modulus of k_z in the argument of F.

The anti-Hermitian part of this tensor, which describes the damping, is

$$\epsilon''_{xx} = \epsilon''_{yy} = \epsilon'_{xy} = \frac{\pi^{1/2}\Omega_e^2}{2^{3/2}\omega|k_z|v_{Te}} \exp\left\{-\frac{(\omega - \omega_{Be})^2}{2v_{Te}^2 k_z^2}\right\}. \tag{55.8}$$

The Hermitian part, in the immediate neighbourhood of $\omega = \omega_{Be}$, has the form

$$\begin{gathered} \epsilon'_{xx} - 1 = \epsilon'_{yy} - 1 = -\epsilon''_{xy} = -\Omega_e^2(\omega - \omega_{Be})/2\omega v_{Te}^2 k_z^2, \\ |\omega - \omega_{Be}|/v_{Te}|k_z| \ll 1. \end{gathered} \tag{55.9}$$

At the point $\omega = \omega_{Be}$ itself, this part passes through zero and changes sign. We see here how the inclusion of spatial dispersion removes the poles of the permittivity (52.11) of a cold plasma: the discontinuous variation shown by the broken line in Fig. 16 is replaced by the continuous line.†

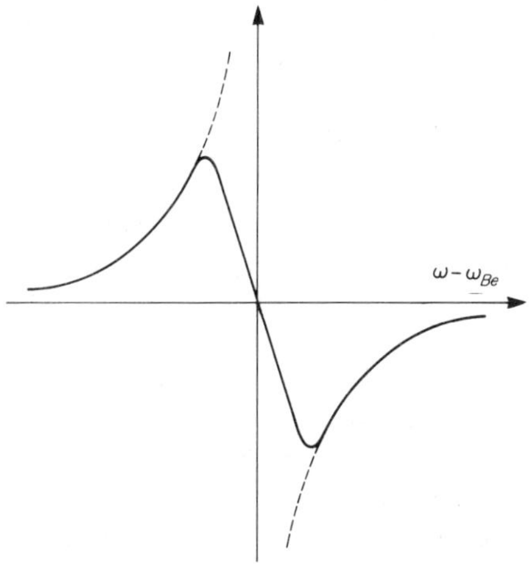

FIG. 16.

In the limit $|k_z| \to 0$, the expression (55.8) reduces to a delta function:

$$\epsilon''_{xx} = \epsilon''_{yy} = \epsilon'_{xy} \to (\pi\Omega_e^2/2\omega)\delta(\omega - \omega_{Be}): \tag{55.10}$$

when $\omega - \omega_{Be} \neq 0$, (55.8) is zero in the limit, and the integral of this function over $d\omega$ is $\pi\Omega_e^2/2\omega$ for any value of k_z. The significance of the result is clear: in the absence of spatial dispersion ($k \to 0$), the width of the absorption line tends to zero, and damping occurs only when ω coincides exactly with ω_{Be}. Formula (55.10) may be used in place of (55.8) in expressions integrated with respect to ω.

The formula (55.10) can also be derived directly from the expressions (52.11) for the permittivity of a cold plasma, by means of Landau's avoidance rule, according

†The expression (55.7) does not have the property (52.1), of course, which would occur only if the absorption line near $\omega = -\omega_{Be}$ were considered as well as that near $\omega = \omega_{Be}$.

to which the frequency ω at a pole is to be taken as $\omega + i0$. Thus the pole factors in (52.11) are really to be understood as follows:

$$\frac{1}{\omega^2 - \omega_{Be}^2} \to \frac{1}{2\omega_{Be}}\left[\frac{1}{\omega - \omega_{Be} + i0} - \frac{1}{\omega + \omega_{Be} + i0}\right],$$

and according to the rule (29.8)

$$\frac{1}{\omega^2 - \omega_{Be}^2} \to \mathrm{P}\frac{1}{\omega^2 - \omega_{Be}^2} - \frac{i\pi}{2\omega_{Be}}[\delta(\omega - \omega_{Be}) - \delta(\omega + \omega_{Be})]. \tag{55.11}$$

Making this change in (52.11), we obtain (55.10).

When $k_z = 0$ (i.e. $\mathbf{k} \perp \mathbf{B}_0$), there is no Landau damping in a magnetoactive plasma: the speed of the particles no longer appears in (55.1) and (55.2), and these conditions cannot be satisfied except when ω coincides exactly with some $n\omega_B$.† This property is due to the non-relativistic approximation; in a relativistic plasma, the Landau (cyclotron) damping can occur even when $k_z = 0$. The frequency of rotation round the direction of $\mathbf{B}_0$ for a relativistic charged particle with energy ϵ is

$$\omega_B mc^2/\epsilon = \omega_B\sqrt{(1 - v^2/c^2)},$$

with ω_B defined as before. This value is to be used in place of ω_B on the right-hand side of the condition (55.2). In particular, for $k_z = 0$ we have

$$\omega = n\omega_B\sqrt{(1 - v^2/c^2)}; \tag{55.12}$$

to allow this condition to be satisfied, it is necessary only that $\omega < n\omega_B$.

Landau damping in a magnetoactive relativistic plasma can occur even in the limit $\mathbf{k} \to 0$ (unlike the cases of a magnetoactive non-relativistic plasma and a relativistic plasma with no magnetic field). It is due to particles in simple cyclotron resonance with a uniform variable field—the condition (55.12) with $n = 1$—and therefore exists at frequencies $\omega < \omega_B$; see Problem 2.

PROBLEMS

PROBLEM 1. Find the permittivity tensor for a magnetoactive plasma at $\omega \lesssim |k_z| v_{Te}$; the conditions (55.4) also are assumed satisfied.

SOLUTION. In the zero-order approximation with respect to the small parameter $k_\perp v_{Te}/\omega_{Be}$, the distribution function for this case (the $s = 0$ term in the Fourier series (53.14), (53.15)) is

$$\delta f = Q_0 \omega_{Be}/i(k_z v_z - \omega),$$

where

$$Q_0 = -\frac{ef_0}{\omega_{Be}T}\frac{1}{2\pi}\int_0^{2\pi} \mathbf{E}\,.\,\mathbf{v}\,d\tau = -\frac{ev_zE_z}{\omega_{Be}T}f_0.$$

†In the limit $B_0 \to 0$, the damping of course reappears because of electrons that satisfy the condition $\omega = \mathbf{k}\,.\,\mathbf{v} \equiv \mathbf{k}_\perp\,.\,\mathbf{v}_\perp$.

With this function δf, the polarization vector $\mathbf{P}$ is in the z-direction, and the only non-zero component of the tensor $\epsilon_{\alpha\beta} - \delta_{\alpha\beta}$ is

$$\epsilon_{zz} - 1 = \frac{4\pi e^2}{\omega T} \int \frac{f_0(p) v_z^2 \, d^3p}{k_z v_z - \omega - i0}.$$

After the identical transformation

$$v_z^2 = \frac{1}{k_z}(k_z v_z - \omega) v_z + \frac{\omega}{k_z} v_z,$$

the integral of the first term over dp_z gives zero, and the second term yields the result

$$\epsilon_{zz} - 1 = \frac{\Omega_e^2}{k_z^2 v_{Te}^2}[F(\omega/\sqrt{2}|k_z|v_{Te}) + 1].$$

The imaginary part of this is

$$\epsilon''_{zz} = \frac{\pi^{1/2}\omega\Omega_e^2}{2^{1/2}|k_z|^3 v_{Te}^3} \exp(-\omega^2/2k_z^2 v_{Te}^2).$$

PROBLEM 2. Find the anti-Hermitian part of the permittivity tensor for an ultra-relativistic magnetoactive electron plasma in the limit $\mathbf{k} \to 0$.

SOLUTION. In the relativistic case, the transport equation (53.5) remains unchanged, but in the transformation to (53.6) the relativistic relation $\mathbf{p} = \epsilon \mathbf{v}/c^2$ (where ϵ is the electron energy) in place of $\mathbf{p} = m\mathbf{v}$ replaces ω_{Be} by $\omega_{Be} mc^2/\epsilon$; the subsequent formulae in §53, with this change, then remain valid. When $k = 0$, damping arises only from the simple cyclotron resonance. Hence, to calculate the anti-Hermitian part of $\epsilon_{\alpha\beta}$, it is sufficient to take the term with $s = 1$ in (53.14) and (53.15). Analogously to (55.5) and (55.6), we find

$$\delta f = -\frac{iep_\perp c^2 e^{i\varphi} f_0}{2T\epsilon(\omega - \omega_{Be} mc^2/\epsilon)}(E_x - iE_y).$$

The ultra-relativistic ($T \gg mc^2$) function f_0 is†

$$f_0 = (N_e c^3/8\pi T^3) e^{-\epsilon/T}.$$

The polarization vector is calculated as

$$\mathbf{P} = \frac{e}{i\omega} \int (\mathbf{p}c^2/\epsilon) \delta f \, d^3p,$$

where d^3p is to be expressed as $p^2 \, dp \, do = p\epsilon \, d\epsilon \, do/c^2$. The integration over do, with $cp = (\epsilon^2 - m^2c^4)^{1/2}$, gives

$$\begin{aligned} \epsilon_{xx} - 1 = \epsilon_{yy} - 1 &= i\epsilon_{xy} \\ &= -\frac{\Omega_e^2 mc^2}{12\omega^2 T^4} \int_{mc^2}^{\infty} \frac{(\epsilon^2 - m^2c^4)^{3/2} e^{-\epsilon/T} \, d\epsilon}{\epsilon - \omega_{Be} mc^2/\omega + i0}. \end{aligned}$$

The integral has an imaginary part if the pole $\epsilon = \omega_{Be} mc^2/\omega$ lies in the range of integration, i.e. if $\omega < \omega_{Be}$. In this case the final result is

$$\epsilon''_{xx} = \epsilon''_{yy} = \epsilon'_{xy} = \frac{\pi\Omega_e^2 \omega_{Be}^3}{12\omega^5}\left(1 - \frac{\omega^2}{\omega_{Be}^2}\right)^{3/2} \exp\left(-\frac{mc^2 \omega_{Be}}{\omega T}\right).$$

§56. Electromagnetic waves in a magnetoactive cold plasma

We can derive a general equation giving the dependence of the frequency on the wave vector (the *dispersion relation*) for free monochromatic waves propagating in a medium with any dielectric tensor $\epsilon_{\alpha\beta}(\omega, \mathbf{k})$.

†In this expression, the normalization factor is written with ultra-relativistic accuracy; we cannot put $\epsilon \approx cp$, on account of the subsequent integration with respect to p from 0 to ∞.

For an electromagnetic field whose dependence on the time and coordinates is given by $\exp(-i\omega t + i\mathbf{k}\cdot\mathbf{r})$, Maxwell's equations (28.2) become†

$$\mathbf{k}\times\mathbf{E} = \omega\mathbf{B}/c, \quad \mathbf{k}\times\mathbf{B} = -\omega\mathbf{D}/c, \tag{56.1}$$

$$\mathbf{k}\cdot\mathbf{B} = 0, \quad \mathbf{k}\cdot\mathbf{D} = 0. \tag{56.2}$$

Substitution of the first equation (56.1) in the second gives

$$\omega^2\mathbf{D}/c^2 = -\mathbf{k}\times(\mathbf{k}\times\mathbf{E}) = \mathbf{E}k^2 - \mathbf{k}(\mathbf{k}\cdot\mathbf{E}),$$

or in components

$$E_\alpha k^2 - k_\alpha k_\beta E_\beta = \omega^2 D_\alpha/c^2 = \omega^2\epsilon_{\alpha\beta}E_\beta/c^2. \tag{56.3}$$

The condition for these linear homogeneous equations to be compatible is the vanishing of the determinant:

$$|k^2\delta_{\alpha\beta} - k_\alpha k_\beta - \omega^2\epsilon_{\alpha\beta}/c^2| = 0. \tag{56.4}$$

This is the required *dispersion relation.*‡ For a given (real) **k**, it determines the frequencies $\omega(\mathbf{k})$ (in general complex), i.e. the *eigenvibration spectrum* of the medium. In the general case where frequency dispersion and spatial dispersion are present, equation (56.4) defines an infinity of branches of the function $\omega(\mathbf{k})$.

Let us consider electromagnetic waves in a magnetoactive cold plasma whose permittivity tensor is given by (52.7) and (52.11).§ Since this tensor is Hermitian, the values of k^2c^2/ω^2 determined by (56.4) are obviously real.

In the absence of spatial dispersion, $\epsilon_{\alpha\beta}$ depends only on ω, and the dispersion relation (56.4) is therefore algebraic with respect to **k**. Expansion of the determinant gives by a simple calculation‖

$$A(kc/\omega)^4 + B(kc/\omega)^2 + C = 0, \tag{56.5}$$

where

$$A = \epsilon_{\alpha\beta}k_\alpha k_\beta/k^2 = \epsilon_\perp \sin^2\theta + \epsilon_\parallel \cos^2\theta \equiv \epsilon_l, \tag{56.6}$$

$$B = -\epsilon_\perp\epsilon_\parallel(1+\cos^2\theta) - (\epsilon_\perp{}^2 - g^2)\sin^2\theta, \tag{56.7}$$

$$C = \epsilon_\parallel(\epsilon_\perp{}^2 - g^2), \tag{56.8}$$

and θ is the angle between **k** and $\mathbf{B}_0$. For given values of ω and θ, equation (56.5)

†The variable magnetic field **B** of the wave is not to be confused with the constant field $\mathbf{B}_0$.

‡Called in crystal optics the *Fresnel equation.*

§Electromagnetic waves in a magnetoactive cold plasma were first studied (the role of ions being neglected) by E. V. Appleton (1928) and H. Lassen (1927).

‖In which it is convenient to take one of the coordinate planes, say xz, to pass through $\mathbf{B}_0$ and **k**.

yields two values of k^2, so that in general two types of wave can propagate in the plasma.†

Let us first consider the cases of wave propagation exactly along ($\theta = 0$) and exactly across ($\theta = \frac{1}{2}\pi$) the magnetic field. These have special features.

When $\theta = 0$, the roots of the dispersion relation are

$$\left(\frac{kc}{\omega}\right)^2 = \epsilon_\perp \pm g = 1 - \frac{\Omega_e^2}{\omega(\omega \pm \omega_{Be})} - \frac{\Omega_i^2}{\omega(\omega \mp \omega_{Bi})}. \tag{56.9}$$

It is easy to see from (56.3) that these waves are transverse ($E_z = 0$) and circularly polarized ($E_y/E_x = \mp i$). The infinity of (56.9) at $\omega = \omega_{Be}$ or $\omega = \omega_{Bi}$ corresponds to resonance: the frequency and direction of rotation of the vector **E** coincide with those of the Larmor rotation of the electrons or ions. Figure 17 shows, as an

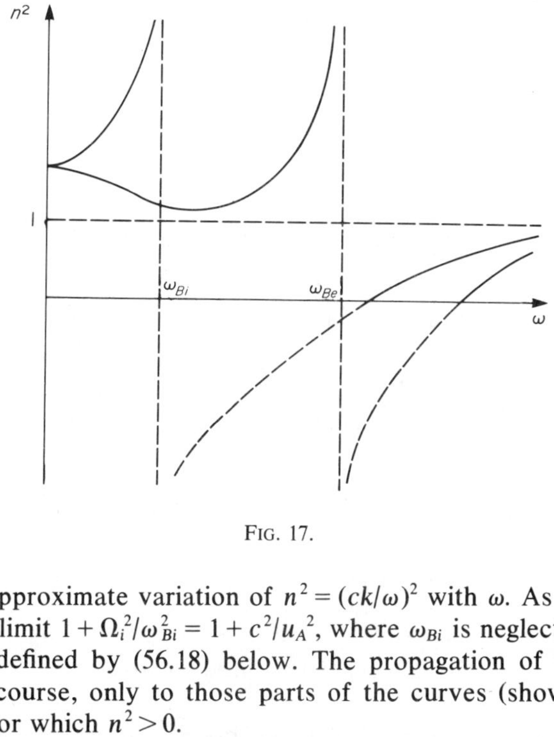

FIG. 17.

illustration, the approximate variation of $n^2 = (ck/\omega)^2$ with ω. As $\omega \to 0$, the values of n^2 tend to the limit $1 + \Omega_i^2/\omega_{Bi}^2 = 1 + c^2/u_A^2$, where ω_{Bi} is neglected in comparison with ω_{Be}; u_A is defined by (56.18) below. The propagation of undamped waves corresponds, of course, only to those parts of the curves (shown by continuous lines in Fig. 17) for which $n^2 > 0$.

When $\theta = 0$, equation (56.5) is satisfied also if $\epsilon_\parallel = 0$, corresponding to ordinary longitudinal plasma waves whose frequency $\omega \approx \Omega_e$ is independent of **k**.

When $\theta = \frac{1}{2}\pi$, the two roots of the dispersion relation are

$$(ck/\omega)^2 = \epsilon_\parallel, \quad (ck/\omega)^2 = \epsilon_\perp - g^2/\epsilon_\perp. \tag{56.10}$$

†The waves concerned are usually distinguished as *ordinary* and *extraordinary*. These terms do not, however, mean the same thing here as in the optics of uniaxial crystals; neither of the waves behaves like a wave in an isotropic medium.

The first corresponds to a wave with dispersion relation independent of $\mathbf{B}_0$:

$$\omega^2 \approx c^2k^2 + \Omega_e^2.$$

This is a transverse wave ($\mathbf{E} \perp \mathbf{k}$) and is linearly polarized, with $\mathbf{E} \parallel \mathbf{B}_0$. The second root (56.10) corresponds to a wave with $\mathbf{E} \perp \mathbf{B}_0$ and having both longitudinal and transverse components relative to $\mathbf{k}$. If the frequency is so high that the ion contribution to $\epsilon_{\alpha\beta}$ is negligible, namely $\omega \gg (\omega_{Be}\omega_{Bi})^{1/2}$, the condition (52.15), then in this wave†

$$\left(\frac{ck}{\omega}\right)^2 = 1 - \frac{\Omega_e^2(\omega^2 - \Omega_e^2)}{\omega^2(\omega^2 - \omega_{Be}^2 - \Omega_e^2)}. \tag{56.11}$$

In the general case of any angle θ (not zero or $\frac{1}{2}\pi$), we note first of all that for each value there exist frequencies for which the coefficient A in (56.5) becomes zero:

$$\begin{aligned}\epsilon_l &\equiv \epsilon_\perp \sin^2\theta + \epsilon_\parallel \cos^2\theta \\ &= 1 - \frac{\Omega_e^2 + \Omega_i^2}{\omega^2}\cos^2\theta - \left[\frac{\Omega_e^2}{\omega^2 - \omega_{Be}^2} + \frac{\Omega_i^2}{\omega^2 - \omega_{Bi}^2}\right]\sin^2\theta = 0.\end{aligned} \tag{56.12}$$

If the frequencies given by this equation, called the *plasma resonance frequencies*, are such as to satisfy also the "slowness" condition $\omega \ll kc$, then from §32 they correspond to longitudinal characteristic oscillations of the plasma. The vanishing of the coefficient of k^4 in equation (56.5), which is quadratic in k^2, signifies that one of its roots becomes infinite; as $A \to 0$, the roots are $-C/B$ and $-B/A$.

Equation (56.12) is cubic in ω^2 and has three real roots. These are easily determined by using the fact that Ω_i/Ω_e and ω_{Bi}/ω_{Be} are small. Two are found by neglecting the ion contribution in (56.12):

$$\omega_{1,2}^2 \approx \tfrac{1}{2}(\Omega_e^2 + \omega_{Be}^2) \pm \tfrac{1}{2}[(\Omega_e^2 + \omega_{Be}^2)^2 - 4\Omega_e^2\omega_{Be}^2\cos^2\theta]^{1/2}. \tag{56.13}$$

The ions must, however, be taken into account in the range $\omega \approx \omega_{Bi}$ where the third root lies; for this root, we easily find

$$\omega_3^2 \approx \omega_{Bi}^2(1 - (zm/M)\tan^2\theta), \tag{56.14}$$

assuming that $\Omega_e \gg \omega_{Bi}$. Formulae (56.13) and (56.14) for $\omega_2(\theta)$ and $\omega_3(\theta)$ are inapplicable when θ is so close to $\frac{1}{2}\pi$ that $\cos\theta \ll m/M$. In this range,

$$\left.\begin{aligned}\omega_2^2 &\equiv \omega_{2h}^2 = \omega_{Be}^2(\Omega_i^2 + \omega_{Bi}^2)/(\Omega_e^2 + \omega_{Be}^2),\\ \omega_3^2 &= \Omega_e^2\omega_{Bi}^2\cos^2\theta/(\Omega_e^2 + \omega_{Be}^2).\end{aligned}\right\} \tag{56.15}$$

The role of the ions is not negligible for either ω_3 or ω_2.

†Plasma oscillations in which the ions play no part are called *high-frequency* oscillations, and those in which the influence of the ions is important are called *low-frequency* oscillations.

Figure 18 shows diagrammatically the dependence of the frequencies ω_1, ω_2, ω_3 on the angle θ.† The curves of $\omega_1(\theta)$ and $\omega_2(\theta)$ nowhere intersect. They respectively begin at $\theta = 0$ from the higher and lower of the frequencies Ω_e and ω_{Be}. At $\theta = \frac{1}{2}\pi$, they reach the respective values

$$\omega_{1h} = (\Omega_e^2 + \omega_{Be}^2)^{1/2} \tag{56.16}$$

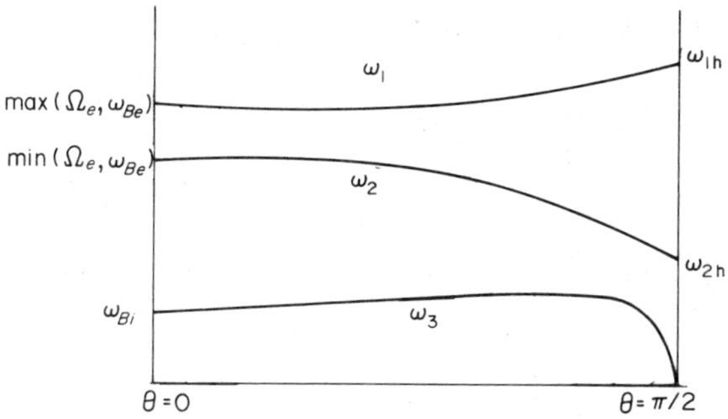

FIG. 18.

and ω_{2h}. The frequencies ω_{1h} and ω_{2h} are called respectively the *upper* and *lower hybrid frequencies*. When $\Omega_e^2 \gg \omega_{Be}^2$(and so certainly $\Omega_i^2 \gg \omega_{Bi}^2$), $\omega_{2h} = (\omega_{Be}\omega_{Bi})^{1/2}$.

The position of the frequencies ω_1, ω_2, ω_3 largely determines the configuration of the various branches of the spectrum governed by the dispersion relation (56.5). Being quadratic in $(ck/\omega)^2$, it has two roots for given ω and θ. If we follow the variation and the infinities of these roots as functions of ω for a fixed θ, we easily obtain Fig. 19, which shows the functions diagrammatically. The points of intersection with the abscissa

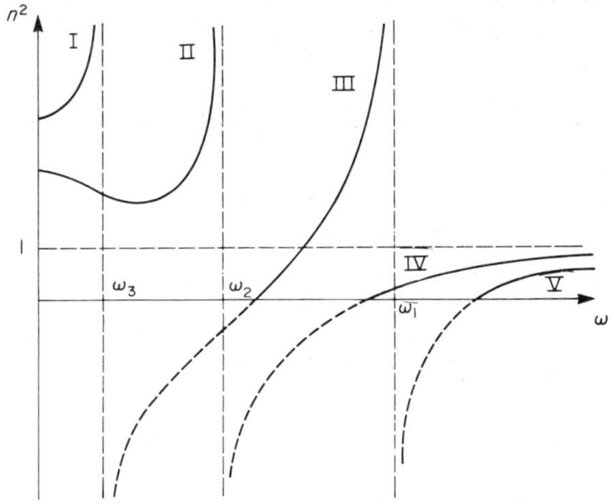

FIG. 19.

†It may be noted immediately that oscillations with the frequency ω_3 exist in practice only in a narrow range of angles close to $\frac{1}{2}\pi$. At other angles, they are strongly damped by the cyclotron absorption at simple ion resonance.

axis are given by $C = 0$, i.e. $\epsilon_\parallel = 0$ or $\epsilon_\perp^2 = g^2$. Their position is independent of θ, and one of them, corresponding to $\epsilon_\parallel = 0$, is always $\omega \approx \Omega_e$.

The spectrum of characteristic oscillations for a magnetoactive cold plasma thus has a total of five branches. Two of these (I and II in Fig. 19) reach the low-frequency oscillation region; the limiting values (as $\omega \to 0$) of the phase velocity in these branches are

$$\left.\begin{aligned}(\omega/k)_{\mathrm{I}} &= u_A|\cos\theta|/(1+u_A^2/c^2)^{1/2},\\ (\omega/k)_{\mathrm{II}} &= u_A/(1+u_A^2/c^2)^{1/2},\end{aligned}\right\} \tag{56.17}$$

where

$$u_A = c\omega_{Bi}/\Omega_i = B_0/(4\pi N_i M)^{1/2} \tag{56.18}$$

is called the *Alfvén velocity*. The expressions (56.17) are easily found from (56.5) by using the limiting values

$$\epsilon_\perp \approx 1 + u_A^2/c^2, \quad \epsilon_\parallel \approx -\Omega_e^2/\omega^2, \quad g \sim \omega.$$

When $u_A \ll c$, the phase velocities (56.17) are respectively $u_A|\cos\theta|$ and u_A. These limiting values correspond to waves which exist in a cold plasma in accordance with the ordinary equations of magnetic fluid dynamics (see *ECM*, §52); the spectrum of hydromagnetic waves has three branches, in all of which $\omega(\mathbf{k})$ is linear but in general depends on the direction of $\mathbf{k}$:

$$\left.\begin{aligned}(\omega/k)_A^2 &= u_A^2\cos^2\theta,\\ (\omega/k)_f^2 &= \tfrac{1}{2}\{u_s^2 + u_A^2 + [(u_s^2+u_A^2)^2 - 4u_s^2u_A^2\cos^2\theta]^{1/2}\},\\ (\omega/k)_s^2 &= \tfrac{1}{2}\{u_s^2 + u_A^2 - [(u_s^2+u_A^2)^2 - 4u_s^2u_A^2\cos^2\theta]^{1/2}\},\end{aligned}\right\} \tag{56.19}$$

where u_s is the speed of sound calculated formally from the adiabatic compressibility of the medium. The phase velocity of the first branch, called *Alfvén waves*, is exactly the same as the limiting value for the first branch (56.17). In order to go to the limit of a cold plasma in the second formula, we must put $u_s = 0$, since in a gas $u_s \sim (T/M)^{1/2}$. Then $(\omega/k)_f$, corresponding to *fast magnetosonic waves*, is equal to the limiting value of $(\omega/k)_{\mathrm{II}}$. The third branch, with $(\omega/k)_s$ (*slow magnetosonic waves*), has a velocity which tends to zero as $u_s \to 0$, and therefore does not appear in a cold plasma. The assumption of a cold plasma enables us to neglect the thermal spread of ion velocities and to describe them in terms of fluid mechanics even in the absence of collisions. The condition $u_A \ll c$ justifies neglecting the displacement currents in the equations of magnetic fluid dynamics.

In the opposite case of high frequencies, the phase velocities of two branches (IV and V) tend to $\omega/k = c$, corresponding to transverse high-frequency waves in an isotropic plasma, as we should expect, since the magnetic field plays no part when $\omega \gg \omega_{Be}$.

Lastly, let us consider the interesting case of waves which can exist when $\Omega_e \gg \omega_{Be}$; the resonance frequency is then $\omega_2 \approx \omega_{Be}\cos\theta$. We shall take the frequency range intermediate (on branch II) between ω_2 and $\omega_3 \approx \omega_{Bi}$, defined by

$$\omega_{Bi} \ll \omega \ll \omega_{Be}\cos\theta, \quad \omega \ll \Omega_e^2/\omega_{Be}. \tag{56.20}$$

The condition $\omega \gg \omega_{Bi}$ enables the ion contribution to g to be neglected, and from the condition $\omega \ll \omega_{Be}$ we have

$$\epsilon_{xy} = ig = -i\Omega_e^2/\omega\omega_{Be}. \tag{56.21}$$

With the conditions (56.20), we also have $\epsilon_\parallel \gg g \gg \epsilon_\perp$.

The required solution of the dispersion relation is found more directly by writing the latter as

$$|k^2\epsilon^{-1}{}_{\alpha\beta} - k_\alpha k_\gamma \epsilon^{-1}{}_{\gamma\beta} - (\omega^2/c^2)\delta_{\alpha\beta}| = 0, \tag{56.22}$$

changing from the tensor $\epsilon_{\alpha\beta}$ in (56.4) to its inverse (i.e. expressing **E** in terms of **D** in equations (57.3)). The components of the inverse tensor are

$$\epsilon^{-1}{}_{xx} = \epsilon^{-1}{}_{yy} \approx -\epsilon_\perp/g^2, \quad \epsilon^{-1}{}_{zz} = 1/\epsilon_\parallel, \quad \epsilon^{-1}{}_{xy} = -\epsilon^{-1}{}_{yx} \approx i/g,$$

and the largest of these is $\epsilon^{-1}{}_{xy}$. Neglecting the other components, and taking the xz-plane to be that of $\mathbf{B}_0$ and $\mathbf{k}$, we obtain the dispersion relation

$$\begin{vmatrix} -\omega^2/c^2 & ik_z^2/g \\ -ik^2/g & -\omega^2/c^2 \end{vmatrix} = 0,$$

whence

$$\begin{aligned} \omega &= k^2c^2(\omega_{Be}/\Omega_e^2)\,|\cos\theta| \\ &= cB_0|\cos\theta|k^2/4\pi e N_e. \end{aligned} \tag{56.23}$$

These are called *helicon waves*;† they are of purely electron origin.

The name of these waves arises from the nature of their polarization. The equation $\mathbf{k}\cdot\mathbf{D} = 0$ (56.2) gives, with the above choice of coordinate axes,

$$D_x \sin\theta + D_z \cos\theta = 0. \tag{56.24}$$

From equations (56.3) written as

$$[k^2\epsilon^{-1}{}_{\alpha\beta} - k_\alpha k_\gamma \epsilon^{-1}{}_{\gamma\beta}]D_\beta = (\omega^2/c^2)D_\alpha, \tag{56.25}$$

we find $D_x = -i|\cos\theta|D_y$. In the same approximation (i.e. retaining only $\epsilon^{-1}{}_{xy}$ among all the $\epsilon^{-1}{}_{\alpha\beta}$), the electric field of the wave lies entirely in the xy plane, which is perpendicular to $\mathbf{B}_0$: $E_z = \epsilon^{-1}{}_{z\beta}D_\beta = 0$. The other field components are

$$E_x = \epsilon^{-1}{}_{xy}D_y, \quad E_y = \epsilon^{-1}{}_{yx}D_x = -\epsilon^{-1}{}_{xy}D_x,$$

and from (56.24)

$$E_y = i|\cos\theta|E_x. \tag{56.26}$$

†In geophysical applications, *whistlers*.

Thus the wave is elliptically polarized in the plane perpendicular to $\mathbf{B}_0$; when $\theta = \frac{1}{2}\pi$, the polarization becomes linear. In coordinates $\xi y \zeta$ with the ζ-axis parallel to $\mathbf{k}$,

$$E_\xi = -i\frac{|\cos\theta|}{\cos\theta}E_y, \quad E_\zeta = E_\xi \tan\theta. \tag{56.27}$$

The vector $\mathbf{E}$ describes a circular cone about the direction of $\mathbf{k}$.

The expression (56.21) for ϵ_{xy} has a simple physical significance. When $\omega_{Be} \gg \omega$, together with the condition (52.17) $k_\perp v_{Te}/\omega_{Be} = k_\perp r_{Be} \ll 1$, which is everywhere assumed satisfied, we can suppose that the transverse (relative to $\mathbf{B}_0$) motion of the electrons takes place in a constant and uniform field $\mathbf{E}$. When a charge moves in constant and uniform crossed fields $\mathbf{E}$ and $\mathbf{B}_0$, its mean transverse velocity (the electric drift velocity) is

$$\bar{\mathbf{v}}_\perp = c\mathbf{E} \times \mathbf{B}_0/B_0^2; \tag{56.28}$$

see *Fields*, § 22. This corresponds to (56.21). Thus helicon waves are associated with the electric drift of electrons in the plasma.

§ 57. Effect of thermal motion on electromagnetic wave propagation in magnetoactive plasmas

When the thermal motion of the particles is taken into account, the dispersion relation in general becomes transcendental, and yields an infinite number of branches of the function $\omega(\mathbf{k})$. The vast majority of these oscillations are, however, strongly damped. Only in exceptional cases is the damping so weak that the oscillations can be propagated as waves. These cases include, first of all, the waves considered in § 56, for which the thermal motion causes (if the conditions (52.17) and (53.17) are satisfied) only small corrections to the dispersion relation, and a small Landau damping rate.

We have seen, however, that for waves in cold plasmas there are frequency ranges in which the ratio ck/ω becomes indefinitely large (the neighbourhoods of plasma resonances). But as $k \to \infty$ the conditions (52.17) are certainly violated, so that it becomes necessary to take account of the thermal motion. We shall now show that doing so, even as a small correction to the permittivity, eliminates the divergence of the roots of the dispersion relation, and generates some qualitatively new properties of the plasma oscillation spectrum (B. N. Gershman, 1956). As we shall see, the conditions for the Landau damping to be exponentially small may still be satisfied, so that the anti-Hermitian part of $\epsilon_{\alpha\beta}$ is still negligible. We shall take the particular case of the neighbourhood of high-frequency plasma resonances, where it is sufficient to include the thermal motion of the electrons only.

The correction terms in $\epsilon_{\alpha\beta}$ are proportional to $(kv_{Te})^2$.† Similar corrections arise in the coefficients A, B, C in the dispersion relation (56.5). To investigate just the

†They are derived from the first-order terms in the expansion of the integrand in (54.5) in powers of k^2.

divergent root of this equation, it is sufficient to include the correction terms in the coefficient A only, which vanishes at the resonance point in the absence of the corrections.

In the neighbourhood of the resonance frequency ω_1 (say), we write this coefficient as

$$A = a_r(\omega - \omega_1) - A_{1r}(v_{Te}k/\omega_1)^2. \tag{57.1}$$

The second term is the correction due to the thermal motion.The coefficients a_r and A_{1r} are taken at the point $\omega = \omega_1$, so that they are independent of the variable ω (but of course depend on the direction of **k**, i.e. on the angle θ). Putting $\omega = \omega_1$ also in the coefficients B and C, and denoting the resulting values by B_r and C_r, we obtain as the dispersion relation in the neighbourhood of the resonance frequency

$$[a_r(\omega - \omega_1) - A_{1r}(v_{Te}^2/c^2)(kc/\omega_1)^2](kc/\omega_1)^4 + B_r(kc/\omega_1)^2 + C_r = 0. \tag{57.2}$$

We are concerned with the root of this equation which as $v_{Te} \to 0$ becomes

$$(kc/\omega_1)^2 \approx -B_r/a_r(\omega - \omega_1),$$

i.e.

$$\omega - \omega_1 = -B_r\omega_1^2/a_rc^2k^2. \tag{57.3}$$

Since $(kc/\omega_1)^2$ in this solution is large, the term C_r which does not contain the large quantity is to be omitted from (57.2) in finding the solution. Then we have the dispersion relation

$$\omega - \omega_1 = (A_{1r}/a_r)(kv_{Te}/\omega_1)^2 - (B_r/a_r)(\omega_1/kc)^2. \tag{57.4}$$

Here two cases are to be distinguished according to the sign of A_{1r} (a_r and B_r are always positive).†

In Fig. 20, the continuous curve shows the dispersion relation (57.4) for $A_{1r} > 0$. The curve meets the abscissa axis at‡

$$k^2 = (\omega_1^2/cv_{Te})\sqrt{(B_r/A_{1r})}. \tag{57.5}$$

As $v_{Te} \to 0$, this point moves to infinity on the right, and we come back to the curve corresponding to the dispersion relation (57.3) for a cold plasma (shown by the broken curve in Fig. 20).

The allowance for thermal motion thus causes the oscillation spectrum branch to be extended into the range $\omega > \omega_1$. In the limit of zero external field, it is this part of the branch which corresponds to ordinary longitudinal plasma oscillations: in the absence of the field, $B_r = 0$, the frequency ω_1 coincides with Ω_e, and the whole

†It is easily seen from (56.6) and (56.7) that B_r is positive: eliminating $\epsilon_\parallel$ by means of the condition $A = 0$, we find $B_r = \epsilon_\perp^2 \tan^2\theta + g^2 \sin^2\theta > 0$. From the expressions (56.6) for A and (52.11) for $\epsilon_\perp$ and $\epsilon_\parallel$ it follows that $\partial A/\partial\omega > 0$, and therefore $a_r = (\partial A/\partial\omega)_{\omega=\omega_1}$ is positive.

‡For this value of k, the ratio kv_{Te}/ω_1 contains $(v_{Te}/c)^{1/2}$, and is therefore small. This is the above-mentioned condition for the Landau damping to be small.

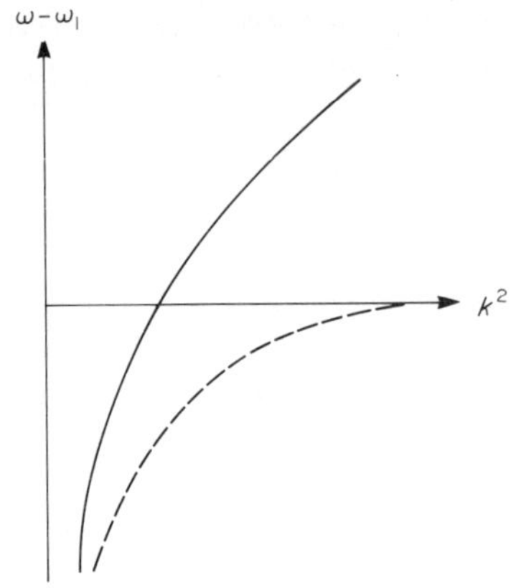

FIG. 20.

curve of $\omega - \Omega_e$ as a function of k^2 becomes a straight line from the origin, whose equation is the same as (32.5).†

When thermal motion is neglected, the oscillations in plasma resonances are longitudinal. This property does not in general occur when spatial dispersion is taken into account: $A = \epsilon_{\alpha\beta}k_\alpha k_\beta/k^2 \equiv \epsilon_l$ becomes a function of k, and the condition $\epsilon_l = 0$ for the oscillations to be longitudinal is incompatible with relation given by the dispersion relation between the same variables ω, k and θ. Both at the plasma resonance points themselves, which cease to be distinctive, and near them, the waves remain almost longitudinal, however: since A is small and the wave is slow (ω/kc is small), the transverse component $E^{(t)}$ is small in comparison with $E^{(l)}$, according to (32.10).

Let us now turn to the case $A_{1r} < 0$. Figure 21 shows how $\omega - \omega_1$ varies with k in

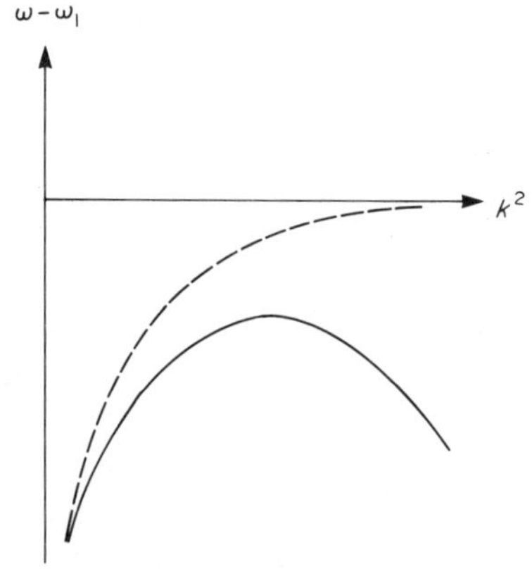

FIG. 21.

†For this reason, the waves which (in a magnetoactive plasma) correspond to the upper part of the continuous curve in Fig. 20 are usually called plasma waves, in contrast to the ordinary or extraordinary waves corresponding to the lower part of the curve. This terminology is conventional, however: there is really just one branch of oscillations, whose point of intersection with the abscissa axis ($\omega = \omega_1$) has no distinctive feature.

this case. The curve does not enter the region $\omega > \omega_1$, but has a maximum at the point

$$k^2 = (\omega_1^2/cv_{Te})(B_r/|A_{1r}|)^{1/2}, \quad \omega - \omega_1 = (2v_{Te}/a_r c)(|A_{1r}|/B_r)^{1/2}. \tag{57.6}$$

When $v_{Te} \to 0$, this point moves to infinity on the right, at the same time approaching the abscissa axis, and we again return to the form (57.3).

As a further example, let us consider the transverse waves near electron cyclotron resonance, propagating along the magnetic field. When thermal motion is neglected, the dispersion relation for these waves is given by (56.9) with the lower signs, and in the neighbourhood of $\omega = \omega_{Be}$†

$$\omega = \omega_{Be}(1 - \Omega_e^2/k^2c^2) \tag{57.7}$$

with $kc \ll \Omega_e$; the whole of this spectrum is at $\omega < \omega_{Be}$.

To examine these waves with allowance for the thermal motion of the electrons, it is necessary to construct the dispersion relation with the permittivity tensor (55.7) which applies to the cyclotron resonance region.‡ Expanding the determinant (56.4) (with the vector **k** parallel to the z-axis), we obtain

$$\frac{k^2c^2}{\omega^2} = 1 + \frac{\Omega_e^2}{\omega(\omega - \omega_{Be})} F\left(\frac{\omega - \omega_{Be}}{\sqrt{2}kv_{Te}}\right). \tag{57.8}$$

Outside the resonance absorption line, i.e. for $|\omega_{Be} - \omega| \gg kv_{Te}$, but of course still $|\omega_{Be} - \omega| \ll \omega_{Be}$, this relation becomes

$$\frac{k^2c^2}{\omega_{Be}^2} = -\frac{\Omega_e^2}{\omega_{Be}(\omega - \omega_{Be})} + i\sqrt{\frac{\pi}{2}}\frac{\Omega_e^2}{\omega_{Be}kv_{Te}}\exp\left(-\frac{(\omega - \omega_{Be})^2}{2k^2v_{Te}^2}\right).$$

Hence we again have the dispersion relation (57.7) for the real part of the frequency, and the expression

$$\gamma = \sqrt{\frac{\pi}{2}}\,\omega_{Be}\frac{\omega_{Be}}{kv_{Te}}\left(\frac{\Omega_e}{ck}\right)^4 \exp\left\{-\frac{1}{2}\left(\frac{\Omega_e}{ck}\right)^4\left(\frac{\omega_{Be}}{kv_{Te}}\right)^2\right\} \tag{57.9}$$

for the Landau damping rate.

As ω approaches closer to ω_{Be}, in the range $|\omega_{Be} - \omega| \ll kv_{Te}$, the damping rate increases and becomes comparable with the frequency ω; in this range, wave propagation does not occur.

§58. **Equations of fluid dynamics in a magnetoactive plasma**

If the characteristic spatial dimensions L in a moving plasma are large compared with the mean free paths,

$$L \gg l, \tag{58.1}$$

†We assume the particular case where not only $\omega_{Be} - \omega \ll \omega_{Be}$ but also $\Omega_e > \omega_{Be}$, so that unity is certainly negligible on the right of (56.9).

‡Formulae (55.7), it will be remembered, also presuppose the fulfilment of the condition (55.4): $\omega_{Be} \gg kv_{Te}$.

we may suppose that thermodynamic equilibrium with local values of the temperature (the same for electrons and ions), pressure, etc., is established by collisions in any small region of the plasma. The movement of the plasma may then be described by the macroscopic equations of fluid dynamics.

The equations of magnetic fluid dynamics have been given in *ECM*, §51, but it was assumed that the transport properties of the medium (viscosity and thermal conductivity) were independent of the magnetic field. For this to be true in a plasma, the following conditions are necessary:

$$\nu_i \gg \omega_{Bi}, \quad \nu_e \gg \omega_{Be}$$

(of which the second follows from the first). These conditions are often too rigorous, and it is therefore necessary to derive equations of fluid dynamics that are free from the restriction mentioned.†

The equation of continuity for the mass density ρ remains, of course, as usual:

$$\partial\rho/\partial t + \operatorname{div}(\rho\mathbf{V}) = 0, \tag{58.2}$$

where $\mathbf{V}$ is the macroscopic velocity. The general form is also the same for the Navier–Stokes equation

$$\rho\left[\frac{\partial V_\alpha}{\partial t} + (\mathbf{V}\,.\,\nabla)V_\alpha\right] + \frac{\partial P}{\partial x_\alpha} - \frac{1}{c}(\mathbf{j}\times\mathbf{B})_\alpha = -\frac{\partial\sigma'_{\alpha\beta}}{\partial x_\beta} \tag{58.3}$$

and for the energy conservation equation

$$\frac{\partial}{\partial t}(\tfrac{1}{2}\rho V^2 + \rho U + B^2/8\pi) = -\operatorname{div}[\rho\mathbf{V}(\tfrac{1}{2}V^2 + W) - \sigma'\,.\,\mathbf{V} + c\mathbf{E}\times\mathbf{B}/4\pi + \mathbf{q}], \tag{58.4}$$

where $\sigma'_{\alpha\beta}$ is the viscous stress tensor, $\sigma'\,.\,\mathbf{V}$ denotes the vector with components $\sigma'_{\alpha\beta}V_\beta$, $\mathbf{q}$ is the energy flux (including both the dissipative part due to thermal conduction and thermoelectric effects, and the convective transfer of energy by the current; see the definition (58.8) below), and U and W the internal energy and heat function of the medium per unit mass. The tensor $\sigma'_{\alpha\beta}$ and the vector $\mathbf{q}$ must be expressed in terms of the gradients of thermodynamic quantities and the velocities; the form of these expressions depends on the magnetic field.

The following remark should be made in connection with equation (58.3). This equation takes into account the force exerted on the plasma by the magnetic field (the last term of the left), but not the force $e(zN_i - N_e)\mathbf{E}$ exerted by the electric field. This treatment is justifiable here, since it follows from the condition (58.1) that *a fortiori* $L \gg a$, and the plasma is therefore quasi-neutral, so that we can put $zN_i = N_e$, and there are no uncompensated charges in the plasma.‡

†Moreover, in *ECM*, §51, the terms representing the thermoelectric effect were omitted from the equations.

‡This argument is based on the inequality $l \gg a$. We are everywhere considering a fully ionized plasma. In a partly ionized one, the inequality $l \gg a$ need not be satisfied, because the mean free path is reduced by collisions with neutral atoms, and then $l \gg a$ is to be regarded as a further necessary condition for the bulk electrical force to be negligible.

Equations (58.2)–(58.4) must be supplemented by Maxwell's equations for a quasi-stationary electromagnetic field (the displacement current being then omitted):

$$\operatorname{curl} \mathbf{E} = -\frac{1}{c}\frac{\partial \mathbf{B}}{\partial t}, \quad \operatorname{div} \mathbf{B} = 0, \quad \operatorname{curl} \mathbf{B} = 4\pi \mathbf{j}/c. \tag{58.5}$$

To say the field is quasi-stationary means that its frequency of variation $\omega \ll c/L$. The electric field induced by the variable magnetic field is $E \sim \omega LB/c \ll B$; for this reason, we need include in (58.4) only the magnetic, and not the electric, field energy density. The neglect of the displacement current is, moreover, in accordance with the assumption that the plasma is quasi-neutral: the last equation (58.5) implies that $\operatorname{div} \mathbf{j} = 0$.

Lastly, the "generalized Ohm's law" equation is needed:

$$\mathbf{E} + \mathbf{V} \times \mathbf{B}/c = \mathbf{F}, \tag{58.6}$$

where $\mathbf{F}$ is some linear combination of the current $\mathbf{j}$ and the gradients of the thermodynamic quantities. The combination of $\mathbf{E}$ and $\mathbf{B}$ on the left of (58.6) arises (cf. *ECM*, §49) from the transformation of $\mathbf{E}$ when we go from the rest frame of a given volume element in the medium to a frame in which the element is moving with velocity $\mathbf{V}$.

In a quasi-neutral plasma, the relative concentration of the electron and ion components is a given constant ($N_e/N_i = z$). Hence only the temperature and the pressure are independent thermodynamic variables; the problem of expressing $\mathbf{F}$ and $\mathbf{q}$ in terms of the gradients of these variables (and the current $\mathbf{j}$) is formally the same as in the theory of thermogalvanomagnetic effects in metals (see *ECM*, §25).†

The relations between $\mathbf{j}$ and $\mathbf{q}$, and between the field and the gradients of thermodynamic quantities, are written as generalizations of (44.12) and (44.13):

$$F_\alpha + \frac{1}{e}\frac{\partial \mu_e}{\partial x_\alpha} = \sigma^{-1}{}_{\alpha\beta} j_\beta + \alpha_{\alpha\beta}\frac{\partial T}{\partial x_\beta}, \tag{58.7}$$

$$q_\alpha = -\frac{\mu_e}{e} j_\alpha + \beta_{\alpha\beta} j_\beta - \kappa_{\alpha\beta}\frac{\partial T}{\partial x_\beta}. \tag{58.8}$$

Here μ_e is the chemical potential of the electrons; the tensors $\sigma'_{\alpha\beta}$, $\alpha_{\alpha\beta}$, $\beta_{\alpha\beta}$ depend on the magnetic field $\mathbf{B}$ as a parameter. The absence of the term $-\varphi\mathbf{j}$ (cf. (44.13)) on the left of (58.8) is due to the fact that this has already been included in (58.4) by means of the Poynting vector in the energy flux, as is easily seen by using Maxwell's equations (58.5) to transform its divergence. In the stationary case (§ 44), we have

$$-\operatorname{div} c\mathbf{E} \times \mathbf{B}/4\pi = \mathbf{j} \cdot \mathbf{E} = -\operatorname{div}(\varphi \mathbf{j}).$$

†We must repeat that the discussion refers to a fully ionized plasma. The presence of several types of heavy particle (various ions, neutral atoms) would make it necessary to consider the corresponding diffusion processes.

Thus the energy flux $\mathbf{q}$ in (58.8) already excludes the transfer of energy $-e\varphi$ by particles.

By Onsager's principle, the coefficients in the relations (58.7) and (58.8) satisfy the relations

$$\sigma_{\alpha\beta}(\mathbf{B}) = \sigma_{\beta\alpha}(-\mathbf{B}), \quad \kappa_{\alpha\beta}(\mathbf{B}) = \kappa_{\beta\alpha}(-\mathbf{B}), \tag{58.9}$$

$$\beta_{\alpha\beta}(\mathbf{B}) = T\alpha_{\beta\alpha}(-\mathbf{B}). \tag{58.10}$$

Since $\mathbf{B}$ is the only available vector parameter, the dependence of the tensors on the direction $\mathbf{b} = \mathbf{B}/B$ may be written in the general form

$$\alpha_{\alpha\beta}(\mathbf{B}) = \alpha_1\delta_{\alpha\beta} + \alpha_2 b_\alpha b_\beta + \alpha_3 e_{\alpha\beta\gamma}b_\gamma, \tag{58.11}$$

and similarly for the other tensors; the scalar coefficients α_1, α_2, α_3 are functions of the field B, satisfying the condition of symmetry under inversion ($\mathbf{B}$ is an axial vector and its components are unaffected by inversion, as is required for the components of the true tensors $\alpha_{\alpha\beta}$, etc.). Expressions of the form (58.11) necessarily satisfy the relations (58.9), and (58.10) becomes

$$\beta_{\alpha\beta}(\mathbf{B}) = T\alpha_{\alpha\beta}(\mathbf{B}). \tag{58.12}$$

In the practical application of the expressions (58.7) and (58.8) in magnetic fluid dynamics, the gradient of the chemical potential is more conveniently expressed in terms of the pressure and temperature gradients by

$$\nabla\mu_e = -s_e\nabla T + (1/N_e)\nabla P_e, \quad \mu_e = w_e - Ts_e,$$

where $P_e = N_eT = Pz/(1+z)$ is the partial pressure of electrons in the plasma, and s_e and w_e are the entropy and heat function per particle in the electron component of the plasma. We finally write (58.7) and (58.8) in vector form as

$$\mathbf{E} + \mathbf{V}\times\mathbf{B}/c + (1/eN_e)\nabla P_e$$
$$= \mathbf{j}_\parallel/\sigma_\parallel + \mathbf{j}_\perp/\sigma_\perp + \mathcal{R}\mathbf{B}\times\mathbf{j} + \alpha_\parallel(\nabla T)_\parallel + \alpha_\perp(\nabla T)_\perp + \mathcal{N}\mathbf{B}\times\nabla T, \tag{58.13}$$

$$\mathbf{q} + (w_e/e)\mathbf{j} = \alpha_\parallel T\mathbf{j}_\parallel + \alpha_\perp T\mathbf{j}_\perp + \mathcal{N}T\mathbf{B}\times\mathbf{j} - \kappa_\parallel(\nabla T)_\parallel - \kappa_\perp(\nabla T)_\perp + \mathcal{L}\mathbf{B}\times\nabla T, \tag{58.14}$$

with a fresh notation for the coefficients (all of which are functions of $\mathbf{B}$), and the suffixes $\parallel$ and $\perp$ denoting vector components longitudinal and transverse with respect to $\mathbf{B}$. The definition of the coefficient $\alpha_\parallel$ in (58.13) differs from that in (58.7) by the inclusion of s_e/e. The coefficients $\mathcal{R}$, $\mathcal{N}$ and $\mathcal{L}$ represent the Hall, Nernst and Leduc–Righi effects. The terms $\mathcal{R}\mathbf{B}\times\mathbf{j}$ in (58.13) and $\mathcal{L}\mathbf{B}\times\nabla T$ in (58.14) are non-dissipative transport effects, which do not appear in the products $\mathbf{E}\cdot\mathbf{j}$ and $\mathbf{q}\cdot\nabla T$ and therefore do not cause an increase of entropy.

The general expression for the viscous stress tensor $\sigma'_{\alpha\beta}$ in terms of the gradients of the macroscopic velocity has already been given in § 13. When applied to a plasma, this expression is somewhat simplified by the vanishing of the two second viscosity coefficients ζ and ζ_1. The vanishing of ζ is a general property of monatomic gases such as plasmas. The reason for the absence of the ζ_1 term is explained in §59.

The remaining terms in (13.18) may conveniently be rearranged for application to a plasma, since in general the magnetic field there has a strong influence on the viscosity, not a weak one as in a neutral gas; there is therefore no sense in distinguishing the ordinary viscosity coefficient η. Here we put $\sigma'_{\alpha\beta}$ in a form differing from (13.18) only in that the η term is replaced by

$$\eta_0(3b_\alpha b_\beta - \delta_{\alpha\beta})(b_\gamma b_\delta V_{\gamma\delta} - \tfrac{1}{3}\operatorname{div}\mathbf{V}), \tag{58.15}$$

where (instead of $\mathbf{h}$) $\mathbf{b} = \mathbf{B}/B$; see the penultimate footnote to §59 regarding the desirability of this definition of η_0.

If the z-axis is taken parallel to $\mathbf{b}$, the stress tensor components are

$$\left.\begin{aligned}
\sigma'_{xx} &= -\eta_0(V_{zz} - \tfrac{1}{3}\operatorname{div}\mathbf{V}) + \eta_1(V_{xx} - V_{yy}) + 2\eta_3 V_{xy},\\
\sigma'_{yy} &= -\eta_0(V_{zz} - \tfrac{1}{3}\operatorname{div}\mathbf{V}) + \eta_1(V_{xx} - V_{yy}) - 2\eta_3 V_{xy},\\
\sigma'_{zz} &= 2\eta_0(V_{zz} - \tfrac{1}{3}\operatorname{div}\mathbf{V}),\\
\sigma'_{xy} &= 2\eta_1 V_{xy} - \eta_3(V_{xx} - V_{yy}),\\
\sigma'_{xz} &= 2\eta_2 V_{xz} + 2\eta_4 V_{yz},\\
\sigma'_{yz} &= 2\eta_2 V_{yz} - 2\eta_4 V_{xz}.
\end{aligned}\right\} \tag{58.16}$$

§59. Transport coefficients of a plasma in a strong magnetic field

To calculate the transport coefficients of a magnetoactive plasma we must, as usual, seek the particle distribution functions in the form $f = f_0 + \delta f$, where δf is a small correction to the local-equilibrium distribution and is proportional to the corresponding gradient of the thermodynamic quantities. On substituting this expression in the transport equation, e.g. for electrons,

$$\frac{\partial f_e}{\partial t} + \mathbf{v}\cdot\frac{\partial f_e}{\partial \mathbf{r}} - e\mathbf{E}\cdot\frac{\partial f_e}{\partial \mathbf{p}} - \frac{e}{c}\mathbf{v}\times\mathbf{B}\cdot\frac{\partial f_e}{\partial \mathbf{p}} = C(f_e), \tag{59.1}$$

we put $f_e = f_{0e}$ in the four terms on the left; the fourth term then vanishes, since $\partial f_{0e}/\partial \mathbf{p}$ is parallel to $\mathbf{v}$, so that the term in δf_e must be retained there, and we find as the equation for δf_e†

$$\frac{\partial f_{0e}}{\partial t} + \mathbf{v}\cdot\frac{\partial f_{0e}}{\partial \mathbf{r}} - e\mathbf{E}\cdot\frac{\partial f_{0e}}{\partial \mathbf{p}} = \frac{e}{c}\mathbf{v}\times\mathbf{B}\cdot\frac{\partial \delta f_e}{\partial \mathbf{p}} + I(\delta f_e), \tag{59.2}$$

where I is the linearized collision integral.

†In the calculation of the permittivity of a plasma in §29, the magnetic-field term in this equation was omitted because it is a second-order small quantity when $\mathbf{E}$ and $\mathbf{B}$ are small. In the present problem, the magnetic field $\mathbf{B}$ (unlike the electric field $\mathbf{E}$) is not assumed to be small.

Let us note first of all that the longitudinal electrical conductivity $\sigma_\parallel$ and thermal conductivity $\kappa_\parallel$ are independent of **B**, and have the same values as in the absence of a magnetic field, i.e. are the ordinary scalars σ and κ. It is obvious from symmetry that, when the vector **E** or ∇T is parallel to **B**, the distribution function δf is independent of the position angle φ of the transverse velocity $\mathbf{v}_\perp$ in the plane perpendicular to **B**. Also, $\mathbf{v}\times\mathbf{B}\,.\,\partial\delta f/\partial\mathbf{p} = -(B/m)\partial\delta f/\partial\varphi$, and so, when $\partial\delta f/\partial\varphi = 0$, the magnetic field does not appear in the transport equation.†

For a similar reason, the viscosity η_0, which determines the viscous stresses $\sigma'_{\alpha\beta}$, is independent of the magnetic field (and therefore is the same as the ordinary viscosity η) when the velocity **V** is parallel to **B** (along the z-axis); it then depends only on z, and in the expressions (58.16) there remain only the terms in $\sigma'_{xx} = \sigma'_{yy} = -\frac{1}{2}\sigma'_{zz} = -\frac{2}{3}\eta_0 dV/dz$.

Lastly, the coefficient ζ_1 would have to be independent of the field. For the velocity distribution mentioned, it would contribute to the stress tensor $\sigma'_{xx} = \sigma'_{yy} = \frac{1}{2}\sigma'_{zz} = \zeta_1 dV/dz$. Since this effect is absent in the absence of the field, $\zeta_1 = 0$ even when the field is present.‡ (Note that this reason does not depend on the plasma being classical; the result $\zeta_1 = 0$ is therefore valid in the relativistic case also, whereas $\zeta \neq 0$ in a relativistic plasma.)

The calculation of the other transport coefficients can be carried out similarly in the limit of strong magnetic fields, when (for each type of particle) the Larmor frequency $\omega_B \gg \nu$. Under these conditions, collisions function as a small correction.§

Electrical conductivity

Let us first calculate the coefficients which determine the electric current in the plasma. This is conveniently done in a frame of reference such that a given plasma volume element is at rest. Neglecting quantities $\sim m/M$, we can regard such a frame as coinciding with the rest frame of the ion component. The electric current is then purely electronic. We have therefore to solve only the transport equation for electrons.

The left-hand side of the transport equation would have to be transformed by means of the equations of fluid dynamics in the same way as was done for an ordinary gas in §6. In the chosen frame of reference, the macroscopic velocity (but not, of course, its derivatives) is zero at the point considered.‖

There is, however, no need to carry out the complete calculations here (for electrons). First of all, the term $\partial\delta f_e/\partial t$ can be omitted altogether. The differentiation with respect to time leads to terms containing the derivatives $\partial T/\partial t$, $\partial P/\partial t$

†We must at once add, however, that these arguments (and similar ones below) assume that the process of particle scattering is independent of the magnetic field. For this to be so, it is necessary that the field should satisfy the inequality (59.10) below.

‡We must again emphasize that all these statements depend on the form of the term in **B** in the transport equation (59.2); they therefore do not apply to an ordinary gas, whose molecules have a magnetic moment through which (not through the particle charge as in a plasma) the interaction with the magnetic field takes place in that case.

§The transport coefficients for a magnetoactive plasma were calculated by R. Landshoff (1949), E. S. Fradkin (1951) and S. I. Braginskiĭ (1952). The analytical method given below is due to I.E. Tamm (1951).

‖This has essentially been assumed already by making use of the parallelism of the vectors $\partial f_0/\partial\mathbf{p}$ and $\mathbf{v}$.

and $\partial\mathbf{V}/\partial t$. Of these, the first two can be expressed in terms of the scalar div $\mathbf{V}$ (cf. (6.16)), and such terms, as we know, cancel in any case for a monatomic gas (such as a plasma). The derivative $\partial\mathbf{V}/\partial t$, expressed by means of the equation (58.3) of fluid dynamics, contains a factor $1/\rho$, or $1/M$; the inclusion of such terms in the transport equation would give only corrections $\sim m/M$, which are of no interest. Next, in (59.2) we can put $\mathbf{E}=0$, since we know that $\mathbf{E}$ can appear in the required current $\mathbf{j}$ only as the sum $\mathbf{E}+(1/eN_e)\nabla P$. Lastly, since we do not propose to calculate the "longitudinal" transport coefficients $\sigma_\parallel$, $\kappa_\parallel$, η_0, which are independent of the magnetic field, all thermodynamic quantities for the plasma may be regarded as depending only on the coordinates in a plane perpendicular to $\mathbf{B}$. Denoting by $\nabla_\perp$ the operator of differentiation in that plane, we can thus write the transport equation as

$$(\mathbf{v}\cdot\nabla_\perp)f_{0e}=(e/c)\mathbf{v}\times\mathbf{B}\,.\,\partial\delta f_e/\partial\mathbf{p}+I(\delta f_e). \tag{59.3}$$

This equation in turn can be solved by successive approximation in powers of $1/\omega_{Be}$. The first approximation, denoted by the superscript (1), entirely neglects the collision integral, so that the equation is

$$\mathbf{v}\times\mathbf{b}\cdot\frac{\partial\delta f_e^{(1)}}{\partial\mathbf{v}}=\frac{1}{\omega_{Be}}(\mathbf{v}\,.\,\nabla_\perp)f_{0e}, \tag{59.4}$$

with $\mathbf{b}=\mathbf{B}/B$. The solution is

$$\delta f_e^{(1)}=-\frac{1}{\omega_{Be}}\mathbf{v}\,.\,\mathbf{b}\times\nabla_\perp f_{0e}, \tag{59.5}$$

as is easily seen by direct substitution. It is evident that this solution can be used to calculate only the non-dissipative transport coefficients: in the absence of collisions, there is no dissipation of energy.

The electric current density is given by the integral

$$\mathbf{j}=-e\int\mathbf{v}\,\delta f_e\,d^3p. \tag{59.6}$$

Substitution of (59.5) gives

$$\begin{aligned}\mathbf{j}^{(1)}&=(mc/B)(\mathbf{b}\times\nabla_\perp\,.\,\langle\mathbf{v}\rangle\mathbf{v}\rangle N_e\\&=(mc/3B)\mathbf{b}\times\nabla_\perp N_e\langle v^2\rangle,\end{aligned}$$

where the averaging is over the Maxwellian distribution. The result is

$$\mathbf{j}^{(1)}=(c/B)\mathbf{b}\times\nabla_\perp P_e,\quad \nabla_\perp P_e=-(B/c)\mathbf{b}\times\mathbf{j}^{(1)}. \tag{59.7}$$

Comparison of this expression with the definition of the coefficient $\mathscr{R}$ in (58.13) shows that

$$\mathscr{R}=-1/N_eec. \tag{59.8}$$

In the next approximation, we seek the solution of equation (59.3) as $\delta f_e = \delta f_e^{(1)} + \delta f_e^{(2)}$, and obtain for $\delta f_e^{(2)}$ the equation

$$\omega_{Be}\mathbf{v}\times\mathbf{b}\,.\,\partial\delta f_e^{(2)}/\partial\mathbf{v} = -I(\delta f_e^{(1)})$$
$$= (1/\omega_{Be})I(\mathbf{v}\cdot\mathbf{b}\times\nabla_\perp f_{0e}); \qquad (59.9)$$

the operator $\nabla_\perp$ cannot be taken outside I, since in the linearized collision integral the integrand contains among its coefficients quantities such as N_i which depend on the coordinates.

As already mentioned, the magnetic field is assumed so strong that $\omega_{Be} \gg \nu_e$. In this section, however, we shall make the further assumption that

$$r_{Be} = v_{Te}/\omega_{Be} \gg a_e, \qquad (59.10)$$

i.e. that $\omega_{Be} \ll \Omega_e$; this places an upper limit on the field. When this condition is satisfied, the field causes almost no curvature of the electron trajectories (and still less of the ion trajectories) in the collision region, and so has no effect on the collision process. The operator I therefore does not depend explicitly on the field. Then, by symmetry, the right-hand side of equation (59.9) must have a vector structure of the form $\mathbf{v}\,.\,\mathbf{b}\times\nabla_\perp\varphi(v^2)$; as regards the variable $\mathbf{v}$, this is of the same kind as on the right of (59.4), but with $\mathbf{b}\times\nabla_\perp$ in place of $\nabla_\perp$. The solution of (59.9) is therefore

$$\delta f_e^{(2)} = -\frac{1}{\omega_{Be}^2}I(\mathbf{v}\,.\,\mathbf{b}\times(\mathbf{b}\times\nabla_\perp)f_{0e}) = \frac{1}{\omega_{Be}^2}I(\mathbf{v}\,.\,\nabla_\perp f_{0e}). \qquad (59.11)$$

In the calculation of the current, a non-zero contribution comes only from *ei* collisions: since, under the conditions assumed, collisions are a small effect, the contributions to the conductivity from *ee* and *ei* collisions may be considered separately, and this means, for example, that contribution from *ee* collisions is calculated from the distribution function obtained by solving the transport equation with only this collision integral on the right, as if the electrons did not collide with the ions at all. In that case, the integral $\int \mathbf{v}\delta f_e^{(2)}\,d^3p$ with $\delta f_e^{(2)}$ in the form (59.11) is zero, because the law of conservation of momentum in collisions with any distribution function f_e gives identically

$$\int \mathbf{v}C_{ee}(f_e)\,d^3p = 0;$$

cf. §5.

Thus, in calculating the electric current, I in (59.11) is to be taken as the electron–ion collision integral. Then†

$$I_{ei}(\mathbf{v}\,.\,\nabla_\perp f_{0e}) = -\nu_{ei}(v)(\mathbf{v}\,.\,\nabla_\perp)f_{0e}, \qquad (59.12)$$

†Cf. (44.1). A formula of this type for $C(f)$ is valid if collisions take place with particles which may be regarded as immovable and if δf has the form $\mathbf{v}\cdot\mathbf{A}g(v)$, with $\mathbf{A}$ a constant vector. In the present case, $\mathbf{A}$ is represented by the vector operator $\nabla_\perp$.

where, from (44.3),

$$\nu_{ei}(v) = 4\pi z e^4 N_e L_e / m^2 v^3.$$

The contribution to the current from the distribution function (59.11), (59.12) is

$$\mathbf{j}^{(2)} = \frac{eN_e}{3\omega_{Be}^2}\nabla_\perp\langle v^2\nu_{ei}(v)\rangle$$
$$= \frac{4\sqrt{(2\pi)}ze^5 L_e N_e}{3m^{3/2}\omega_{Be}^2}\nabla_\perp(P_e/T^{3/2}). \tag{59.13}$$

To calculate the required transport coefficients, we must substitute the current $\mathbf{j}_\perp = \mathbf{j}^{(1)} + \mathbf{j}^{(2)}$ in equation (58.13):

$$\frac{1}{eN_e}\nabla_\perp P_e = \frac{\mathbf{j}_\perp}{\sigma_\perp} + \mathcal{R}B\mathbf{b}\times\mathbf{j}_\perp + \alpha_\perp\nabla_\perp T + \mathcal{N}B\mathbf{b}\times\nabla_\perp T, \tag{59.14}$$

which determines these coefficients. First of all putting $\nabla T = 0$ and collecting terms of order $1/\omega_{Be}$, we find that

$$\mathbf{j}^{(1)}/\sigma_\perp + \mathcal{R}B\mathbf{b}\times\mathbf{j}^{(2)} = 0,$$

whence

$$\sigma_\perp = 3\pi^{1/2}e^2N_e/2^{1/2}m\nu_{ei}, \tag{59.15}$$

where ν_{ei} (without argument) denotes

$$\nu_{ei} = \nu_{ei}(v_{Te}) = 4\pi z e^4 L_e N_e / m^{1/2}T^{3/2}. \tag{59.16}$$

The quantity (59.15) is of the same order as the conductivity (43.8) in the absence of the field, which here is equal to $\sigma_\parallel$.

Similarly, putting $\nabla P_e = 0$ in (59.14) and collecting terms $\sim 1/\omega_{Be}$, we find

$$\mathcal{R}B\mathbf{b}\times\mathbf{j}^{(2)} + \mathcal{N}B\mathbf{b}\times\nabla T = 0,$$

whence

$$\mathcal{N} = -\nu_{ei}/(2\pi)^{1/2}mc\omega_{Be}^2 = -3cN_e/2\sigma_\perp B^2. \tag{59.17}$$

The coefficient $\alpha_\perp$ appears only in the next approximation with respect to $1/\omega_{Be}$; its value for $z = 1$ is

$$\alpha_\perp = 0.36(\nu_{ei}/\omega_{Be})^2. \tag{59.18}$$

Electronic thermal conductivity

The heat flux in the plasma consists of electronic and ionic parts; let us first consider the former. It is calculated as

$$\mathbf{q}_e = \tfrac{1}{2}m\int v^2\mathbf{v}\,\delta f_e\,d^3p. \tag{59.19}$$

In the first approximation with respect to $1/\omega_{Be}$, substitution of (59.5) gives

$$\mathbf{q}_e^{(1)} = -(m/2\omega_{Be})(\mathbf{b}\times\nabla_\perp . \langle\mathbf{v}\rangle\mathbf{v}v^2\rangle N_e$$
$$= -(m/6\omega_{Be})\mathbf{b}\times\nabla_\perp N_e\langle v^4\rangle,$$

whence

$$\mathbf{q}_e^{(1)} = -(5c/2eB)\mathbf{b}\times\nabla_\perp P_e T$$
$$= -(w_e/e)\mathbf{j}^{(1)} - (5cP_e/2eB)\mathbf{b}\times\nabla_\perp T, \qquad (59.20)$$

where $w_e = 5T/2$ is the electronic heat function per electron. Comparison with the definition of the coefficient $\mathscr{L}$ in (58.14) shows that

$$\mathscr{L}_e = -5cN_eT/2eB^2. \qquad (59.21)$$

In the next approximation, the integral (59.19) is to be calculated with the distribution function (59.11). However, both *ei* and *ee* collisions contribute to the heat flux. For the former, we again use (59.11) and (59.12), obtaining

$$\mathbf{q}_e^{(ei)} = -(mN_e/6\omega_{Be}^2)\nabla_\perp\langle v^4\nu_{ei}(v)\rangle,$$

whence

$$\mathbf{q}_e^{(ei)} = -\frac{4\sqrt{(2\pi)}}{3}\frac{ze^4N_eL_e}{m^{3/2}\omega_{Be}^2}\nabla_\perp(P_e/\sqrt{T}). \qquad (59.22)$$

To find from this the corresponding part of the thermal conductivity $\kappa_\perp$, however, we must also use the condition $\mathbf{j} = \mathbf{j}^{(1)} + \mathbf{j}^{(2)} = 0$, since by (58.14)$\kappa_\perp$ is defined in terms of the heat flux in the absence of a current. With (59.7) and (59.13), we find that this condition leads to the following relation between the pressure and temperature gradients:

$$(c/B)\mathbf{b}\times\nabla_\perp P_e = \frac{eN_e\nu_{ei}}{\sqrt{(2\pi)}m\omega_{Be}^2}\nabla_\perp T;$$

in the calculations, we everywhere neglect terms above the second order in $1/\omega_{Be}$. Using this relation to calculate the sum $\mathbf{q}_e^{(1)} + \mathbf{q}_e^{(ei)}$, we find

$$\kappa_{\perp e}^{(ei)} = \frac{13}{6\sqrt{(2\pi)}}\frac{N_eT\nu_{ei}}{m\omega_{Be}^2}. \qquad (59.23)$$

This formula has a simple physical significance. In order of magnitude, the thermal conductivity must be $\kappa_\perp \sim C_eD_\perp$, where $C_e \sim N_e$ is the specific heat of the electrons per unit volume, and $D_\perp$ the electron diffusion coefficient across the magnetic field. The latter in turn is estimated as $\langle(\Delta x)^2\rangle/\delta t$, where $\langle(\Delta x)^2\rangle$ is the mean square displacement in the time δt. In a magnetic field, the transverse displacement is due only to collisions, and the electron moves a distance $\sim r_{Be}$. Hence $D_\perp \sim \nu_{ei}r_{Be}^2$, which leads to (59.23).

Let us now deal with the contribution from *ee* collisions. The calculations here are more laborious, and we shall only outline them.

In (59.11), I is now to be understood as the linearized Landau collision integral:

$$I_{ee}(\delta f_e) = -\operatorname{div}_{\mathbf{p}}\mathbf{s}^{(ee)},$$

where

$$s_\alpha^{(ee)} = 2\pi e^4 L_e \int \frac{w^2\delta_{\alpha\beta} - w_\alpha w_\beta}{w^3}\left\{f_{0e}\frac{\partial \delta f'_e}{\partial p'_\beta} + \delta f_e \frac{\partial f'_{0e}}{\partial p'_\beta} - f'_{0e}\frac{\partial \delta f_e}{\partial p_\beta} - \delta f'_e \frac{\partial f_{0e}}{\partial p_\beta}\right\} d^3p' \quad (59.24)$$

and $\mathbf{w} = \mathbf{v} - \mathbf{v}'$. The integral (59.19) with this distribution function becomes, on integrating by parts,

$$\mathbf{q}_e^{(ee)} = (1/2\omega_{Be}^2)\int \{v^2\mathbf{s}^{(ee)} + 2\mathbf{v}(\mathbf{v}\cdot\mathbf{s}^{(ee)})\}\, d^3p. \quad (59.25)$$

The coefficient here is written so that δf_e in (59.24) is to be taken as $(\mathbf{v}\cdot\nabla_\perp)f_{0e}$. The differentiation $\nabla_\perp$ need only be applied to the temperature T in the exponent of the Maxwellian function f_{0e}:

$$(\mathbf{v}\cdot\nabla_\perp)f_{0e} \to f_{0e}(mv^2/2T^2)\,\mathbf{v}\cdot\nabla_\perp T;$$

the terms arising from differentiation of the coefficient of the exponential cancel.†

After a simple though fairly lengthy calculation, the integral (59.25) is brought to the form $-\kappa_{\perp e}^{(ee)}\nabla_\perp T$, where‡

$$\kappa_{\perp e}^{(ee)} = \frac{\pi L_e e^4}{3T^2\omega_{Be}^2}\int\!\!\int\left\{wV^2 + \frac{(\mathbf{w}\cdot\mathbf{V})^2}{w} + \dots\right\} f_{0e}(p)f_{0e}(p')\, d^3p\, d^3p',$$

$\mathbf{w} = \mathbf{v} - \mathbf{v}'$, $\mathbf{V} = \frac{1}{2}(\mathbf{v} + \mathbf{v}')$, and the dots in the braces stand for terms containing odd powers of $\mathbf{w}\cdot\mathbf{V}$, which vanish on integration. Noting that

$$f_{0e}(p)f_{0e}(p') \propto \exp(-mV^2/T - mw^2/4T),$$

and carrying out the integration over $d^3p\, d^3p'$, we have finally

$$\kappa_{\perp e}^{(ee)} = \frac{2}{3\sqrt{\pi}}\frac{N_e T\nu_{ee}}{m\omega_{Be}^2}, \quad (59.26)$$

where

$$\nu_{ee} = 4\pi e^4 N_e L_e/m^{1/2}T^{3/2}. \quad (59.27)$$

Thus the total electronic contribution to the transverse thermal conductivity is

$$\kappa_{\perp e} = \frac{2N_e T\nu_{ee}}{3\sqrt{\pi} m\omega_{Be}^2}\left(1 + \frac{13}{4}z\right). \quad (59.28)$$

†This is evident from the general property noted in §6: the collision integral for like particles is zero for functions of the form $\mathbf{v}f_0$.

‡The pressure gradient does not appear here, and so there is no need to eliminate it by means of the condition $\mathbf{j} = 0$.

IONIC THERMAL CONDUCTIVITY

First of all, note that the condition $\omega_{Bi} \gg \nu_{ii}$ for the approximation under consideration to be applicable to *ii* collisions is stronger than for electrons. Since $\nu_{ii} \sim \nu_{ee}(m/M)^{1/2}$, and $\omega_{Bi} \sim \omega_{Be} m/M$, it follows from $\omega_{Bi} \gg \nu_{ii}$ that $\omega_{Be} \gg \nu_{ee}(M/m)^{1/2}$, which is stronger than $\omega_{Be} \gg \nu_{ee}$; the condition $r_{Bi} \gg a$ is certainly satisfied, being weaker than (59.10).

The transport equation for ions is analogous to (59.2):

$$\frac{\partial f_{0i}}{\partial t} + \mathbf{v} \cdot \frac{\partial f_{0i}}{\partial \mathbf{r}} + ze\mathbf{E} \cdot \frac{\partial f_{0i}}{\partial \mathbf{p}} = -\frac{ze}{c} \mathbf{v} \times \mathbf{B} \cdot \frac{\partial \delta f_i}{\partial \mathbf{p}} + I(\delta f_i). \tag{59.29}$$

As regards the transformation of the left-hand side, however, there is a difference from the electronic case. Substituting

$$f_{0i} = \frac{N_i}{(2\pi TM)^{3/2}} \exp\{-M(\mathbf{v} - \mathbf{V})^2/2T\},$$

we must now differentiate $\mathbf{V}$ with respect to t (and then again assume that the frame of reference is so chosen that $\mathbf{V} = 0$). With $\mathbf{V} = 0$ we have from the equation of motion in fluid dynamics

$$\frac{\partial \mathbf{V}}{\partial t} = -\frac{1}{\rho} \nabla P + \frac{1}{\rho c} \mathbf{j} \times \mathbf{B},$$

where the pressure $P = P_e + P_i$ and the density $\rho = N_i M$. The transport equation then becomes

$$\mathbf{v} \,.\, \nabla_\perp f_{0i} - \frac{f_{0i}}{N_i T} \mathbf{v} \cdot \left(\nabla_\perp P - \frac{1}{c} \mathbf{j} \times \mathbf{B}\right) = -\frac{ze}{c} \mathbf{v} \times \mathbf{B} \cdot \frac{\partial \delta f_i}{\partial \mathbf{p}} + I(\delta f_i), \tag{59.30}$$

where we have again, as in (59.3), put $\mathbf{E} = 0$ and written $\nabla_\perp$ in place of ∇.†

We can solve equation (59.30) by successive approximation with respect to $1/\omega_{Bi}$. In the first approximation we have, analogously to (59.5),

$$\delta f_i^{(1)} = \frac{1}{\omega_{Bi}} \mathbf{v} \,.\, \mathbf{b} \times \left(\nabla_\perp f_{0i} - \frac{f_{0i}}{N_i T} \nabla_\perp P + \frac{f_{0i}}{c N_i T} \mathbf{j} \times \mathbf{B}\right).$$

In this approximation, from (59.7), $\nabla_\perp P_e = \mathbf{j} \times \mathbf{B}/c$, whence

$$\delta f_i^{(1)} = \frac{1}{\omega_{Bi}} \mathbf{v} \,.\, \mathbf{b} \times \left(\nabla_\perp f_{0i} - \frac{f_{0i}}{P_i} \nabla_\perp P_i\right). \tag{59.31}$$

This distribution function gives, of course, no contribution to the

†For electrons, the second term on the left would contain not M/ρ but $m/\rho = m/MN_i$, thus being negligible.

current, $\int \delta f_i^{(1)}\mathbf{v}\, d^3p = 0$, as we should expect in a frame of reference where the ionic component of the plasma is at rest. The heat flux is then

$$\begin{aligned}\mathbf{q}_i^{(1)} &= \tfrac{1}{2}M\int v^2\mathbf{v}\,\delta f_i^{(1)}\,d^3p\\ &= (M/6\omega_{Bi})\mathbf{b}\times[\nabla_\perp(N_i\langle v^4\rangle) - (\langle v^4\rangle/T)\nabla_\perp P_i]\\ &= (5cP_i/2zeB)\mathbf{B}\times\nabla T,\end{aligned}$$

whence

$$\mathscr{L}_i = 5cN_iT/2zeB^2 = -\mathscr{L}_e/z^2. \tag{59.32}$$

To calculate the heat flux in the next approximation, only *ii* collisions are important: *ei* collisions make a contribution that is smaller by a factor $\sim (m/M)^{1/2}$, because the ion momentum change in collisions with electrons is small. The calculations are exactly similar to those given above for *ee* collisions.† The ionic part of the thermal conductivity is therefore obtained from (59.26) on replacing the electronic quantities by ionic ones:

$$\kappa_{\perp i} = 2N_iT\nu_{ii}/3\sqrt{\pi}M\omega_{Bi}^2, \quad \nu_{ii} = 4\pi z^2e^4L_iN_i/M^{1/2}T^{3/2}. \tag{59.33}$$

A comparison of (59.33) and (59.23) shows that (when $z\sim 1$)$\kappa_{\perp i}\to\kappa_{\perp e}(M/m)^{1/2}$. Thus, in fields so strong that $\omega_{Bi}\gg\nu_{ii}$, the transverse thermal conductivity is almost entirely ionic. The electronic conductivity becomes comparable when $\omega_{Bi}\sim (m/M)^{1/4}\nu_{ii}$; in making the comparison, it is to be taken into account that in such fields the effect of the magnetic field on κ_i is negligible. In still weaker fields, the ionic contribution to $\kappa_\perp$ becomes unimportant; in that case, if $\omega_{Be}\gg\nu_{ee}$, $\kappa_\perp$ is given by (59.28).

Viscosity

The momentum of a moving plasma is concentrated mainly in the ions, and the viscosity is therefore determined by the ion distribution function. Since collisions between an ion and electrons do not greatly change its momentum, only ion–ion collisions need be taken into account in the transport equation.

The left-hand side of the transport equation (59.29) is transformed in the same way as in §§6 and 8, and takes the same form as there.‡ Thus the transport equation for the viscosity problem is

$$(M/T)v_\alpha v_\beta(V_{\alpha\beta} - \tfrac{1}{3}\delta_{\alpha\beta}\,\mathrm{div}\,\mathbf{V})f_{0i} = -(ze/cM)\mathbf{v}\times\mathbf{B}\,.\,\partial\delta f_i/\partial\mathbf{v} + I_{ii}(\delta f_i). \tag{59.34}$$

The solution of this equation is to be sought in the form

$$\delta f_i = \sum_{n=0}^{4} g_n(v^2)V_{\gamma\delta}^{(n)}v_\gamma v_\delta, \tag{59.35}$$

†The term in ∇P_i which distinguishes (59.31) from (59.5) is unimportant here: this part of the distribution function is $\propto \mathbf{v}f_{0i}$, and has zero collision integral; cf. the last-but-one footnote.

‡Here it must be noted that the plasma pressure $P = (N_i + N_e)T = N_i(1+z)T$, and the heat capacity per ion is $3(1+z)/2$.

where the $V^{(n)}_{\gamma\delta}$ are linear combinations of the components of the tensor $V_{\alpha\beta}$, which appear in the expression

$$\sigma'_{\alpha\beta} = \sum_{n=1}^{4} \eta_n V^{(n)}_{\alpha\beta} \tag{59.36}$$

for the viscous stress tensor, according to the definitions (13.18) and (58.15); it will be recalled that all the $V^{(n)}_{\alpha\alpha} = 0$. The stress tensor is calculated as the integral

$$-\sigma'_{\alpha\beta} = \int M v_\alpha v_\beta \, \delta f_i \, d^3p .$$

Substituting (59.35), averaging over the direction of $\mathbf{v}$ by means of the formula

$$\langle v_\alpha v_\beta v_\gamma v_\delta \rangle = \tfrac{1}{15} v^4 (\delta_{\alpha\beta}\,\delta_{\gamma\delta} + \delta_{\alpha\gamma}\,\delta_{\beta\delta} + \delta_{\alpha\delta}\,\delta_{\beta\gamma}),$$

and comparing with (59.36), we find

$$\eta_n = -(2M/15) \int v^4 g_n(v^2)\, d^3p . \tag{59.37}$$

The equations which determine the functions g_n are obtained by substituting (59.35) in (59.34) and equating the coefficients of the various tensors $V^{(n)}_{\alpha\beta}$ on the two sides of the equation. We shall omit the details of these fairly laborious calculations, and give only the final results.

Non-zero viscosity coefficients η_3 and η_4 arise even when the collision integral is neglected, and are therefore proportional to $1/\omega_{Bi}$. The coefficients η_1 and η_2 arise only in the next approximation, when collisions are taken into account, and are therefore proportional to $1/\omega^2_{Bi}$:†

$$\left.\begin{aligned} \eta_1 &= \tfrac{1}{4}\eta_2 = 2\pi^{1/2}(ze)^4 L_i N_i^2 / 5(MT)^{1/2}\omega^2_{Bi}, \\ \eta_3 &= \tfrac{1}{2}\eta_4 = N_i T/2\omega_{Bi}. \end{aligned}\right\} \tag{59.38}$$

Lastly, it may be noted that all the expressions derived in this section for the "transverse" transport coefficients remain meaningful even under conditions less stringent than the general formula (58.1). It is easy to see that the correction to the distribution function is small provided that the characteristic dimensions of the problem are large in comparison with the Larmor radius r_B of the corresponding particles; this ensures that the above expressions are applicable. The condition is sufficient also for the applicability of the equations of fluid dynamics themselves, if the pressure and temperature gradients are everywhere transverse to the magnetic field.

In this discussion, we have everywhere considered a plasma in which the electron and ion temperatures are equal. However, "two-temperature" conditions

†The suitability of defining the viscosity η_0 for a magnetoactive plasma as in (58.15) is due to the fact that all the other coefficients η then tend to zero as $B \to \infty$.

often arise, because of the large mass difference between electrons and ions. In such a case, we can again formulate equations of the fluid dynamics type, and calculate the transport coefficients that occur in them.†

PROBLEM

A plasma inhomogeneous in the x-direction is confined by a magnetic field in the z-direction. With the condition $\omega_{Be} \gg \nu_{ei}$, determine the density and magnetic field distributions in the plasma, assuming the temperature distribution given (I.E. Tamm 1951).

SOLUTION. The gradients of the temperature T and the pressure P are in the x-direction, and so is the electric field $\mathbf{E}$ (a potential field in the stationary case) which results from the inhomogeneity of the plasma. The confinement signifies that there is no movement of the plasma or electric current in the x-direction: $V_x = 0$, $j_x = 0$.

Using these results and Maxwell's equation curl $\mathbf{B} = 4\pi\mathbf{j}/c$, we obtain by taking the y-component of (58.13)

$$(c/4\pi)dB/dx = -j_y = \mathcal{N}\sigma_\perp B\, dT/dx.$$

Substitution of (59.17) for $\mathcal{N}\sigma_\perp$ gives

$$\frac{d}{dx}\left(\frac{B^2}{8\pi}\right) = -\frac{3}{2}N_e\frac{dT}{dx}. \tag{1}$$

The magnetic field is "expelled" from the hotter parts of the plasma.

By taking the x-component of (58.3) and neglecting the viscosity terms which make a contribution of a higher order of smallness in $1/B$, we obtain a second equation,

$$d(P_e + P_i)/dx = j_y B/c,$$

which by means of the same Maxwell's equation can be converted (when $z = 1$) to

$$2N_eT + B^2/8\pi = \text{constant}. \tag{2}$$

Equation (1) may be put in a more convenient form by eliminating the magnetic field by means of equation (2). After integration, we have

$$N_eT^{1/4} = \text{constant}. \tag{3}$$

The formulae (2) and (3) give the solution. The temperature distribution comes from the equation of thermal conduction.

§60. The drift approximation

In examining the transport coefficients of a plasma in a strong magnetic field (§59), we used the Landau collision integral, which assumes the inequality $r_{Be} \gg a$ (59.10). We shall now show how this limitation may be removed, i.e. how formulae may be obtained which are suitable even for fields so strong that the opposite inequality holds for electrons:

$$r_{Be} \ll a. \tag{60.1}$$

Here it is convenient to make use of a particular approximation called the *drift approximation*, made in the transport equation itself and not only in the solving of

†This topic is dealt with by S. I. Braginskiĭ, Transport processes in a plasma, *Reviews of Plasma Physics* 1, 205, Consultants Bureau, New York, 1965.

it. The drift approximation is valid if the magnetic and electric fields vary sufficiently slowly in space and time: the field frequency ω and the effective collision frequency ν must be small in comparison with the Larmor frequency, and the characteristic distance $1/k$ over which the fields vary must be large compared with the Larmor radius. These conditions must be satisfied for each type of particle to which the drift approximation is applied. In the present section, we shall write all formulae for the particular case of electrons; the corresponding formulae for ions are obtained, as usual, by making the changes $e \to -ze$, $\omega_{Be} \to -\omega_{Bi}$, $m \to M$. Thus we shall assume that

$$\omega,\ \nu_{ei} \ll \omega_{Be}, \quad 1/k \gg r_{Be}. \tag{60.2}$$

The method in question is based on approximately solving the equations of motion of charged particles in specified fields $\mathbf{E}(t, \mathbf{r})$ and $\mathbf{B}(t, \mathbf{r})$, taking account of the slowness of variation of these with t and $\mathbf{r}$. The motion of particles in such fields is a combination of a rapid rotation (with frequency ω_{Be}) in circular Larmor orbits and a slow movement of the *guiding centres*, i.e. the centres of these orbits. The method of solution is to separate the rapidly oscillating component of the motion and average over it.

The position vector and the velocity of the electron may be written as

$$\mathbf{r} = \mathbf{R}(t) + \boldsymbol{\zeta}(t), \quad \mathbf{v} = \mathbf{V} + \dot{\boldsymbol{\zeta}}, \quad \mathbf{V} = \dot{\mathbf{R}}, \tag{60.3}$$

where $\mathbf{R}$ is the position vector of the guiding centre, and $\boldsymbol{\zeta}$ the oscillatory position vector of the electron relative to it.† In the zero-order approximation, where the space and time variations of the field, and collisions, are entirely neglected, we have simply a motion in crossed uniform and constant fields $\mathbf{E}$ and $\mathbf{B}$. As we know (see *Fields*, § 22), the vector $\boldsymbol{\zeta}$ in this case is exactly in the plane perpendicular to $\mathbf{B}$, and rotates in that plane with a constant angular velocity $\omega_{Be} = eB/mc$, remaining constant in magnitude. The radius $|\boldsymbol{\zeta}|$ of the circle is related to the constant speed $|\dot{\boldsymbol{\zeta}}| \equiv v_\perp$ by $|\boldsymbol{\zeta}| = v_\perp/\omega_{Be}$; the vector relation between $\boldsymbol{\zeta}$ and $\dot{\boldsymbol{\zeta}}$ is

$$\boldsymbol{\zeta} = -\mathbf{b} \times \dot{\boldsymbol{\zeta}}/\omega_{Be}, \tag{60.4}$$

where $\mathbf{b} = \mathbf{B}/B$. The centre of the orbit moves with velocity

$$\dot{\mathbf{R}} = \mathbf{V}_0 = v_{0\parallel}\mathbf{b} + \mathbf{w}_0,$$

where $v_{0\parallel}$ is the speed of the uniformly accelerated motion along the magnetic field, which satisfies the equation

$$m\dot{v}_{0\parallel} = -e\mathbf{b} \cdot \mathbf{E} \tag{60.5}$$

and

$$\mathbf{w}_0 = \dot{\mathbf{R}}_\perp = (c/B)\mathbf{E} \times \mathbf{b} \tag{60.6}$$

is the *electric drift* velocity in the plane perpendicular to $\mathbf{B}$.‡

†The quantity $\mathbf{V}$ in this section is not to be confused with the macroscopic velocity denoted by $\mathbf{V}$ in § 59.

‡Here we assume, of course, that $E/B \ll 1$, so that $w \ll c$ and relativistic effects may be neglected.

From now on, we shall use this approximation, neglecting terms arising from the non-constancy of the fields **E** and **B**, i.e. shall regard these fields as constant. Accordingly, the suffix 0 will be omitted from all quantities.

The essence of the drift approximation is to change, in the transport equation, to the slowly varying quantities **R**, $v_\parallel$ and $v_\perp = |\dot{\boldsymbol{\zeta}}|$. These together constitute five independent variables in the distribution function.

The phase volume element in the new variables is

$$\begin{aligned} d^3x\, d^3p &= d^3R\,.\,2\pi m^3\, dv_\parallel\,.\,v_\perp\, dv_\perp \\ &= 2\pi m^3\, d^3R\, dv_\parallel dJ, \end{aligned} \tag{60.7}$$

where the quantity

$$J = \tfrac{1}{2}v_\perp^2 \tag{60.8}$$

will be convenient later. As regards the derivation of (60.7) it must be remembered that in the approximation used the fields may be regarded as constant.

The electron current density can be expressed in terms of the new variables. For one electron, the current density is $-e\mathbf{v}\delta(\mathbf{r}-\mathbf{r}_e)$, where **r** denotes variable coordinates in space, and $\mathbf{r}_e$ the position of the electron. Putting $\mathbf{v} = \mathbf{V} + \dot{\boldsymbol{\zeta}}$ and $\mathbf{r}_e = \mathbf{R} + \boldsymbol{\zeta}$, we write

$$-e\mathbf{v}\delta(\mathbf{r}-\mathbf{r}_e) \approx -e(\mathbf{V}+\dot{\boldsymbol{\zeta}})[\delta(\mathbf{r}-\mathbf{R}) - \boldsymbol{\zeta}\cdot\nabla_\mathbf{r}(\mathbf{r}-\mathbf{R})].$$

We average this expression over the angle of rotation by means of the obvious relation

$$\omega_{Be}\langle \dot{\zeta}_\alpha(\mathbf{b}\times\boldsymbol{\zeta})_\beta\rangle = \langle \dot{\zeta}_\alpha\dot{\zeta}_\beta\rangle = \tfrac{1}{2}v_\perp^2\delta_{\alpha\beta},$$

where α and β are two-dimensional vector suffixes in the plane perpendicular to the magnetic field. The result is

$$-e\mathbf{V}\delta(\mathbf{r}-\mathbf{R}) + (mcJ/B)\mathbf{b}\times\nabla_\mathbf{r}\delta(\mathbf{r}-\mathbf{R}).$$

Multiplying this by the electron distribution function f_e and integrating over $d^3p = 2\pi m^3\, dv_\parallel dJ$, we find the current density in **R**-space:†

$$\mathbf{j}_e = -e\int \mathbf{V} f_e\, d^3p - (mc/B)\,\mathrm{curl}\Big(\mathbf{b}\int Jf_e\, d^3p\Big). \tag{60.9}$$

The first term here corresponds to charge transfer with the moving Larmor orbits; the second term takes account of the rotation of particles in these orbits,‡ and has a simple physical significance: if it is written as $c\,\mathrm{curl}\,\mathbf{M}$, the vector

$$\mathbf{M} = -(m\mathbf{b}/B)\int f_e J\, d^3p \tag{60.10}$$

†With integration by parts in the second term, which transfers the operator $\nabla_\mathbf{r}$ to $\mathbf{b}f_e$.

‡A similar averaging of the charge density $-e\delta(\mathbf{r}-\mathbf{r}_e)$ gives the usual expression $-e\int f_e\, d^3p$; correction terms due to the rotation of the particles would appear here only when second-order small quantities (second derivatives with respect to the coordinates) were taken into account.

is the magnetization of the plasma due to the rotation of the charges. The magnetic moment (60.10) is independent of the sign of the charges and is in the direction opposite to the magnetic field, i.e. corresponds to diamagnetism.

Let us now transform the transport equation to the new variables. Since the distribution function f_e relates to the same element of phase space as before, which is simply put in the different form (60.7), the transport equation again has the form $df_e/dt = C(f_e)$, or, expanding the left-hand side in terms of the new variables,

$$\frac{\partial f_e}{\partial t} + v_\parallel \frac{\partial f_e}{\partial R_\parallel} + \frac{c}{B}\mathbf{E}\times\mathbf{b}\cdot\frac{\partial f_e}{\partial \mathbf{R}_\perp} = C(f_e), \tag{60.11}$$

where we have used an obvious notation for the components of the vectors, and taken account of (60.5) and (60.6). In this approximation there is no term in $\dot{v}_\perp$, since $v_\perp$ does not vary during the drift.

Next, let us express the collision integral in terms of the drift variables.† Note first of all that a collision in these variables is an "instantaneous" change in the velocities $v_\parallel$ and $v_\perp$ and the components $\mathbf{R}_\perp$, perpendicular to the magnetic field, of the position vector of the guiding centre. The parallel component $R_\parallel$ is almost equal to the corresponding coordinate of the particle itself, and is unchanged in the collision.

Collisions occur only between particles which pass at impact parameters ρ not exceeding the screening length a: $\rho \lesssim a$. If ρ is much less than the Larmor radii of the colliding particles, the magnetic field has no effect on the scattering process, since at such distances the field causes no appreciable curvature of the particle trajectories. It is not natural to describe such collisions in terms of the drift variables. The use of the collision integral expressed with these variables is therefore appropriate only when for at least one of the colliding particles $r_B \ll a$.

In a Coulomb interaction of particles, with or without a magnetic field present, distant collisions, and accordingly small changes in all the variables, are important. The derivation of the collision integral in **p**-space given in §41 therefore remains valid in the space of the variables $\mathbf{R}_\perp = (X, Y)$, $v_\parallel$, J (with the z-axis along the magnetic field), if we now replace the momentum components by the four variables $g_k\{X, Y, v_\parallel, J\}$ and take Δg_1, $\Delta g_2, \ldots$ to be the changes in these variables in collisions.

The collision integral is again brought to the form

$$\begin{aligned} C(f) &= -\sum_{k=1}^{4} \partial s_k/\partial g_k \\ &= -\partial \mathbf{s}_\perp/\partial \mathbf{R}_\perp - \partial s_\parallel/\partial v_\parallel - \partial s_J/\partial J \end{aligned} \tag{60.12}$$

(the flux $\mathbf{s}_\perp$ has, by definition, components only in the plane perpendicular to **B**). Here it is important that the volume element in the space of the variables g_k reduces simply to the product of their differentials; the collision integral is therefore an ordinary divergence. The proof given in §41 needs only slight

†This was done by E. M. Lifshitz (1937) for an electron gas, and the result was generalized to plasmas by S. T. Belyaev (1955).

changes. First of all, in (41.2) we have used the fact that $\Delta\mathbf{p} \equiv \mathbf{q} = -\Delta\mathbf{p}'$, by the law of conservation of momentum. For the drift variables considered here, there is of course no such relation. Repeating the derivation without this assumption, we find for electron–ion collisions (for example)

$$s_k^{(ei)} = \sum_{l=1}^{4} \frac{1}{2} \int \{\langle \Delta g_{ek} \Delta g_{el} \rangle f_i \partial f_e / \partial g_{el} + [\Delta g_{ek} \Delta g_{il} \rangle f_e \partial f_i / \partial g_{il}\} \, d^3 p_i, \qquad (60.13)$$

where $d^3 p_i = 2\pi M^3 \, dJ_i \, dv_{i\parallel}$, the Δg_k are the changes in the quantities g_k in a collision, and the angle brackets denote averaging over collisions.

An important point in the derivation of (60.13) is the possibility of interchanging the initial and final states in the collision integral, after which the terms linear in Δg_k evidently cancel; this also allows the integration to be taken over all g-space. In §41 this transformation was effected by virtue of the symmetry under time reversal, which relates the probabilities of direct and reverse collisions. When a magnetic field is present, this symmetry exists only if the direction of the field $\mathbf{B}$ is reversed, and therefore relates the collision probabilities in essentially different fields. However, we shall see that in this case the symmetry under time reversal is restored by integration over impact parameters.

Lastly, in (60.13) we have used the fact that mutual scattering of Larmor orbits occurs only when they pass at distances not exceeding the screening length a. Assuming that the distribution function varies only slightly over such distances, we have put approximately $f_i(\mathbf{R}_i, v_{i\parallel}, J_i) \approx f_i(\mathbf{R}_e, v_{i\parallel}, J_i)$ and integrated over $d^3 R_i$. This has left in (60.13) only the integration over $d^3 p_i$; the averaging over collisions includes integration over the positions $\mathbf{R}_i$. In specific cases below, this averaging will be expressed by means of the appropriate scattering cross-section. Here we shall simply note that the mean values $\langle \Delta \mathbf{R}_\perp \Delta J \rangle$, $\langle \Delta \mathbf{R}_\perp \Delta v_\parallel \rangle$ are zero, as is seen from the fact that the products $\Delta X \Delta J$, $\Delta Y \Delta J$ (and the same with $\Delta v_\parallel$ in place of ΔJ) form a vector in the xy-plane. Since there are no preferred directions for the Larmor orbits in that plane, this vector must give zero on averaging.

An important property of the collision integral in the drift variables is that when it is included in the transport equation there is a change in the expression for the particle flux (in ordinary space) in terms of the distribution function. To see this, we write the transport equation as

$$\frac{\partial f_e}{\partial t} + \frac{\partial (\mathbf{V} f_e)}{\partial \mathbf{R}_\perp} + \frac{\partial}{\partial v_\parallel}(v_\parallel f_e) = -\frac{\partial \mathbf{s}_{e\perp}}{\partial \mathbf{R}_\perp} - \frac{\partial s_{e\parallel}}{\partial v_\parallel} - \frac{\partial s_{eJ}}{\partial J}; \qquad (60.14)$$

$\mathbf{V}$ can be taken under the differentiation, since $\mathbf{B}$ and $\mathbf{E}$ are assumed constant. Integration of this equation over $d^3 p$ gives

$$\partial N_e / \partial t + \operatorname{div}_{\mathbf{R}} \int (\mathbf{V} f_e + \mathbf{s}_{e\perp}) \, d^3 p = 0, \quad N_e = \int f_e \, d^3 p, \qquad (60.15)$$

where the suffix e to the electronic variables is omitted for brevity; N_e is the spatial number density of orbits, and the expression acted upon by $\operatorname{div}_{\mathbf{R}}$ is therefore the flux of orbits. We see that the ordinary expression $\int \mathbf{V} f_e \, d^3 p$ is augmented by the collision term $\int \mathbf{s}_{e\perp}\, d^3p$, which is essentially the diffusion flux transverse to the magnetic field. With this description, in contrast to the ordinary description of diffusion, it appears directly in the transport equation.

When these expressions are used we must, of course, take account of the fact that the electric current density is related to the flux of actual particles, not of orbits. According to (60.9), the flux of particles differs from the flux of orbits by a curl term which represents the magnetization. The final expression for the electron current density is therefore

$$\mathbf{j}_e = -e\int \mathbf{V} f_e\, d^3p - (mc/B)\,\mathrm{curl}\left(\mathbf{b}\int f_e J\, d^3p\right) - \int e\mathbf{s}_{e\perp}\, d^3p. \tag{60.16}$$

The significance of the expression (60.13) can be appreciated only when the mean values in it have been evaluated. We shall show how this done for the case of the electron integral in electron–ion collisions.

The calculations are done in different ways in two ranges of values of the impact parameters ρ, specified by the inequalities

$$\text{(I)}\ \rho \ll r_{Be}, \quad \text{(II)}\ r_{Be} \ll \rho \ll a. \tag{60.17}$$

The integrations with respect to the parameter ρ will be logarithmic, as is usual for Coulomb scattering. With logarithmic accuracy, no distinction is necessary between strong ($\gg$) and weak ($>$) inequalities. The ranges (60.17) therefore cover essentially the whole variation of the impact parameter; in accordance with (60.1), it is of course assumed that $r_{Be} \ll a$. For the existence of range I, it is also necessary that

$$r_{Be} \gg \rho_{\min} = ze^2/mv_{Te}^2, \tag{60.18}$$

where $\rho_{\min}$ is the impact parameter for which the scattering angle becomes ~ 1; we are here considering only the quasi-classical case $e^2/\hbar v_{Te} \gg 1$.

We shall also suppose that $r_{Bi} \gtrsim a$. Then, for all impact parameters $\rho \lesssim a$, the influence of the magnetic field on the motion of the ions (in a collision) is unimportant: the ion path is only slightly curved by the field at distances $\sim \rho$. We can neglect the ion recoil in the limit $m/M \to 0$, i.e. take as zero the changes in all the ion characteristics $R_\perp$, $v_\parallel$, J.† Then the second term in the braces in (60.13) vanishes, so that the electron–ion part of the electron current becomes

$$s_\alpha^{(ei)} = -\tfrac{1}{2}N_i\langle \Delta X_\alpha \Delta X_\beta\rangle^{(ei)}\, \partial f_e/\partial X_\beta. \tag{60.19}$$

The quantities $\langle \Delta X_\alpha \Delta X_\beta\rangle$ form a spatial tensor transverse to the field, which we write in the explicitly transverse form

$$\langle \Delta X_\alpha \Delta X_\beta\rangle = \tfrac{1}{2}\langle(\Delta \mathbf{R}_\perp)^2\rangle(\delta_{\alpha\beta} - b_\alpha b_\beta); \tag{60.20}$$

†This cannot be done if there are impact parameters such that $a \gg \rho \gg r_{Bi}$. In such collisions, the ion drifts in the field of the electron, and its large mass does not take effect.

the flux (60.19) then becomes

$$\mathbf{s}^{(ei)} = -\tfrac{1}{4}N_i\langle(\Delta\mathbf{R}_\perp)^2\rangle^{(ei)}\nabla_\perp f_e, \tag{60.21}$$

where $\nabla_\perp = \nabla_\mathbf{R} - \mathbf{b}(\mathbf{b}\,.\,\nabla_\mathbf{R})$ is the operator of differentiation in the directions transverse to **b**.

The expressions for the "velocity fluxes" analogous to (60.19) are

$$\left.\begin{aligned} s_\parallel^{(ei)} &= -\tfrac{1}{2}N_i\{\langle(\Delta v_\parallel)^2\rangle^{(ei)}\partial f_e/\partial v_\parallel + \langle\Delta v_\parallel\langle\Delta J\rangle^{(ei)}\partial f_e/\partial J\}, \\ s_J^{(ei)} &= -\tfrac{1}{2}N_i\{\langle\Delta v_\parallel\Delta J\rangle^{(ei)}\partial f_e/\partial v_\parallel + \langle(\Delta J)^2\rangle^{(ei)}\partial f_e/\partial J\}. \end{aligned}\right\} \tag{60.22}$$

In equilibrium, i.e. for a Maxwellian distribution

$$f_e = \text{constant}\times\exp\left\{-\frac{m}{T}(\tfrac{1}{2}v_\parallel^2 + J)\right\}, \tag{60.23}$$

the collision integral must be zero. Substituting (60.23) in (60.22) and equating the fluxes to zero, we have

$$\langle\Delta v_\parallel\Delta J\rangle^{(ei)} = -v_\parallel\langle(\Delta v_\parallel)^2\rangle^{(ei)} = -(1/v_\parallel)\langle(\Delta J)^2\rangle^{(ei)}. \tag{60.24}$$

Let us first calculate the contribution from range I, where the magnetic field can be supposed not to influence the scattering process, since at such distances there is no appreciable curvature of the path for ions or electrons. The natural variable to describe the collision is then the ordinary momentum **p** of the electron, and the drift variables must be expressed in terms of this. According to (60.3), (60.4) and (60.8),

$$\mathbf{r} = \mathbf{R} - (1/m\omega_{Be})\mathbf{b}\times\mathbf{p}_\perp, \quad v_\parallel = p_\parallel/m, \quad J = p_\perp{}^2/2m^2.$$

Since the coordinates **r** of the particle, unlike the coordinates **R** of the orbit centre, are unaffected by the collision, we hence find

$$\Delta\mathbf{R}_\perp = (1/m\omega_{Be})\mathbf{b}\times\mathbf{q}_\perp, \quad \Delta v_\parallel = q_\parallel/m, \quad \Delta J = (1/m^2)\mathbf{p}_\perp\,.\,\mathbf{q}_\perp, \tag{60.25}$$

where **q** is the small change in the momentum **p**.

Denoting by the suffix I the contribution from this class of collisions, we now write

$$\langle(\Delta\mathbf{R}_\perp)^2\rangle_\text{I}^{(ei)} = \int(\Delta\mathbf{R}_\perp)^2 v\,d\sigma = (1/m^2\omega_{Be}^2)\int q_\perp{}^2 v\,d\sigma, \tag{60.26}$$

where $d\sigma$ is the cross-section for scattering of an electron by an ion at rest. With $d\sigma$ given by (41.6), the integration gives

$$\langle(\Delta\mathbf{R}_\perp)^2\rangle_\text{I}^{(ei)} = \frac{8\pi z^2 e^4 L_\text{I}}{m^2\omega_{Be}^2 v}\cdot\tfrac{1}{2}(1+\cos^2\theta), \tag{60.27}$$

where θ is the angle between $\mathbf{v}$ and $\mathbf{b}$, and

$$L_{\rm I} = \log(mr_{Be}v_{Te}^2/ze^2) = \log(mv_{Te}^3/ze^2\omega_{Be}) \tag{60.28}$$

is the Coulomb logarithm cut off at a maximum impact parameter $\rho \sim r_{Be}$ (the upper limit of range I). Finally, expressing this result in terms of the drift variables, we have

$$\langle(\Delta\mathbf{R}_\perp)^2\rangle_{\rm I}^{(ei)} = \frac{8\pi z^2e^2c^2L_{\rm I}}{B^2}\frac{v_\parallel^2+J}{(v_\parallel^2+2J)^{3/2}}. \tag{60.29}$$

A similar calculation gives

$$\langle\Delta v_\parallel \Delta J\rangle_{\rm I}^{(ei)} = -\frac{8\pi z^2e^4L_{\rm I}}{m^2}\frac{Jv_\parallel}{(v_\parallel^2+2J)^{3/2}}, \tag{60.30}$$

and the remaining two quantities are determined from (60.24).

Let us now turn to range II. Here the drift variables are the natural ones, and the collision is described as a drift deflection of an orbit moving in the direction of $\mathbf{b}$ (the z-direction) in the Coulomb field of the ion at rest. The speed $v_\perp$ and therefore J are not changed by the drift; this in turn implies the conservation of $v_\parallel$, because of the conservation of energy in scattering by a heavy ion. Hence range II makes no contribution to the quantities (60.24).

The contribution to $\langle(\Delta\mathbf{R}_\perp)^2\rangle$ is calculated as

$$\langle(\Delta\mathbf{R}_\perp)^2\rangle_{\rm II}^{(ei)} = (\Delta\mathbf{R}_\perp)^2|v_\parallel|d\sigma = \int(\Delta\mathbf{R}_\perp)^2|v_\parallel|d^2\rho, \tag{60.31}$$

where $\boldsymbol{\rho}$ is the position vector of the orbit centre (the value of $\mathbf{R}_\perp$) before the collision. The change in $\mathbf{R}_\perp$ as the orbit travels in a constant uniform magnetic field $\mathbf{B}$ and a constant electric field $\mathbf{E} = ez\mathbf{R}/R^3$ (the field of the ion) is determined by the drift equation

$$d\mathbf{R}_\perp/dt = (c/B)\mathbf{b}\times\mathbf{E} = (zec/B)\mathbf{b}\times\mathbf{R}_\perp/(R_\parallel^2+R_\perp^2)^{3/2}; \tag{60.32}$$

see (60.6). In the first approximation, we can put on the right of this equation $\mathbf{R}_\perp \approx \boldsymbol{\rho}$, $R_\parallel = v_\parallel t$. The total change in $\mathbf{R}_\perp$ in a collision is found by integrating (60.32) with respect to t from $-\infty$ to ∞, and is

$$\Delta\mathbf{R}_\perp = (2zec/B|v_\parallel|)\mathbf{b}\times\boldsymbol{\rho}/\rho^2. \tag{60.33}$$

Substituting this expression in (60.31) and carrying out the integration (with logarithmic accuracy, corresponding to the limits of range II), we find

$$\langle(\Delta\mathbf{R}_\perp)^2\rangle_{\rm II}^{(ei)} = 8\pi z^2e^2c^2L_{\rm II}/B^2|v_\parallel|, \quad L_{\rm II} = \log(a/r_{Be}). \tag{60.34}$$

The contributions (60.29) and (60.34) are in general of the same order of magnitude:

$$N_i\langle(\Delta\mathbf{R}_\perp)^2\rangle \sim \nu_{ei}r_{Be}^2,$$

where ν_{ei} is the mean electron–ion collision frequency. The particular feature of (60.34) is that it becomes infinite as $v_{\parallel}\to 0$, whatever the value of $v_{\perp}$. The physical significance of this divergence is that when the speed $v_{\parallel}$ is small the orbit spends a long time in the field of the ion, during which the drift carries it to a great distance.

In reality, of course, formula (60.34) becomes invalid when $v_{\parallel}$ is small, for a variety of reasons: (1) if $r_{Bi} \gg a$, then for $|v_{\parallel}| \ll v_{Ti}$ the ion can leave the electron during the collision time, and this mechanism cuts off the divergence at $|v_{\parallel}| \sim v_{Ti}$; (2) in deriving the formula, it is always assumed that $|\Delta\mathbf{R}_{\perp}| \ll \rho$; (3) the orbit can leave the ion in question because of drift in the field of other particles (a three-body collision).

The above formulae solve the problem of constructing the transport equation in the drift approximation. In particular, it enables us to find the transport coefficients for the plasma in the first non-vanishing approximation with respect to $1/B$; see Problem 1.

Lastly, it has to be explained how the integration over $d^2\rho$ formally restores the symmetry under time reversal, as already utilized in writing (60.13). The loss of this symmetry is shown by the change in sign of the deflection $\Delta\mathbf{R}_{\perp}$ in (60.33) when the direction of $\mathbf{B}$ is reversed. The previous sign can be restored, however, by changing the sign of the integration variable, $\boldsymbol{\rho}\to-\boldsymbol{\rho}$, so that the change in the sign of $\mathbf{B}$ can have no effect in this approximation. (In range I, the magnetic field never affects the scattering process.)

PROBLEMS

PROBLEM 1. In the drift approximation, determine the Hall coefficient $\mathcal{R}$ and the transverse conductivity $\sigma_{\perp}$ of the plasma (S. T. Belyaev, 1955).

SOLUTION. Considering a plasma with an electron density gradient (but with no electric field or temperature gradient), we assume the distribution function f_e Maxwellian in (60.16) and (60.21), obtaining

$$\mathbf{j} = (cT/B)\mathbf{b}\times\nabla N_e + eD_{\perp}\nabla_{\perp}N_e,$$

the transverse diffusion coefficient being

$$D_{\perp} = \tfrac{1}{4}N_i\overline{\langle(\Delta\mathbf{R}_{\perp})^2\rangle},$$

where the bar denotes averaging over the Maxwellian distribution of the electrons. On comparing with the general expression (58.13), we find in the first approximation with respect to $1/B$ the previous expression (59.8) for $\mathcal{R}$. In the next approximation,

$$\sigma_{\perp} = T/e^2N_e\mathcal{R}B^2D_{\perp}. \qquad (1)$$

In range II (see (60.17)), we take $\langle(\Delta\mathbf{R}_{\perp})^2\rangle$ from (60.34). With logarithmic accuracy,

$$\overline{|v_{\parallel}|^{-1}} = (m/2\pi T)^{1/2}\int_{-\infty}^{\infty}\exp(-mv_{\parallel}^2/2T)dv_{\parallel}/|v_{\parallel}|$$
$$\approx 2(m/2\pi T)^{1/2}\log(v_{Te}/v_{\min}),$$

where $v_{\min}$ is determined by one of the mechanisms mentioned at the end of the section. For example, putting $v_{\min}\sim v_{Ti}$, we find

$$D_{\perp}^{\text{II}} = [(2\pi m)^{1/2}z^2e^2c^2N_i/T^{1/2}B^2]\log(M/m)\log(a/r_{Be}). \qquad (2)$$

Similarly, taking $\langle(\Delta\mathbf{R}_\perp)^2\rangle$ from (60.27), we find the contribution to the diffusion coefficient from range I:

$$D_\perp{}^{\mathrm{I}} = [4(2\pi m)^{1/2}z^2e^2c^2N_i/3T^{1/2}B^2]\log(mv_{Te}^3/ze^2\omega_{Be}). \qquad (3)$$

If the inequality (59.10) opposite to (60.1) is assumed to be satisfied, then range II does not exist, and the logarithm in (3) is replaced by its ordinary Coulomb value (41.10). Then substitution of (3) in (1) gives formula (59.15) for $\sigma_\perp$.

PROBLEM 2. Determine the transverse diffusion coefficient $D_\perp$ for collisions between electrons and neutral atoms.

SOLUTION. Because the electron–atom interaction is short-range, there is only range I, in which a is to be taken as the size of the atom.† Formula (60.26) also remains valid, but we must now substitute in it the cross-section for electron scattering by a neutral atom. After integration over angles, $D_\perp$ is expressed in terms of the transport cross-section σ_t for this scattering:

$$D_\perp = \tfrac{1}{4}N_a\overline{\langle(\Delta R_\perp)^2\rangle} = (N_a/2\omega_{Be}^2)\,\overline{v^3\sigma_t},$$

where N_a is the number density of atoms. With the cross-section σ_t independent of the electron speed, we have on averaging over the Maxwellian distribution

$$D_\perp = \frac{2}{3\sqrt{\pi}}\left(\frac{2T}{m}\right)^{3/2}\frac{N_a\sigma_t}{\omega_{Be}^2}.$$

†There may be some doubt as to the applicability of the collision integral (60.12) for scattering by a short-range potential, which of course occurs with angles of the order of unity. It is easy to see, however, that in this problem only the change in position of the orbit centre, $\Delta R_\perp \sim r_{Be}$, need be small in comparison with the characteristic distances over which the electron density varies; this corresponds to the condition for the equation of transverse diffusion to be applicable (cf. the end of §59).

CHAPTER VI

INSTABILITY THEORY

§61. Beam instability

ACCORDING to the results of §34, the amplitude of a perturbation with wave vector **k** in a homogeneous unbounded medium has the asymptotic form (as $t \to \infty$)

$$e^{-i\omega(\mathbf{k})t}, \tag{61.1}$$

where $\omega(\mathbf{k})$ is the frequency of waves propagating in the medium. In particular, for longitudinal waves in a plasma the frequencies $\omega(\mathbf{k})$ are the roots of the equation†

$$\epsilon_l(\omega, \mathbf{k}) = 0. \tag{61.2}$$

The frequencies $\omega(k)$ are in general complex. If the imaginary part im $\omega \equiv -\gamma < 0$, the perturbation is damped in the course of time. If, however, $\gamma < 0$ in some range of **k**, such perturbations grow: the medium is unstable with respect to oscillations in that range of wavelengths, and $|\gamma|$ is then called the *instability growth rate.* We should emphasize immediately that, in referring to an "unlimited" increase of the perturbation, as $\exp(|\gamma|t)$, we are considering here and subsequently only the behaviour in the linear approximation. In reality, of course, the increase is limited by non-linear effects.

In a collisionless plasma, the imaginary part of the frequency is due to Landau damping. The thermodynamic equilibrium state of the plasma, corresponding to the absolute maximum of the entropy, is stable with regard to any perturbation. However, it has already been noted in §30 that, for non-equilibrium distributions in plasmas, the absorption of energy of the oscillations may be replaced by amplification. This is shown by the appearance of a range of values of the independent variables **k** and ω ($\omega > 0$) in which the imaginary part of the permittivity is negative: $\epsilon_l''(\omega, \mathbf{k}) < 0$. It must be emphasized, however, that the existence of such ranges does not in itself necessarily signify that the plasma is unstable (at least in the linear approximation); some branch of the plasma oscillation spectrum must also actually fall in this range.

A typical instance of instability is afforded by a directed beam of electrons passing through a plasma at rest (A. I. Akhiezer and Ya. B. Faĭnberg 1949, D. Bohm and E. P. Gross 1949). The beam is assumed to be electrically compensated:

†For an anisotropic plasma, this dispersion relation refers to quasi-longitudinal "slow" waves (see §32).

the sum of the electron charge densities in the plasma and the beam is equal to the ion charge density in the plasma. The system is homogeneous and unbounded, i.e. both the beam and the plasma extend throughout space, and the directed velocity **V** of the beam is everywhere the same. We shall assume that **V** is non-relativistic.

Let us first suppose that both the beam and the plasma are cold, i.e. that the thermal motion of their particles is negligible. The necessary condition for this will be ascertained later.

In the electron oscillation frequency range, the longitudinal permittivity of the plasma-beam system has the form

$$\epsilon_l(\omega, \mathbf{k}) - 1 = -\frac{\Omega_e^2}{\omega^2} - \frac{\Omega_e'^2}{(\omega - \mathbf{k}\cdot\mathbf{V})^2}. \tag{61.3}$$

The first term on the right corresponds to the plasma at rest; $\Omega_e = (4\pi e^2 N_e/m)^{1/2}$ is the corresponding electron plasma frequency. The second term is due to the beam electrons. In a frame of reference K' moving with the beam, the contribution of the beam electrons to $\epsilon_l - 1$ is $-(\Omega_e'/\omega')^2$, where ω' is the oscillation frequency in that frame, and $\Omega_e' = (4\pi e^2 N_e'/m)^{1/2}$ (N_e' being the electron density in the beam). On return to the original frame K, the frequency ω' is replaced by

$$\omega' = \omega - \mathbf{k}\cdot\mathbf{V}, \tag{61.4}$$

and we have (61.3).†

We shall assume the beam density to be small, in the sense that

$$N_e' \ll N_e, \tag{61.5}$$

and so $\Omega_e' \ll \Omega_e$. Then the presence of the beam changes only slightly the principal branch of the spectrum of longitudinal oscillations of the plasma, i.e. the root of the dispersion relation $\epsilon_l = 0$ for which $\omega \approx \Omega_e$. As well as this branch, however, another branch appears, on account of the presence of the beam, which we have now to consider.

In order that the term with the small numerator $\Omega_e'^2$ should not disappear from the dispersion relation

$$\frac{\Omega_e^2}{\omega^2} + \frac{\Omega_e'^2}{(\omega - \mathbf{k}\cdot\mathbf{V})^2} = 1, \tag{61.6}$$

this smallness must be compensated by that of the denominator. We therefore seek the solution in the form $\omega = \mathbf{k}\cdot\mathbf{V} + \delta$, where δ is small. The equation then becomes

$$\frac{\Omega_e^2}{(\mathbf{k}\cdot\mathbf{V})^2} + \frac{\Omega_e'^2}{\delta^2} = 1, \tag{61.7}$$

†The law of transformation for the frequency is easily found by transforming the phase factor of the wave. The position vector of a point in the frame K' is $\mathbf{r}' = \mathbf{r} - \mathbf{V}t$. Hence

$$\mathbf{k}\cdot\mathbf{r} - \omega t = \mathbf{k}\cdot\mathbf{r}' - (\omega - \mathbf{k}\cdot\mathbf{V})t = \mathbf{k}\cdot\mathbf{r}' - \omega' t.$$

whence

$$\delta = \pm \frac{\Omega_e'}{[1-(\Omega_e/\mathbf{k}\cdot\mathbf{V})^2]^{1/2}};\qquad (61.8)$$

the condition $\delta < \mathbf{k}\cdot\mathbf{V}$ requires that $|\mathbf{k}\cdot\mathbf{V}|$ should not be too close to Ω_e. The assumption that the plasma is cold implies that $kv_{Te} \ll \omega$, and in the present case therefore that $v_{Te} \ll V$: the speed of the beam is much greater than the thermal speed of the plasma electrons.

If $(\mathbf{k}\cdot\mathbf{V})^2 > \Omega_e^2$, then both roots (61.8) are real, and the oscillations do not grow. If, however,

$$(\mathbf{k}\cdot\mathbf{V})^2 < \Omega_e^2,\qquad (61.9)$$

the two values of δ are imaginary, and the one for which $\operatorname{im}\omega = \operatorname{im}\delta > 0$ corresponds to growing oscillations. The system is thus unstable with respect to oscillations having sufficiently small values of $\mathbf{k}\cdot\mathbf{V}$.

A different situation occurs when the thermal motion of the electrons is taken into account. In the general case, we have in place of (61.3)

$$\epsilon_l(\omega, \mathbf{k}) = \epsilon_l^{(\mathrm{pl})}(\omega, \mathbf{k}) - \Omega_e'^2/(\omega - \mathbf{k}\cdot\mathbf{V})^2,\qquad (61.10)$$

where $\epsilon_l^{(\mathrm{pl})}$ pertains to the plasma in the absence of the beam. Solving the equation $\epsilon_l = 0$ by the same method, we now find

$$\delta = \pm\Omega_e'/[\epsilon_l^{(\mathrm{pl})}(\mathbf{k}\cdot\mathbf{V}, \mathbf{k})]^{1/2}.\qquad (61.11)$$

Because of the Landau damping, $\epsilon_l^{(\mathrm{pl})}$ always has an imaginary part (for any $\mathbf{k}$). Consequently, δ is always complex, and by virtue of the double sign in (61.11) $\operatorname{im}\delta > 0$ for one branch of the oscillations, i.e. these are unstable. For large $\mathbf{V}$, corresponding to the cold-plasma case discussed above, the part of $\operatorname{im}\epsilon_l$ due to the Landau damping becomes exponentially small, and we return to (61.8).

In the above analysis, the thermal spread of electron speeds in the beam has been neglected. This is justifiable if the amount of it

$$v'_{Te} \ll |\delta|/k.\qquad (61.12)$$

PROBLEMS

PROBLEM 1. Determine the boundary of the beam instability region in a cold plasma for values of $\mathbf{k}\cdot\mathbf{V}$ close to Ω_e.

SOLUTION. For small values of $(\mathbf{k}\cdot\mathbf{V})^2 - \Omega_e^2$, (61.7) is insufficiently accurate. Retaining the term of the next order in δ in the equation $\epsilon_l = 0$, with ϵ_l from (61.3), we obtain

$$\frac{\Omega_e'^2}{\delta^2} \approx 1 - \frac{\Omega_e^2}{(\mathbf{k}\cdot\mathbf{V})^2} + \frac{2\Omega_e^2\delta}{(\mathbf{k}\cdot\mathbf{V})^3} \approx \frac{2(\mathbf{k}\cdot\mathbf{V}-\Omega_e)}{\Omega_e} + \frac{2\delta}{\Omega_e}.$$

With new variables ξ and τ defined by

$$\delta = \xi(\tfrac{1}{2}\Omega_e'^2\Omega_e)^{1/3},\quad \tau = (2/\Omega_e'^2\Omega_e)^{1/3}(\mathbf{k}\cdot\mathbf{V}-\Omega_e),$$

we reduce this equation to

$$\xi^3 + \tau\xi^2 = 1 \tag{1}$$

(taking the particular case where $\mathbf{k}\cdot\mathbf{V}$ is close to $+\Omega_e$, not to $-\Omega_e$). All three roots of equation (1) are real if $\tau > 3.2^{-2/3}$, and this determines the instability region. Two of them correspond to the two roots of (61.6), and the other corresponds to the oscillation frequency of the plasma at rest, which is close to them if $\Omega_e \approx \mathbf{k}\cdot\mathbf{V}$.

PROBLEM 2. Investigate the stability of ion-sound waves in a two-temperature plasma ($T_e \gg T_i$) in which the electron component moves relative to the ion component with a macroscopic velocity $\mathbf{V}$, and $V \ll v_{Te}$.

SOLUTION. With the condition $V \ll v_{Te}$, the directed motion of the electrons has little effect on the dispersion relation for the ion-sound waves, which is again given by (33.4):

$$\frac{\omega}{k} = \left(\frac{zT_e}{M}\right)^{1/2} \frac{1}{(1+k^2a_e^2)^{1/2}}. \tag{2}$$

The electronic part of the damping rate is found from (33.6) by the change (61.4):

$$\gamma = (\mathbf{k}\cdot\mathbf{V} - \omega)(\pi zm/8M)^{1/2}. \tag{3}$$

The instability condition is $\mathbf{k}\cdot\mathbf{V} > \omega$; for this to be so, we must always have $V > \omega/k$. Near the instability limit, the factor $\mathbf{k}\cdot\mathbf{V} - \omega$ in (3) is small, and it may then be necessary to take account in γ of the ionic part of the damping, which in ordinary conditions is small.

§62. Absolute and convective instabilities

If the dispersion relation has roots in the upper half-plane of ω, a small initial perturbation in the form of a plane wave will grow, i.e. the system is unstable with respect to such a perturbation. In reality, however, any initial perturbation is a wave packet of finite extent in space, and the plane waves are only its separate Fourier components. In the course of time, the packet spreads out, and its amplitude (in an unstable system) increases. Simultaneously it moves in space, like any wave packet. Here there are two possibilities.

In one case, regardless of the movement of the packet, the perturbation increases without limit at every point in space. This is called *absolute instability*. In the other case, the packet moves so quickly that the perturbation tends to zero as $t \to \infty$ at any fixed point; this is *convective instability*.

We must emphasize immediately that this difference is a relative one, in the sense that the nature of the instability has to be defined in relation to a particular frame of reference, and may be different in different frames: an instability convective in one frame becomes absolute in a frame moving with the wave packet, and an absolute instability becomes convective in a frame that moves sufficiently rapidly away from the packet.

This, however, does not mean that the difference between the two types of instability has no physical significance. In actual problems, there is always an experimentally preferred frame of reference relative to which the instability is to be considered. The permissibility of regarding a physical system as infinite in extent does not exclude the fact that it really has boundaries such as walls, which constitute a laboratory frame of reference. Moreover, the actual boundedness of the system may have the result that in convective instability the perturbation is unable to develop before the packet is carried out of the system (for example, by the flow of liquid in a pipe).

The theory below, which leads to criteria distinguishing the two types of instability, is a very general one.† The system may be any that is homogeneous and infinite in at least one (the x) direction. We shall therefore not specify here the nature of the medium and the perturbation in it, denoting the latter by some $\psi(t, \mathbf{r})$ and considering only the case of a one-dimensional packet. For a three-dimensional system, this means that the perturbations considered have the form

$$\psi(t, \mathbf{r}) = \psi(t, x)e^{i(k_y y + k_z z)}$$

with given k_y and k_z.

Let us express $\psi(t, x)$ as a one-sided Fourier expansion with respect to the time from $t = 0$ (the instant when the perturbation arises) to $t = \infty$. The components will be denoted by $\varphi(\omega, x)$:

$$\varphi(\omega, x) = \int_0^\infty \psi(t, x)e^{i\omega t}\, dt. \tag{62.1}$$

We shall have to consider perturbations which increase as $t \to \infty$. We shall assume, what is in fact true, that this increase is not more rapid than an exponential $\exp(\sigma_0 t)$. Then the integral (62.1) can be made to converge by regarding ω as a complex quantity with $\operatorname{im} \omega = \sigma > \sigma_0$. In this region, $\varphi(\omega, x)$ has no singularity as a function of the complex variable ω. In the region $\operatorname{im} \omega < \sigma_0$, however, $\varphi(\omega, x)$ is to be treated as an analytical continuation, and here it has singularities, of course.

The inverse expression for $\psi(t, x)$ in terms of its Fourier transform is

$$\psi(t, x) = \int_{-\infty+i\sigma}^{\infty+i\sigma} e^{-i\omega t}\varphi(\omega, x)\frac{d\omega}{2\pi}, \tag{62.2}$$

with $\sigma > \sigma_0$, so that the contour of integration (which we shall call the ω-contour) passes above all singularities of $\varphi(\omega, x)$ in the upper half-plane of ω.

The function $\varphi(\omega, x)$ in turn can be expanded as a Fourier integral with respect to the coordinate x:

$$\varphi(\omega, x) = \int_{-\infty}^{\infty} \psi_{\omega k}^{(+)}\, e^{ikx}\frac{dk}{2\pi}; \tag{62.3}$$

for brevity, the suffix in k_x is omitted.

The function $\psi_{\omega k}^{(+)}$ is obtained in each specific case by solving the linearized "equations of motion" of the system concerned:

$$\psi_{\omega k}^{(+)} = g_{\omega k}/\Delta(\omega, k), \tag{62.4}$$

†Such a criterion was first established by P. A. Sturrock (1958). The formulation given below is due to R. J. Briggs (1964), whose analysis we shall largely follow in §§62–64.

where $g_{\omega k}$ is determined by the initial perturbation, and $\Delta(\omega, k)$ is a characteristic of the system itself. For example, in a plasma the "equation of motion" is the transport equation, $\Delta(\omega, k)$ is the longitudinal permittivity of the plasma, and $g_{\omega k}$ is expressed in terms of the Fourier component of the initial perturbation by (34.12).

We shall assume that $g_{\omega k}$ as a function of the complex variables ω and k has no singularity for finite values of these variables, i.e. is an entire function of them.† Then all the singularities of $\psi_{\omega k}^{(+)}$ are singularities of the factor $1/\Delta(\omega, k)$. The equation

$$\Delta(\omega, k) = 0 \tag{62.5}$$

is the dispersion relation of the system. Its roots $\omega(k)$ give the frequencies of oscillations with specified (real) values of the wave number k. As we have seen in § 34, it is these frequencies which determine the asymptotic (as $t \to \infty$) law of time variation of the Fourier component of a perturbation with a specified value of k:

$$\psi_k(t) \propto e^{-i\omega(k)t} = e^{-i\omega'(k)t + \omega''(k)t}.$$

On this basis, the determination of the asymptotic law of variation of the perturbation at a given point in space would require the investigation of the integral

$$\psi(t, x) \propto \int e^{-i\omega'(k)t} e^{\omega''(k)t} e^{ikx}\, dk. \tag{62.6}$$

In the presence of instability, when $\omega''(k) > 0$ for some range of values of k, one factor in the integrand increases without limit as $t \to \infty$, and the other oscillates indefinitely rapidly. These opposite tendencies make it difficult to estimate the integral.

Instead, let us return to the form (62.2) for $\psi(t, x)$, before the integration with respect to ω is carried out. We move the ω-contour downwards until it "catches" on the first (the highest, i.e. with the largest ω'') singularity of $\varphi(\omega, x)$; let this point be at $\omega = \omega_c$ (it will become clear that ω_c is independent of x). Evidently, the asymptotic value of the integral is determined by the neighbourhood of that point, so that

$$\psi(t, x) \propto e^{-i\omega_c t} = \exp(-i\omega_c' t + \omega_c'' t). \tag{62.7}$$

If $\omega_c'' > 0$, the perturbation increases at any fixed point x, i.e. the instability is absolute, but if $\omega_c'' < 0$ the perturbation tends to zero at a fixed point, i.e. the instability is convective. Thus the required criterion involves simply a determination of ω_c.

The function $\varphi(\omega, x)$ is given by the integral (62.3) with $\psi_{\omega k}^{(+)}$ from (62.4):

$$\varphi(\omega, x) = \int_{-\infty}^{\infty} \frac{g_{\omega k}}{\Delta(\omega, k)} e^{ikx} \frac{dk}{2\pi}. \tag{62.8}$$

†For this to be so, it is always necessary that the initial wave packet should diminish in space sufficiently rapidly (faster than $\exp(-\alpha|x|)$).

Since $g_{\omega k}$ is assumed to be an entire function of k, the singularities of the integrand as a function of the complex variable k are at the singularities of $1/\Delta(\omega, k)$; these are usually poles, the roots $k(\omega)$ of (62.5).

For some value of ω, a point on the ω-contour, with a sufficiently large positive imaginary part $\omega'' = \sigma$, let the singularities lie in the k-plane as shown in Fig. 22: some in the upper and some in the lower half-plane. The contour of integration with respect to k in (62.8), which we call the k-contour, lies along the real axis. We now modify ω by gradually reducing ω''. The singularities move in the k-plane and may reach the real axis for some ω.† These values of ω are not singularities of the function $\varphi(\omega, k)$: there is no objection to moving the k-contour in such a way as to remove it from the neighbourhood of singularities that have crossed the real axis, as shown in Fig. 22b. A singularity of the integral occurs, however, if two moving singularities come close together and pinch the contour of integration between them, thereby eliminating the possibility of removing this contour from their neighbourhood (Fig. 22c).

Thus the value of ω which determines the nature of the stability is chosen from among those such that two roots $k(\omega)$ of the dispersion relation fall together. Only cases such that the two roots approach from opposite sides of the k-contour are to be considered; that is, as $\omega'' \to 0$ these roots must lie on opposite sides of the real axis. Note, incidentally, that the values of ω_c must be independent of x, since they are determined solely by the properties of the function $1/\Delta(\omega, k)$.

When two simple roots of the equation coincide, a double root is formed, near which the dispersion relation is

$$\Delta(\omega, k) \approx (\omega - \omega_c)(\partial \Delta/\partial \omega)_c + \tfrac{1}{2}(k - k_c)^2(\partial^2\Delta/\partial k^2)_c = 0, \tag{62.9}$$

so that $k - k_c \propto \pm(\omega - \omega_c)^{1/2}$.‡ At the point $\omega = \omega_c$, the function $\omega(k)$ satisfies the

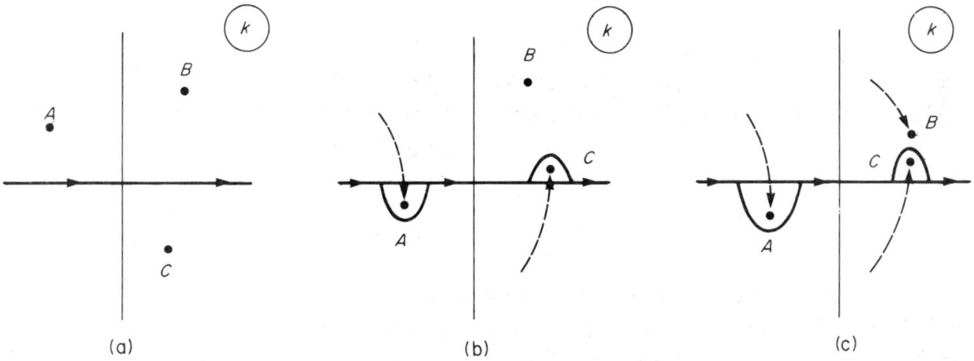

FIG. 22.

†For an unstable system, the singularity must reach the real axis of k while $\omega'' > 0$, at least for some range of values of ω', since there are certainly roots of $\Delta(\omega, k) = 0$ such that $\omega'' > 0$ for real k.

‡In some cases there may be a coincidence of a still larger number of roots, forming a multiple root of higher order. Such cases, however, can in general occur only for particular values of the parameters of the system, since they impose additional restrictions on the points ω_c, k_c: in the expansion of $\Delta(\omega, k)$, other coefficients besides $(\partial\Delta/\partial k)_c$ must be zero.

condition

$$d\omega/dk = 0, \tag{62.10}$$

i.e. ω_c is a saddle point of the analytic function $\omega(k)$.

The integral (62.8) taken over the neighbourhood of the point $k = k_c$ is

$$\varphi(\omega, x) \propto \frac{e^{ik_c x}}{\sqrt{(\omega - \omega_c)}}; \tag{62.11}$$

the function $\varphi(\omega, x)$ thus has a square-root pole at $\omega = \omega_c$. The integral (62.2), taken now over the neighbourhood of the point $\omega = \omega_c$, as a function of t and x, has the form

$$\psi(t, x) \propto \frac{1}{\sqrt{t}} e^{-i(\omega_c t - k_c x)}; \tag{62.12}$$

since this asymptotic expression has been derived for $t \to \infty$ and fixed x, it is valid only if $|k_c x| \ll |\omega_c t|$.

Although the coalescence of roots of the dispersion relation is the principal source of singularities of $\varphi(\omega, x)$ (and usually determines the nature of the instability), another type of singularity may be mentioned, which occurs at a frequency for which the root $|k| \to \infty$.† The imaginary part of such a frequency ω_c, however, is in practice always negative and therefore certainly cannot cause absolute instability (if ω_c'' were positive, the system under consideration would be unstable with respect to oscillations of infinitesimal wavelength). We shall meet such a case later; see (63.10).

As has already been emphasized, an instability convective in one frame of reference (the laboratory frame) may be absolute in another frame. Let us seek the speed V of the frame in which the instability is absolute and has the greatest growth rate.

The change from the laboratory frame to one moving with speed V is made by changing ω to $\omega - kV$ in all formulae. As we have seen, the value ω_c corresponds to the instant when, as ω'' decreases on the ω-contour, two poles of the function $1/\Delta(\omega, k)$ in the k-plane coalesce, and these poles must approach from opposite sides of the real axis, so that one of them must first cross that axis. Let ω''_{max} denote the maximum value of ω'' (independent of V) for real k. Since $\omega_c''(k, V)$ is certainly less than the value of ω'' at which the pole crossed the real axis, we have $\omega_c''(k, V) \leq \omega''_{max}$ for all V. This means that the highest value of ω_c'' is reached if the poles coalesce on the real axis at the maximum of $\omega''(k)$. Replacing $\omega(k)$ by $\omega(k) - kV$ in (62.10), and separating the real and imaginary parts of the equation

†Such a root causes an essential singularity of $\varphi(\omega, x)$. For example, if $|k| \to \infty$ according to $k^{-n} = C(\omega - \omega_c)$, the contribution from the neighbourhood of the singularity to the integral (62.8) is

$$\varphi(\omega, x) \propto \exp\left\{\frac{ix}{[C(\omega - \omega_c)]^{1/n}}\right\}.$$

(for real k), we find two equations:

$$d\omega''/dk = 0, \tag{62.13}$$

$$V = d\omega'/dk. \tag{62.14}$$

Thus the greatest growth rate of the instability is given by the maximum value of $\omega''(k)$ as a function of real k. The speed of the frame of reference in which such an instability occurs is determined by the corresponding value of $d\omega'/dk$. This value of V may naturally be taken to define the group velocity of the wave packet in a convectively unstable medium.

§63. Amplification and non-transparency

So far, we have considered instability problems as regards the development with time of a perturbation specified in space at some initial instant. The Fourier expansion of such a perturbation includes components with real values of the wave vectors **k**, and their time dependence is governed by the frequencies $\omega(\mathbf{k})$, the complex roots of the dispersion relation.

There is, however, another possible formulation of the instability problem, in which we consider a perturbation set up with a specified time variation in some region of space. The Fourier expansion of such a perturbation contains components with real frequencies ω, and their propagation in space is governed by the wave numbers $k(\omega)$ found by solving the dispersion relation, this time for k; correspondingly, the wave numbers, not the frequencies, are complex. (As in §62, we are considering a one-dimensional problem, and therefore write $k \equiv k_x$ in place of the vector **k**.)

The fact that the wave numbers are complex may have various meanings. In some cases it may simply mean that the relevant waves cannot propagate in the medium (*non-transparency*): in other cases it may signify *amplification* of the waves by the medium as they propagate from the source. We must emphasize at once that the sign of $\operatorname{im} k$ certainly cannot be the criterion for distinguishing between these two possibilities: the waves can propagate in the positive and the negative x-direction, and a change in the direction of propagation is equivalent to a change in the sign of k.

It is physically obvious that only an unstable medium can amplify. Hence, for example, it is clear immediately that there is non-transparency for transverse electromagnetic waves in a plasma with dispersion relation $\omega^2 = \Omega_e^2 + c^2k^2$ (see §32, Problem 1) at frequencies $\omega < \Omega_e$, for which $k(\omega)$ is imaginary: the function $\omega(k)$ given by this equation is real for all real k, so that the system is certainly stable.

For an exact formulation of the problem let us consider a point source as regards the coordinate x, called a *signal*, which starts at $t = 0$ and thereafter creates a monochromatic perturbation ψ (with some frequency ω_0), called the *response* of the system to the signal. The source strength is then

$$\left.\begin{aligned} g(t, x) &= 0 && \text{for} \quad t < 0, \\ &= \text{constant} \times \delta(x) e^{-i\omega_0 t} && \text{for} \quad t > 0. \end{aligned}\right\} \tag{63.1}$$

We shall not specify the physical nature of the perturbation ψ, nor therefore that of the source strength g. The only important point is that the ωk-components of the perturbation are determined from the source by

$$\psi_{\omega k} = g_{\omega k}/\Delta(\omega, k). \tag{63.2}$$

This expression is derived from the inhomogeneous linearized "equation of motion" of the system, in which $g(t, x)$ acts as the "right-hand side", just as (62.4) was the solution of the homogeneous equation with the initial condition specified by the function $g(0, x)$. The source (63.1) is†

$$g_{\omega k} = \text{constant}/i(\omega - \omega_0). \tag{63.3}$$

The function $\psi(t, x)$ is then found from the inversion formula

$$\psi(t, x) = \text{constant} \times \int_{-\infty+i\sigma}^{\infty+i\sigma} \Phi(\omega, x) \frac{e^{-i\omega t}}{i(\omega - \omega_0)} \frac{d\omega}{2\pi}, \tag{63.4}$$

$$\Phi(\omega, x) = \int_{-\infty}^{\infty} \frac{e^{ikx}}{\Delta(\omega, k)} dk. \tag{63.5}$$

This expression necessarily satisfies the equation $\psi(t, x) = 0$ for $t < 0$ in accordance with the conditions of the problem: the perturbation occurs only after the source comes in at $t = 0$.

The problem is now to find the asymptotic expression for $\psi(t, x)$ far from the source ($|x| \to \infty$) in steady conditions, i.e. at a long time after the source begins to operate ($t \to \infty$). If the perturbation then tends to zero as $x \to \pm\infty$, we have non-transparency: if it increases in one or other direction from the source, there is amplification. In either case, we can evidently speak of only a convectively unstable (or a stable) system. With absolute instability, the perturbation increases without limit in the course of time at every point in space, so that no steady conditions can possibly be reached.

To find the required asymptotic form, we note first of all that the asymptotic limit $t \to \infty$ must be taken before $|x| \to \infty$: since the perturbation cannot propagate to infinity in a finite time, $\psi \to 0$ as $|x| \to \infty$ for finite t.

As in §62, we move the contour of integration with respect to ω in (63.4) downwards in order to get the asymptotic expression as $t \to \infty$. The analytical properties of $\Phi(\omega, x)$ are similar to those of $\varphi(\omega, x)$ in §62. Since the system is assumed only convectively unstable, $\Phi(\omega, x)$ has no singularity in the upper half-plane of ω, and the highest singularity of the integrand in (63.4) is the pole $\omega = \omega_0$ on the real axis. Hence the asymptotic form as $t \to \infty$ is

$$\psi(t, x) \propto e^{-i\omega_0 t} \Phi(\omega_0, x). \tag{63.6}$$

†In calculating $g_{\omega k}$, it must be remembered that the integration in the inverse transformation formula is taken along a contour with im $\omega > 0$; hence $e^{i\omega t} \to 0$ as $t \to \infty$.

To find the asymptotic form of $\Phi(\omega_0, x)$ as $|x| \to \infty$, we must now move the path of integration with respect to k upwards for $x > 0$ or downwards for $x < 0$, until it catches on the pole of the integrand in (63.5), i.e. the root of the equation $\Delta(\omega_0, k) = 0$.

Let $k_+(\omega)$ and $k_-(\omega)$ denote the poles which as im $\omega \to \infty$ are respectively in the upper and lower half-planes of k. As im ω decreases, the poles move, and for a real $\omega = \omega_0$ they may either remain in their original half-plane or enter the other half-plane. In the first case, the contour of integration in $\Phi(\omega_0, x)$ remains on the real axis as in Fig. 22a; in the second case, it is deformed as shown in Fig. 22b, so as to embrace the poles $k_+(\omega_0)$ and $k_-(\omega_0)$ (points A and C) that have "escaped" into the other half-plane. In either case, when the contour is moved up or down, it catches on the poles k_+ and k_- respectively. The asymptotic form of $\psi(t, x)$ as $x \to +\infty$ is determined by the contribution from the lowest pole $k_+(\omega_0)$; that as $x \to -\infty$ is determined by the highest pole $k_-(\omega_0)$. The pole concerned is thus the closest to the real axis (if all poles of a given class are still in their original half-planes), or the farthest from the real axis among those which have moved into the other half-plane. With these values of k_+ and k_-, we have

$$\left.\begin{aligned} \psi(t, x) \propto \exp\{ik_+(\omega_0)x - i\omega_0 t\} \quad &\text{for} \quad x > 0, \\ \exp\{ik_-(\omega_0)x - i\omega_0 t\} \quad &\text{for} \quad x < 0. \end{aligned}\right\} \tag{63.7}$$

For a stable system, all poles remain in their original half-planes when $\omega = \omega_0$, since the absence of oscillation branches with im $\omega(k) > 0$ (for real k) means that a pole $k(\omega)$ can cross the real axis only with im $\omega < 0$. Hence, in (63.7),

$$\text{im}\, k_+(\omega_0) > 0, \quad \text{im}\, k_-(\omega_0) < 0,$$

so that the waves are damped in both directions from the source.

In the case of convective instability, the poles $k(\omega)$ reach the real axis with im $\omega > 0$. There are therefore certainly poles k_+ or k_- which have entered the other half-plane for $\omega = \omega_0$, i.e. which have im $k_+(\omega_0) < 0$ or im $k_-(\omega_0) > 0$. The presence of such a pole $k_+(\omega_0)$ or $k_-(\omega_0)$ amplifies the wave to the right or left of the source respectively.

From the preceding arguments, we arrive at the following criterion for distinguishing cases of non-transparency and amplification for waves from a source with frequency ω_0 in a convectively unstable system: a wave with complex $k(\omega_0)$ and real ω_0 is amplified if im $k(\omega)$ changes sign when im ω varies from $+\infty$ to 0 with a given re $\omega = \omega_0$; if im $k(\omega)$ does not change sign, there is non-transparency.

The criterion has its origin in the requirements of causality. When the source comes into action instantaneously, the perturbation must always decrease as $x \to \pm\infty$, simply because it cannot propagate to an infinite distance in a finite time. On the other hand, this "infinitely rapid" onset of the source can take place as $e^{-i\omega t}$ with im $\omega \to +\infty$. It is therefore clear that waves that are amplified (for real ω) in one or other direction from the source must be damped in that direction when im $\omega \to \infty$, and this leads to the criterion formulated above.

The results obtained have a further facet which allows us to determine the direction of wave propagation in a medium with absorption or amplification. In a transparent medium (i.e. when ω and k are real), the physical direction of propagation is that of the group velocity vector. In particular, in the one-dimensional case a wave with a positive or negative derivative $d\omega/dk$ moves in the positive or negative x-direction respectively. In a medium with absorption or amplification, however, we can say that waves of the k_+ and k_- groups propagate in the positive and negative directions respectively. For real ω and k, this general formulation is the same as the previous one: small changes in ω and k are related by

$$\delta k = \frac{\delta\omega}{d\omega/dk},$$

from which we see that, if ω acquires an imaginary part im $\omega > 0$, k moves into the upper or lower half-plane according as $d\omega/dk > 0$ or < 0.

As a simple example of the application of the criteria derived in §§62 and 63, let us consider the instability of a cold electron beam in a cold plasma, discussed in §61. The dispersion relation for this system is

$$\frac{\Omega_e^2}{\omega^2} + \frac{\Omega_e'^2}{(\omega - kV)^2} = 1; \tag{63.8}$$

see (61.6) (for waves propagating in the direction of the beam, $\mathbf{k}\cdot\mathbf{V} = kV$). The roots $k(\omega)$ of this equation have, as $|\omega| \to \infty$, the form†

$$k = (\omega \pm \Omega_e')/V. \tag{63.9}$$

When im $\omega \to \infty$, the two roots are in the same (upper) half-plane, i.e. both are in the $k_+(\omega)$ class. In their movement, therefore, they cannot pinch the k-contour as im ω decreases, so that the instability is convective. The asymptotic behaviour of a perturbation created at the initial instant is governed by the frequency $\omega = \Omega_e$, near which the roots of equation (63.8) tend to infinity according to

$$k^2 = \Omega_e\Omega_e'^2/2V^2(\omega - \Omega_e). \tag{63.10}$$

Thus, as $t \to \infty$, only undamped plasma waves remain from the perturbation.

For real values of $\omega < \Omega_e$, equation (63.8) has two complex-conjugate roots $k(\omega)$. The one for which im $k(\omega) < 0$ has moved from the upper to the lower half-plane. Thus, when waves are propagated from a source with frequency $\omega_0 < \Omega_e$, they are amplified in the direction of $x > 0$, i.e. "down" the beam.

§64. Instability with weak coupling of the two branches of the oscillation spectrum

Let us apply the general method developed in §§62 and 63 to investigate the instability which results from the "interaction" of oscillations with neighbouring

†Note that (63.9) is the same as the dispersion relation for the beam itself in the absence of the plasma at rest.

values of ω and k and belonging to different branches of the oscillation spectrum of a non-dissipative system (here denoting one in which both true dissipation and Landau damping are absent).

If the two branches $\omega = \omega_1(k)$ and $\omega = \omega_2(k)$ were completely independent, the dispersion relation would separate into two factors:

$$[\omega - \omega_1(k)][\omega - \omega_2(k)] = 0. \tag{64.1}$$

Near a point of intersection of such branches, the functions $\omega_1(k)$ and $\omega_2(k)$ would have the general form

$$\left.\begin{aligned}\omega_1(k) &= \omega_0 + v_1(k - k_0),\\ \omega_2(k) &= \omega_0 + v_2(k - k_0),\end{aligned}\right\} \tag{64.2}$$

where v_1 and v_2 are some constants, ω_0 and k_0 the (real) values of ω and k at the point of intersection.

Such a case is, however, not realistic in general. The coupling between the two branches could be completely absent for (at best) some specific values of the system parameters, and would appear when these were very slightly changed.† To represent an actual situation, therefore, it would be necessary to take into account the presence of a weak coupling between branches. This has the effect of substituting a small quantity ϵ for zero on the right of (64.1). The dispersion relation near the point then becomes

$$[\omega - \omega_0 - v_1(k - k_0)][\omega - \omega_0 - v_2(k - k_0)] = \epsilon. \tag{64.3}$$

The solution for ω is

$$\omega(k) - \omega_0 = \tfrac{1}{2}[(v_1 + v_2)(k - k_0) \pm [(k - k_0)^2(v_1 - v_2)^2 + 4\epsilon^2]^{1/2}\}, \tag{64.4}$$

and the solution for k is

$$k(\omega) - k_0 = \frac{1}{2v_1v_2}\{(v_1 + v_2)(\omega - \omega_0) \pm [(\omega - \omega_0)^2(v_1 - v_2)^2 + 4\epsilon v_1 v_2]^{1/2}\}. \tag{64.5}$$

The existence of coupling between the branches shifts their point of intersection into the complex region. The functions $\omega(k)$ for real ω and k vary in form according to the sign of the constant ϵ and the relative signs of the constants v_1 and

†An exception is the case where the interaction is absent for reasons of symmetry, for example, if one branch relates to longitudinal waves and the other to transverse waves in an isotropic medium. Since in such a medium the longitudinal current cannot induce a transverse field and vice versa, such waves do not interact. The situation here is analogous to what is found in quantum mechanics for the interaction of terms with different symmetry; see *QM*, §79.

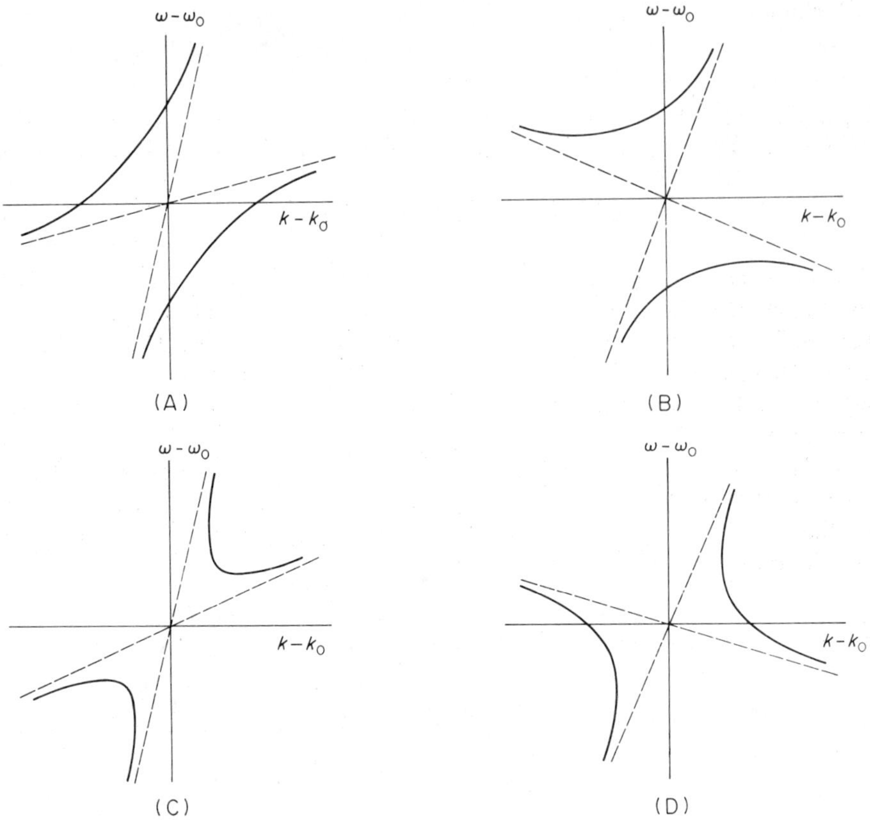

FIG. 23.

v_2. These functions are shown in Fig. 23 for four cases:

$$\left.\begin{array}{ll}\text{(A) } \epsilon > 0, & v_1v_2 > 0,\\ \text{(B) } \epsilon > 0, & v_1v_2 < 0,\\ \text{(C) } \epsilon < 0, & v_1v_2 > 0,\\ \text{(D) } \epsilon < 0, & v_1v_2 < 0,\end{array}\right\} \tag{64.6}$$

which we shall consider in turn.

(A) Here the functions $\omega(k)$ are real for all (real) k, and the system is therefore stable. The functions $k(\omega)$ are also real for all ω, so that the waves propagate without amplification for all ω.

(B) The functions $\omega(k)$ are real for all k, and the system is therefore stable. The functions $k(\omega)$ are complex in the frequency range

$$(\omega - \omega_0)^2 < 4|\epsilon v_1 v_2|/(v_1 - v_2)^2. \tag{64.7}$$

Since the system is stable, there is non-transparency in this range.

(C) When

$$(k - k_0)^2 < 4|\epsilon|/(v_1 - v_2)^2 \tag{64.8}$$

the functions $\omega(k)$ are complex, and for one of them $\text{im}\,\omega(k) > 0$, i.e. there is instability, and it is convective instability, since when $|\omega| \to \infty$ the roots $k(\omega)$ are

$$k \approx \omega/v_1, \quad k \approx \omega/v_2, \tag{64.9}$$

and when $\text{im}\,\omega \to \infty$ they lie in the same half-plane of k. Let v_1 and $v_2 > 0$. Then this is the upper half-plane, and the roots belong to the $k_+(\omega)$ class. For real ω, in the range (64.7), the roots $k(\omega)$ form a complex conjugate pair. The one for which $\text{im}\,k(\omega) < 0$ has moved from the upper to the lower half-plane. In the frequency range (64.7), therefore, there is amplification of waves propagating in the positive x-direction.

It is also easy to find for this case the "group velocity" of the waves, defined by (62.14), i.e. the speed of the frame of reference in which there is absolute instability with the maximum growth rate. Differentiating (64.3) with respect to k and substituting $d\omega/dk = V$ in accordance with (62.13) and (62.14), we obtain

$$\frac{V - v_1}{V - v_2} = -\frac{\omega - \omega_0 - v_1(k - k_0)}{\omega - \omega_0 - v_2(k - k_0)}. \tag{64.10}$$

Since the left-hand side is real, the right-hand side must be so, even if ω is complex. This condition shows that $k = k_0$; then from (64.10)

$$V = \tfrac{1}{2}(v_1 + v_2). \tag{64.11}$$

and from (64.3) the corresponding maximum growth rate is

$$(\text{im}\,\omega)_{\text{max}} = |\epsilon|^{1/2}. \tag{64.12}$$

(D) The functions $k(\omega)$ are real for all (real) ω, but the $\omega(k)$ are complex in the range (64.8), so that the system is unstable. To determine the nature of this instability, we note that from (64.9) (with opposite signs of v_1 and v_2), as $\text{im}\,\omega \to \infty$, the roots $k(\omega)$ are in opposite half-planes. These two roots coalesce at a point in the upper half-plane of ω given by

$$\omega = \omega_c = \omega_0 + 2i\sqrt{(v_1 v_2 \epsilon)}/|v_1 - v_2|. \tag{64.13}$$

The instability is therefore absolute, with growth rate $\text{im}\,\omega_c$. For $v_1 = -v_2$, corresponding to the perturbation in a frame of reference moving with the speed (64.11), the growth rate reaches its maximum value (64.12).

PROBLEM

Determine the nature of the instability of low-frequency ($\omega \sim \omega_{Bi}$) "slow" ($\omega/k \ll c$) transverse electromagnetic waves propagating along a constant magnetic field in a cold magnetoactive plasma, with a low-density cold electron beam moving through the plasma in the same direction.

SOLUTION. To establish the dispersion relation, we first write it for the beam electrons only, in a frame of reference for which the beam is at rest. According to (56.9), we have in this frame

$$k^2c^2 - \omega^2 = -\Omega_e'^2\omega/(\omega \pm \omega_{Be}),$$

where Ω_e' is the plasma frequency corresponding to the beam density. On returning to the laboratory frame, in which the beam moves with velocity **V** (which we take to be in the x-direction), we must replace ω by $\omega - kV$ on the right of the equation; the difference $k^2c^2 - \omega^2$ is invariant with respect to a change in the frame of reference. Now adding, in the laboratory frame, the terms due to the plasma electrons and ions, we obtain

$$k^2c^2 - \omega^2 = -\frac{\Omega_e'^2(\omega - kV)}{\omega - kV \pm \omega_{Be}} - \frac{\omega\Omega_e^2}{\omega \pm \omega_{Be}} - \frac{\omega\Omega_i^2}{\omega \mp \omega_{Bi}}.$$

Neglecting here (in accordance with the conditions of the problem) ω in comparison with ck and ω_{Be}, and noting also that $\Omega_e^2/\omega_{Be} = \Omega_i^2/\omega_{Bi}$, we bring the dispersion relation to the form

$$[k^2c^2 - \Omega_i^2\omega^2/\omega_{Bi}(\omega_{Bi} \mp \omega)](\omega - kV \pm \omega_{Be}) = -\Omega_e'^2(\omega - kV). \tag{1}$$

The first factor on the left corresponds to the "principal" oscillation branch, and the second to the beam branch; the right-hand side describes the "interaction" of these branches.

With the upper signs in (1), the dispersion relations for the two independent branches are shown by the continuous curves in Fig. 24; it is, as always, sufficient to consider the branches with $\omega > 0$. Near the point ω_0, k_0, where they intersect, the expansion of equation (1) is

$$2k_0c^2[k - k_0 - (\omega - \omega_0)/v_1][\omega - \omega_0 - V(k - k_0)] = \Omega_e'^2\omega_{Be}$$

with a positive coefficient v_1 (as is clear from the slope of the curves in Fig. 24). Comparison with (64.3) shows that we have case C, convective instability. The broken curves in Fig. 24 show the form of the branches when their interaction is taken into account.

Similar diagrams for the lower signs in (1) are given in Fig. 25. Near the point of intersection, the dispersion relation is

$$2k_0c^2[k - k_0 + (\omega - \omega_0)/v_1][\omega - \omega_0 - V(k - k_0)] = -\Omega_e'^2\omega_{Be},$$

where again $v_1 > 0$. We now have case D, absolute instability. The second intersection in this case is seen from the diagram to occur at $\omega \gtrsim \omega_{Be}$, which contradicts the conditions of the problem.

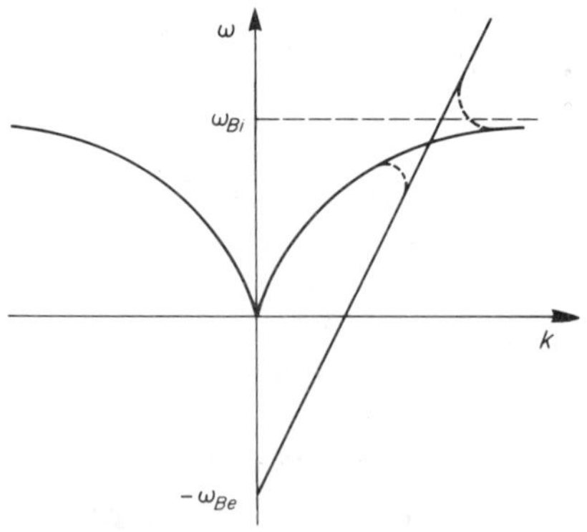

FIG. 24.

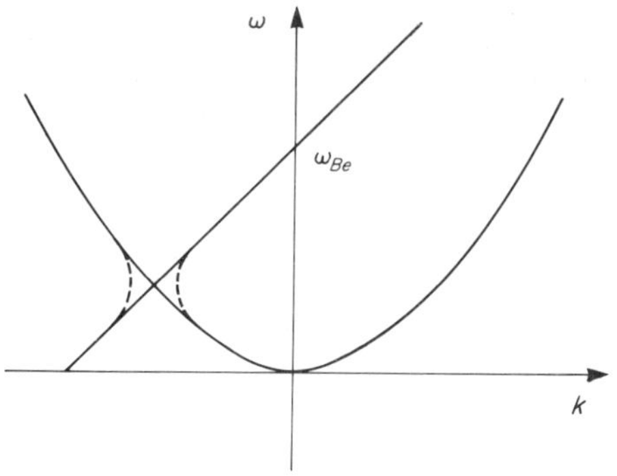

FIG. 25.

§65. Instability in finite systems

The whole of the theory in §§61–63 related to homogeneous media of unlimited extent in at least one direction (the x-axis). For applications to actual bounded systems, this means that effects due to the reflection of waves from the boundaries are neglected; that is, such a theory is limited to times of the order of the time taken by the perturbation to propagate the length of the system.

Let us now consider stability in the opposite situation where the finiteness of the system is important and the spectrum of its characteristic oscillations is governed by the boundary conditions at the ends; as before, we shall investigate only the one-dimensional case, and the length of the system in the x-direction will be denoted by L. The frequency spectrum of a finite system is discrete, and if one or more of the characteristic frequencies has a positive imaginary part the system is unstable. The distinction between absolute and convective instability has no meaning in this case.

Thus the problem of determining the stability or instability of a finite system is equivalent to that of finding its (complex) eigenfrequencies. The dispersion relation which gives these frequencies can be derived in a general form for a system with finite but sufficiently large dimensions L, such that $\operatorname{im}|k|.\ L \gg 1$ (A. G. Kulikovskiĭ, 1966).

Let $k(\omega)$ be solutions of the dispersion relation for an infinite medium. We again divide the branches of this many-valued function into two groups, $k_+(\omega)$ and $k_-(\omega)$, as defined in §63. The characteristic oscillations of a finite system may be regarded as resulting from the superposition of travelling waves reflected by the two boundaries (in a medium without absorption and amplification, they would be ordinary stationary waves). The reflection is in general accompanied by a mutual transformation of waves belonging to different branches of the spectrum. The travelling wave of a given frequency is therefore a superposition of all branches. Far from the boundaries, however, the main contribution to each wave comes from only one term in the superposition. For example, in a wave propagating from the left-hand boundary $x = 0$ (Fig. 26) in the positive x-direction, the asymptotic

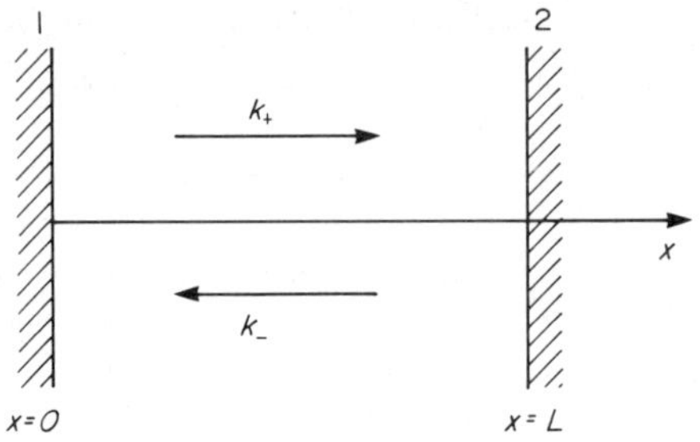

FIG. 26.

expression far from that boundary is

$$\psi = a\exp\{i[k_+(\omega)x - \omega t]\}, \tag{65.1}$$

and $k_+(\omega)$ must be taken as the branch in this group for which $\operatorname{im} k_+(\omega)$ has its algebraically least value for the given real ω.†

After reflection from the right-hand boundary $x = L$, the wave propagates to the left, and at sufficiently great distances from that boundary has the asymptotic form

$$\psi = R_2 a\exp\{ik_+(\omega)L\}\exp\{i[k_-(\omega)(x-L) - \omega t]\}, \tag{65.2}$$

where $k_-(\omega)$ is the branch in this group for which $\operatorname{im} k_-(\omega)$ has its algebraically greatest value. The coefficient R_2 depends on the wave transformation law at a particular boundary.

Lastly, after a second reflection, this time at the left-hand boundary, we again have a wave propagating to the right:

$$\psi = R_1 R_2 a e^{i(k_+-k_-)L} e^{i(k_+x-\omega t)}. \tag{65.3}$$

Since $\psi(t, x)$ is one-valued, this must coincide with (65.1), so that

$$R_1 R_2 \exp\{i[k_+(\omega) - k_-(\omega)]L\} = 1. \tag{65.4}$$

This determines the frequency spectrum for the finite system, i.e. is its dispersion relation.

Taking the modulus of each side, we have

$$|R_1 R_2| \exp\{-\operatorname{im}(k_+ - k_-)L\} = 1. \tag{65.5}$$

When $L \to \infty$, the exponential factor tends to zero or infinity, depending on the sign

†That is, the least positive value if all $\operatorname{im} k_+(\omega) > 0$, and the greatest (in absolute magnitude) negative value if there are branches for which $\operatorname{im} k_+(\omega) < 0$. In the first case, (65.1) is the wave least rapidly damped (with increasing x); in the second case, it is the one most rapidly amplified.

of $\operatorname{im}(k_+ - k_-)$. Hence, for sufficiently long systems, equation (65.5) is possible only if

$$\operatorname{im}[k_+(\omega) - k_-(\omega)] = 0. \tag{65.6}$$

In this case, therefore, the dispersion relation reduces to a form depending only on the properties of the medium itself and independent of the specific conditions at the boundaries. Equation (65.6) defines a curve in the ω-plane, on which the discrete eigenfrequencies lie very close together (for large L). If the curve lies even partly in the upper half-plane, the system is unstable. Since this instability is due to the properties of the system as a whole, it is called *global instability*.

Some further comments may be made on the relation between global instability of a finite system and the instability of an infinite medium. First of all, it is easy to see that in the presence of global instability the infinite medium is certainly unstable, since there exist real values of k for which $\operatorname{im}\omega(k) > 0$. For, by the definition of $k_+(\omega)$ and $k_-(\omega)$, their values for $\operatorname{im}\omega \to \infty$ lie in different half-planes of k. The condition (65.6) signifies that, as $\operatorname{im}\omega$ decreases, the points $k_+(\omega)$ and $k_-(\omega)$ may come into the same half-plane, and do so (in the case of global instability) while $\operatorname{im}\omega > 0$. Consequently, at least one of these points crosses the real axis even earlier, i.e. certainly while $\operatorname{im}\omega > 0$; this proves the above statement.

The converse statement, however, is valid only for absolute (not convective) instability of the infinite medium: the presence of absolute instability is sufficient to cause global instability of the finite system. The condition for absolute instability is that there exists a branch point of $k(\omega)$ with $\operatorname{im}\omega > 0$, and the coalescing branches belong to the k_+ and k_- groups. At such a point, the condition (65.6) is certainly satisfied also.

A convectively unstable medium may be either stable or unstable when boundaries are present.

CHAPTER VII

INSULATORS

§66. Interaction of phonons

THE physical nature of transport processes such as thermal and electrical conduction in gases consists in transfer by the thermal motion of the gas particles; in solids, the particles are replaced by quasi-particles. In going on to study these processes, we shall begin with thermal conduction in non-magnetic insulators. The relative simplicity of the physical picture here, as compared with transport processes in solids of other kinds, arises from the presence of quasi-particles of only one sort, namely *phonons*.

The concept of free phonons is the result of quantization of the vibrational motion of atoms in the crystal lattice in the harmonic approximation, i.e. with only the quadratic terms (in the displacements of the atoms) included in the Hamiltonian; see *SP* 1, § 72. The various phonon interaction processes result when terms of higher orders of smallness are considered: the anharmonic terms of the third and subsequent orders in the displacements.†

The first anharmonic (the cubic) terms in the classical lattice energy are

$$H^{(3)} = \frac{1}{6} \sum_{(\mathbf{n},\, s)} \Lambda^{s_1 s_2 s_3}_{\alpha\beta\gamma} (\mathbf{n}_1 - \mathbf{n}_3, \mathbf{n}_2 - \mathbf{n}_3) U_{s_1\alpha}(\mathbf{n}_1) U_{s_2\beta}(\mathbf{n}_2) U_{s_3\gamma}(\mathbf{n}_3). \tag{66.1}$$

Here $\mathbf{U}_s(\mathbf{n})$ are the atomic displacement vectors in the lattice; α, β, γ are vector suffixes taking the values x, y, z; s_1, s_2, s_3 number the atoms in the unit cell; $\mathbf{n}_1, \mathbf{n}_2, \mathbf{n}_3$ are integral "vectors" giving the position of the cell in the lattice; $(\mathbf{n}, s)$ under the summation sign denotes summation over all $\mathbf{n}$ and s. Because the crystal is homogeneous, the functions Λ depend only on the relative positions $\mathbf{n}_1 - \mathbf{n}_3$, $\mathbf{n}_2 - \mathbf{n}_3$ of the cells, not on their absolute positions in the lattice.

The second-quantized Hamiltonian is found by replacing the displacement vectors in (66.1) by the operators $\hat{\mathbf{U}}_s(\mathbf{n})$ expressed in terms of the creation and annihilation operators $\hat{c}^+_{\mathbf{k}g}$, $\hat{c}_{\mathbf{k}g}$ for phonons of type g (i.e. branch g of the phonon spectrum) with quasi-momentum $\mathbf{k}$:

$$\hat{\mathbf{U}}_s(\mathbf{n}) = \sum_{g,\mathbf{k}} [2M\mathcal{N}\omega_g(\mathbf{k})]^{-1/2} \{\hat{c}_{\mathbf{k}g}\mathbf{e}^{(g)}_s(\mathbf{k})e^{i\mathbf{k}\cdot\mathbf{r}_\mathbf{n}} + \hat{c}^+_{\mathbf{k}g}\mathbf{e}^{(g)*}_s(\mathbf{k})e^{-i\mathbf{k}\cdot\mathbf{r}_\mathbf{n}}\}, \tag{66.2}$$

†The need to take account of the anharmonicity of atomic vibrations in the lattice in considering thermal conduction in a crystal was first noted by P. Debye (1914) and M. Born (1914).

where $\mathcal{N}$ is the number of cells in the lattice, M the total mass of the atoms in the cell, $\mathbf{e}_s^{(g)}(\mathbf{k})$ the phonon polarization vectors, and $\omega_g(\mathbf{k})$ the energy of phonons of type g.† The substitution gives rise to terms which contain the operators $\hat{c}$ and $\hat{c}^+$ in sets of three. These terms represent processes involving three phonons: products of the form $\hat{c}^+\hat{c}^+\hat{c}$ for the decay of one phonon into two, and $\hat{c}^+\hat{c}\hat{c}$ for the coalescence of two colliding phonons into one; terms $\hat{c}\hat{c}\hat{c}$ and $\hat{c}^+\hat{c}^+\hat{c}^+$ would correspond to processes that are prohibited by the law of conservation of energy.

Les us write down, for example, the terms corresponding to the decay of a phonon $\mathbf{k}_1$, g_1 into two phonons $\mathbf{k}_2$, g_2 and $\mathbf{k}_3$, g_3. Changing in (66.1) from summation over $\mathbf{n}_1$, $\mathbf{n}_2$, $\mathbf{n}_3$ to summation over $\boldsymbol{\nu}_1 = \mathbf{n}_1 - \mathbf{n}_3$, $\boldsymbol{\nu}_2 = \mathbf{n}_2 - \mathbf{n}_3$, $\mathbf{n}_3$, we can put these terms in the form

$$\hat{H}^{(3)}_{\text{dec}} = \Omega \frac{\hat{c}_1\hat{c}_2{}^+\hat{c}_3{}^+}{\mathcal{N}^{3/2}(\omega_1\omega_2\omega_3)^{1/2}} \sum_{\mathbf{n}_3} \exp\{i(\mathbf{k}_1 - \mathbf{k}_2 - \mathbf{k}_3)\cdot \mathbf{r}_{\mathbf{n}_3}\}, \tag{66.3}$$

where

$$\Omega = (2M)^{-3/2} \sum_{\boldsymbol{\nu}, s} \Lambda^{s_1s_2s_3}_{\alpha\beta\gamma}(\boldsymbol{\nu}_1, \boldsymbol{\nu}_2) e_{1\alpha} e^*_{2\beta} e^*_{3\gamma} \exp\{i(\mathbf{k}_1 \cdot \mathbf{r}_{\boldsymbol{\nu}_1} - \mathbf{k}_2 \cdot \mathbf{r}_{\boldsymbol{\nu}_2})\}, \tag{66.4}$$

$$\hat{c}_1 \equiv \hat{c}_{\mathbf{k}_1 g_1}, \quad \omega_1 \equiv \omega_{g_1}(\mathbf{k}_1), \quad \mathbf{e}_1 \equiv \mathbf{e}^{(g_1)}_{s_1}(\mathbf{k}_1), \ldots$$

The exponential factor is separated in (66.3) which depends on the absolute position $\mathbf{n}_3$ of the cell in the lattice. Summation of this factor over all $\mathbf{n}_3$ gives $\mathcal{N}$ if $\mathbf{k}_1 - \mathbf{k}_2 - \mathbf{k}_3$ is equal to any reciprocal lattice period $\mathbf{b}$; otherwise, it is zero. Hence

$$\hat{H}^{(3)}_{\text{dec}} = \Omega \frac{\hat{c}_1\hat{c}_2{}^+\hat{c}_3{}^+}{\mathcal{N}^{1/2}(\omega_1\omega_2\omega_3)^{1/2}}, \tag{66.5}$$

and the phonon quasi-momenta satisfy the conservation law

$$\mathbf{k}_1 = \mathbf{k}_2 + \mathbf{k}_3 + \mathbf{b}. \tag{66.6}$$

The condition (66.6) is to be regarded as an equation giving the value of, say, the quasi-momentum $\mathbf{k}_3$ from specified values of $\mathbf{k}_1$ and $\mathbf{k}_2$. The latter have to be taken within some one chosen unit cell of the reciprocal lattice (including all the physically different values of the quasi-momentum), and we have to verify that $\mathbf{k}_3$ is also in that cell. This last condition determines the necessary value of $\mathbf{b}$ in (66.6), and does so unambiguously. For, if with specified $\mathbf{k}_1$, $\mathbf{k}_2$ and $\mathbf{b}$ the vector $\mathbf{k}_3$ lies in the chosen cell, then any change in $\mathbf{b}$ will certainly bring $\mathbf{k}_3$ outside that cell. Processes (in this case, phonon decay) for which the law of conservation of quasi-momentum involves a non-zero vector $\mathbf{b}$ are called *Umklapp processes*, in contrast to *normal processes*, for which $\mathbf{b} = 0$. Note that the difference between these two classes of processes is to some extent conventional: any particular process may be of either class, depending on the choice of the base cell. It is important, however,

†In this chapter, we use units such that $\hbar = 1$. The dimensions of the momentum and the wave vector are then the same, and those of the energy and the frequency are the same.

that no choice can make **b** zero simultaneously for all possible processes. It is convenient to choose the base cell of the reciprocal lattice so that the point $\mathbf{k} = 0$ (infinite wavelength) is at its centre; this will be assumed henceforward. With this choice, low values of the quasi-momentum ($k \ll 1/d$, where d is the lattice constant) correspond to all low-frequency phonons, and all processes involving only low-frequency phonons are normal processes.† Large values of the quasi-momentum ($k \sim 1/d$) correspond to short-wavelength phonons with high energy (of the order of the Debye temperature Θ).

Let us return to the phonon decay process. According to the general principles of quantum mechanics (see *QM* (43.1)), the probability of a decay in which the quasi-momentum of one of the two newly formed phonons lies in the range d^3k_2 is given by the square of the corresponding matrix element of the perturbation operator (66.5):

$$dW = 2\pi|\langle N_1 - 1, N_2 + 1, N_3 + 1|H^{(3)}|N_1, N_2, N_3\rangle|^2 \times \\ \times \delta(\omega_1 - \omega_2 - \omega_3)\mathcal{V}\, d^3k_2/(2\pi)^3, \tag{66.7}$$

where $N_1 \equiv N_{\mathbf{k}_1 g_1}$, N_2, N_3 are the phonon occupation numbers in the initial state of the crystal. The matrix elements of the phonon creation and annihilation operators are given by

$$\langle N - 1|\hat{c}|N\rangle = \langle N|\hat{c}^+|N - 1\rangle = \sqrt{N}. \tag{66.8}$$

We thus obtain the decay probability in the form

$$dW = wN_1(N_2 + 1)(N_3 + 1)\delta(\omega_1 - \omega_2 - \omega_3)d^3k_2/(2\pi)^3, \tag{66.9}$$

where

$$w = w(g_2\mathbf{k}_2, g_3\mathbf{k}_3; g_1\mathbf{k}_1) = 2\pi v|\Omega|^2/\omega_1\omega_2\omega_3, \tag{66.10}$$

and $v = \mathcal{V}/\mathcal{N}$ is the volume of the crystal lattice cell. The probability of the processes is therefore proportional to the number N_1 of initial phonons in the initial state of the crystal, and also to the numbers $N_2 + 1$ and $N_3 + 1$ of final phonons in the final state. The latter property is related to the Bose statistics obeyed by the phonons, and is true of all processes involving bosons.‡

The process inverse to decay is the coalescence of two photons $\mathbf{k}_2$ and $\mathbf{k}_3$ to form one phonon $\mathbf{k}_1$. We can easily show that the terms in the Hamiltonian that are responsible for this process differ from (66.5) in that the c-operators in the numerator are replaced by $\hat{c}_1^+\hat{c}_2\hat{c}_3$, and Ω by Ω^*. The probability of this process is therefore given by a formula which differs from (66.9) only as regards the

†If, on the other hand, the base cell is chosen so that the point $\mathbf{k} = 0$ is at one of its vertices, for example, low frequencies will also correspond to the neighbourhoods of the other vertices, near which $\mathbf{k}$ is not small.

‡The phonon distribution function $N_\mathbf{k}$ or $N(\mathbf{k})$ will be defined as the occupation numbers of quantum states with various values of the quasi-momentum $\mathbf{k}$. The number of states belonging to an element d^3k in $\mathbf{k}$-space is $d^3k/(2\pi)^3$, and so the distribution relative to d^3k is $N_\mathbf{k}/(2\pi)^3$.

N-factors:

$$dW = wN_2N_3(N_1+1)\delta(\omega_1-\omega_2-\omega_3)\, d^3k_2/(2\pi)^3. \tag{66.11}$$

The functions w here and in (66.9) are the same, in accordance with the general rule that in the Born approximation (the first approximation of perturbation theory) the probabilities of direct and reverse scattering events are equal; see *QM*, § 126.

The branches of the phonon spectrum always include three *acoustic* branches, for which the energy tends to zero as $\mathbf{k}\to 0$; for long-wavelength (small $\mathbf{k}$) acoustic phonons, the function $\omega(\mathbf{k})$ is linear. The behaviour of the function w in (66.10) for such phonons will be important below. It can be determined by noting the property of the coefficients Λ in the Hamiltonian (66.1) which expresses the fact that a simple displacement of the crystal as a whole leaves its energy unchanged, whether or not the crystal is deformed. This means that the energy $H^{(3)}$ must be unaffected if any of the factors $\mathbf{U}_s(\mathbf{n})$ in it is replaced by $\mathbf{U}_s+\mathbf{a}$ with a vector $\mathbf{a}$ that is independent of $\mathbf{n}$ and s. For this to be so, we must have

$$\sum_{\mathbf{n}_1,\, s_1} \Lambda^{s_1s_2s_3}_{\alpha\beta\gamma}(\mathbf{n}_1,\mathbf{n}_2,\mathbf{n}_3)=0, \tag{66.12}$$

where the summation is over at least one pair of variables $\mathbf{n}_1$, s_1.

Of the three phonons involved in the process, either one or all three may be long-wavelength acoustic phonons: if there are two such phonons and a short-wavelength one, the momentum and energy conservation laws cannot be satisfied. For an acoustic phonon in the limit $\mathbf{k}\to 0$, the polarization vectors $\mathbf{e}_s(\mathbf{k})$ tend to a constant independent of s, since all the atoms in the cell vibrate in unison, and the factors $\exp(i\mathbf{k}\cdot\mathbf{r_n})$ tend to unity. Because of the property (66.12), the quantity Ω (66.4) therefore tends to zero, and for small $\mathbf{k}$ it is proportional to k or (the same thing for an acoustic phonon) to ω. The result is

$$w \propto k_1 \tag{66.13}$$

if there is one long-wavelength phonon, or

$$w \propto k_1k_2k_3 \tag{66.14}$$

if there are three.

The results (66.13) and (66.14) can also be reached in a more obvious way by noting that the long-wavelength acoustic phonons correspond to macroscopic sound waves, which can be treated by macroscopic elasticity theory. Then the energy of the deformed crystal is expressed in terms of the strain tensor

$$U_{\alpha\beta}=\frac{1}{2}\left(\frac{\partial U_\alpha}{\partial x_\beta}+\frac{\partial U_\beta}{\partial x_\alpha}\right), \tag{66.15}$$

where $\mathbf{U}(\mathbf{r})$ is the macroscopic displacement vector for points in the elastic medium. The components of this tensor are the small quantities used in the expansion of the elastic energy. In second quantization, the vector $\mathbf{U}$ is replaced by the operator $\hat{\mathbf{U}}$ analogous to (66.2). The differentiation of $\hat{\mathbf{U}}$ with respect to the coordinates to obtain the operators $\hat{U}_{\alpha\beta}$ gives the additional factor k which leads to the results (66.13) and (66.14).

§67. The transport equation for phonons in an insulator

In a solid crystal, the phonons form a rarefied gas, and their transport equation is obtained in a similar way to that for an ordinary gas.

Let $N \equiv N_g(t, \mathbf{r}, \mathbf{k})$ be the distribution function for phonons of type g. The transport equation for each type of phonon is written as

$$\frac{\partial N}{\partial t} + \mathbf{u} \cdot \frac{\partial N}{\partial \mathbf{r}} = C(N), \tag{67.1}$$

where $\mathbf{u} = \partial\omega/\partial\mathbf{k}$ is the phonon velocity.

An important difference from the case of ordinary gases, however, is that in collisions in the phonon gas neither the number of phonons nor (because of Umklapp processes) their total quasi-momentum is in general conserved. The only remaining conservation law is that of energy, expressed by

$$\sum_g \int \omega C(N) d^3k/(2\pi)^3 = 0. \tag{67.2}$$

Multiplying (67.1) by ω, integrating over d^3k, and summing over g, we obtain the law of conservation of energy in the form

$$\partial E/\partial t + \operatorname{div} \mathbf{q} = 0, \tag{67.3}$$

where the thermal energy density E of the crystal and the energy flux $\mathbf{q}$ are given by the obvious expressions

$$E = \sum_g \int \omega N\, d^3k/(2\pi)^3, \quad \mathbf{q} = \sum_g \int \omega \mathbf{u} N\, d^3k/(2\pi)^3. \tag{67.4}$$

The collision integral in (67.1) must in principle include all processes that can occur as a result of the interaction of phonons of type g with all other phonons. In practice, however, the chief contribution to it comes from the three-phonon processes discussed in §66. Processes involving a greater number of phonons arise from subsequent terms in the expansion of the Hamiltonian in powers of the displacements of the atoms: these terms decrease rapidly as their order increases. The reason for the decrease is that the ratio of the vibration amplitude ξ to the lattice constant d is small; in solid crystals, it remains small at all temperatures up to the melting-point.† For a rough estimate, we can begin from the classical relation

†Except in the "quantum crystal", solid helium.

$M\omega^2\xi^2 \sim T$; estimating the characteristic frequency as $\omega \sim u/d$,† we find

$$(\xi/d)^2 \sim T/Mu^2 \ll 1. \tag{67.5}$$

The collision integral is, as always, the difference between the numbers of processes (per unit time) which create phonons in a given state g, $\mathbf{k}$, and which remove phonons from that state. Taking account of three-phonon processes only, we have

$$\begin{aligned} C(N) = \int \Big\{ & \frac{1}{2} \sum_{g_1, g_2} w(\mathbf{k}_1, \mathbf{k}_2; \mathbf{k}) \delta(\omega - \omega_1 - \omega_2) \times \\ & \times [(N+1)N_1N_2 - N(N_1+1)(N_2+1)] \\ & + \sum_{g_1, g_3} w(\mathbf{k}, \mathbf{k}_1; \mathbf{k}_3) \delta(\omega_3 - \omega - \omega_1) \times \\ & \times [(N+1)(N_1+1)N_3 - NN_1(N_3+1)] \Big\} \, d^3k_1/(2\pi)^3, \end{aligned} \tag{67.6}$$

where $N_1 \equiv N_{g_1}(\mathbf{k}_1)$, $\omega_1 \equiv \omega_{g_1}(\mathbf{k}_1), \ldots$ The first term in the braces corresponds to the direct and reverse processes

$$(g, \mathbf{k}) \rightleftarrows (g_1, \mathbf{k}_1) + (g_2, \mathbf{k}_2), \quad \mathbf{k}_2 = \mathbf{k} - \mathbf{k}_1 - \mathbf{b}; \tag{67.7}$$

the factor $\frac{1}{2}$ in this term takes into account the fact that, because of the identity of the phonons, we have to sum only over half of the final states. The second term in the braces corresponds to processes

$$(g_3, \mathbf{k}_3) \rightleftarrows (g, \mathbf{k}) + (g_1, \mathbf{k}_1), \quad \mathbf{k}_3 = \mathbf{k} + \mathbf{k}_1 + \mathbf{b}; \tag{67.8}$$

the factor $\frac{1}{2}$ is not needed in this term, since one of the two phonons formed by the decay is specified. In the integrand of (67.6), it should be noted, the triple products NN_1N_2 and NN_1N_3 cancel.

The collision integral is identically zero for the equilibrium phonon distribution, the Planck distribution

$$N_0 = (e^{\omega/T} - 1)^{-1}. \tag{67.9}$$

This is easily shown for the integral (67.6) by direct calculation: multiplication of the factors gives

$$N_0(N_{01}+1)(N_{02}+1) = (N_0+1)N_{01}N_{02} \exp[(\omega_1 + \omega_2 - \omega)/T], \tag{67.10}$$

and by the law of conservation of energy the exponential factor on the right is equal to unity.

If Umklapp processes were absent, not only the total energy but also the total quasi-momentum of the phonons would be zero. Then not only the distribution

†In the estimates, we shall regard u as the speed of sound, although this is of course literally correct only for long-wavelength acoustic phonons.

function (67.9), but also the functions

$$N_0 = \left[\exp\frac{\omega - \mathbf{k}\,.\,\mathbf{V}}{T} - 1\right]^{-1} \tag{67.11}$$

corresponding to the translational motion (*drift*) of the phonon gas as a whole relative to the lattice with any velocity $\mathbf{V}$, would be equilibrium functions. This result is in accordance with the general principles of statistical physics. It can also be proved directly: with the functions (67.11) as N_0, a further factor $\exp[\mathbf{V}\cdot(\mathbf{k}-\mathbf{k}_1-\mathbf{k}_2)/T]$ appears on the right of (67.10), and is equal to unity for non-Umklapp processes, where $\mathbf{k}=\mathbf{k}_1+\mathbf{k}_2$.

The distribution (67.11) leads, of course, to a non-zero energy flux $\mathbf{q}$. Thus, in the absence of Umklapp processes, a heat flux could exist in the crystal although the temperature were constant throughout the body; that is, the crystal would have an infinite thermal conductivity. A finite conductivity arises only because Umklapp processes exist.†

To calculate the thermal conductivity, we have to write the transport equation for a crystal in which the temperature varies slowly through the volume. As usual, we seek the phonon distribution functions in the form

$$N(\mathbf{r},\mathbf{k}) = N_0(\mathbf{k}) + \delta N(\mathbf{r},\mathbf{k}), \tag{67.12}$$

where δN is a small correction to the equilibrium function. The transport equations are then

$$(\mathbf{u}\,.\,\nabla T)\partial N_0/\partial T = I(\delta N), \tag{67.13}$$

where $I(\delta N)$ is the linearized collision integral.

The functions δN must also satisfy the further condition

$$\sum_g \int \omega\,\delta N\, d^3k/(2\pi)^3 = 0, \tag{67.14}$$

which signifies that the perturbed distribution functions give the same value of the lattice energy density as do the equilibrium functions. As already noted in §6, this condition essentially specifies the definition of the temperature in a non-equilibrium body. The other conditions imposed on δN in §6 do not apply to a phonon gas, in contrast to an ordinary gas. The number of particles in the phonon gas is not a fixed quantity, but depends on the temperature. The total actual momentum (not quasi-momentum) of the phonons in the crystal is necessarily zero, since otherwise there would be a flow of the solid, which is certainly impossible for an ideal (defect-free) crystal lattice. Each atom in the lattice executes only a finite motion, the oscillation about the lattice site; the mean momentum of such a motion is identically zero.

†The quantum theory of thermal conduction in insulators, based on the transport equation for phonons, is due to R. E. Peierls (1929), who also first drew attention to the role of Umklapp processes in transport processes in solids.

Thus the phonon flux (associated with the energy flux) in a solid crystal is not accompanied by a transfer of mass.†

Let us write explicitly the linearized collision integral (67.6). Here, it is convenient to use instead of δN new unknown functions χ defined by

$$\delta N = -(\partial N_0/\partial\omega)\chi = N_0(N_0+1)\chi/T. \qquad (67.15)$$

The process of linearization is simplified by noting that

$$\delta\frac{N}{1+N} = \frac{N_0}{1+N_0}\frac{\chi}{T}. \qquad (67.16)$$

The expression in square brackets in the first integral in (67.6), for example, may be written

$$(N+1)(N_1+1)(N_2+1)\left[\frac{N_1}{N_1+1}\frac{N_2}{N_2+1} - \frac{N}{N+1}\right].$$

In the factors taken outside the brackets, we can put immediately $N = N_0$. The difference in the brackets gives

$$\frac{1}{T}\frac{N_0}{N_0+1}(\chi_1+\chi_2-\chi),$$

where we have used the formula

$$\frac{N_{01}}{N_{01}+1}\frac{N_{02}}{N_{02}+1} = \frac{N_0}{N_0+1}.$$

The collision integral is thus brought to the form

$$\begin{aligned} C(N) \approx I(\chi) = \frac{1}{T}\int \Big\{ &\frac{1}{2}\sum_{g_1,g_2} w(\mathbf{k}_1,\mathbf{k}_2;\mathbf{k})N_0(N_{01}+1)(N_{02}+1)\times \\ &\times \delta(\omega_1+\omega_2-\omega)(\chi_1+\chi_2-\chi) \\ &+\sum_{g_1,g_2} w(\mathbf{k},\mathbf{k}_1;\mathbf{k}_3)N_0N_{01}(N_{03}+1)\times \\ &\times \delta(\omega+\omega_1-\omega_3)(\chi_3-\chi_1-\chi)\Big\}\, d^3k_1/(2\pi)^3. \end{aligned} \qquad (67.17)$$

Note that $\chi(\mathbf{k})$ appears in the integrands as simple sums of its values for various $\mathbf{k}$, as in the classical collision integral for gases (6.4), (6.5).

†Unlike a liquid, where the phonon momentum is the actual momentum and the phonon flux does involve a transfer of mass. In a liquid, the atoms execute an infinite motion: in a sufficiently long time, any atom can reach any point in the volume.

To a solution of (67.13), we can always add the obvious solution of the homogeneous equation,

$$\chi = \text{constant} \times \omega, \tag{67.18}$$

which makes the integral (67.17) identically zero because of the conservation of energy in collisions. As already explained in §6, this "extra" solution corresponds simply to a small constant change in the temperature, and is excluded by the further condition (67.14).

A second "extra" solution,

$$\chi = \mathbf{k} \,.\, \delta\mathbf{V}, \tag{67.19}$$

with $\delta\mathbf{V}$ a constant, corresponds to a small change in the velocity of the phonon gas as a whole (cf. (6.6)), and is excluded by the presence of Umklapp processes, which have the effect that the total phonon quasi-momentum is not conserved.

§68. Thermal conduction in insulators. High temperatures

Equation (67.13) allows us to determine immediately the temperature dependence of the thermal conductivity of an insulator at temperatures much greater than the Debye temperature $\Theta \sim u/d$ (or $\hbar u/d$ in ordinary units).

The maximum phonon energy in all branches of the spectrum is of the order of Θ. Hence, when $T \gg \Theta$, the energies of all phonons $\omega \ll T$, and for most of them $\omega \sim \Theta$. The equilibrium distribution function (67.9) then becomes

$$N_0 \approx T/\omega \gg 1. \tag{68.1}$$

In the collision integral (67.17), the temperature separates as a factor T^2; the function w for frequencies $\omega \sim \Theta$ does not affect the temperature dependence of the integral. On the left-hand side of (67.13), $\partial N_0/\partial T \approx 1/\omega$ does not involve the temperature. Hence

$$\chi \propto \nabla T/T^2, \quad \delta N = -(\partial N_0/\partial\omega)\chi \propto \nabla T/T,$$

and so the heat flux†

$$\mathbf{q} = \sum_g \int \omega \mathbf{u}\, \delta N\, d^3k/(2\pi)^3 \propto \nabla T/T.$$

The thermal conductivity is thus inversely proportional to the temperature:

$$\kappa \propto 1/T, \quad T \gg \Theta; \tag{68.2}$$

†The obvious vanishing of $\mathbf{q}$ in equilibrium follows formally from the vanishing of the integral over d^3k because the integrand is an odd function of $\mathbf{k}$: the frequency $\omega(\mathbf{k})$, and therefore $N_0(\omega)$, are even functions of $\mathbf{k}$, and the velocity $\mathbf{u} = \partial\omega/\partial\mathbf{k}$ is an odd function. The function $\omega(\mathbf{k})$ is even because of the symmetry under time reversal, whatever the symmetry of the crystal lattice (see *SP* 1, § 69).

in the classical theory, this result was obtained by Debye. In an anisotropic crystal, the directions of $\mathbf{q}$ and ∇T are in general not the same, and the conductivity is therefore a tensor of rank two, not a scalar; we shall not take account of this when considering its temperature dependence.

Let us estimate the phonon mean free path in the temperature range concerned. According to the elementary relation (7.10) in the kinetic theory of gases, $\kappa \sim C\bar{v}l$, where C is the specific heat per unit volume, $\bar{v}$ the mean speed of the energy carriers, and l their mean free path. The specific heat of the crystal is constant at high temperatures, and so is the speed of the phonons, which may be estimated as the speed of sound u. We then see that the mean free path $l \propto 1/T$. This would have to become of the order of the lattice constant d at temperatures so high that the vibration amplitudes of the atoms are also of the order of d. According to the estimate (67.5), such a temperature is $\sim Mu^2$, and we have for the mean free path and the effective collision frequency $\nu \sim u/l$ the estimates

$$l \sim Mu^2 d/T, \quad \nu \sim T/Mud. \tag{68.3}$$

From this we see that $l \gg d$ at almost all temperatures below the melting-point.

In this analysis, it has essentially been assumed that the three-phonon mechanism of thermal resistance in a crystal lattice is operative for all phonons. The energy fluxes carried by the various groups of phonons are additive, and therefore so are their contributions to the thermal conductivity. If that mechanism were insufficient for even one group of phonons, it would be insufficient to provide a finite conductivity. The long-wavelength acoustic phonons need special consideration here.

Let us first consider processes involving only such phonons with small quasi-momenta of comparable magnitude, denoted by $\mathbf{f}$ with the appropriate suffix. We shall estimate for these processes the collision integral (67.17) as regards its dependence on f. According to (66.14), in this case $w \propto ff_1f_2 \propto f^3$. The factors $N_0 \sim T/\omega \propto 1/f$. The integration in k-space is over a volume $\sim f^3$, but the delta function separates within the volume a surface with area $\sim f^2$. We thus find for the collision integral

$$I(\chi) \propto f^2\chi \propto f^4 \delta N,$$

where the last expression uses the fact that $\delta N \propto \chi/f^2$ by (67.15); the result can also be written in terms of the effective collision frequency as

$$\nu(f) \propto f^4. \tag{68.4}$$

On the left-hand side of the transport equation (67.13), the factor $\mathbf{u}$ is (for acoustic phonons) independent of f, and $\partial N_0/\partial T \propto 1/f$. Hence

$$\delta N \propto 1/f\nu.$$

The contribution of the long-wavelength phonons to the energy flux $\mathbf{q}$ is given by the integral (67.4) taken over a volume $\sim f^3$. This integral,

$$\int \omega u \, \delta N \, d^3f/(2\pi)^3 \propto \int d^3f/\nu(f), \tag{68.5}$$

diverges as $1/f$ for small f, however. Thus the three-phonon processes between only long-wavelength acoustic phonons would lead to an infinite thermal conductivity; to arrive at a finite conductivity, collisions between these and short-wavelength phonons are necessary (I. Ya. Pomeranchuk 1941).

Let a short-wavelength phonon with quasi-momentum $\mathbf{k}$ decay into a long-wavelength acoustic phonon $\mathbf{f}$ and a short-wavelength phonon $\mathbf{k}-\mathbf{f}-\mathbf{b}$ belonging to the same branch of the spectrum $\omega(\mathbf{k})$ as the phonon $\mathbf{k}$; in the following analysis, the absolute value of k is less important than the fact that $k \gg f$. Since $\omega(\mathbf{k})$ is periodic in the reciprocal lattice, we have $\omega(\mathbf{k}-\mathbf{f}-\mathbf{b}) = \omega(\mathbf{k}-\mathbf{f})$, and the law of conservation of energy states that

$$\omega(\mathbf{k}) = \omega(\mathbf{k}-\mathbf{f}) + u(\mathbf{n})f. \tag{68.6}$$

The second term on the right, the acoustic phonon frequency, is a linear function of f; $u(\mathbf{n}) = \omega(\mathbf{f})/f$ is the phase velocity of sound, which depends on the direction $\mathbf{n} = \mathbf{f}/f$. Expanding $\omega(\mathbf{k}-\mathbf{f})$ in powers of the small vector $\mathbf{f}$, we can put this equation in the form

$$\mathbf{f} \,.\, \partial\omega/\partial\mathbf{k} = fu(\mathbf{n}). \tag{68.7}$$

It can be satisfied only if the speed of the short-wavelength phonon exceeds that of sound:

$$|\partial\omega/\partial\mathbf{k}| > u(\mathbf{n}). \tag{68.8}$$

In this sense, the most "dangerous" acoustic branch is that with the highest speed of sound, and this is the branch we shall have in mind when referring to acoustic phonons.†

Other possibilities for three-phonon processes occur when there are points of degeneracy in $\mathbf{k}$-space, where the energies of two or more branches of the phonon spectrum coincide (C. Herring 1954); the presence of such points (either isolated, or forming a line or plane) is in many cases a necessary consequence of the crystal lattice symmetry. The resulting possibilities are illustrated by a graphical construction which we shall first give for the case already discussed, that of emission by a "supersonic" short-wavelength phonon.

For a given direction of $\mathbf{f}$, we take that direction as the x-axis; in Fig. 27a, the continuous curve represents the function $\omega(k_x)$ (with given k_y and k_z) for short-

†In an isotropic solid, one branch of the acoustic spectrum corresponds to longitudinal vibrations, and the other two to transverse ones; the speed of longitudinal sound waves exceeds that of transverse waves. In an anisotropic crystal, the division of the waves into longitudinal and transverse has in general no meaning. However, in the literature the branch with the highest speed of sound is often arbitrarily called longitudinal.

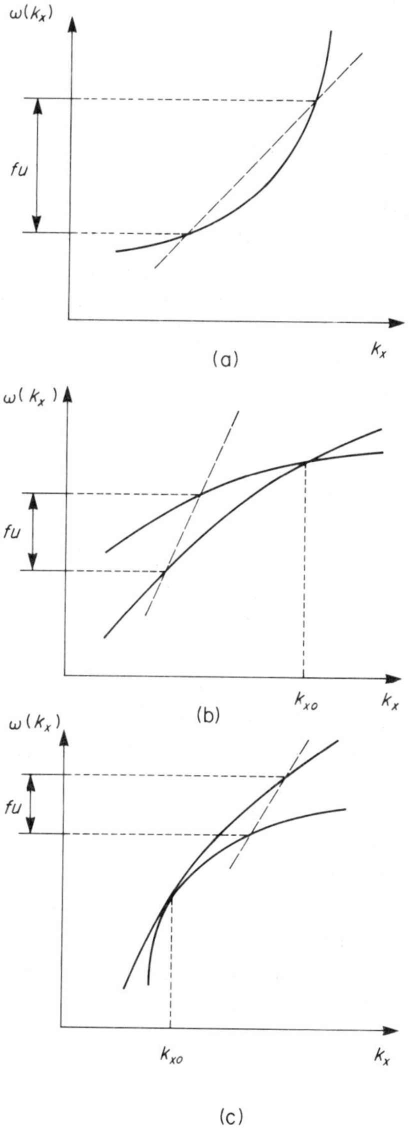

FIG. 27.

wavelength phonons. Writing the condition (68.7) in the form

$$v_x \equiv \partial\omega/\partial k_x = u(\mathbf{n}_x),$$

we see that the emission of an acoustic phonon is possible if at some point the slope of the curve is equal to the speed of sound. The frequencies $\omega(\mathbf{k})$ and $\omega(\mathbf{k}-\mathbf{f})$ of the short-wavelength phonons near this point are then given by the intersections of the curve with the broken line, whose slope is $u(\mathbf{n}_x)$; the difference between the ordinates of these points gives the frequency fu.

If, however, the curves of two branches $\omega(k_x)$ intersect at a point $k_x = k_{x0}$, a three-phonon process is always possible near that point, whatever the slopes of the curves, and whether or not k_{x0} is a simple intersection (Fig. 27b) or a point of contact (Fig. 27c). The two short-wavelength phonons then belong to different branches of the spectrum.

Let us now estimate the effective number of collisions of a long-wavelength acoustic phonon when there are points of degeneracy. These must be processes (67.8) of absorption and emission of the phonon; the decay of the phonon by the processes (67.7) would lead to two long-wavelength phonons, i.e. the case already discussed. We have therefore to estimate the second term in (67.17) with the assumption that

$$\omega_1, \omega_3 \gg \omega \propto f \to 0.$$

Here we use the facts that $w \propto f$, $N_0 \propto 1/f$, and the remaining factors in the integrand may be replaced by mean values independent of f, since the integration is taken only over the neighbourhood of the degeneracy points. Again using $\delta N \propto \chi/f^2$, we obtain an estimate of the dependence of the collision integral on f in the form $I(\chi) \propto \nu(f)\delta N$, where

$$\nu(f) \propto f^2 \int \delta[\omega_1(\mathbf{k}-\mathbf{f}) + u(\mathbf{n})f - \omega_3(\mathbf{k})]\, d^3k. \tag{68.9}$$

The integral can be transformed into a surface integral in **k**-space, the surface being defined by

$$\omega_1(\mathbf{k}-\mathbf{f}) + u(\mathbf{n})f - \omega_3(\mathbf{k}) = 0, \tag{68.10}$$

by means of the formula†

$$\int \delta(F)\, d^3k = \oint dS/|\nabla_{\mathbf{k}} F|, \tag{68.11}$$

the integral being taken over the surface $F(\mathbf{k}) = 0$. We then have

$$\nu(f) \propto f^2 \Delta S(f) \left\langle \left| \frac{\partial \omega_3(\mathbf{k})}{\partial \mathbf{k}} - \frac{\partial \omega_1(\mathbf{k}-\mathbf{f})}{\partial \mathbf{k}} \right|^{-1} \right\rangle, \tag{68.12}$$

where $\Delta S(f)$ is the area of the surface (68.10), and the angle brackets denote averaging over the surface.

Let us consider a typical case, in which the degeneracy points form a curve in **k**-space. Then, as $f \to 0$, the surface (68.10) contracts to a curve on which the

†This is derived immediately by noting that

$$d^3k = dS\, dl = dS\, dF/|\nabla_{\mathbf{k}} F|,$$

where l is the distance along the normal to the surface.

degeneracy points lie, and for small f it is a narrow tube surrounding this line; the dependence of the area ΔS on f is therefore the same as the f-dependence of the tube diameter.

If the surfaces $\omega(\mathbf{k})$ meet on the degeneracy curve but do not touch (Fig. 27b), the distance of the point $\mathbf{k}$ from the degeneracy point varies linearly with f, and so $\Delta S \propto f$ also. Since the difference of the derivatives in this case is finite at the point of intersection, we have

$$\nu(f) \propto f^3. \tag{68.13}$$

The integral (68.5) now diverges only logarithmically. This divergence is to be removed in the same way as when there is no degeneracy (see below). Because the divergence is weak, it usually does not cause any significant change in the law (68.2).

Now let the surfaces $\omega(\mathbf{k})$ have a quadratic contact at the degeneracy point. Then, as we see from Fig. 27c, f is proportional to the square of the distance from the point of contact. The area ΔS, being proportional to this distance itself, is $\Delta S \propto f^{1/2}$. In the case concerned, the same dependence on f occurs for the difference of the derivatives in (68.12), since the derivative curves intersect without contact. In this case, therefore,

$$\nu(f) \propto f^2, \tag{68.14}$$

and there is no divergence in the thermal conductivity.

Other types of degeneracy may be treated similarly.†

If there are no degeneracy points in the phonon spectrum, the condition (68.6) must be satisfied for all directions $\mathbf{n}$ in at least one branch of the spectrum $\omega(\mathbf{k})$ in order to ensure a finite thermal conductivity from three-phonon processes. Otherwise, a finite conductivity results only from higher-order (four-phonon) processes, and the law (68.2) does not hold. At low temperatures the mean free path increases and may become comparable with the size L of the body; the divergence of the integral (68.5) can then be cut off at $f \sim 1/L$, which would make the thermal conductivity dependent on L.

§69. Thermal conduction in insulators. Low temperatures

At low temperatures ($T \ll \Theta$), heat transfer in insulators becomes quite different. The reason is that under such conditions the number of Umklapp processes becomes exponentially small, as is clear from the following arguments.

The conservation of quasi-momentum in a three-phonon process with Umklapp, expressed by $\mathbf{k} = \mathbf{k}_1 + \mathbf{k}_2 + \mathbf{b}$, requires that at least one of the three quasi-momenta should be large; let this one be $k_1 \sim b$. Then the energy $\omega_1 \sim \Theta$, and the conservation of energy ($\omega = \omega_1 + \omega_2$) requires that the energy $\omega \sim \Theta$ should also be large. When

†A discussion of them may be found in the original paper by C. Herring, *Physical Review* **95**, 954, 1954.

$T \ll \Theta$, however, the majority of the phonons have energies $\sim T$, and the number with energies $\sim \Theta$ is exponentially small. Thus, both for the phonon decay process and for the inverse process of coalescence of two phonons, the numbers of initial phonons, and therefore the numbers of processes, are exponentially small. It is easy to see that the fact that it is a three-phonon process is unimportant in these arguments, which apply equally to processes involving a greater number of phonons.

In this situation, the physical picture of heat transfer is as follows. Numerous normal collisions of phonons, in which the total quasi-momentum is conserved, establish only "internal" equilibrium in the phonon gas, which may still be moving relative to the lattice with any velocity **V**. The small number of collisions with Umklapp change the distribution function only slightly, but establish a definite value of **V**, proportional to the temperature gradient; this in turn determines the heat flux. We shall now show how this picture is represented in the mathematical solution of the problem.†

The transport equation is written as

$$(\partial N_0/\partial T)\mathbf{u}\cdot\nabla T = I_N(\chi) + I_U(\chi), \tag{69.1}$$

the collision integral being separated into parts associated with normal and Umklapp collisions. The equilibrium distribution function corresponding to the movement of the gas as a whole with velocity **V** is obtained from $N_0(\omega)$ on replacing the argument ω by $\omega - \mathbf{k}\,.\,\mathbf{V}$; when **V** is small, we have

$$N_0(\omega - \mathbf{k}\,.\,\mathbf{V}) \approx N_0(\omega) - \mathbf{k}\,.\,\mathbf{V}\,\partial N_0/\partial\omega. \tag{69.2}$$

In accordance with the picture described above, we seek the solution of equation (69.1) in the form

$$\chi = \chi_N + \chi_U, \quad \chi_N = \mathbf{k}\,.\,\mathbf{V}; \tag{69.3}$$

χ_U is the part of the change in the distribution function that is due to Umklapp processes, and is small in comparison with χ_N. If ν_U and ν_N denote the orders of magnitude of the effective collision frequencies with and without Umklapp ($\nu_U \ll \nu_N$), then

$$\chi_U/\chi_N \sim \nu_U/\nu_N. \tag{69.4}$$

Substitution in (69.1) gives

$$(\partial N_0/\partial T)\mathbf{u}\cdot\nabla T = I_N(\chi_U) + I_U(\chi_N), \tag{69.5}$$

where the linear operators acting on the functions χ are defined by (67.17). In (69.5) we have used the fact that $I_N(\chi_N) = 0$, and omitted $I_U(\chi_U)$ as a small quantity; the

†It should be noted that the unambiguous separation of Umklapp processes as a small effect is achieved by precisely the choice described in § 66 for the base cell in the reciprocal lattice, as a result of which all collisions between long-wavelength phonons only, with low energies, are normal ones.

two remaining terms on the right are of the same order of magnitude if (69.4) is valid.

Let us emphasize first of all that, when Umklapp processes are neglected and the temperature gradient is not zero, the transport equation has no solution. For, let us multiply equation (69.5) by **k**, integrate over $d^3k/(2\pi)^3$, and sum over all branches of the phonon spectrum. Since normal collisions conserve the total quasi-momentum, the term $I_N(\chi_U)$ becomes zero, leaving

$$\sum_g \int \mathbf{k}(\mathbf{u}\cdot\nabla T)\frac{\partial N_0}{\partial T}\frac{d^3k}{(2\pi)^3} = \sum_g \int \mathbf{k} I_U(\chi_N)\frac{d^3k}{(2\pi)^3}. \tag{69.6}$$

When Umklapp processes are neglected, the right-hand side is zero, whereas the left-hand side is certainly not zero, the integrand being an even function, since $\omega(\mathbf{k})$ is even and $\mathbf{u} = \partial\omega/\partial\mathbf{k}$ is odd. This contradiction means that the transport equation has no solution.

With Umklapp processes taken into account, however, equation (69.6) determines the unknown quantity **V** in the solution (69.3). To simplify the notation, we shall suppose that the crystal has cubic symmetry. The anisotropy of the crystal then does not appear† in the integrals in (69.6), and this equation becomes, after substitution of χ_N from (69.3),

$$\beta_1 \nabla T = -\nu_U \beta_2 T \mathbf{V}, \tag{69.7}$$

with the notation

$$\left.\begin{aligned}
\beta_1 &= \frac{1}{3}\frac{\partial}{\partial T}\sum_g \int \mathbf{k}\cdot\mathbf{u} N_0 \frac{d^3k}{(2\pi)^3},\\
\beta_2 &= \frac{1}{3}\frac{\partial}{\partial T}\sum_g \int \frac{k^2}{\omega} N_0 \frac{d^3k}{(2\pi)^3},\\
T\nu_U\beta_2 &= -\frac{1}{3}\sum_g \int \mathbf{k}\cdot I_U(\mathbf{k})\frac{d^3k}{(2\pi)^3};
\end{aligned}\right\} \tag{69.8}$$

the factor β_2 is separated in order to simplify the later formulae.

Equation (69.7) determines **V**, and the energy flux is then calculated as the integral (67.4), in which N is to be replaced by

$$\delta N_N = -\mathbf{k}\cdot\mathbf{V}\,\partial N_0/\partial\omega = \mathbf{k}\cdot\mathbf{V}(T/\omega)\,\partial N_0/\partial T.$$

Then $\mathbf{q} = T\beta_1\mathbf{V}$, and together with (69.7) this gives $\mathbf{q} = -\kappa\nabla T$, with the thermal conductivity

$$\kappa = \beta_1^2/\nu_U\beta_2. \tag{69.9}$$

It is noteworthy that in this case the calculation of κ does not require the transport equation (69.5) to be solved, but only the integrals (69.8) to be calculated.

†For cubic symmetry, any tensor of rank two reduces to a scalar: $a_{\alpha\beta} = \frac{1}{3}a\delta_{\alpha\beta}$, $a \equiv a_{\alpha\alpha}$.

The integrals β_1 and β_2 are governed by the frequency range $\omega \sim T$, which contains the majority of the phonons. They depend on T only by a power law. Since only acoustic phonons can have low energies, β_1 and β_2 need in practice be summed only over the three acoustic branches of the spectrum. It is easy to see that we then have

$$\beta_1, \beta_2 \propto T^3. \tag{69.10}$$

The exponential dependence is contained in the integral ν_U. Its specific form can be obtained by means of (67.17). For Umklapp processes,

$$\chi_{N_1} + \chi_{N_2} - \chi_N = \mathbf{V} \cdot (\mathbf{k}_1 + \mathbf{k}_2 - \mathbf{k}) = \mathbf{V} \cdot \mathbf{b}.$$

For the majority of phonons, $\omega \sim T$ and the distribution function $N_0 \sim 1$; for phonons with $\omega \gg T$, however, $N_0 \ll 1$. The factors $N_0 + 1 \sim 1$ therefore need not be considered in estimating the integral. The functions

$$N_0 = e^{-\omega/T}(N_0 + 1)$$

contain factors $e^{-\omega/T}$, which may be exponentially small, and which have a decisive effect on the estimate of the integral.

Thus, if we consider only the exponential temperature dependence of ν_U, we have

$$\nu_U \propto \sum_{(g, \mathbf{b})} e^{-\omega/T} \delta(\omega - \omega_1 - \omega_2)\, d^3k\, d^3k_1; \tag{69.11}$$

the summation is over all branches g, g_1, g_2 of the spectrum and over all non-zero values of $\mathbf{b}$ occurring in Umklapp processes. The equation

$$\omega_g(\mathbf{k}) = \omega_{g_1}(\mathbf{k}_1) + \omega_{g_2}(\mathbf{k} - \mathbf{k}_1) \tag{69.12}$$

defines a five-dimensional surface in six-dimensional $\mathbf{kk}_1$-space. Let $\Delta(g, g_1, g_2)$ be the minimum value of $\omega_g(\mathbf{k})$ on this hypersurface; since the energies of phonons involved in Umklapp processes are large, these values are $\sim \Theta$. Each of the integrals in the sum over (g) in (69.11) is proportional to $\exp[-\Delta(g, g_1, g_2)/T]$. Retaining only the largest of them, we have

$$\nu_U \propto \exp(-\Delta_{\min}/T), \tag{69.13}$$

where $\Delta_{\min}$ is the smallest of the $\Delta(g, g_1, g_2)$.

We thus conclude that the thermal conductivity depends on the temperature essentially according to the exponential relation

$$\kappa \propto \exp(\Delta_{\min}/T), \tag{69.14}$$

with $\Delta_{\min} \sim \Theta$ (R. E. Peierls, 1929).

The higher-order processes, involving a larger number of phonons, lead to a temperature dependence of a similar type, with Δ the lowest possible value of the energy of the initial phonons in each process (or, equivalently, half the least total energy of all phonons, initial and final, taking part in the process). In principle, it may happen that this value is less than for three-phonon processes, in which case the contribution of higher-order processes to the thermal conductivity may become predominant, despite the fact that the coefficient of the exponential decreases, of course, as the order of the process increases.

Unlike the Umklapp process frequency ν_U, the effective frequency ν_N of normal collisions decreases as a power of the temperature; we shall determine the relationship, in order to use it in §71.

Normal collisions take place between acoustic phonons with $\omega \sim T$, which form the majority. Their quasi-momenta $k \sim \omega/u \sim T/u$. In the collision integral (67.17), the integration is over a surface with area $\sim k^2$ distinguished by the delta function in a volume $\sim k^3$. In this region, the functions $N_0 \sim 1$ and $\omega \propto k^3$, according to (66.14). Hence $\nu_N \propto T^5$. The proportionality coefficient is most simply determined from the condition that when $T \sim \Theta$ this expression and the estimate (68.3) must give the same result, so that

$$\nu_N \sim T^5/\Theta^4 Mud. \tag{69.15}$$

§70. Phonon scattering by impurities

In §§68 and 69 we have assumed that the crystal lattice is ideal and free from defects. Let us now consider the possible significance of phonon scattering by impurity atoms as regards thermal conduction in insulators.

In relation to the long-wavelength acoustic phonons, the impurity atom is a point defect in the lattice. A characteristic property of scattering by such defects is that it is elastic (the phonon frequency is unchanged), and the scattering cross-section decreases rapidly with the frequency, i.e. with the wave number, as k^4.†

The collision integral for phonon scattering by impurities is

$$C(N_{\mathbf{k}}) = N_{\text{imp}} \int w(\mathbf{k}, \mathbf{k}')\{N_{\mathbf{k}'}(1+N_{\mathbf{k}}) - N_{\mathbf{k}}(1+N_{\mathbf{k}'})\}\delta(\omega'-\omega)d^3k'/(2\pi)^3. \tag{70.1}$$

As usual, the first term in the braces gives the number of scattering events per unit time which bring a phonon into a state with a given quasi-momentum $\mathbf{k}$ from states with any other values $\mathbf{k}'$ that correspond to the same energy. Similarly, the second term gives the number of scattering events that take phonons from that state into any other. If the impurity atoms are arranged randomly, and the mean distance between them is much greater than the scattering amplitude, then different atoms scatter independently and the probabilities are additive. Under these conditions, which have been assumed in (70.1), the total number of scattering events is

†This is a general property of sound wave scattering by obstacles small compared with the wavelength; cf. *FM*, §76. Compare also the corresponding case in the scattering of long electromagnetic waves (*Fields*, §79).

proportional to the impurity atom concentration N_{imp}. For scattering in an anisotropic medium, the function $w(\mathbf{k}, \mathbf{k}')$ depends on the directions of both vectors $\mathbf{k}$ and $\mathbf{k}'$, but its dependence on the magnitude k is $w \propto k^4$. In (70.1) we have put $w(\mathbf{k}, \mathbf{k}') = w(\mathbf{k}', \mathbf{k})$. In the Born approximation, this follows from the unitarity condition and the smallness of the scattering amplitude when second-order terms are neglected (see *QM*, § 126). The Born approximation is not in general applicable to phonon scattering by an impurity atom. At low temperatures, however, when phonons with small $\mathbf{k}$ are concerned, the scattering amplitude is small for another reason, being proportional to k^2; if terms $\propto k^4$ are neglected, we again have the same equation.

The products $N_\mathbf{k} N_{\mathbf{k}'}$ in the braces in (70.1) cancel, and after the substitution $N = N_0 + \delta N$ the collision integral is immediately linearized:

$$C(N) \equiv I_{imp}(\delta N) = N_{imp} \int w(\delta N_{\mathbf{k}'} - \delta N_\mathbf{k}) \delta(\omega' - \omega)\, d^3k'/(2\pi)^3. \qquad (70.2)$$

This integral is, like w, proportional to k^4. Since $\partial N_0/\partial T \propto 1/\omega \propto 1/k$ when $\omega \ll T$, in this frequency range we have

$$\delta N \propto k^{-5}. \qquad (70.3)$$

A similar case has already occurred in § 68; cf. (68.4). The relation (70.3) leads to divergence of the integral for the heat flux. Thus the presence of impurities in the crystal cannot in itself ensure a finite thermal conductivity of an insulator.

This does not mean, however, that impurities play no part in determining the conductivity. The reason is that the scattering by impurity atoms does not conserve the quasi-momentum of the phonons, and in this sense may take the place of Umklapp processes. In sufficiently pure samples, there may exist a range of low temperatures in which the effective frequency ν_{imp} of scattering by impurities (for phonons with $\omega \sim T$) is intermediate between the frequencies of normal and Umklapp collisions between phonons:

$$\nu_N \gg \nu_{imp} \gg \nu_U. \qquad (70.4)$$

Under such conditions, the role of the Umklapp processes is taken by the impurity scattering, and equations (69.6)–(69.8) remain valid if I_U is replaced by I_{imp}. The thermal conductivity is then given by (69.9) with ν_{imp} in place of ν_U:

$$\kappa = \beta_1^2/\beta_2 \nu_{imp}.$$

According to (70.2), $\nu_{imp} \propto \omega^4 \sim T^4$. The quantities β_1 and β_2 for acoustic phonons are proportional to T^3; see (69.10). We thus have $\kappa \propto 1/T$ in this case.

§ 71. Phonon gas dynamics in insulators

The approximate conservation of quasi-momentum when the mean free path (l_N) for normal collisions is small compared with that (l_U) for Umklapp processes,

$$l_N/l_U \sim \nu_U/\nu_N \ll 1, \tag{71.1}$$

makes the phonon system in the crystal at low temperatures similar in many respects to an ordinary gas. The normal collisions establish internal equilibrium in each volume element of the gas (large compared with l_N), which may still be moving with any velocity $\mathbf{V}$. If $\mathbf{V}$ and the temperature T vary appreciably only over distances large compared with l_N (and over times long compared with $1/\nu_N$), a system of "hydrodynamic" equations can be derived for them. We shall construct these in the linear approximation with respect to the velocity $\mathbf{V}$ and the temperature gradient, which will be regarded as small quantities of the same order. Moreover, to simplify the formulae we shall again (as in §69) suppose that the crystal has cubic symmetry.

One of the required equations expresses the law of conservation of energy. It is obtained by substituting the distribution function (69.2) in (67.3) and (67.4). The integrals of $\omega(\mathbf{k}\cdot\mathbf{V})\partial N_0/\partial\omega$ and of $\omega\mathbf{u}N_0$ are zero when the integration over the directions of $\mathbf{k}$ is carried out; cf. the first footnote to §68. The function $N_0(\omega)$ depends on the coordinates and the time only through T. Neglecting terms which contain the product $\mathbf{V}\cdot\nabla T$, we find

$$\beta_3\partial T/\partial t + \beta_1 T \operatorname{div}\mathbf{V} = 0, \tag{71.2}$$

where

$$\beta_3 = \partial E_0/\partial T, \tag{71.3}$$

E_0 is the equilibrium energy density, and β_1 is defined in (69.8).

The second equation expresses the (approximate) conservation of quasi-momentum. It is obtained from the transport equation

$$\partial N/\partial t + \mathbf{u}\,.\,\nabla N = C_N(N) + C_U(N) \tag{71.4}$$

by substituting N in the form (69.2), multiplying by $\mathbf{k}$, integrating over d^3k, and summing over the types of phonon. The integral of $\mathbf{k}C_N(N)$ is zero by the conservation of quasi-momentum in normal collisions. The result is

$$\beta_2 T\,\partial\mathbf{V}/\partial t + \beta_1\nabla T = -\nu_U\beta_2 T\mathbf{V}, \tag{71.5}$$

with β_2 and ν_U given by (69.8). Equations (71.2) and (71.5) are the hydrodynamic equations for a phonon gas in an insulator.

The exponentially small (like ν_U) term on the right of (71.5) represents the effect of Umklapp processes. When this term is neglected, the quasi-momentum is exactly conserved. Under such conditions, undamped waves can propagate in the phonon gas, analogous to second sound waves in a superfluid (V. P. Peshkov 1946): eliminating $\mathbf{V}$ from (71.2) and (71.5), we have in that case

$$\partial^2 T/\partial t^2 = (\beta_1^2/\beta_2\beta_3)\triangle T, \tag{71.6}$$

i.e. the wave equation describing the propagation of temperature oscillations with speed

$$u_2 = (\beta_1^2/\beta_2\beta_3)^{1/2}. \tag{71.7}$$

As already mentioned, contributions to the integrals β_1, β_2, β_3 at low temperatures come almost entirely from the acoustic branches of the spectrum. For linear dispersion relations $\omega(\mathbf{k})$, these integrals are proportional to T^3; the speed (71.7) is then independent of the temperature, and of the same order as the speed of sound.†

So far we have assumed the crystal to be of infinite size. At low temperatures, when the phonon mean free path rapidly increases, a situation may actually occur in which the mean free path becomes comparable with or even much greater than the size L of the crystal. This applies in particular to the exponentially increasing l_U.

Let us consider heat transfer in an insulator with $l_U \gg L$ (the condition will be more precisely specified below), but still $l_N \ll L$; the latter inequality enables us to use the equations of phonon hydrodynamics (J. A. Sussmann and A. Thellung 1963, R. N. Gurzhi 1964).

Because of the microscopic unevennesses of the crystal surface, phonons are usually reflected from it randomly (or *diffusely*); this means that the macroscopic velocity $\mathbf{V}$ of the phonon gas is zero at the surface. Equations (71.2) and (71.5), however, do not allow such a boundary condition; their solutions can satisfy only the condition that the normal component of the velocity is zero at the surface. As in the hydrodynamics of ordinary liquids, the boundary condition that the tangential velocity component is zero demands that the viscosity be taken into account.

In the steady state, equation (71.2) gives $\operatorname{div} \mathbf{V} = 0$. The inclusion of the viscosity adds a term in $\triangle\mathbf{V}$ to the right-hand side of (71.5), similar to the corresponding term in the Navier–Stokes equation for an ordinary viscous liquid. In the steady state, this equation is

$$(\beta_1/\beta_2 T)\nabla T = \mu\triangle\mathbf{V} - \nu_U\mathbf{V}. \tag{71.8}$$

The quantity μ has the dimensions $[L^2/T]$, and acts as the kinematic viscosity of the phonon gas.‡ Its calculation requires in principle the solution of the corresponding transport equation. However, for an order-of-magnitude estimate we can use the ordinary formula from the kinetic theory of gases, whereby

$$\mu \sim l_N\bar{v} \sim u^2/\nu_N. \tag{71.9}$$

Size effects are predominant when the term $\nu_U\mathbf{V}$ in (71.8) is negligible in comparison with $\mu\triangle\mathbf{V}$. For example, let us consider heat transfer along a cylindrical rod with diameter R, which is the characteristic length as regards the variation

†In an isotropic liquid with a phonon energy spectrum (superfluid helium at low temperatures) there is only one acoustic branch, in which $\omega = uk$. Then $\beta_1/\beta_2 = u^2$, $\beta_1/\beta_3 = \frac{1}{3}$, and the speed of second sound is $u_2 = u/\sqrt{3}$.

‡Considering the problem purely qualitatively, we here neglect entirely the anisotropy of the crystal. It should be remembered that, even with cubic symmetry, the viscosity is described not by a scalar coefficient but by a tensor of rank four having more than one independent component.

of the velocity **V**, so that $\triangle \mathbf{V} \sim \mathbf{V}/R^2$. We see that the term $\nu_U \mathbf{V}$ is negligible if $\mu/R^2 \gg \nu_U$. With the estimate (71.9), this condition becomes $l_U \gg l_{\text{eff}}$, where

$$l_{\text{eff}} \sim R^2/l_N \tag{71.10}$$

acts as an effective phonon mean free path in the finite body. If $l_{\text{eff}} \gg l_U$, on the other hand, the size of the body is unimportant, and (69.14) is valid.

The process of heat transfer along the rod when $l_U \gg l_{\text{eff}}$ is a Poiseuille flow of a viscous phonon gas, and may be described by an effective thermal conductivity which determines the energy flux as $-\kappa_{\text{eff}}\nabla T$, where ∇T is the temperature gradient along the rod. This flux may be estimated by substituting (71.10) in the expression $\kappa_{\text{eff}} \sim Cul_{\text{eff}}$. At low temperatures, the lattice specific heat $C \propto T^3$, and $l_N \sim u/\nu_N \propto T^{-5}$ according to (69.15). The effective thermal conductivity is therefore

$$\kappa_{\text{eff}} \propto R^2 T^8 \quad \text{when} \quad R^2/l_U \ll l_N \ll R; \tag{71.11}$$

it decreases with falling temperature.

Lastly, at still lower temperatures, when also $l_N \gg R$, collisions between phonons become unimportant, as in the Knudsen case for highly rarefied ordinary gases. The role of the mean free path is then taken by the size R of the body, and the effective thermal conductivity is

$$\kappa_{\text{eff}} \sim CuR \propto T^3 R \tag{71.12}$$

(H. B. G. Casimir 1938).

§72. **Sound absorption in insulators. Long waves**

The nature of sound absorption in insulator crystals depends greatly on the relation between the wavelength and the mean free path l of thermal phonons. If the wavelength is much greater than l ($fl \ll 1$, where **f** is the sound wave vector), the macroscopic theory based on the equations of elasticity theory is valid (see *TE*, §35), according to which the sound absorption coefficient comprises two terms which are respectively determined by the thermal conductivity and the viscosity of the medium. Both terms are proportional to the square of the frequency. Our aim here is to find their dependence on the temperature.

The thermal conductivity contribution to the sound absorption coefficient is given in order of magnitude by†

$$\gamma_{\text{th}} \sim \omega^2 \kappa T \alpha^2 \rho / uC^2, \tag{72.1}$$

where α is the thermal expansion coefficient of the body, C the specific heat per unit volume and ρ the density. At high temperatures $T \gg \Theta$, the thermal conductivity

†We give the absorption coefficient per unit path length. The frequency and temperature dependences are the same for the coefficient per unit time, since the two coefficients differ only by a constant factor, the velocity of sound.

$\kappa \propto 1/T$, while C and α are independent of the temperature; see *SP* 1, §§ 65, 67. In this range, therefore, γ_{th} is independent of the temperature. At low temperatures, its temperature dependence is governed mainly (in an ideal lattice) by the thermal conductivity, which increases exponentially as T decreases.

Let us now determine the viscosity part of the sound absorption coefficient (A. I. Akhiezer 1938). The external sound field alters the phonon dispersion relation by causing a macroscopic deformation of the crystal lattice. The wavelength of thermal phonons is small in comparison with the wavelength of sound; hence the deformation may be regarded as uniform in relation to a thermal phonon, i.e. the latter may be regarded as being in a lattice that is still regular but has slightly altered periods. In the first approximation with respect to the small deformation, the phonon frequency $\omega(\mathbf{k})$ in such a lattice is related to its value $\omega^{(0)}(\mathbf{k})$ in the undeformed lattice by

$$\omega(\mathbf{k}) = \omega^{(0)}(\mathbf{k})(1 + \lambda_{\alpha\beta} U_{\alpha\beta}), \tag{72.2}$$

where

$$U_{\alpha\beta} = \frac{1}{2}\left(\frac{\partial U_\alpha}{\partial x_\beta} + \frac{\partial U_\beta}{\partial x_\alpha}\right)$$

is the strain tensor and $\mathbf{U}$ the displacement vector. The characteristic tensor $\lambda_{\alpha\beta}$ of the crystal in general depends on $\mathbf{k}$; for long-wavelength acoustic phonons with a linear dispersion relation, however, it does not depend on the magnitude of $\mathbf{k}$.

The parentheses in (72.2) should also contain a term of the form $\lambda \operatorname{curl} \mathbf{U}$ expressing the trivial fact that, if the deformation causes rotation of a lattice volume element ($\operatorname{curl} \mathbf{U} \neq 0$), this changes the direction of the axes (of the reciprocal lattice) with respect to which the quasi-momentum of a phonon in the dispersion relation is to be defined; the term $\lambda \operatorname{curl} \mathbf{U}$ would represent the corresponding change in $\mathbf{k}$. We have not written this term in (72.2), since it is evident *a priori* that it cannot influence the energy dissipation in the sound wave, which concerns us here: the actual physical effect (dissipation) cannot depend on the vector $\operatorname{curl} \mathbf{U}$, which differs from zero even for a mere rotation of the body as a whole.

The change in the phonon distribution function due to the lattice deformation is given by the transport equation

$$(\partial N/\partial \omega)\dot{\omega} + (\partial N/\partial T)\dot{T} = C(N), \tag{72.3}$$

where $C(N)$ is the phonon–phonon collision integral (67.6), and $\dot{T}$ the rate of variation of the temperature at a given point in the crystal that necessarily results from the deformation. Linearizing this equation in the usual way, and using the function χ defined by (67.15), we can reduce it to the form

$$\omega \frac{\partial N_0}{\partial \omega}(\lambda_{\alpha\beta}\dot{U}_{\alpha\beta} - \dot{T}/T) = I(\chi), \tag{72.4}$$

where $I(\chi)$ is the linearized collision integral (67.17). On the left-hand side, the derivative $\dot{\omega}$ has been transformed by means of (72.2); the superscript (0) to the unperturbed frequency is omitted, here and henceforward.

The derivative $\dot{T}$ can in principle be expressed in terms of the same tensor $\lambda_{\alpha\beta}$. Multiplying both sides of equation (72.4) by ω, integrating over **k**-space, and summing over all branches of the phonon spectrum, reduces the right-hand side to zero by virtue of the conservation of energy in collisions. The left-hand side gives

$$\dot{T}/T = \bar{\lambda}_{\alpha\beta}\dot{U}_{\alpha\beta}, \tag{72.5}$$

where $\bar{\lambda}_{\alpha\beta}$ is the tensor $\lambda_{\alpha\beta}$ averaged over $\omega^2\partial N_0/\partial\omega$. In both the limiting cases of high and low temperature, $\bar{\lambda}_{\alpha\beta}$ is independent of temperature: when $T \gg \Theta$, the important phonons in the averaging are those with the temperature-independent quasi-momentum $k \sim k_{max} \sim 1/d$, and when $T \ll \Theta$ the long-wavelength acoustic phonons are the important ones, with $\lambda_{\alpha\beta}$ independent of k, so that the averaging again causes no dependence on the temperature.

With $\lambda_{\alpha\beta} - \bar{\lambda}_{\alpha\beta} = \tilde{\lambda}_{\alpha\beta}$, we write the transport equation as

$$\omega(\partial N_0/\partial\omega)\tilde{\lambda}_{\alpha\beta}\dot{U}_{\alpha\beta} = I(\chi). \tag{72.6}$$

Next, let us derive a formula for the dissipation of energy in a non-equilibrium phonon gas. We start from the expression for the entropy per unit volume of a Bose gas:

$$S = \sum_g \int [(N+1)\log(N+1) - N\log N]\, d^3k/(2\pi)^3 \tag{72.7}$$

(see *SP* 1, §55). Differentiation with respect to time gives

$$\dot{S} = \sum_g \int \dot{N} \log\frac{N+1}{N}\, d^3k/(2\pi)^3. \tag{72.8}$$

Replacing $\dot{N}$ here by the integral $C(N)$ (cf. §4) and renaming appropriately the variables $\mathbf{k}$, $\mathbf{k}_1$ and $\mathbf{k}_2$ in the two terms in (67.6), we can put $\dot{S}$ in the form

$$\dot{S} = \tfrac{1}{2}\sum_{g_1, g_2, g_3}\int w(\mathbf{k}_2, \mathbf{k}_3; \mathbf{k}_1)\delta(\omega_1 - \omega_2 - \omega_3)\log\frac{(N_1+1)N_2N_3}{N_1(N_2+1)(N_3+1)} \times$$
$$\times [(N_1+1)N_2N_3 - N_1(N_2+1)(N_3+1)]\, d^3k_1\, d^3k_2/(2\pi)^6.$$

Multiplication by T gives the dissipative function, i.e. the energy dissipated per unit time and volume. Substituting $N = N_0 + \delta N$, with δN in the form (67.15), and keeping the first (quadratic) terms in the expansion in powers of δN, we find

$$T\dot{S} = \frac{1}{2T}\sum_{g_1, g_2, g_3}\int w(\mathbf{k}_2, \mathbf{k}_3 : \mathbf{k}_1)\delta(\omega_1 - \omega_2 - \omega_3) \times$$
$$\times (N_{01}+1)N_{02}N_{03}(\chi_1 - \chi_2 - \chi_3)^2\, d^3k_1\, d^3k_2/(2\pi)^6. \tag{72.9}$$

The above formulae are sufficient to determine the temperature dependence of the sound absorption coefficient. Let us first consider the range of high temperatures.

In this case, the collision integral $I(\chi)$ contains the temperature as a factor T^2 (see the beginning of §68). On the left-hand side of the transport equation (72.6), we have $\omega\partial N_0/\partial\omega \approx -T/\omega$, and for the majority of the phonons the frequency $\omega \sim \Theta$ is independent of the temperature. For these frequencies, therefore,

$$\chi \sim (1/T)\lambda_{\alpha\beta}\dot{U}_{\alpha\beta}.$$

From (72.9), in which we must put $N_0 \approx T/\omega \gg 1$, we now find that the dissipative function is independent of the temperature. The same is true of the absorption coefficient, obtained by dividing the dissipative function by the energy flux in the sound wave, which is independent of the temperature. When $T \gg \Theta$, therefore, both the viscosity part and the thermal-conduction part of the sound absorption coefficient are independent of the temperature.

At low temperatures there is, first of all, a fundamental difference from the problem of thermal conduction: the sound absorption coefficient is finite even when Umklapp processes (whose frequency is small at low temperatures) are neglected. In the thermal conduction case, the absence of any solution of the transport equation when Umklapp processes are neglected was shown by the contradiction arising on multiplication of this equation by $\mathbf{k}$ and integration over the whole phonon spectrum: the right-hand side is then zero, but the left-hand side is certainly not zero; cf. (69.6). For equation (72.6), however, the contradiction does not occur: since its left-hand side is an even function of $\mathbf{k}$, it becomes an odd function on multiplication by $\mathbf{k}$, and vanishes on integration over d^3k. Here we assume that the integral of the term containing the Umklapp process operator, i.e. of $\mathbf{k}I_U(\chi)$, is also zero. Since this is not ensured by any conservation law, a certain condition is thereby imposed on the solution of the transport equation: the function $\chi(\mathbf{k})$ must be even in $\mathbf{k}$ (and $\mathbf{k}I_U(\chi)$ is then an odd function, since it is easily shown that the operator I does not change the parity of χ). This condition eliminates the arbitrariness due to the existence (in contrast to Umklapp processes) of an "extra" solution of the form $\chi = \mathbf{k}\cdot\delta\mathbf{V}$, an odd function of $\mathbf{k}$, and ensures a correct passage to the limiting case where these processes are absent.

When $T \ll \Theta$, phonons with energy $\omega \sim T$ are the most important in the collision integral (and in the dissipative function). These are long-wavelength phonons in the acoustic branches of the spectrum; their frequency varies linearly with $\mathbf{k}$, and they therefore have $k \sim T/u$. According to (66.14), the function w in the integral (67.17) for collisions of such phonons is $w \propto kk_1k_2$. The distribution function N_0 depends only on the ratio ω/T, so that $N_0 \sim 1$ when $\omega \sim T$. The integration is over $d^3k_1 = k_1^2\,dk_1\,do_1$, and for k_1 over a region $\sim T$. Each factor k, k_1, k_2 therefore contributes a factor T to the integral, and the delta function gives a factor $1/T$. Thus the whole integral, as regards its temperature dependence, is estimated as χT^4. The left-hand side of the transport equation (72.6) is independent of the temperature when $\omega \sim T$. Hence we have, when $\omega \sim T$,

$$\chi \propto T^{-4}\tilde{\lambda}_{\alpha\beta}\dot{U}_{\alpha\beta}.$$

A corresponding estimate of the integral (72.9) then leads to the result that the dissipative function, and therefore the viscosity part of the sound absorption coefficient, are inversely proportional to T. Thus

$$\gamma_{\text{vi}} \propto \omega^2/T \quad \text{when} \quad T \ll \Theta. \tag{72.10}$$

The absence of any need for Umklapp processes has the result that this part of the absorption coefficient increases only by a power law with decreasing temperature, not exponentially.

The use of the dissipative function in the foregoing analysis has made it possible to avoid expressing the viscous stress tensor in the crystal in terms of the phonon distribution function. This is not a trivial problem, because the actual momentum flux tensor is involved, and this momentum is not the same as the quasi-momentum of the phonons. We shall show how this expression can in turn be derived from the form of the dissipative function.

To do so, we again start from the integral (72.8), and now write $\dot{N}$ in it as the expression on the left of the transport equation (72.6). The logarithm in the integrand is written in the form (see (67.16))

$$-\log\frac{N}{N+1} = -\log\left[\frac{N_0}{N_0+1}\left(1+\frac{\chi}{T}\right)\right] \approx \frac{\omega-\chi}{T}.$$

The result is

$$T\dot{S} = \sum_g \int \omega\tilde{\lambda}_{\alpha\beta}\delta N \frac{d^3k}{(2\pi)^3}\dot{U}_{\alpha\beta}, \tag{72.11}$$

where $\delta N = -\chi \partial N_0/\partial\omega$; the term with the factor ω in place of χ is identically zero, by the definition of $\tilde{\lambda}_{\alpha\beta}$. Instead of $\tilde{\lambda}_{\alpha\beta} = \lambda_{\alpha\beta} - \bar{\lambda}_{\alpha\beta}$, we can here put simply $\lambda_{\alpha\beta}$, since the integral containing the constant factor $\bar{\lambda}_{\alpha\beta}$ is zero by the further condition (67.14) imposed on δN.

The dissipative function (per unit volume) can be expressed in terms of the viscous stress tensor $\sigma'_{\alpha\beta}$ as $\sigma'_{\alpha\beta}\dot{U}_{\alpha\beta}$; cf. *TE*, §34. A comparison with (72.11) thus gives the following expression for the viscous stress tensor:

$$\sigma'_{\alpha\beta} = \sum_g \int \omega\lambda_{\alpha\beta}\,\delta N\, d^3k/(2\pi)^3 \tag{72.12}$$

(V. L. Gurevich 1980).

§73. Sound absorption in insulators. Short waves

In the opposite case of short wavelengths, $fl \gg 1$, the process of sound wave damping may be regarded as the result of absorption of individual sound quanta when they collide with thermal phonons (L. D. Landau and Yu. B. Rumer, 1937). For this treatment to be permissible, the energy and momentum of the thermal phonons must be defined with sufficient precision: when changed by the absorption of a sound quantum, they must come into a range outside the quantum uncertainty due to the finite mean free path, and this is ensured by the inequality $fl \gg 1$. In practice, such a situation can occur only at low temperatures, when the mean free path becomes sufficiently long.

In the first approximation, i.e. when processes involving the smallest number of phonons are considered, we have three-phonon processes:

$$\mathbf{k}_1 + \mathbf{f} = \mathbf{k}_2, \quad \omega_1 + \omega = \omega_2, \tag{73.1}$$

where ω and $\mathbf{f}$ are the energy and quasi-momentum of the sound quantum, while $\omega_1, \mathbf{k}_1$ and $\omega_2, \mathbf{k}_2$ belong to thermal phonons. The latter are $\omega_1, \omega_2 \sim T$; $k_1, k_2 \sim T/u$. We shall assume that

$$\hbar\omega \ll T. \tag{73.2}$$

Then ω_1, ω_2 and k_1, k_2 are large compared with ω and f respectively.

As we have seen in §68, the conservation laws (73.1) can be obeyed only if the speed of the thermal phonon exceeds that of the sound quanta absorbed (or emitted). Without entering into a discussion of various possible cases, we shall suppose that the sound wave is not "longitudinal" (i.e. does not correspond to the acoustic branch of the phonon spectrum for which the speed is greatest), and that the condition stated may therefore be satisfied. Since ω and f are small, the initial and final thermal phonons belong in general to the same acoustic branch of the phonon spectrum; at low temperatures, they are long-wavelength phonons.

The probabilities of phonon emission or absorption in a three-phonon process are given by (66.9) or (66.11). The occupation numbers $N_1 \equiv N(\mathbf{k}_1)$ and $N_2 \equiv N(\mathbf{k}_2)$ are given by the Planck equilibrium distribution function (67.9). A macroscopic sound wave corresponds to a very large occupation number for a given phonon state $\mathbf{f}$; in comparison with it, unity is of course negligible. Omitting the factor $N(\mathbf{f})$, we obtain the probability per sound quantum.

Thus the probability of absorption of a sound quantum in its collisions with thermal phonons having all possible values of $\mathbf{k}_1$ is given by the integral

$$\int Ak_1k_2fN_1(N_2+1)\delta(\omega_1+\omega-\omega_2)\, d^3k_1/(2\pi)^3. \tag{73.3}$$

The probability of the inverse process of emission of a phonon f by all possible phonons $\mathbf{k}_2$ is

$$\int Ak_1k_2fN_2(N_1+1)\delta(\omega_1+\omega-\omega_2)\, d^3k_1/(2\pi)^3. \tag{73.4}$$

The function w in (66.9) and (66.11) is written, in accordance with (66.14), in the form Ak_1k_2f, all three phonons being long-wavelength ones (A is a function of the directions of all the phonons).

The phonon absorption (the relative rate of decrease of the number of phonons) is determined by the difference of these two probabilities. Since the frequency ω is small in comparison with ω_1 and ω_2, we have

$$N_1(N_2+1)-(N_1+1)N_2 = N_1 - N_2 = -(\partial N_1/\partial \omega_1)\omega.$$

The absorption coefficient is therefore

$$\gamma \propto \omega f \int Ak_1k_2|\partial N_1/\partial \omega_1|\delta(\omega_1+\omega-\omega_2)\, d^3k_1. \tag{73.5}$$

We are concerned with the dependence of this quantity on the sound frequency ω and the crystal temperature T. It is governed entirely by the fact that all the frequencies in (73.5) are first-order homogeneous functions of the wave vectors. To simplify the discussion, it is sufficient to take $\omega = Uf$, $\omega_1 = uk_1$, $\omega_2 = uk_2$, where U and u are speeds independent of direction.

Since f is small, we can put $k_1 \approx k_2$. For the same reason,

$$\omega_2 - \omega_1 \approx (\partial \omega_1/\partial \mathbf{k}_1) \cdot \mathbf{f} = uf \cos\theta = \omega(u/U)\cos\theta,$$

where θ is the angle between $\mathbf{f}$ and $\mathbf{k}$. Then

$$\delta(\omega_1+\omega-\omega_2) = \frac{1}{\omega}\delta\left(1-\frac{u}{U}\cos\theta\right),$$

and the integral (73.5) becomes

$$\gamma \propto \omega \int Ak_1^2\left|\frac{\partial N_1}{\partial \omega}\right| \delta\left(1-\frac{u}{U}\cos\theta\right)k_1^2\, dk_1\, d\cos\theta, \tag{73.6}$$

or, after removing the delta function,

$$\gamma \propto \omega \int k_1^4|\partial N_1/\partial k_1|dk_1.$$

Since N_1 is a function only of the ratio $\omega_1/T = uk_1/T$ (because of the rapid convergence, the integration with respect to k_1 can be extended to infinity), the remaining integral is proportional to T^4. Thus

$$\gamma \propto \omega T^4, \tag{73.7}$$

Here, the sound absorption coefficient varies linearly with the frequency.

With the condition (73.2) assumed above, the sound attenuation mechanism in question is exactly analogous to Landau damping in a plasma. The "resonance electrons" are here represented by phonons moving in phase with the sound wave. There is therefore naturally a resenblance between (73.6) and the Landau damping formula (30.1).

CHAPTER VIII

QUANTUM LIQUIDS

§74. Transport equation for quasi-particles in a Fermi liquid

THE transport equation for quasi-particles in a normal Fermi liquid has already been discussed (*SP* 2, §§4 and 5) in connection with the propagation of oscillations in it; the collision integral in the equation was there unimportant. We shall now continue the discussion of the transport equation with a view to its application to dissipative processes relating specifically to collisions.

The quasi-particles in a Fermi liquid have a spin of $\frac{1}{2}$. Accordingly, their distribution function is in general a matrix with regard to the spin variables. However, there is a wide range of problems in which it is sufficient to consider a distribution independent of the spin variables, reducing to a scalar function $n(\mathbf{r}, \mathbf{p})$ normalized so that $n\, d^3p/(2\pi\hbar)^3$ is the number of quasi-particles per unit volume with momenta in the range d^3p and a given spin component. This will be assumed in §§74–76.

The characteristic property of the spectrum of a Fermi liquid is that the energy ϵ of the quasi-particles is a functional of the distribution function. When the latter changes by a small amount:

$$n(\mathbf{r}, \mathbf{p}) = n_0(\mathbf{p}) + \delta n(\mathbf{r}, \mathbf{p}), \tag{74.1}$$

where n_0 is the equilibrium distribution, the energy changes by

$$\delta\epsilon(\mathbf{r}, \mathbf{p}) = \int f(\mathbf{p}, \mathbf{p}')\delta n(\mathbf{r}, \mathbf{p}')\, d^3p'/(2\pi\hbar)^3, \tag{74.2}$$

where $f(\mathbf{p}, \mathbf{p}')$ is the quasi-particle interaction function. Thus the distribution (74.1) corresponds to the quasi-particle energy

$$\epsilon(\mathbf{r}, \mathbf{p}) = \epsilon_0(\mathbf{p}) + \delta\epsilon(\mathbf{r}, \mathbf{p}), \tag{74.3}$$

where $\epsilon_0(\mathbf{p})$ is the energy corresponding to the equilibrium distribution.

The transport equation is

$$\frac{\partial n}{\partial t} + \frac{\partial \epsilon}{\partial \mathbf{p}} \cdot \frac{\partial n}{\partial \mathbf{r}} - \frac{\partial \epsilon}{\partial \mathbf{r}} \cdot \frac{\partial n}{\partial \mathbf{p}} = C(n). \tag{74.4}$$

Its characteristic feature is that in an inhomogeneous liquid the left-hand side of the equation contains a term involving the derivative $\partial\epsilon/\partial\mathbf{r}$ even in the absence of an external field, because of the coordinate dependence of ϵ (74.3).

The collision integral on the right of (74.4) has the form

$$C(n) = \int w(\mathbf{p}, \mathbf{p}_1; \mathbf{p}', \mathbf{p}_1')[n'n_1'(1-n)(1-n_1) - nn_1(1-n')(1-n_1')] \times$$
$$\times\, \delta(\epsilon + \epsilon_1 - \epsilon' - \epsilon_1')\, d^3p_1\, d^3p'/(2\pi\hbar)^6, \tag{74.5}$$

where n, n_1, n', n_1' are functions of the momenta $\mathbf{p}, \mathbf{p}_1, \mathbf{p}', \mathbf{p}_1'$ of the colliding quasi-particles. The law of conservation of momentum in collisions is assumed already taken into account, so that $\mathbf{p} + \mathbf{p}_1 = \mathbf{p}' + \mathbf{p}_1'$; the integration in (74.5) is therefore taken over only two momenta, not three. The conservation of energy is ensured by the explicitly written delta function. Lastly, w is a function of the momenta which gives the collision probability. The two terms in the square brackets give respectively the numbers of quasi-particles entering and leaving a particular quantum state as a result of collisions. They differ from the corresponding terms in the Boltzmann gas collision integral by the factors $(1-n)$, etc. The presence of these factors is due to the Fermi statistics, whereby collisions can take quasi-particles only into unoccupied states.

The Born approximation is in general not applicable to collisions of quasi-particles in a Fermi liquid. Nevertheless, the probabilities of direct and reverse scattering processes may be assumed to be the same. We are considering quantities already averaged over the directions of the quasi-particle spins. Under these conditions, the scattering probability depends only on the initial and final momenta of the colliding quasi-particles. This enables us to apply the same arguments as were used in §2 when deriving the principle of detailed balancing in the form (2.8). Here it is important that in a Fermi liquid there is again invariance under spatial inversion. We thus arrive at the equation

$$w(\mathbf{p}', \mathbf{p}_1'; \mathbf{p}, \mathbf{p}_1) = w(\mathbf{p}, \mathbf{p}_1; \mathbf{p}', \mathbf{p}_1'),$$

already used in the collision integral (74.5). The function w depends in general on the state occupation numbers, and therefore on the temperature. However, since the temperature is low (an important point in the whole theory of Fermi liquids), w in the collision integral is to be taken as the function calculated for $T = 0$.

The integral (74.5) vanishes identically, as it should, when we substitute for n the Fermi equilibrium distribution function

$$n_0(\epsilon) = \left[\exp\left(\frac{\epsilon-\mu}{T}\right) + 1\right]^{-1}. \tag{74.6}$$

For, since

$$\frac{n_0}{1-n_0} = \exp\left(-\frac{\epsilon-\mu}{T}\right),$$

we see at once that the law of conservation of energy leads to the equation

$$\frac{n_0 n_{01}}{(1-n_0)(1-n_{01})} = \frac{n_0' n_{01}'}{(1-n_0')(1-n_{01}')}. \tag{74.7}$$

Let us use the transport equation to express the mass, energy and momentum conservation laws in a Fermi liquid in terms of the distribution function. The dependence of the energy of the quasi-particles on their distribution makes this a fairly specific problem.

We integrate both sides of (74.4) over $2d^3p/(2\pi\hbar)^3$; the factor 2 takes account of the two possible directions of the spin. Because of the conservation of number of quasi-particles in collisions, the integral of $C(n)$ is zero. On the left-hand side, the term $-(\partial n/\partial \mathbf{p}) \,.\, (\partial \epsilon/\partial \mathbf{r})$ is integrated by parts, and the equation then becomes

$$\partial N/\partial t + \operatorname{div} \mathbf{i} = 0,$$

where N is the number density of quasi-particles,

$$\mathbf{i} = \langle \mathbf{v} \rangle, \tag{74.8}$$

and $\mathbf{v} = \partial \epsilon/\partial \mathbf{p}$ is the quasi-particle velocity.† This is the continuity equation for quasi-particles, and $\mathbf{i}$ is therefore the quasi-particle flux. Since the number of quasi-particles in a Fermi liquid is the same as the number of actual particles, $\mathbf{i}$ is also the flux of actual particles, so that $\mathbf{i} = \langle \mathbf{p}/m \rangle$.

Let us now apply the same operations to (74.4) after first multiplying both sides by $\mathbf{p}$. The integral of $\mathbf{p}C(n)$ is zero, because of the conservation of the total momentum of quasi-particles in collisions. The left-hand side in vector components is

$$\frac{\partial \langle p_\alpha \rangle}{\partial t} + \int p_\alpha \left(\frac{\partial n}{\partial x_\beta} \frac{\partial \epsilon}{\partial p_\beta} - \frac{\partial n}{\partial p_\beta} \frac{\partial \epsilon}{\partial x_\beta} \right) \frac{2d^3p}{(2\pi\hbar)^3}.$$

The integrand in the second term can be rewritten as

$$\frac{\partial}{\partial x_\beta} \left(p_\alpha \frac{\partial \epsilon}{\partial p_\beta} n \right) + n \frac{\partial \epsilon}{\partial x_\alpha} - \frac{\partial}{\partial p_\beta} \left(p_\alpha \frac{\partial \epsilon}{\partial x_\beta} n \right).$$

On integration, the third term gives zero, and the second term gives the derivative $\partial E/\partial x_\alpha$ of the energy density E of the liquid, the quasi-particle energy in a Fermi liquid being determined by the variation of the internal energy,

$$\delta E = \int \epsilon \delta n \,.\, 2d^3p/(2\pi\hbar)^3. \tag{74.9}$$

Thus we have the equation of conservation of momentum in the form

$$\frac{\partial}{\partial t} \langle p_\alpha \rangle + \frac{\partial \Pi_{\alpha\beta}}{\partial x_\beta} = 0,$$

†Here and in the rest of § 74, $\langle \ldots \rangle$ denotes integration over the distribution n:

$$\langle \ldots \rangle = \int \ldots n \,.\, 2d^3p/(2\pi\hbar)^3.$$

where the momentum flux tensor is

$$\Pi_{\alpha\beta} = \langle p_\alpha v_\beta \rangle + \delta_{\alpha\beta}(\langle \epsilon \rangle - E). \tag{74.10}$$

Finally, multiplying both sides of (74.4) by ϵ and integrating, we similarly obtain the equation of conservation of energy:

$$\partial E/\partial t + \operatorname{div} \mathbf{q} = 0,$$

where the energy flux is

$$\mathbf{q} = \langle \epsilon \mathbf{v} \rangle. \tag{74.11}$$

In equilibrium, all the fluxes **i**, **q** and $\Pi_{\alpha\beta}$ are zero. We can derive expressions for them that are linear in the small correction δn in the perturbed distribution (74.1).

The equilibrium function n_0 depends only on the energy of the quasi-particle, which in turn corresponds to the equilibrium distribution. Denoting this fact by the suffix zero to ϵ, we write the definition (74.1) in the more precise form

$$n(\mathbf{r}, \mathbf{p}) = n_0(\epsilon_0) + \delta n(\mathbf{r}, \mathbf{p}). \tag{74.12}$$

If n_0 is expressed as a function of the actual quasi-particle energy ϵ, we must put

$$n_0(\epsilon_0) = n_0(\epsilon) - \delta\epsilon \,.\, \partial n_0/\partial\epsilon$$

and the perturbed distribution function then becomes

$$n(\mathbf{r}, \mathbf{p}) = n_0(\epsilon) + \delta\tilde{n}(\mathbf{r}, \mathbf{p}), \tag{74.13}$$

$$\begin{aligned} \delta\tilde{n} &= \delta n - \delta\epsilon \,.\, \partial n_0/\partial\epsilon \\ &= \delta n - \frac{\partial n_0}{\partial \epsilon}\int f(\mathbf{p}, \mathbf{p}')\delta n(\mathbf{r}, \mathbf{p}')\, d^3p'/(2\pi\hbar)^3. \end{aligned}$$

Since, in the integrals (74.8)–(74.11), ϵ and $\mathbf{v} = \partial\epsilon/\partial\mathbf{p}$ are the actual energy and velocity of the quasi-particle, it is sufficient to substitute n in them in the form (74.13), which gives immediately

$$\left.\begin{aligned} \mathbf{i} &= \int \mathbf{v}\delta\tilde{n} \,.\, 2d^3p/(2\pi\hbar)^3, \\ \mathbf{q} &= \int \epsilon\mathbf{v}\delta\tilde{n} \,.\, 2d^3p/(2\pi\hbar)^3, \\ \Pi_{\alpha\beta} &= \int p_\alpha v_\beta \delta\tilde{n} \,.\, 2d^3p/(2\pi\hbar)^3; \end{aligned}\right\} \tag{74.14}$$

in the last expression, we have also used (74.9). Now, as the first-order terms in $\delta\tilde{n}$ have been separated, we can of course treat ϵ as $\epsilon_0(\mathbf{p})$ in the integrals (74.14).

As in previous cases, we express δn as

$$\delta n = -\psi \partial n_0 / \partial \epsilon. \tag{74.15}$$

In this case, the separation of the factor $\partial n_0/\partial \epsilon$ has a special significance. The perturbation δn is concentrated in the blurred region of the Fermi distribution. The derivative $\partial n_0/\partial \epsilon$ too is appreciably different from zero in just that region; when this factor has been separated, the remaining function ψ is a slowly varying one. Together with (74.15), we shall write

$$\begin{aligned}\delta \tilde{n} &= -\varphi\, \partial n_0/\partial \epsilon \\ &= n_0(1-n_0)\varphi/T,\end{aligned} \tag{74.16}$$

where

$$\varphi = \psi - \int f(\mathbf{p}, \mathbf{p}') \frac{\partial n_0(\epsilon')}{\partial \epsilon'} \psi(\mathbf{r}, \mathbf{p}') \frac{d^3p'}{(2\pi\hbar)^3}. \tag{74.17}$$

In the zero-order approximation with respect to the small ratio T/ϵ_F, the function $n_0(\epsilon)$ may be replaced by a step function cut off at the limiting energy ϵ_F. Then

$$\partial n_0/\partial \epsilon = -\delta(\epsilon - \epsilon_F), \tag{74.18}$$

and the integration over d^3p reduces to an integration over the Fermi surface $\epsilon = \epsilon_F$. The volume element between two infinitely close constant-energy surfaces in momentum space is

$$dS\, d\epsilon / |\partial \epsilon / \partial p|, \tag{74.19}$$

where dS is an area element on the constant-energy surface. The integration over d^3p thus becomes one over the Fermi surface according to the formula

$$\int \ldots \delta(\epsilon - \epsilon_F)\, d^3p = \int \ldots dS_F / v_F, \tag{74.20}$$

where v_F is the velocity on the Fermi surface. This formula does not assume that the Fermi surface is spherical; on a sphere, $dS_F = p_F^2\, do$ with constant p_F.

After this transformation, the definition (74.17) becomes

$$\varphi(\mathbf{r}, \mathbf{p}) = \psi(\mathbf{r}, \mathbf{p}) + \int f(\mathbf{p}, \mathbf{p}_F')\psi(\mathbf{r}, \mathbf{p}_F')\, dS_F'/v_F'(2\pi\hbar)^3, \tag{74.21}$$

where $\mathbf{p}_F$ denotes the momentum (with variable direction) on the Fermi surface. The particle flux is

$$\mathbf{i} = \int (\mathbf{v}_F/v_F)\varphi \,.\, 2dS_F/(2\pi\hbar)^3 \tag{74.22}$$

and the momentum flux is given by a similar expression. In the energy flux, the approximation (74.18) is certainly inadequate: it would reduce $\mathbf{q}$ simply to the convective energy transfer $\epsilon_F \mathbf{i}$, the first term in the expression

$$\mathbf{q} = \epsilon_F \mathbf{i} - \int \mathbf{v}(\epsilon - \epsilon_F) \frac{\partial n_0}{\partial \epsilon} \varphi \frac{2d^3p}{(2\pi\hbar)^3}. \tag{74.23}$$

To linearize the collision integral, it must be noted that the equilibrium distribution $n_0(\epsilon)$ as a function of the actual energy ϵ makes the collision integral zero.† The linearization is thus carried out by substituting n in the form (74.13) and (74.16). The calculations are similar to those in going from (67.6) to (67.17). The expression in the square brackets in (74.5) is written in the form

$$(1-n)(1-n_1)(1-n')(1-n_1')\left[\frac{n'}{1-n'}\frac{n_1'}{1-n_1'} - \frac{n}{1-n}\frac{n_1}{1-n_1}\right],$$

and we use the fact that

$$\delta \frac{n}{1-n} = \frac{n_0}{1-n_0}\frac{\varphi}{T}.$$

The result is

$$\begin{aligned} C(n) &\equiv I(\varphi) \\ &= \frac{1}{T}\int w n_0 n_{01}(1-n_0')(1-n_{01}')(\varphi' + \varphi_1' - \varphi - \varphi_1) \times \\ &\quad \times \delta(\epsilon' + \epsilon_1' - \epsilon - \epsilon_1)\, d^3p_1\, d^3p'/(2\pi\hbar)^6. \end{aligned} \tag{74.24}$$

Note that the required perturbation of the distribution function (to be found by solving the transport equation) occurs in the collision integral as the same $\delta\tilde{n}$ that appears in the fluxes (74.14). If the terms in δn can be omitted on the left of the transport equation (as in calculating the thermal conductivity and the viscosity; see §75), then the quasi-particle distribution function $f(\mathbf{p}, \mathbf{p}')$ does not occur explicitly in the resulting equations: the equations with f for the unknown $\delta\tilde{n}$ are the same as those with $f \equiv 0$ for the unknown δn. In such problems, therefore, the Fermi-liquid effects do not appear, and the situation is formally identical with that for a Fermi gas.

We shall show that a similar case occurs in a particular class of problems where the first-order terms in δn have to be retained on the left-hand side of the transport equation. If the function n_0 is independent of the coordinates, these terms are

$$\begin{aligned} &\frac{\partial \delta n}{\partial t} + \frac{\partial \delta n}{\partial \mathbf{r}} \cdot \frac{\partial \epsilon_0}{\partial \mathbf{p}} - \frac{\partial n_0}{\partial \mathbf{p}} \cdot \frac{\partial \delta\epsilon}{\partial \mathbf{r}} \\ &= \frac{\partial \delta n}{\partial t} + \mathbf{v} \cdot \frac{\partial \delta n}{\partial \mathbf{r}} - \mathbf{v}\frac{\partial n_0}{\partial \epsilon} \cdot \frac{\partial}{\partial \mathbf{r}} \int f(\mathbf{p}, \mathbf{p}')\delta n(\mathbf{r}, \mathbf{p}') \frac{d^3p'}{(2\pi\hbar)^3}. \end{aligned}$$

†This is a general remark, which applies to any collision integral involving Fermi quasi-particles, not only to (74.5).

With $\delta\tilde{n}$ from (74.13), they become

$$\frac{\partial \delta n}{\partial t} - \mathbf{v} \cdot \frac{\partial \delta \tilde{n}}{\partial \mathbf{r}}. \tag{74.25}$$

If the time derivative may be neglected, again only $\delta\tilde{n}$ occurs here.

These statements remain valid not only for an electrically neutral Fermi liquid, discussed here, but also for the electron liquid in metals, which will be considered in Chapter IX. For this reason and in order not to have to return to the topic, some additional comments will be made here.

If the quasi-particles carry an electric charge $-e$, then in the presence of an electromagnetic field the derivative $\dot{\mathbf{p}} = -\partial\epsilon/\partial\mathbf{r}$ contains a further term, the Lorentz force on the charge. Accordingly, the left-hand side of the transport equation contains a term

$$-e\left(\mathbf{E} + \frac{1}{c}\frac{\partial \epsilon}{\partial \mathbf{p}} \times \mathbf{B}\right) \cdot \frac{\partial n}{\partial \mathbf{p}}.$$

The electric field is generally assumed to be weak, and in the term $-e\mathbf{E} \,.\, \partial n/\partial \mathbf{p}$ it is sufficient to put $n = n_0$. The magnetic field term is identically zero for a function $n_0(\epsilon)$ that depends only on ϵ. If the field is strong, however, it may be necessary to retain also the first-order terms in δn. These are

$$-\frac{e}{c}\mathbf{v} \times \mathbf{B} \cdot \frac{\partial \delta n}{\partial \mathbf{p}} - \frac{e}{c}\frac{\partial \delta \epsilon}{\partial \mathbf{p}} \times \mathbf{B} \cdot \frac{\partial n_0}{\partial \mathbf{p}} = -\frac{e}{c}\mathbf{v} \times \mathbf{B} \cdot \left\{\frac{\partial \delta n}{\partial \mathbf{p}} - \frac{\partial n_0}{\partial \epsilon}\frac{\partial \delta \epsilon}{\partial \mathbf{p}}\right\},$$

where $\mathbf{v} = \partial\epsilon_0/\partial\mathbf{p}$. The factor $\partial n_0/\partial\epsilon$ which depends only on ϵ can be taken under $\partial/\partial\mathbf{p}$ in the braces; its derivative is parallel to $\mathbf{v}$ and gives zero on multiplication by $\mathbf{v} \times \mathbf{B}$. These terms are thus brought to the form

$$-\frac{e}{c}\mathbf{v} \times \mathbf{B} \cdot \frac{\partial \delta \tilde{n}}{\partial \mathbf{p}}. \tag{74.26}$$

which again contains $\delta\tilde{n}$ only.

§75. Thermal conductivity and viscosity of a Fermi liquid

The temperature dependences of the viscosity and thermal conductivity of a Fermi liquid can be established by simple qualitative arguments (I. Ya. Pomeranchuk 1950).

According to the elementary formula (8.11) in the kinetic theory of gases, the viscosity is $\eta \sim mN\bar{v}l$, where m is the particle mass, N the particle number density, $\bar{v}$ the mean thermal speed and l the mean free path. In the present case, the particles of the kinetic theory are quasi-particles, but since the numbers of each are the same the product mN is a quantity independent of the temperature, namely the density of the liquid.† The speed $\bar{v} \sim v_F$, where v_F is the temperature-independent speed on the Fermi surface. The mean free path $l \sim v_F\tau$, where τ is the time between quasi-particle collisions. This time varies with temperature as T^{-2} (see *SP* 2, § 1), so that for the viscosity also

$$\eta \propto T^{-2}. \tag{75.1}$$

The thermal conductivity is estimated from (7.10): $\kappa \sim cN\bar{v}l$, where c is the specific heat per particle. For a Fermi liquid $c \propto T$, and so

$$\kappa \propto T^{-1}. \tag{75.2}$$

For an exact determination of η and κ, we have to use the transport equation. The sequence of calculation for the conductivity is as follows.

The left-hand side of the transport equation (74.4) is transformed in a similar way to the procedure in § 7 for the thermal conductivity of a classical gas.

Let there be a temperature gradient in the liquid, which is macroscopically at rest. The latter condition implies that the pressure is constant throughout the liquid, and the temperature distribution is steady. On the left of (74.4), we substitute for n and ϵ their local equilibrium values with a temperature varying through the liquid. Then $\partial\epsilon/\partial\mathbf{r} = 0$, and only the term $\mathbf{v} \cdot \partial n_0/\partial\mathbf{r}$ remains (we omit the suffix 0 to ϵ and $\mathbf{v}$). The function n involves only the combination $(\epsilon - \mu)/T$, and since we shall seek only the limiting forms as $T \to 0$, the chemical potential $\mu(T)$ may be taken to have its value at $T = 0$ (which is the same as the limiting energy ϵ_F). Then

$$\begin{aligned}\mathbf{v} \cdot \partial n_0/\partial\mathbf{r} &= (\partial n_0/\partial T)\mathbf{v} \cdot \nabla T \\ &= \frac{n_0(1-n_0)}{T} \cdot \frac{\epsilon-\mu}{T}\mathbf{v} \cdot \nabla T,\end{aligned}$$

and the transport equation becomes

$$n_0(1-n_0)\frac{\epsilon-\mu}{T^2}\mathbf{v} \cdot \nabla T = I(\varphi), \tag{75.3}$$

with $I(\varphi)$ from (74.24). The solution of this equation must be subjected to a further condition which expresses the absence of any macroscopic mass transfer:

$$\int \mathbf{v}\varphi \frac{\partial n_0}{\partial\epsilon} \frac{d^3p}{(2\pi\hbar)^3} = 0. \tag{75.4}$$

Because of this condition, only the second term remains in the energy flux (74.23).

As already noted in § 74, equations (75.3) and (75.4) do not explicitly contain the quasi-particle interaction function, so that the problem of thermal conduction in a Fermi liquid (and the same applies to the viscosity problem) is formally identical with that for a Fermi gas.

†Since we shall seek the limiting form of the function $\eta(T)$ at low temperatures, this limit is of course meant for all quantities which tend to a finite value as $T \to 0$.

In all the integrals, the most important region is that where $\epsilon - \mu \sim T$ and the Fermi distribution is blurred; the quasi-particle momenta are close to the radius p_F of the Fermi sphere, and in this range $\epsilon - \mu = v_F(p - p_F)$. Wherever the momenta occur other than as the difference $p - p_F$, we can put $p = p_F$, and the speed can everywhere be equated to v_F. In particular, this can be done in w, which then becomes a function only of the angles which describe the relative orientation of the vectors $\mathbf{p}$, $\mathbf{p}_1$, $\mathbf{p}'$, $\mathbf{p}_1'$. For given $\mathbf{p}$ and $\mathbf{p}_1$, the law of conservation of momentum fixes the angle between the vectors $\mathbf{p}'$ and $\mathbf{p}_1' = \mathbf{p} + \mathbf{p}_1 - \mathbf{p}'$; the integration with respect to this angle removes the delta function from the collision integral. There remain integrations over the magnitudes p_1 and p' (and over the other angle variables). The integration over these magnitudes is replaced by one over $T^2\, du_1\, du'$, where $u = (\epsilon - \mu)/T = v_F(p - p_F)/T$ are variables on which the distribution functions n_0 depend; in view of the rapid convergence, these integrations can be taken from $-\infty$ to ∞. We then find that the whole integral $I(\varphi)$ is proportional to T, and the solution of (75.3) is

$$\varphi = -T^{-2} g(u) \mathbf{v} \cdot \nabla T.$$

When this is substituted in (74.23), integration over the directions of $\mathbf{v}$ puts the heat flux in the form $\mathbf{q} = -\kappa \nabla T$, with

$$\kappa = \frac{8\pi v_F p_F^2}{3T} \int_{-\infty}^{\infty} u g(u) \left| \frac{\partial n_0}{\partial u} \right| du.$$

Hence we see again that $\kappa \propto T^{-1}$.

The above simplifications of the collision integral are sufficient to solve the transport equation exactly (and the same is true of the viscosity problem). The formulae obtained for κ and η express them in terms of the parameters p_F and v_F and the function w suitably averaged over directions.†

§76. **Sound absorption in a Fermi liquid‡**

It has been shown in *SP* 2 (§4) that the nature of waves propagating in a Fermi liquid depends essentially on the value of the product $\omega\tau$, where τ is the mean free time.

When $\omega\tau \ll 1$, we have ordinary hydrodynamic sound waves. The frequency and temperature dependences of the coefficient γ for the absorption of these waves per unit distance can be found from the familiar formula $\gamma \sim \omega^2\eta/\rho u^3$, where η is the viscosity, ρ the density of the liquid and u the speed of sound; see *FM*, §77. Since in a Fermi liquid $\eta \propto T^{-2}$, we have

$$\gamma \propto \omega^2/T^2. \tag{76.1}$$

†See G. A. Brooker and J. Sykes, *Physical Review Letters* **21**, 279. 1968.
‡The results in this section are due to L. D. Landau (1957).

This result can be more formally derived by noting that the absorption is described by the first correction term (with respect to the small parameter) in the sound dispersion relation:

$$k = (\omega/u)(1 + i\alpha\omega\tau), \tag{76.2}$$

where α is a constant. The imaginary part of this expression (for a real frequency) gives γ; since $\tau \propto T^{-2}$, we return to (76.1).

When $\omega\tau \sim 1$ the absorption becomes very strong, and the propagation of sound waves cannot occur.

When $\omega\tau \gg 1$ the propagation of weakly damped waves (*zero sound*) again becomes possible. The absorption is described by a correction term in the dispersion relation, in this case involving the small parameter $1/i\omega\tau$:

$$k = \frac{\omega}{u_0}\left(1 + \frac{i\alpha}{\omega\tau}\right), \tag{76.3}$$

where u_0 is the speed of propagation of zero sound. The absorption coefficient is accordingly proportional to the collision frequency: $\gamma \propto 1/\tau$, and the latter is in turn proportional to the squared width of the blurred region of the quasi-particle distribution. When $\hbar\omega \ll T$, this width is governed by the temperature, so that $1/\tau \propto T^2$, and the absorption coefficient is

$$\gamma = aT^2, \quad T \gg \hbar\omega \gg \hbar/\tau. \tag{76.4}$$

If, however, $\hbar\omega \gg T$ (but still $\hbar\omega \ll \epsilon_F$ as the necessary condition for the whole theory to be applicable), the distribution is blurred in a region of width $\sim\hbar\omega$. The absorption of zero sound is then

$$\gamma = b\omega^2, \quad \hbar\omega \gg T. \tag{76.5}$$

This case includes, in particular, zero sound of all frequencies at $T = 0$. It will be shown below that there is a relation between the constants a and b in (76.4) and (76.5).

The difference in the nature of the absorption of ordinary and zero sound is due to a difference in their physical nature. In an ordinary sound wave, in any volume element small compared with the wavelength, the quasi-particle distribution corresponds, in the first approximation, to equilibrium for a given local temperature and velocity of the liquid. In this approximation, there is no dissipation, and sound absorption occurs only when we take into account the effect of the temperature and velocity gradients on the quasi-particle distribution. In a zero sound wave, however, the vibrations themselves cause the distribution function to depart from equilibrium in every volume element, and the collisions of quasi-particles cause absorption of sound.

According to the basic ideas of normal Fermi liquid theory, a quasi-particle in such a liquid may be regarded in one sense as a particle in the self-consistent field of the surrounding particles. In a zero sound wave, this field is periodic in time and space. According to the general rules of quantum mechanics, a collision of two quasi-particles in such a field is accompanied by a change in their total energy and momentum by $\hbar\omega$ and $\hbar\mathbf{k}$ respectively; we may say that in the collision a "zero sound quantum" is emitted or absorbed.† The overall effect of such collisions is to reduce the total number of sound quanta; the sound absorption coefficient is proportional to the rate of this decrease.

With this approach, the absorption coefficient of zero sound is

$$\gamma = \int W\{n_1 n_2(1-n_1')(1-n_2') - n_1' n_2'(1-n_1)(1-n_2)\} \times \\ \times \delta(\epsilon_1' + \epsilon_2' - \epsilon_1 - \epsilon_2 - \hbar\omega)\delta(\mathbf{p}_1' + \mathbf{p}_2' - \mathbf{p}_1 - \mathbf{p}_2 - \hbar\mathbf{k}) \times \\ \times d^3p_1\, d^3p_2\, d^3p_1'\, d^3p_2'/(2\pi\hbar)^{12}. \tag{76.6}$$

In the integrand, the delta functions which provide for the conservation of energy and momentum in collisions are shown explicitly. The first term in the braces corresponds to collisions $\mathbf{p}_1, \mathbf{p}_2 \to \mathbf{p}_1', \mathbf{p}_2'$ with absorption of a quantum, the second to $\mathbf{p}_1', \mathbf{p}_2' \to \mathbf{p}_1, \mathbf{p}_2$ with emission of a quantum. The function W, which is related to the probability of "radiative" collisions, is determined by the properties of the zero sound wave; this wave itself may be regarded as propagating at $T = 0$ (see *SP* 2, §4), and W is then independent of the temperature.‡

It is, however, not necessary to know the function W if we seek only to express the absorption coefficient in terms of its value in the limiting case $\hbar\omega \ll T$. To do so, we note that in the integral (76.6) the only important values of the quasi-particle energies are those in the region of blurring of the Fermi distribution. In this region, the only factors in the integrand which vary rapidly are those containing the functions $n(\epsilon)$. Moreover, the angle integrals in (76.6) are almost unchanged when we go from $\hbar\omega \ll T$ to $\hbar\omega \gg T$. It is therefore sufficient to calculate the integral

$$J = \int \{n_1 n_2(1-n_1')(1-n_2') - n_1' n_2'(1-n_1)(1-n_2)\} \times \\ \times \delta(\epsilon_1' + \epsilon_2' - \epsilon_1 - \epsilon_2 - \hbar\omega)\, d\epsilon_1\, d\epsilon_2\, d\epsilon_1'\, d_2', \tag{76.7}$$

taken only with respect to the energies. The proportionality factor between γ and J depends only on ω, not on T, so that it can be found from the limiting value of γ when $\hbar\omega/T \ll 1$.

In the integral (76.7) we can, of course, neglect the slight distortion of the distribution function in the wave, putting

$$n(\epsilon) = [e^{(\epsilon-\mu)/T} + 1]^{-1}.$$

With the notation

$$x = (\epsilon - \mu)/T, \quad \xi = \hbar\omega/T,$$

†Such emission or absorption by one quasi-particle is impossible, since the speed of zero sound exceeds the Fermi speed v_F.

‡To avoid misunderstanding, it should be emphasized that W is not the same as w in the collision integral (74.5).

we have

$$J = T^3 \int_{-\infty}^{\infty} \frac{(1-e^{-\xi})\delta(x_1'+x_2'-x_1-x_2-\xi)\,dx_1\,dx_2\,dx_1'\,dx_2'}{(e^{x_1}+1)(e^{x_2}+1)(1+e^{-x_1'})(1+e^{-x_2'})}.$$

Because of the rapid convergence of the integral, the range of integration can be extended from $-\infty$ to ∞.

To carry out the integration, we change to variables y_1, y_2, u_1, u_2, where $y = x - x'$, $u = e^x$. The integration with respect to u_1 and u_2 is elementary, and gives

$$\begin{aligned} T^{-3}J &= (1-e^{-\xi})\int_{-\infty}^{\infty}\int_{-\infty}^{\infty}\int_0^{\infty}\int_0^{\infty} \frac{\delta(y_1+y_2+\xi)\,du_1\,du_2\,dy_1\,dy_2}{(u_1+1)(u_2+1)(u_1+e^{y_1})(u_2+e^{y_2})} \\ &= (1-e^{-\xi})\int_{-\infty}^{\infty}\int_{-\infty}^{\infty} \frac{y_1 y_2 \delta(y_1+y_2+\xi)}{(1-e^{y_1})(1-e^{y_2})}\,dy_1\,dy_2 \\ &= \int_{-\infty}^{\infty} y(\xi+y)\left\{\frac{1}{e^y-1} - \frac{1}{e^{y+\xi}-1}\right\} dy. \end{aligned}$$

To calculate this difference of two divergent integrals, we first put in a finite lower limit $-\Lambda$, writing

$$\begin{aligned} T^{-3}J &= \int_{-\Lambda}^{\infty} \frac{y(\xi+y)}{e^y-1}\,dy - \int_{-\Lambda+\xi}^{\infty} \frac{y(y-\xi)}{e^y-1}\,dy \\ &= 2\xi\int_{-\Lambda}^{\infty}\frac{y}{e^y-1}\,dy - \int_{-\Lambda+\xi}^{-\Lambda}\frac{y(y-\xi)}{e^y-1}\,dy. \end{aligned}$$

Intending to take the limit $\Lambda\to\infty$, we neglect e^y in the denominator of the second integral. The first integral is transformed as follows:

$$\begin{aligned} \int_{-\Lambda}^{\infty}\frac{y}{e^y-1}\,dy &= \int_0^{\infty}\frac{y}{e^y-1}\,dy + \int_{-\Lambda}^{0}\frac{y}{e^y-1}\,dy \\ &= \tfrac{1}{6}\pi^2 + \int_{-\Lambda}^{0}\left(\frac{y}{1-e^{-y}} - y\right) dy \\ &= \tfrac{1}{6}\pi^2 + \int_0^{\Lambda}\frac{y}{e^y-1}\,dy + \tfrac{1}{2}\Lambda^2. \end{aligned}$$

Cancelling terms and then taking the limit $\Lambda\to\infty$, we have finally

$$J = \tfrac{2}{3}\pi^2\xi T^3(1+\xi^2/4\pi^2).$$

The proportionality factor between γ and J is found, as already mentioned, from the condition that when $\xi \ll 1$ we have $\gamma = aT^2$ from (76.4). This gives

$$\gamma = a[T^2 + (\hbar\omega/2\pi)^2]. \tag{76.8}$$

In particular, in the limit of high frequencies $\hbar\omega \gg T$, we hence obtain

$$\gamma = (a/4\pi^2)(\hbar\omega)^2, \tag{76.9}$$

which establishes the relation between the coefficients in (76.4) and (76.5).

§77. Transport equation for quasi-particles in a Bose liquid

If the mean free path of quasi-particles in a Bose superfluid is small in comparison with the characteristic dimensions of the problem, the motion of the liquid is described by the Landau equations of two-velocity hydrodynamics (see *FM*, Chapter XVI). The dissipative terms in these equations contain several transport coefficients (the thermal conductivity and four viscosity coefficients). Their calculation requires a detailed discussion of various scattering processes, the multiplicity of which is due to the existence of two types of quasi-particles (phonons and rotons). Actually, in liquid helium, the situation is further complicated by the instability of the initial part of the phonon spectrum. Such topics will not be discussed here.

The mean free paths of the quasi-particles increase as the temperature falls, if only because of the decrease in their number density. Hence, at sufficiently low temperatures, there can easily be a considerable disequilibrium of the quasi-particle system. Under these conditions, the equations of two-velocity hydrodynamics are not applicable. The concepts of the temperature and of the normal velocity $\mathbf{v}_n$ also cease to be meaningful (they can be defined only in terms of an equilibrium distribution of quasi-particles), and along with $\mathbf{v}_n$ so does the separation of the liquid density into superfluid and normal parts. The total density ρ and the superfluid velocity $\mathbf{v}_s$, however, retain their meaning, and in this respect are essentially mechanical variables. The whole set of equations describing a superfluid must then consist of the transport equation for the quasi-particle distribution function $n(t, \mathbf{r}, \mathbf{p})$, the continuity equation for the density ρ, and the equation for the velocity $\mathbf{v}_s$.

The transport equation has the usual form†

$$\frac{\partial n}{\partial t} + \frac{\partial n}{\partial \mathbf{r}} \cdot \frac{\partial \tilde{\epsilon}}{\partial \mathbf{p}} - \frac{\partial n}{\partial \mathbf{p}} \cdot \frac{\partial \tilde{\epsilon}}{\partial \mathbf{r}} = C(n), \tag{77.1}$$

where $\tilde{\epsilon}$ is the quasi-particle energy, depending on the superfluid velocity $\mathbf{v}_s$ as a parameter; the symbol ϵ is retained for the quasi-particle energy in a fluid at rest. The relation between ϵ and $\tilde{\epsilon}$ is established as follows.

By definition, $\epsilon(p)$ is the dispersion relation for quasi-particles in a frame of reference K_0 such that $\mathbf{v}_s = 0$. That is, in the presence of only one quasi-particle the energy of the liquid (relative to that at $T = 0$) is $\epsilon(p)$, and its momentum is equal to the momentum $\mathbf{p}$ of the quasi-particle. We make a Galilean transformation to a

†It is, of course, assumed that the quasi-classicality condition is satisfied: all quantities vary only slightly over distances of the order of the quasi-particle wavelength $\hbar/p$.

frame of reference K at rest, in which the superfluid velocity is $\mathbf{v}_s$. In this frame, the energy and momentum of a mass M of the liquid are

$$E = \epsilon(p) + \mathbf{p}\cdot\mathbf{v}_s + \tfrac{1}{2}Mv_s^2,\quad \mathbf{P} = \mathbf{p} + M\mathbf{v}_s. \tag{77.2}$$

From this we see that, in a liquid in superfluid motion, the energy of a quasi-particle is

$$\tilde{\epsilon}(\mathbf{p}) = \epsilon(p) + \mathbf{p}\cdot\mathbf{v}_s; \tag{77.3}$$

cf. the arguments in the derivation of the superfluidity condition (*SP* 2, §23).

The derivatives which occur in the transport equation are therefore†

$$\left.\begin{aligned} \frac{\partial\tilde{\epsilon}}{\partial\mathbf{p}} &= \frac{\partial\epsilon}{\partial\mathbf{p}} + \mathbf{v}_s, \\ \frac{\partial\tilde{\epsilon}}{\partial\mathbf{r}} &= \frac{\partial\epsilon}{\partial\mathbf{r}} + \frac{\partial}{\partial\mathbf{r}}(\mathbf{p}\cdot\mathbf{v}_s) = \frac{\partial\epsilon}{\partial\rho}\nabla\rho + (\mathbf{p}\cdot\nabla)\mathbf{v}_s. \end{aligned}\right\} \tag{77.4}$$

In the second equation, we have used the facts that the energy ϵ depends on the variable density ρ, and so may depend on the coordinates; and (in transforming the derivative of $\mathbf{p}\cdot\mathbf{v}_s$) that the superfluid flow is always a potential flow:

$$\operatorname{curl}\mathbf{v}_s = 0. \tag{77.5}$$

The continuity equation for the density is

$$\partial\rho/\partial t + \operatorname{div}\mathbf{i} = 0, \tag{77.6}$$

where $\mathbf{i}$ is by definition the momentum of the liquid per unit volume. An expression for $\mathbf{i}$ can be found directly from the second formula (77.2) by summation over all the quasi-particles in unit volume:

$$\mathbf{i} = \rho\mathbf{v}_s + \langle\mathbf{p}\rangle. \tag{77.7}$$

Here and in the rest of §77, the angle brackets denote integration over the momentum distribution:

$$\langle\ldots\rangle = \int \ldots n\, d^3p/(2\pi\hbar)^3.$$

It remains to derive an equation for the superfluid velocity. To do so, we start from the law of conservation of momentum, expressed by

$$\frac{\partial i_\alpha}{\partial t} + \frac{\partial\Pi_{\alpha\beta}}{\partial x_\beta} = 0, \tag{77.8}$$

†Formula (77.2) has been derived, strictly speaking, for a homogeneous superfluid flow ($\mathbf{v}_s$ = constant). In an inhomogeneous flow, the energy may contain terms in the spatial derivatives of $\mathbf{v}_s$. However, if $\mathbf{v}_s$ is assumed to vary slowly, these terms would lead to corrections of higher orders of smallness in the transport equation.

where $\mathbf{i}$ is given by (77.7) and $\Pi_{\alpha\beta}$ is the momentum flux tensor. Let $\Pi_{\alpha\beta}^{(0)}$ be the value of this tensor in the frame of reference K_0. Transformation to the frame K yields†

$$\begin{aligned}\Pi_{\alpha\beta} &= \Pi_{\alpha\beta}^{(0)} + \rho v_{s\alpha}v_{s\beta} + v_{s\alpha}i_{\beta}^{(0)} + v_{s\beta}i_{\alpha}^{(0)} \\ &= \Pi_{\alpha\beta}^{(0)} + \rho v_{s\alpha}v_{s\beta} + v_{s\alpha}\langle p_\beta\rangle + v_{s\beta}\langle p_\alpha\rangle; \qquad (77.9)\end{aligned}$$

$\mathbf{i}^{(0)} = \langle \mathbf{p}\rangle$ is the momentum per unit volume of the liquid in the frame K_0. This determines the dependence of the tensor $\Pi_{\alpha\beta}$ on the velocity $\mathbf{v}_s$.

For a further transformation of equation (77.8), we go back to the transport equation (77.1), multiply it by p_α, and integrate over $d^3p/(2\pi\hbar)^3$. Because the total momentum of the quasi-particles is conserved in collisions, the right-hand side of the equation becomes zero. The integral on the left-hand side is transformed exactly as in the derivation of (74.10), giving

$$\frac{\partial}{\partial t}\langle p_\alpha\rangle + \frac{\partial}{\partial x_\beta}\left\langle p_\alpha \frac{\partial \tilde{\epsilon}}{\partial p_\beta}\right\rangle + \left\langle \frac{\partial \tilde{\epsilon}}{\partial x_\alpha}\right\rangle = 0. \qquad (77.10)$$

We now substitute in (77.8) the expressions (77.7) and (77.9) for $\mathbf{i}$ and $\Pi_{\alpha\beta}$, and then eliminate $\partial\rho/\partial t$ and $\partial\langle \mathbf{p}\rangle/\partial t$ by means of (77.6) and (77.10). The result is

$$\frac{\partial v_{s\alpha}}{\partial t} + \frac{\partial}{\partial x_\alpha}\frac{v_s^2}{2} + \frac{1}{\rho}\frac{\partial \Pi_{\alpha\beta}^{(0)}}{\partial x_\beta} - \frac{1}{\rho}\left\langle\frac{\partial \epsilon}{\partial \rho}\right\rangle\frac{\partial \rho}{\partial x_\alpha} - \frac{1}{\rho}\frac{\partial}{\partial x_\beta}\left\langle p_\alpha\frac{\partial \epsilon}{\partial p_\beta}\right\rangle = 0.$$

From the condition curl $\mathbf{v}_s = 0$ (which has already been used in the second term) it follows that the sum of the last three terms must be the gradient of some function. Moreover, the tensor $\Pi_{\alpha\beta}^{(0)}$ in the absence of quasi-particles must be equal to $P_0\delta_{\alpha\beta}$, where $P_0(\rho)$ is the pressure of the liquid at $T = 0$. These arguments give as the only possible form of $\Pi_{\alpha\beta}^{(0)}$

$$\Pi_{\alpha\beta}^{(0)} = \langle p_\alpha\, \partial\epsilon/\partial p_\beta\rangle + \delta_{\alpha\beta}[P_0 + \rho\langle \partial\epsilon/\partial\rho\rangle]. \qquad (77.11)$$

The equation for $\mathbf{v}_s$ now becomes

$$\frac{\partial \mathbf{v}_s}{\partial t} + \nabla\left[\tfrac{1}{2}v_s^2 + \frac{\mu_0}{m} + \left\langle\frac{\partial \epsilon}{\partial \rho}\right\rangle\right] = 0, \qquad (77.12)$$

where μ_0 is the chemical potential of the liquid (at $T = 0$), related to the pressure P_0 by the thermodynamic formula $d\mu_0 = m\,dP_0/\rho$ (where m is the mass of a liquid particle and m/ρ the molecular volume).

Equations (77.1), (77.6) and (77.12) for a complete set for the description of a superfluid in the non-equilibrium state (I. M. Khalatnikov 1952).

For completeness, let us also consider the law of conservation of energy. This is expressed by

$$\partial E/\partial t + \operatorname{div}\mathbf{q} = 0, \qquad (77.13)$$

†The Galilean transformation formula for $\Pi_{\alpha\beta}$ is easily found by considering a classical system of particles, for which $\Pi_{\alpha\beta} = \Sigma\, p_\alpha v_\beta = \Sigma\, m v_\alpha v_\beta$, where the summation is over all the particles in unit volume.

where $\mathbf{q}$ is the energy flux in the liquid. According to (77.2),

$$E = E_0(\rho) + \langle\epsilon\rangle + \mathbf{v}_s \cdot \langle\mathbf{p}\rangle + \tfrac{1}{2}\rho v_s^2, \tag{77.14}$$

where $E_0(\rho)$ is the energy at $T = 0$, related to the chemical potential by $dE_0 = \mu_0 d\rho/m$. By differentiating (77.14) with respect to time and using the equations already available for the various quantities, the energy flux can be found. Passing over the calculations, we shall give the final result:

$$\mathbf{q} = (\langle\mathbf{p}\rangle + \rho\mathbf{v}_s)\left[\frac{\mu_0}{m} + \left\langle\frac{\partial\epsilon}{\partial\rho}\right\rangle + \tfrac{1}{2}v_s^2\right] + \left\langle(\epsilon + \mathbf{p}\cdot\mathbf{v}_s)\left(\frac{\partial\epsilon}{\partial\mathbf{p}} + \mathbf{v}_s\right)\right\rangle. \tag{77.15}$$

The equilibrium quasi-particle distribution function in a frame of reference where the "quasi-particle gas" is at rest as a whole (i.e. the normal velocity $\mathbf{v}_n = 0$) is the ordinary Bose distribution with the quasi-particle energy $\tilde{\epsilon}$ given by (77.3). The distribution in a frame where the normal velocity is not zero is obtained on replacing $\tilde{\epsilon}$ by $\tilde{\epsilon} - \mathbf{p}\cdot\mathbf{v}_n$. Thus the equilibrium distribution of quasi-particles when both motions are present is

$$n(\mathbf{p}) = \left[\exp\frac{\epsilon + (\mathbf{v}_s - \mathbf{v}_n)\cdot\mathbf{p}}{T} - 1\right]^{-1}. \tag{77.16}$$

By averaging the above equations over this distribution, the equations of two-velocity hydrodynamics (without the dissipative terms, in this approximation) can be derived, but we shall not pause to do so here.

PROBLEM

Determine the sound absorption coefficient in a Bose liquid at frequencies $\omega \gg \nu$, where ν is the quasi-particle collision frequency. The temperature is assumed so low that almost all the quasi-particles are phonons (A. F. Andreev and I. M. Khalatnikov 1963).

SOLUTION. Under the conditions stated, we can neglect the collision integral in (77.1). We put $\rho = \rho_0 + \delta\rho$, $n = n_0 + \delta n$ (where $\delta\rho$ and δn are small corrections to the equilibrium density of the liquid and the equilibrium phonon distribution function), and linearize equations (77.1), (77.6) and (77.12) with respect to the small quantities $\delta\rho$, δn and $\mathbf{v}_s$. Assuming all these proportional to $\exp(-i\omega t + i\mathbf{k}\cdot\mathbf{r})$, we find

$$(\mathbf{k}\cdot\mathbf{v} - \omega)\delta n = \frac{\partial n}{\partial\epsilon}\mathbf{v}\cdot\mathbf{k}\left(\frac{\partial\epsilon}{\partial\rho}\delta\rho + \mathbf{p}\cdot\mathbf{v}_s\right), \tag{1}$$

$$\omega\delta\rho - \mathbf{k}\cdot\mathbf{v}_s\rho = \int \mathbf{k}\cdot\mathbf{p}\,\delta n\, d^3p/(2\pi\hbar)^3, \tag{2}$$

$$\omega\mathbf{v}_s - \mathbf{k}u_0^2\delta\rho/\rho = \mathbf{k}\int \{n(\partial^2\epsilon/\partial\rho^2)\delta\rho + (\partial\epsilon/\partial\rho)\delta n\}\, d^3p/(2\pi\hbar)^3. \tag{3}$$

Here we have used the thermodynamic relations

$$d(\mu_0/m) = dP_0/\rho = u_0^2 d\rho/\rho,$$

where u_0 is the speed of sound at $T = 0$; the subscript 0 to ρ and n is omitted here and henceforward.

Since the number of phonons is small at temperatures near zero, the expressions on the right of equations (1)–(3) are small corrections. Omitting them altogether, we have from (2) and (3)

$$\omega = u_0 k, \quad \mathbf{v}_s = u_0(\delta\rho/\rho)\mathbf{k}/k. \tag{4}$$

In the next approximation we substitute these on the right of (1):

$$\delta n = \frac{\partial n}{\partial \epsilon}\frac{v\cos\theta}{v\cos\theta-\omega/k}\left(\frac{\partial \epsilon}{\partial \rho}+\frac{pu_0}{\rho}\cos\theta\right)\delta\rho, \tag{5}$$

where θ is the angle between **p** and **k**. The phonon dispersion relation is written

$$\epsilon(p) = u_0 p(1+\alpha p^2), \quad v = \partial\epsilon/\partial p = u_0(1+3\alpha p^2),$$

including the next term after the linear one in the expansion; for liquid helium at ordinary pressures, $\alpha > 0$, which means that the phonons are unstable with respect to spontaneous decay.

The presence of a "resonance" denominator in (5) leads (see below) to a large logarithmic factor in the integration. We use only logarithmic accuracy and neglect on the right of (3) the term in $\delta\rho$, which does not have such a denominator. Then, eliminating $\mathbf{v}_s$ from (2) and (3), we finally arrive at the dispersion relation

$$\frac{\omega^2}{k^2} - u_0^2 = A\frac{u_0^2}{\rho}\int\frac{p^2}{\cos\theta - 1 + 3\alpha p^2 - i0}\frac{\partial n}{\partial\epsilon}\frac{d^3p}{(2\pi\hbar)^3}, \tag{6}$$

where

$$A = \left(1+\frac{\rho}{u_0}\frac{du_0}{d\rho}\right)^2.$$

The imaginary part of the integral with respect to $\cos\theta$ is determined by passing round the pole (which is in the range of integration $\alpha > 0$). The real part is calculated with logarithmic accuracy by cutting off the integration at the lower limit $1-\cos\theta \sim \alpha p^2 \sim \alpha T^2/u_0^2$ and at the upper limit $1-\cos\theta \sim 1$. The left-hand side of equation (6) is written as $2u_0(\delta u - u_0\gamma/\omega)$, where γ is the absorption coefficient and δu the correction to the speed of sound ($u = u_0 + \delta u$). Calculation of the integral gives

$$\delta u = (3\rho_n u_0 A/4\rho)\log(u_0^2/\alpha T^2), \quad \gamma = 3\pi\omega\rho_n A/4\rho, \tag{7}$$

where $\rho_n = 2\pi^2 T^4/45\hbar^3 u_0^5$ is the phonon part of the normal density of the liquid. The frequency and temperature dependences of γ are, of course, the same as those found in §73.

CHAPTER IX

METALS

§78. Residual resistance

THE transport properties of metals are considerably more complex than those of insulators, if only because they contain quasi-particles of different kinds (conduction electrons and phonons).

The electric charge is, of course, transported by the conduction electrons. Heat transfer, on the other hand, is by both electrons and phonons. In practice, however, the electrons are also predominant in thermal conduction in metals of sufficiently high purity, mainly because their speeds (v_F on the Fermi surface) are much greater than those of the phonons (the speed of sound). Moreover, at low temperatures the electron specific heat considerably exceeds the phonon specific heat.

The conduction electrons undergo collisions of various types: with one another, with phonons and with impurity atoms (and other lattice defects). The collision frequency for the first two types decreases with the temperature. At sufficiently low temperatures, therefore, the scattering of electrons by impurities is the determining factor in transport phenomena. This range is called the *residual resistance* range, and we shall consider it as the first topic in the kinetics of metals.

The relations between the electric current **j** and the dissipative energy flux **q**′ in a metal, and the electric field **E** and the temperature gradient, have the form (44.12), (44.13):

$$\mathbf{E} + \nabla(\mu/e) = \mathbf{j}/\sigma + \alpha\nabla T, \tag{78.1}$$

$$\mathbf{q}' = \mathbf{q} - (\varphi - \mu/e)\mathbf{j} = \alpha T\mathbf{j} - \kappa\nabla T. \tag{78.2}$$

In this form they apply to crystals with cubic symmetry, and for simplicity this symmetry will everywhere be assumed. For crystals that do not have cubic symmetry, the coefficients σ, κ and α are replaced by tensors of rank two. The relation (78.2) is more convenient to use if **j** in it is expressed in terms of **E** by means of (78.1):

$$\mathbf{q}' = \sigma\alpha T[\mathbf{E} + \nabla(\mu/e)] - (\kappa + T\sigma\alpha^2)\nabla T. \tag{78.3}$$

The discussion in §74 about the transport equation for a Fermi liquid remains largely valid for an electron liquid in a metal. The momentum of the quasi-particles is here replaced by their quasi-momentum, and the form of the Fermi surface is in general complex and different for each individual metal.

The transport coefficients for a metal are in principle calculated by means of the linearized transport equation

$$-e\mathbf{E}\,.\,\mathbf{v}\partial n_0/\partial\epsilon + \mathbf{v}\,.\,\partial n_0/\partial\mathbf{r} = I(\delta\tilde{n}),$$

where $\mathbf{v} = \partial\epsilon/\partial\mathbf{p}$, and the collision integral is linearized with respect to the required small function $\delta\tilde{n}$ defined by (74.13). Differentiation of n_0 with respect to $\mathbf{r}$ can be arbitrarily carried out with μ = constant, since the gradient of μ would still enter in the combination $e\mathbf{E} + \nabla\mu$, as it should by (78.1). Then

$$\frac{\partial n_0}{\partial T} = -\frac{\epsilon-\mu}{T}\frac{\partial n_0}{\partial\epsilon},$$

and the transport equation takes the form

$$-\left(e\mathbf{E} + \frac{\epsilon-\mu}{T}\nabla T\right)\cdot\mathbf{v}\frac{\partial n_0}{\partial\epsilon} = I(\delta\tilde{n}). \tag{78.4}$$

The current density and the dissipative energy flux are given by the integrals

$$\mathbf{j} = -e\int \mathbf{v}\delta\tilde{n}\,.\,2d^3p/(2\pi\hbar)^3, \quad \mathbf{q}' = \int(\epsilon-\mu)\mathbf{v}\delta\tilde{n}\,.\,2d^3p/(2\pi\hbar)^3; \tag{78.5}$$

when $\mathbf{q}'$ is calculated as the flux of the kinetic energy $\epsilon - \mu$, there is no need to subtract the convective transport of potential energy, $\varphi\mathbf{j}$.

A characteristic of conduction electron scattering by impurity atoms is that it is elastic. Because the atoms have a large mass and are "bound" in the lattice, the electron energy may be regarded as unchanged in a collision. We shall show that the assumption of elastic scattering is by itself sufficient to give a simple relation between the electrical and thermal conductivities of the metal.

To obtain this, we note that the elastic collision operator does not affect the dependence of the function $\delta\tilde{n}$ on the energy ϵ; the collisions simply move the particles on the constant-energy surface. This means that any factor in $\delta\tilde{n}$ which depends only on ϵ can be taken outside the operator I. We can therefore seek the solution of the transport equation in the form

$$\delta\tilde{n} = \frac{\partial n_0}{\partial\epsilon}\left(e\mathbf{E} + \frac{\epsilon-\mu}{T}\nabla T\right)\cdot\mathbf{l}(\mathbf{p}), \tag{78.6}$$

where $\mathbf{l}(\mathbf{p})$ satisfies the equation

$$I(\mathbf{l}) = -\mathbf{v}. \tag{78.7}$$

The current density calculated from the distribution (78.6) is

$$\mathbf{j} = -e\int\left\{e(\mathbf{E}\,.\,\mathbf{l})\mathbf{v} + \frac{\epsilon-\mu}{T}(\mathbf{l}\,.\,\nabla T)\mathbf{v}\right\}\frac{\partial n_0}{\partial\epsilon}\frac{2d^3p}{(2\pi\hbar)^3}. \tag{78.8}$$

The first term gives the conductivity tensor

$$\sigma_{\alpha\beta} = -e^2 \int v_\alpha l_\beta \frac{\partial n_0}{\partial \epsilon} \frac{2d^3p}{(2\pi\hbar)^3}. \tag{78.9}$$

In a crystal with cubic symmetry, $\sigma_{\alpha\beta} = \sigma\delta_{\alpha\beta}$, and the conductivity is

$$\sigma = -\tfrac{1}{3}e^2 \int \mathbf{l} \cdot \mathbf{v} \frac{\partial n_0}{\partial \epsilon} \frac{2d^3p}{(2\pi\hbar)^3},$$

or, transforming the integral as in (74.18)–(74.20),

$$\sigma = \tfrac{2}{3}e^2 J_F, \quad J = \int \mathbf{l} \cdot \mathbf{v}\, dS/v(2\pi\hbar)^3. \tag{78.10}$$

The integration in J_F is taken over all sheets of the Fermi surface within one unit cell of the reciprocal lattice.

Similarly, the second term in (78.8) gives on comparison with (78.1)

$$\alpha\sigma = \frac{2e}{3T} \int \eta (\mathbf{v} \cdot \mathbf{l}) \frac{\partial n_0}{\partial \epsilon} \frac{d^3p}{(2\pi\hbar)^3},$$

where $\eta \equiv \epsilon - \mu$. The integration over d^3p is replaced by integration over the constant-energy surfaces $\eta =$ constant and over η. Again with J as in (78.10), we find

$$\alpha\sigma = \frac{2e}{3T} \int J\eta \frac{\partial n_0}{\partial \eta} d\eta. \tag{78.11}$$

The function

$$\frac{\partial n_0}{\partial \epsilon} = -\frac{1}{T(e^{\eta/T}+1)(e^{-\eta/T}+1)}$$

decreases exponentially as $\eta \to \pm\infty$; the integration with respect to η can therefore be extended from $-\infty$ to $+\infty$. The integral is governed mainly by the range $|\eta| \sim T$; $J(\eta)$, on the other hand, varies significantly only in the range $\eta \sim \mu \gg T$. It is therefore sufficient to write

$$J \approx J_F + \eta\, dJ/d\epsilon_F.$$

On substitution in (78.11), the integral of the first term is zero, because the integrand is an odd function of η; the second term gives

$$\alpha\sigma = \frac{2e}{3T}\frac{dJ}{d\epsilon_F} 2 \int_0^\infty \eta^2 \frac{\partial n_0}{\partial \eta} d\eta$$

$$= -\frac{8e}{3T}\frac{dJ}{d\epsilon_F} \int_0^\infty \eta n_0\, d\eta.$$

The integral

$$\int_0^\infty \frac{\eta\, d\eta}{e^{\eta/T}+1} = \frac{\pi^2}{12}T^2;$$

using also (78.10), we find

$$\alpha = -\frac{\pi^2 T}{3e}\frac{d\log J}{d\epsilon_F}. \tag{78.12}$$

In order of magnitude, $|\alpha| \sim T/e\epsilon_F$.

Let us now put $\mathbf{E} = 0$ and calculate the energy flux. Again using the cubic symmetry, we find

$$\mathbf{q}' = \frac{2\nabla T}{3T}\int_{-\infty}^{\infty} J\eta^2 \frac{\partial n_0}{\partial \eta}\, d\eta.$$

Here it is sufficient to put $J = J_F$, which gives

$$\mathbf{q}' = -\frac{2\pi^2}{9} T J_F \nabla T.$$

Comparison with (78.3) and (78.10) shows that

$$\kappa + T\sigma\alpha^2 = \pi^2\sigma T/3e^2.$$

From the estimate of α given above, the term $T\sigma\alpha^2$ on the left is small compared with the right-hand side, in the ratio $(T/\epsilon_F)^2$. Neglecting this term, we have finally the following relation between the thermal and electrical conductivities:

$$\kappa = (\pi^2 T/3e^2)\sigma, \tag{78.13}$$

the *Wiedemann–Franz law*.†

We must again emphasize that the proof of this relation uses only the fact that the scattering of conduction electrons is elastic. An examination of the proof also readily shows that the assumption of cubic symmetry merely simplifies the formulae. In the general case where the crystal has any symmetry, a similar relation (78.13) exists between the tensors $\kappa_{\alpha\beta}$ and $\sigma_{\alpha\beta}$.

To find the temperature dependence of the coefficients κ and σ separately, the collision integral is needed. For collisions with impurity atoms, its form is exactly analogous to the integral (70.3) for phonon scattering by impurities:

$$C(n) = N_{\text{imp}}\int w(\mathbf{p},\mathbf{p}')[n'(1-n) - n(1-n')]\delta(\epsilon-\epsilon')\,.\,2d^3p'/(2\pi\hbar)^3. \tag{78.14}$$

†A formula of the type (78.13) was derived qualitatively by P. Drude (1900), who first formulated the concept of conduction electrons participating in the thermal equilibrium of the metal. The quantitative result in classical statistics was given by H. A. Lorentz (1905), and in Fermi statistics by A. Sommerfeld (1928).

The factors $1-n$ and $1-n'$ take account of the Pauli principle (a transition can take place only to an unoccupied state); the factors n' and n signify that scattering can occur only from an occupied state. As in (70.3), it is assumed in the integral (78.14) that the impurity atoms are randomly distributed and that the mean distance between them is much greater than the scattering amplitude; the various atoms then scatter independently. The equation $w(\mathbf{p}, \mathbf{p}') = w(\mathbf{p}', \mathbf{p})$ has already been used in (78.14). The Born approximation is in general not applicable to the scattering of conduction electrons by impurity atoms. The equation as given can be justified by the arguments used in deriving the principle of detailed balancing in the form (2.8). Here, however, it is implied that the positions occupied by the impurity atoms in the metal lattice have a symmetry that allows inversion.

The linearization of the collision integral amounts to replacing the difference $n'(1-n) - n(1-n') = n' - n$ by $\delta\tilde{n}' - \delta\tilde{n}$. Equation (78.7) then becomes

$$N_{\text{imp}} \int w(\mathbf{p}, \mathbf{p}')(\mathbf{l}' - \mathbf{l})\delta(\epsilon - \epsilon') \,.\, 2d^3p'/(2\pi\hbar)^3 = -\mathbf{v}. \tag{78.15}$$

This does not involve the temperature. The solution $\mathbf{g}(\mathbf{p})$ is therefore also independent of the temperature, as is the conductivity σ, by (78.10). Thus, at sufficiently low temperatures, when the scattering by impurities is the chief mechanism of electrical resistance, the resistance tends to a constant (residual) value. Accordingly, the thermal conductivity κ in this range is proportional to T.†

For a rough quantitative estimate of the residual resistance, we can use the elementary formula (43.7), putting (for electrons in a metal) $p \sim p_F$:

$$\sigma \sim e^2 N l / p_F, \tag{78.16}$$

where N is the electron density. For scattering by impurities, the mean free path $l \sim 1/N_{\text{imp}}\sigma_t$, where σ_t is the transport scattering cross-section. Hence the residual resistance $\rho_{\text{res}} = 1/\sigma$ is

$$\rho_{\text{res}} \sim N_{\text{imp}}\sigma_t p_F / e^2 N. \tag{78.17}$$

A further comment should be added to the above discussion. The general condition for the applicability of the transport equation to a Fermi liquid requires that the quantum uncertainty of the electron energy should be small in comparison with the width ($\sim T$) of the thermal blurring of the Fermi distribution. This uncertainty is $\sim \hbar/\tau$, where $\tau \sim l/v_F$ is the mean free time. For scattering by impurities $l \sim 1/N_{\text{imp}}\sigma_t$; the uncertainty $\hbar/\tau$ is independent of the temperature, and therefore blurs the Fermi boundary even when $T = 0$. At first sight, it follows from this that the whole of the above discussion is subject to the very severe condition

$$T \gg \hbar v_F \sigma_t N_{\text{imp}}, \tag{78.18}$$

†In this analysis it is assumed that equation (78.15) does not contain quantities that vary rapidly near $\epsilon = \epsilon_F$, and so $\mathbf{l}$ can be replaced by $\mathbf{l}_F$ in (78.9). This is true for scattering by ordinary impurities, but not for scattering by paramagnetic atoms.

depending on the impurity concentration. In reality, however, there is no such limitation (L. D. Landau, 1934).

The reason is that, because of the fixed positions of the impurity atoms and the elasticity of electron scattering by them, the whole problem of calculating the electric current can in principle be formulated as the quantum-mechanical problem of the motion of an electron in a given complex external field having a potential. For electron states determined as stationary states in this field, the energy has no uncertainty; at $T = 0$, the electrons occupy a range of states bounded by a sharp Fermi surface, but in the space of quantum numbers for motion in the field, not in momentum space. With this formulation of the problem, conditions of the type (78.18) do not arise.

§79. Electron–phonon interaction

In sufficiently pure metals, the chief mechanism for the establishment of equilibrium over a wide temperature range is the interaction between conduction electrons and phonons.

The condition for an electron to be able to emit (or absorb) a phonon requires that the speed of the electron be greater than that of the phonon; compare the analogous result in §68 for the emission of a phonon by a phonon. The speed of electrons at the Fermi surface is, however, usually large in comparison with that of the phonons; the condition is therefore satisfied, and the main contribution to the electron–phonon collision integral comes from just these one-phonon processes.

The collision integral then has the following form, analogous to the phonon–phonon integral (67.6):†

$$\begin{aligned} C_{e,\mathrm{ph}}(n_{\mathbf{p}}) = \int & w(\mathbf{p}', \mathbf{k}; \mathbf{p})\{n_{\mathbf{p}'}(1-n_{\mathbf{p}})N_{\mathbf{k}} - n_{\mathbf{p}}(1-n_{\mathbf{p}'})(1+N_{\mathbf{k}})\}\times \\ & \times\delta(\epsilon_{\mathbf{p}} - \epsilon_{\mathbf{p}'} - \omega_{\mathbf{k}})\, d^3k/(2\pi)^3 \\ & + \int w(\mathbf{p}'; \mathbf{p}, \mathbf{k})\{n_{\mathbf{p}'}(1-n_{\mathbf{p}})(1+N_{\mathbf{k}}) - n_{\mathbf{p}}(1-n_{\mathbf{p}'})N_{\mathbf{k}}\}\times \\ & \times\delta(\epsilon_{\mathbf{p}} + \omega_{\mathbf{k}} - \epsilon_{\mathbf{p}'})\, d^3k/(2\pi)^3. \end{aligned} \tag{79.1}$$

The first term corresponds to processes with emission of a phonon having quasi-momentum $\mathbf{k}$ by an electron having a given quasi-momentum $\mathbf{p}$, and the reverse processes with absorption of a phonon $\mathbf{k}$ by electrons $\mathbf{p}'$ with return to the quasi-momentum $\mathbf{p}$:

$$\mathbf{p} = \mathbf{p}' + \mathbf{k} + \mathbf{b}; \tag{79.2a}$$

in these processes, the transitions take place between an electron state with given energy $\epsilon_{\mathbf{p}}$ and states of lower energy. The second term corresponds to processes

†In §§79–83 the units used are such that $\hbar = 1$.

with absorption of a phonon by an electron $\mathbf{p}$ and the reverse processes of its emission by electrons $\mathbf{p}'$:

$$\mathbf{p}+\mathbf{k}=\mathbf{p}'+\mathbf{b}; \tag{79.2b}$$

in these processes, the transitions take place between a specified electron state and ones of higher energy. For the same reasons as in the case of phonon emission by phonons (§66), the value of $\mathbf{b}$ in the equations (79.2) is uniquely determined by specifying the values of $\mathbf{k}$ and $\mathbf{p}$ with the requirement that $\mathbf{p}'$ should be in the same selected cell of the reciprocal lattice. The delta-function factors in (79.1) express the law of conservation of energy; $\epsilon_\mathbf{p}$ is the electron energy and $\omega_\mathbf{k}$ the phonon energy. As in Chapter VII, the phonon distribution function (numbers of occupied states) is denoted by $N_\mathbf{k}$; the electron distribution function is denoted by $n_\mathbf{p}$. The subscripts marking the branch of the phonon spectrum, and the signs of summation with respect to these, will be omitted, for brevity. It is assumed that the transition probabilities are independent of the electron spin, which is unchanged in the transition.

There is a similar expression for the phonon–electron collision integral which is to be added to the phonon–phonon integral on the right-hand side of the transport equation for the phonon distribution function:

$$C_{\mathrm{ph},e}(N_\mathbf{k})=\int w(\mathbf{p};\mathbf{p}',\mathbf{k})\{n_\mathbf{p}(1-n_{\mathbf{p}'})(1+N_\mathbf{k})-n_{\mathbf{p}'}(1-n_\mathbf{p})N_\mathbf{k}\}\times$$
$$\times\,\delta(\epsilon_{\mathbf{p}'}+\omega_\mathbf{k}-\epsilon_\mathbf{p})\,.\,2d^3p/(2\pi)^3, \tag{79.3}$$

with $\mathbf{p}=\mathbf{p}'+\mathbf{k}+\mathbf{b}$. This is the difference between the number of phonons $\mathbf{k}$ emitted by electrons with any quasi-momenta $\mathbf{p}$ and the number absorbed by electrons with any $\mathbf{p}'$. The factor 2 allows for the two possible spin directions of the emitting or absorbing electron.

In first-order perturbation theory, the probabilities of phonon emission or absorption by an electron which occur in these integrals are determined by the electron–phonon interaction operator linear in the phonon operators $\hat{\mathbf{U}}_s(\mathbf{n})$ (66.2); the linearity corresponds to the fact that these operators are responsible for transitions in which only one of the phonon state occupation numbers changes by unity. Without repeating the discussion in §66, we may note that, in the limit as the phonon quasi-momentum $\mathbf{k}$ tends to zero, the phonon emission or absorption probability is proportional to k:

$$w \propto k. \tag{79.4}$$

According to a general property of transition probabilities in the Born approximation, the probabilities of the direct and reverse transitions are equal, and so†

$$w(\mathbf{p}',\mathbf{k};\mathbf{p})=w(\mathbf{p};\mathbf{p}',\mathbf{k}). \tag{79.5}$$

†The quantum numbers i and f of the initial and final states are always written in the order fi in the notation for the probability.

This property has already been made use of in the integrals (79.1) and (79.3).

A further simplification is achieved by taking into account the symmetry (expressed by the fact that the operators $\hat{U}_s$ are real) with which the phonon creation and annihilation operators appear in the electron–phonon interaction operator. Because of this, the emission of a phonon with quasi-momentum $\mathbf{k}$ is equivalent to the absorption of one with quasi-momentum $-\mathbf{k}$. We shall also use the fact that the electron energies $\epsilon_\mathbf{p}$ and $\epsilon_{\mathbf{p}'}$ are close to the Fermi energy ϵ_F. Let $\mathbf{p}_F$ and $\mathbf{p}'_F$ be vectors in the directions of $\mathbf{p}$ and $\mathbf{p}'$, ending on the Fermi surface, and let the functions w be expressed in terms of the directions of $\mathbf{p}_F$ and $\mathbf{p}'_F$ and the differences $\eta_\mathbf{p} = \epsilon_\mathbf{p} - \epsilon_F$, $\eta_{\mathbf{p}'} = \epsilon_{\mathbf{p}'} - \epsilon_F$ which represent the closeness of the electron energy to ϵ_F. As regards these variables, w is a slowly varying function, which changes appreciably only in ranges $\sim \epsilon_F \gg T$. Neglecting quantities $\sim \eta \sim T$, we can put $\eta_\mathbf{p} = \eta_{\mathbf{p}'} = 0$ in these functions. The equivalence mentioned above is then expressed by the equation

$$w(\mathbf{p}'_F, \mathbf{k}; \mathbf{p}_F) = w(\mathbf{p}'_F; \mathbf{p}_F, -\mathbf{k}), \tag{79.6}$$

the w being functions only of the directions of $\mathbf{p}_F$ and $\mathbf{p}'_F$. If now we change the variable of integration $\mathbf{k}$ to $-\mathbf{k}$ in the second term in (79.1), the coefficients w in the two integrals become equal; since $\omega_{-\mathbf{k}} = \omega_\mathbf{k}$, the change simply replaces $N_\mathbf{k}$ by $N_{-\mathbf{k}}$.

The integrals (79.1) and (79.3) are, of course, zero when the equilibrium electron and phonon distribution functions are substituted. The linearization of these integrals for small deviations from equilibrium is carried out simultaneously with respect to both distribution functions, which we write as

$$\left.\begin{aligned} n &= n_0(\epsilon) + \delta\tilde{n}, \quad N = N_0(\omega) + \delta N,\\ \delta\tilde{n} &= -\frac{\partial n_0}{\partial \epsilon}\varphi = \frac{n_0(1-n_0)}{T}\varphi,\\ \delta N &= -\frac{\partial N_0}{\partial \omega}\chi = \frac{N_0(1+N_0)}{T}\chi. \end{aligned}\right\} \tag{79.7}$$

The transformation is exactly similar to those in §§67 and 74. For example, the expression in the braces in the first term in (79.1), rewritten as

$$(1-n)(1-n')(1+N)\left[\frac{n'}{1-n'}\frac{N}{1+N} - \frac{n}{1-n}\right],$$

is put into the form

$$n_0(1-n_0')(1+N_0)\frac{1}{T}(\varphi' - \varphi + \chi).$$

This is conveniently transformed further by means of the equation

$$n_0(\epsilon)[1-n_0(\epsilon')] = [n_0(\epsilon) - n_0(\epsilon')]N_0(\epsilon - \epsilon'), \tag{79.8}$$

which is easily verified by direct calculation. We then find

$$(n_0 - n_0')\frac{N_0(1+N_0)}{T}(\varphi' - \varphi + \chi) = -\frac{\partial N_0}{\partial \omega}(n_0 - n_0')(\varphi' - \varphi + \chi).$$

The other terms are transformed similarly, leading to the following linearized collision integrals:

$$C_{e,\mathrm{ph}}(n) = I_{e,\mathrm{ph}}(\varphi, \chi) = -\int \frac{\partial N_0}{\partial \omega} w(n_0' - n_0)\{(\varphi_{\mathbf{p}'} - \varphi_{\mathbf{p}} + \chi_{\mathbf{k}})\delta(\epsilon_{\mathbf{p}} - \epsilon_{\mathbf{p}'} - \omega_{\mathbf{k}}) - (\varphi_{\mathbf{p}'} - \varphi_{\mathbf{p}} - \chi_{-\mathbf{k}})\delta(\epsilon_{\mathbf{p}} - \epsilon_{\mathbf{p}'} + \omega_{\mathbf{k}})\}\, d^3k/(2\pi)^3, \qquad (79.9)$$

$$C_{\mathrm{ph},e}(N) = I_{\mathrm{ph},e}(\chi, \varphi) = \frac{\partial N_0}{\partial \omega}\int w(n_0' - n_0)(\varphi_{\mathbf{p}'} - \varphi_{\mathbf{p}} + \chi_{\mathbf{k}})\delta(\epsilon_{\mathbf{p}} - \epsilon_{\mathbf{p}'} - \omega_{\mathbf{k}})\,.\, 2d^3p/(2\pi)^3; \qquad (79.10)$$

in both integrals, $\mathbf{p} = \mathbf{p}' + \mathbf{k} + \mathbf{b}$.

These integrals fall naturally into two parts, the linear integral operators acting on φ and χ respectively. For instance,

$$I_{e,\mathrm{ph}}(\varphi, \chi) = I^{(1)}_{e,\mathrm{ph}}(\varphi) + I^{(2)}_{e,\mathrm{ph}}(\chi). \qquad (79.11)$$

An important property of the operator $I^{(1)}_{e,\mathrm{ph}}$ is that it does not change the parity of the function $\varphi(\eta, \mathbf{p}_F)$ with respect to the variable η, i.e., it leaves even and odd functions the same: as regards its effect on a function of η,

$$I^{(1)}_{e,\mathrm{ph}}(\varphi(\eta)) \sim \int K(\eta, \eta')[\varphi(\eta') - \varphi(\eta)]\, d\eta,$$

where

$$K(\eta, \eta') = [n_0(\eta') - n_0(\eta)][\delta(\eta - \eta' - \omega) - \delta(\eta' - \eta - \omega)].$$

Since

$$n_0(\eta) = \tfrac{1}{2}[1 - \tanh(\eta/2T)], \qquad (79.12)$$

and so

$$n_0(\eta') - n_0(\eta) = \tfrac{1}{2}[\tanh(\eta/2T) - \tanh(\eta'/2T)],$$

we see that

$$K(\eta, \eta') = K(-\eta, -\eta'),$$

and this immediately yields the above-mentioned property of the operator, which will be used in §§80 and 82.

The collision integrals (79.9) and (79.10) are identically zero for the functions

$$\varphi = \text{constant} \times \epsilon, \quad \chi = \text{constant} \times \omega, \tag{79.13}$$

with the same constant. This "spurious" solution of the transport equation corresponds, like the solution (67.18) in the phonon–phonon equation, to a change in the temperature of the system by a small constant amount. The integrals (79.9) and (79.10) are, however, also zero when

$$\varphi = \text{constant} \tag{79.14}$$

and $\chi = 0$. This solution is due to the constancy of the total number of electrons (unlike the total number of phonons); formally, it corresponds to a change in the chemical potential of the electrons by a small constant amount.

To proceed to quantitative estimates, we note that the orders of magnitude of the parameters of the electron spectrum in a metal can be expressed in terms of the lattice constant d and the electron effective mass m^* only; for example, the Fermi momentum (in ordinary units) is $p_F \sim \hbar/d$, the speed $v_F \sim p_F/m^* \sim \hbar/m^*d$, and the energy $\epsilon_F \sim v_F p_F \sim \hbar^2/m^*d^2$. The parameters of the phonon spectrum and the electron–phonon interaction also contain the mass M of the atoms. The density of the substance $\rho \propto M$, and the speed of sound $u \propto \rho^{-1/2} \propto M^{-1/2}$; making the dimensions right by means of $\hbar$, d and m^* (which can be done in only one way), we obtain the estimate

$$u \sim v_F(m^*/M)^{1/2}. \tag{79.15}$$

Hence the Debye temperature is

$$\Theta \sim \hbar\omega_{\max} \sim \hbar u/d \sim \epsilon_F(m^*/M)^{1/2}. \tag{79.16}$$

The mass M appears in the electron–phonon interaction operator only through the displacement operators $\hat{\mathbf{U}}_s$ (66.2) of the atoms; this interaction involves no other small terms in $1/M$, its energy being $\sim\epsilon_F$ when $U_s \sim d$. The matrix elements of the operators $\hat{\mathbf{U}}_s$, and therefore those of the electron–phonon interaction operator, are $\propto (M\omega)^{-1/2} \propto M^{-1/4}$; for a given quasi-momentum k, the frequency $\omega \sim uk \propto M^{-1/2}$. The scattering probability is given by the square of the matrix element. Hence the function w in the collision integral is proportional to $M^{-1/2}$, or, making the dimensions right,

$$w \sim \Theta v_F d^2. \tag{79.17}$$

This estimate needs modification in relation to the emission or absorption of a long-wavelength acoustic phonon. The fact that w is then proportional to k means that the estimate must include an extra factor $k/k_{\max} \sim kd$:

$$w \sim \Theta v_F k d^3. \tag{79.18}$$

§80. Transport coefficients in metals. High temperatures

At high temperatures $T \gg \Theta$, phonons with all possible quasi-momenta are excited in the crystal, up to the maximum value, which has the same order of magnitude as the electron Fermi momentum: $k_{max} \sim p_F \sim 1/d$. By the definition of the Debye temperature, the maximum phonon energy $\omega_{max} \sim \Theta$, and so $\omega \ll T$ for all phonons.

Under these conditions, therefore, the phonon energies are small compared with the width of the blurred region in the Fermi distribution of electrons. This enables us to treat phonon emission or absorption approximately as elastic scattering of an electron. The scattering angles are not small, since the electron and phonon quasi-momenta under these conditions are of the same order of magnitude.

At high temperatures, when the phonon state occupation numbers are large, the establishment of equilibrium in each volume element of the phonon gas (phonon–phonon relaxation) takes place very quickly. We can therefore regard the phonon distribution function as being the equilibrium one when considering the electrical and thermal conductivities, i.e. take $\chi = 0$ in the collision integrals (a quantitative estimate of χ will be made at the end of this section). That is, it is sufficient to deal with the transport equation for electrons only.

It may be noted at once that, in an approximation which assumes the electron scattering elastic, the results of §78 remain valid that were based only on this approximation, including the Wiedemann–Franz law (78.13) which gives the ratio σ/κ. To determine the temperature dependence of σ and κ separately, however, it is necessary to examine in more detail the electron–phonon collision integral (79.9).

Under the conditions in question, this integral is greatly simplified. Because the phonon energy $\omega = \pm(\epsilon' - \epsilon)$ is small, we can expand the difference $n_0' - n_0$ in powers of ω:†

$$n_0' - n_0 \approx \pm \omega \partial n_0 / \partial \epsilon.$$

We can then put $\omega = 0$ in the arguments of the delta functions, obtaining

$$I_{e,ph}(\varphi) = 2 \int w \frac{\partial N_0}{\partial \omega} \frac{\partial n_0}{\partial \epsilon} \delta(\epsilon' - \epsilon)(\varphi' - \varphi) \omega \frac{d^3k}{(2\pi)^3}.$$

When $\omega \ll T$ the phonon distribution function $N_0 \approx T/\omega$, so that $\partial N_0/\partial \omega \approx -T/\omega^2$. The derivative $\partial n_0/\partial \epsilon \sim -1/T$. The integral is governed by the range $k \sim k_{max}$ in which $\omega \sim \Theta$. When the delta functions are taken into account, the integration over d^3k adds a factor k_{max}^2/v_F to the estimate of the integral:

$$I_{e,ph}(\varphi) \sim -w(T/\Theta)(k_{max}^2/v_F)\varphi/T.$$

With (79.17) this gives

$$I_{e,ph}(\varphi) \sim -\varphi \sim -T\,\delta\tilde{n}. \tag{80.1}$$

†The presence of ω in this difference is consistent with the approximation that the electron scattering is elastic. It is necessary because, in bringing the collision integral to the form (79.9), we used equation (79.8), the right-hand side of which becomes indeterminate when $\epsilon = \epsilon'$.

This means that the electron–phonon collision frequency $\nu_{e,\mathrm{ph}} \sim T$ ($T/\hbar$ in ordinary units), the mean free path $l \sim v_F/T$, and (78.16) gives for the electrical conductivity (in ordinary units)†

$$\sigma \sim Ne^2\hbar/m^*T. \tag{80.2}$$

The electrical conductivity of the metal is thus inversely proportional to the temperature when $T \gg \Theta$. The Wiedemann–Franz law then shows that the thermal conductivity is constant:

$$\kappa \sim N\hbar/m^*. \tag{80.3}$$

Let us now estimate the correction functions φ and χ in the electron and phonon distributions in order to justify neglecting χ in the collision integral. We can do this, for instance, in the case where there is an electric field but no temperature gradient.

Since the electric field does not affect the motion of the phonons, the left-hand side of the transport equation for phonons is zero. The equation therefore reduces to the vanishing of the sum of the collision integrals for phonons with electrons and phonons with phonons:

$$I^{(1)}_{\mathrm{ph},e}(\varphi) + I^{(2)}_{\mathrm{ph},e}(\chi) + I_{\mathrm{ph,ph}}(\chi) = 0; \tag{80.4}$$

the superscripts (1) and (2) distinguish the two parts of the integral (79.10) in the same way as was done in (79.11).

The integral $I_{\mathrm{ph},e}$ is estimated similarly to $I_{e,\mathrm{ph}}$ above. Here, however, we must take into account that the integration over the electron quasi-momenta $\mathbf{p}$ is in practice taken only near the Fermi surface, over the volume of a layer with thickness $\sim T/v_F$ and area $\sim p_F^2$. The delta function adds a factor $1/\epsilon_F$ to the estimate of the integral. The result is

$$I^{(2)}_{\mathrm{ph},e}(\chi) \sim -w\frac{\chi}{T}\frac{T}{\Theta}\frac{Tp_F^2}{v_F\epsilon_F} \sim -\chi T/\epsilon_F, \quad I^{(1)}_{\mathrm{ph},e}(\varphi) \sim -\varphi T/\epsilon_F. \tag{80.5}$$

The phonon–phonon collision integral is estimated as

$$I_{\mathrm{ph,ph}}(\chi) \sim -\nu_{\mathrm{ph,ph}}\delta N \sim -\nu_{\mathrm{ph,ph}}(T/\Theta^2)\chi,$$

with the effective collision frequency from (68.3):

$$\nu_{\mathrm{ph,ph}} \sim T/Mud \sim T\surd(m^*/M).$$

†Note that the quantum uncertainty of the electron energy, $\sim \hbar\nu_{e,\mathrm{ph}} \sim T$, is of the order of the width of the blurred region in the electron distribution. This fact, however, does not make the results inapplicable, for a reason similar to that given at the end of §78 in connection with scattering by impurities. Because of the relative slowness of the vibrations of atoms in the lattice, and the elasticity of electron scattering, the problem can in principle be formulated as that of electrons moving in the given potential field of the deformed lattice.

Thus

$$I_{\text{ph,ph}}(\chi) \sim -(T^2/\Theta^2)\sqrt{(m^*/M)}\chi \sim T^2\chi/\Theta\epsilon_F. \tag{80.6}$$

Comparison of (80.5) and (80.6) shows, first of all, that

$$I^{(2)}_{\text{ph},e}(\chi)/I_{\text{ph,ph}}(\chi) \sim \Theta/T \ll 1:$$

the effective frequency of phonon–electron collisions (for equilibrium electrons, i.e. with $\varphi = 0$) is small relative to the phonon–phonon collision frequency. We can therefore neglect the second term in (80.4). Comparison of the two remaining terms gives

$$\chi/\varphi \sim \Theta/T \ll 1, \tag{80.7}$$

and this justifies neglecting χ in the electron–phonon collision integral. It is easily seen that the same result (80.7) is obtained when a temperature gradient is present.

The neglect of the function χ in the electron transport equation may, however, be impermissible in the treatment of thermoelectric phenomena.

According to (78.12), the derivation of which was based only on the assumption of elastic scattering of electrons, the thermoelectric coefficient is

$$\alpha^{\text{I}} \sim T/e\epsilon_F; \tag{80.8}$$

the meaning of the superscript I will be explained later. This quantity is "anomalously" small in the sense that the order of magnitude of the integral in (78.8) (the second term) was reduced in the ratio T/ϵ_F because

$$\varphi^{\text{I}} = -(\eta/T)\mathbf{l}\,.\,\nabla T \tag{80.9}$$

is an odd function of $\eta = \epsilon - \mu$. This property is in a sense accidental, and may have the result that a comparatively small addition to φ due to the non-equilibrium of the phonons yields a contribution to α that is comparable with (80.8).

We shall seek the solution of the electron transport equation

$$\frac{\partial n_0}{\partial T}\mathbf{v}\,.\,\nabla T = -\frac{\partial n_0}{\partial \epsilon}\frac{\eta}{T}\mathbf{v}\,.\,\nabla T = I^{(1)}_{e,\text{ph}}(\varphi) + I^{(2)}_{e,\text{ph}}(\chi) \tag{80.10}$$

as a sum $\varphi = \varphi^{\text{I}} + \varphi^{\text{II}}$, where φ^{I} is the solution of the equation without the second term on the right, and φ^{II} is the solution of the equation

$$I^{(1)}_{e,\text{ph}}(\varphi) + I^{(2)}_{e,\text{ph}}(\chi) = 0. \tag{80.11}$$

Here φ^{I} is the "major" part of φ; because the operator $I^{(1)}_{e,\text{ph}}$ is even with respect to the variable η (§79), this part has the form (80.9) and is an odd function of η. Equation (80.11) shows that $\varphi^{\text{II}} \sim \chi$, and therefore

$$\varphi^{\text{II}}/\varphi^{\text{I}} \sim \chi/\varphi^{\text{I}} \sim \Theta/T \ll 1.$$

Unlike φ^{I}, however, φ^{II} is not zero when $\epsilon = \mu$. In the calculation of the corresponding contribution to the current density, therefore, the leading term is not cancelled, and the result is small only in the sense that φ^{II} is relatively small. This means that the latter's contribution to the thermoelectric coefficient is

$$\alpha^{\mathrm{II}} \sim \alpha^{\mathrm{I}}(\epsilon_F/T)(\Theta/T) \sim \Theta/eT. \tag{80.12}$$

At the lower end of the temperature range considered, where $T \sim \Theta$, we have $e\alpha^{\mathrm{II}} \sim 1$ in place of the small quantity $e\alpha^{\mathrm{I}} \sim \Theta/\epsilon_F$.

The thermoelectric coefficient is thus composed of two additive parts. These may be of the same order of magnitude, but they vary differently with the temperature. The physical origin of the second term in α is that heat transfer in the crystal causes a flux of phonons (a "phonon wind"), which carries the electrons with it.†

§81. **Umklapp processes in metals**

The nature of electron–phonon scattering at low temperatures is quite different from that when $T \gg \Theta$. When $T \ll \Theta$, phonons are excited with energies $\omega \sim T$ in the crystal (and belong in general to the acoustic branches of the spectrum). When such a phonon is emitted or absorbed, the electron energy changes by an amount $\sim T$, i.e. by an amount of the order of the total width of the blurred region in the Fermi distribution. The change in the electron quasi-momentum is equal to the phonon quasi-momentum. Since $k \sim T/u \ll k_{\mathrm{max}}$, and $k_{\mathrm{max}} \sim p_F$, it follows that the electron quasi-momentum changes only by a relatively small amount. At low temperatures, therefore, there is a limiting case which is the opposite of elastic scattering: the electron relaxation in energy takes place considerably more rapidly than as regards the quasi-momentum direction.

The energy relaxation is a rapid "mixing" in the blurred region of the Fermi distribution. The relaxation as regards direction is an equalization of the distribution over this surface; it takes place in small amounts ($\sim T/u$), i.e. it is a slow diffusion over the surface.

Before going on to a detailed consideration of the transport phenomena under these conditions, we shall make some general comments about the role of Umklapp processes. As in insulator crystals, the finiteness of the transport coefficients in an ideal metal crystal (without impurities or defects) is due to the occurrence of these processes. With only the normal processes that conserve the total quasi-momentum of electrons and phonons, the transport equations would have spurious solutions corresponding to the movement of the electron and phonon systems as a whole relative to the lattice. These are solutions of the type

$$\varphi = \mathbf{p} \cdot \delta\mathbf{V}, \quad \chi = \mathbf{k} \cdot \delta\mathbf{V} \tag{81.1}$$

with a constant vector $\delta\mathbf{V}$; cf. (67.19). They reduce to zero the collision integrals

†The role of phonon drag on electrons as regards transport effects in metals was elucidated by L. É. Gurevich (1946).

(79.9), (79.10) if the emission or absorption of phonons by electrons takes place with conservation of quasi-momentum ($\mathbf{p} = \mathbf{p}' + \mathbf{k}$).

At high temperatures, when the quasi-momenta of both electrons and phonons are large ($\sim 1/d$), Umklapp processes take place, in general, with the same frequency as normal processes. The need to take account of them therefore does not give rise to any specific features of the transport phenomena.

The electron quasi-momenta lie near the Fermi surface, and in this sense are almost independent of the temperature. At low temperatures, however, the phonon quasi-momenta become small, and Umklapp processes may therefore be impeded. In this respect the situation is substantially different for closed and open Fermi surfaces.

An open Fermi surface, for any choice of the unit cell in $\mathbf{p}$ space (the reciprocal lattice), crosses the cell boundaries. In this case, clearly, Umklapp processes are always possible with emission or absorption of a phonon with arbitrarily low energy: even a small change in the electron quasi-momentum near the cell boundary can transfer the electron to an adjacent cell. In the course of their diffusion over the Fermi surface, all the electrons will ultimately reach the cell boundaries and may thus take part in Umklapp processes. Consequently, in this case also the probability of such processes contains no additional small factor in comparison with normal processes. Indeed, the classification into normal and Umklapp processes depends on the choice of the reciprocal lattice cell, and is to that extent arbitrary. With an open Fermi surface, the property that there is no additional small factor in the frequency of Umklapp processes exists for any choice of cell. It is then desirable to avoid any distinction of two types of scattering event, and to regard all of them as normal (i.e. conserving quasi-momentum) but allow electron quasi-momentum values anywhere in the reciprocal lattice. For the phonons, the unit cell is chosen so that the point $\mathbf{k} = 0$ is at its centre; then all the long-wavelength phonons (the only ones that need be considered when $T \ll \Theta$) are in a small part of the volume of one cell, near its centre. In this treatment, the spurious solution (81.1) is excluded by applying to the electron distribution function the condition of periodicity in the reciprocal lattice:

$$n(\mathbf{p} + \mathbf{b}) = n(\mathbf{p}). \tag{81.2}$$

The equilibrium distribution, depending only on the electron energy $\epsilon(\mathbf{p})$, necessarily satisfies this condition, since $\epsilon(\mathbf{p})$ is periodic. As well as $n_0(\mathbf{p})$, the derivative $\partial n_0/\partial \epsilon$ is periodic, and so therefore is the factor $\varphi(\mathbf{p})$ in $\delta\tilde{n}$; this requirement eliminates the solution (81.1), which does not satisfy it.

Let us now consider a closed Fermi surface. In this case, we can choose the basic reciprocal lattice cell in such a way that the Fermi surface nowhere crosses its boundaries.† Then Umklapp processes correspond to electron transitions between any points on the Fermi surface in the basic cell and its replica in the

†If, however, the Fermi surface consists of a number of closed cavities, it may be necessary to define the basic cell otherwise than as a parallelepiped with plane faces. This is illustrated schematically in Fig. 28 for the case of a plane lattice with two non-equivalent closed cavities forming the "Fermi surface". The broken line shows the basic cell, which does not intersect these cavities. Intersections could not be avoided by any choice of a rectangular cell.

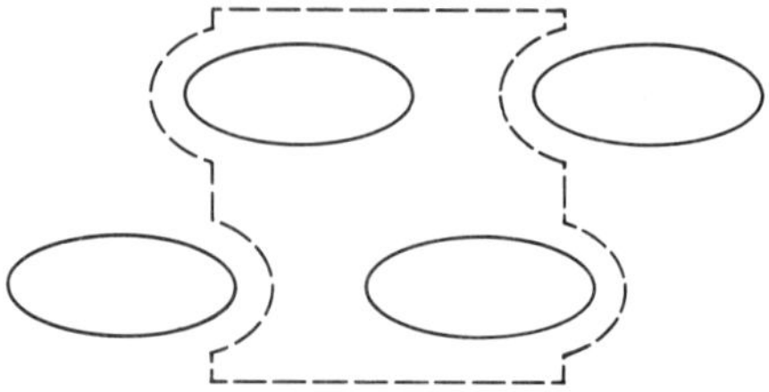

FIG. 28.

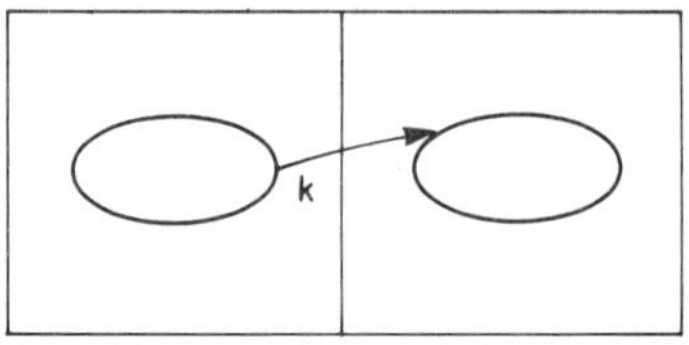

FIG. 29.

adjacent cell, as shown schematically in Fig. 29. The vector $\mathbf{k}$ joining these points is the quasi-momentum of the emitted or absorbed phonon. The distance k is in general large ($\sim 1/d$), and at low temperatures the number of phonons with energy $\omega(\mathbf{k})$ is exponentially small, being proportional to $\exp[-\omega(\mathbf{k})/T]$. The effective frequency of Umklapp scattering events then depends on the temperature according to

$$\nu_U \propto \exp[-\omega(\mathbf{k}_{\min})/T], \tag{81.3}$$

where $\mathbf{k}_{\min}$ is the value of the phonon quasi-momentum (among all vectors of the type concerned) for which the energy $\omega(\mathbf{k})$ has its minimum value. Here it is important, of course, that the electron speed is much greater than the phonon speed ($v_F \gg u$). It is for this reason that we cannot reduce the exponential in (81.3) by changing the length of the vector $\mathbf{k}$ by moving away from the Fermi surface. Although the phonon energy may decrease by an amount $\sim u\delta k$, the energy of the electron involved in the process would simultaneously increase by a much larger amount $\sim v_F\delta k$, thus reducing ν_U instead of increasing it. To find $\mathbf{k}_{\min}$, it is sufficient therefore to consider the Fermi surface as such, without taking into account the blurring of the distribution near it. The points important in practice are usually those near the closest approach of the Fermi surface to its replica in the adjacent cell.

The solution (81.1) implies that there is a macroscopic flux of electrons in the absence of an electric field, i.e. an infinite electrical conductivity. The exponentially small frequency of Umklapp processes causes an exponentially large electrical conductivity (R. E. Peierls).

The thermal conductivity of a metal having a closed Fermi surface remains finite even when Umklapp processes are neglected. This is because, by (78.2), the thermal conductivity κ defines the heat flux in the absence of an electric current, and the condition $\mathbf{j}=0$ necessarily excludes the spurious solution (81.1). The inclusion of Umklapp processes can alter the value of κ only if it is small. The same applies to the thermoelectric coefficient α, which, by the definition (78.1), relates the temperature gradient and the electric field, again with the condition $\mathbf{j}=0$; see §82, Problem.

The above remarks, however, do not apply to compensated metals having closed electron and hole Fermi surfaces, i.e. to metals with equal numbers of electrons and holes, $N_e = N_h$ (see *SP* 2, §61). The reason is that in this case the solution (81.1) is not dependent on the presence of an electric current. The current density corresponding to this solution is

$$\begin{aligned}\mathbf{j} &= e\int \mathbf{v}\frac{\partial n_0}{\partial \epsilon}\mathbf{p}\,.\,\delta\mathbf{V}\frac{2d^3p}{(2\pi)^3}\\ &= e\int \frac{\partial n_0}{\partial \mathbf{p}}\mathbf{p}\,.\,\delta\mathbf{V}\frac{2d^3p}{(2\pi)^3}\\ &= e\int \frac{\partial n_0^{(e)}}{\partial \mathbf{p}}\mathbf{p}\,.\,\delta\mathbf{V}\frac{2d^3p}{(2\pi)^3} - e\int \frac{\partial n_0^{(h)}}{\partial \mathbf{p}}\mathbf{p}\,.\,\delta\mathbf{V}\frac{2d^3p}{(2\pi)^3}.\end{aligned}$$

The two integrals are taken over the electron and hole cavities respectively of the Fermi surface; in the second, the hole distribution used is $n^{(h)} = 1 - n$. We can now integrate by parts; the resulting integrals over the surfaces of the cell faces are zero because of the rapid decrease of $n_0^{(e)}$ and $n_0^{(h)}$ away from the respective Fermi surfaces. The result is

$$\mathbf{j} = e\delta\mathbf{V}(N_h - N_e). \tag{81.4}$$

For a compensated metal, $\mathbf{j}=0$.

This means that the electrical conductivity of a compensated metal is finite even when Umklapp processes are not taken into account. The thermal conductivity and the thermoelectric coefficient, on the other hand, are governed by the Umklapp processes and would be infinite if these were ignored, since the condition $\mathbf{j}=0$ then does not exclude the spurious solution (81.1).

In the arguments and estimates in §§81 and 82, we are essentially implying the simplest assumptions as to the form of the Fermi surface, namely that it is either closed or open, with all its characteristic dimensions of the order of $1/d$. The Fermi surfaces of actual metals, however, in general have a complex form, and may consist of several different sheets; we shall not pause to analyse the resulting complications in the behaviour of the transport coefficients. For example, the sheets of open Fermi surfaces in different cells of the reciprocal lattice may be connected by narrow bridges (of width $\Delta p \ll p_F$). The presence of the small parameter $\Delta p/p_F$ in the problem may create new intermediate temperature ranges with a different temperature dependence of the transport coefficients. The sheets of closed Fermi surfaces may come "anomalously" near together, and this may move the exponential law (81.3) into the range of "anomalously" low temperatures.

§82. Transport coefficients in metals. Low temperatures

In the quantitative study of transport phenomena at low temperatures, we shall have in mind the case of open Fermi surfaces, and therefore pay no special attention to Umklapp processes.

First of all, we shall show that relaxation in the phonon system takes place (when $T \ll \Theta$) mainly by phonon–electron (not phonon–phonon) collisions.

To estimate the phonon–electron collision integral (79.10), we note that at low temperatures $\omega \sim T$, $\epsilon - \mu \sim T$, and therefore $N_0 \sim n_0 \sim 1$, $\partial N_0/\partial\omega \sim 1/T$. The integration over d^3p is taken over the volume of a layer with thickness $\sim T/v_F$ along the Fermi surface. Since k/p is small, the argument of the delta function can be expressed as

$$\epsilon(\mathbf{p}) - \epsilon(\mathbf{p}-\mathbf{k}) - \omega(\mathbf{k}) \approx \mathbf{k}\,.\,\partial\epsilon/\partial\mathbf{p} - \omega \approx \mathbf{v}_F\,.\,\mathbf{k} - \omega. \tag{82.1}$$

The delta function is removed by integration over the directions of $\mathbf{p}$ (or, equivalently, over those of $\mathbf{v}_F$) for a given $\mathbf{k}$, adding a factor $1/v_Fk$ to the integrand. Lastly, w is estimated by means of (79.18). The result is

$$I_{\text{ph},e}(\chi) \sim -\chi(m^*/M)^{1/2} \sim -T(m^*/M)^{1/2}\delta n,$$

so that the effective collision frequency is

$$\nu_{\text{ph},e} \sim T\sqrt{(m^*/M)}. \tag{82.2}$$

The effective frequency of phonon–phonon collisions at low temperatures is, according to the estimate (69.15),

$$\nu_{\text{ph,ph}} \sim T\sqrt{(m^*/M)}(T/\Theta)^4 \ll \nu_{\text{ph},e}, \tag{82.3}$$

and this proves the above statement.

We shall henceforward neglect phonon–phonon collisions. The phonon transport equation is then

$$\mathbf{u}\cdot\frac{\partial N_0}{\partial\mathbf{r}} = -\frac{\omega}{T}\frac{\partial N_0}{\partial\omega}\mathbf{u}\,.\,\nabla T = I_{\text{ph},e}(\chi,\varphi). \tag{82.4}$$

This equation can be solved explicitly for the phonon function χ. Since $\mathbf{k}$ in this equation is given, $\chi_\mathbf{k}$ can be taken outside the integral, and we find

$$\chi_\mathbf{k} = -\frac{\omega}{T\nu_{\text{ph},e}}\mathbf{u}\,.\,\nabla T + \frac{1}{\nu_{\text{ph},e}}\int w(n_0' - n_0)\delta(\epsilon - \epsilon' - \omega)(\varphi - \varphi')\frac{2d^3p}{(2\pi)^3} \equiv \chi_1 + \chi_2, \tag{82.5}$$

where

$$\nu_{\text{ph},e} = \int w(n_0' - n_0)\delta(\epsilon - \epsilon' - \omega)\,.\,2d^3p/(2\pi)^3. \tag{82.6}$$

It is easy to see that $\chi_2 \gg \chi_1$: from the definition of χ_2, it follows that $\chi_2 \sim \varphi$ (the integrals in the numerator and denominator differ only by the factor $\varphi - \varphi'$ in the integrand), and the order of magnitude of φ is governed by the electron transport equation,

$$\mathbf{v} \,.\, \nabla T \partial n_0 / \partial T = I_{e,\mathrm{ph}}(\varphi) \sim -\nu_{e,\mathrm{ph}} \delta \tilde{n} \sim -\nu_{e,\mathrm{ph}} \varphi / T,$$

whence

$$\varphi \sim v_F |\nabla T| / \nu_{e,\mathrm{ph}}.$$

The effective electron–phonon collision frequency is estimated in the same way as $\nu_{\mathrm{ph},e}$ above, the only difference being that the integration over d^3k in $I_{e,\mathrm{ph}}$ is taken over a volume $\sim (T/u)^3$ in momentum space, instead of $\sim p_F^2 T / v_F$ in the integration over d^3p in $I_{\mathrm{ph},e}$:

$$\nu_{e,\mathrm{ph}} \sim T^3 / \Theta^2. \tag{82.7}$$

Finally, since $\chi_1 \sim |\nabla T| u / \nu_{\mathrm{ph},e}$, we have

$$\chi_1 / \chi_2 \sim u \nu_{e,\mathrm{ph}} / v_F \nu_{\mathrm{ph},e} \sim T^2 / \Theta^2 \ll 1, \tag{82.8}$$

as was to be proved.

In calculating the electrical and thermal conductivities (but not the thermoelectric coefficient; see below) we can neglect the small quantity χ_1. Substituting therefore $\chi \approx \chi_2$ from (82.5) in the linearized electron–phonon collision integral in the form (79.11), we find

$$I_{e,\mathrm{ph}}(\varphi, \chi) = I^{(1)}_{e,\mathrm{ph}}(\varphi) + I_{e,\mathrm{ph},e}(\varphi), \tag{82.9}$$

where $I_{e,\mathrm{ph},e}(\varphi)$ denotes the result of substituting χ_2 in the integral $I^{(2)}_{e,\mathrm{ph}}(\chi)$. The first term in (82.9) is the collision integral for electrons and equilibrium phonons; the second term may be called the collision integral between electrons via phonons.

As in §79, we take as independent variables in the function $\varphi(\mathbf{p})$ the quantity $\eta = \epsilon - \mu$ and the vector $\mathbf{p}_F$ which has the direction of $\mathbf{p}$ and ends on the Fermi surface. Both terms in (82.9) have in their integrands the difference

$$\varphi(\eta, \mathbf{p}_F) - \varphi(\eta', \mathbf{p}_F'), \tag{82.10}$$

with

$$\eta - \eta' = \pm \omega, \quad \mathbf{p}_F - \mathbf{p}_F' = \boldsymbol{\kappa},$$

where $\boldsymbol{\kappa}$ is the projection of $\mathbf{k}$ on the tangent plane to the Fermi surface at the point $\mathbf{p}_F$.

With regard to the variable $\mathbf{p}_F$, the function $\varphi(\eta, \mathbf{p}_F)$ varies considerably over ranges $\sim p_F$; the difference $\kappa \sim k \ll p_F$. In this sense, φ varies slowly with $\mathbf{p}_F$, and in a first approximation we can take $\mathbf{p}'_F = \mathbf{p}_F$ in the difference (82.10), i.e. replace it by

$$\varphi(\eta, \mathbf{p}_F) - \varphi(\eta', \mathbf{p}_F). \tag{82.11}$$

The dependence on η, however, is strong, in the sense that $|\eta - \eta'| = \omega \sim T$ is of the same order of magnitude as the range over which φ varies considerably.

Let L_0 be the operator obtained from $I_{e,\mathrm{ph}}$ (82.9) on replacing (82.10) by (82.11); then

$$I_{e,\mathrm{ph}}(\varphi) = L_0(\varphi) + L_1(\varphi),$$

and $L_0 \gg L_1$. The electron transport equation (in the presence of both an electric field and a temperature gradient) is

$$-\left(e\mathbf{E} + \frac{\eta}{T}\nabla T\right) \cdot \mathbf{v}\frac{\partial n_0}{\partial \epsilon} = L_0(\varphi) + L_1(\varphi). \tag{82.12}$$

The two terms on the right have quite different physical meanings: the first causes rapid energy relaxation, the second causes slow diffusive relaxation with respect to the direction of the quasi-momentum.

There are two obvious properties of the operator L_0. First, it is zero for any function of $\mathbf{p}_F$ only, since the difference (82.11) is zero. Second, the integral

$$\int L_0(\varphi)\, d\eta = 0; \tag{82.13}$$

L_0 describes collisions in which only the energy changes, and (82.13) simply states the conservation of the number of electrons having a given direction of $\mathbf{p}$.

We shall seek the solution of the transport equation in the form

$$\varphi(\eta, \mathbf{p}_F) = a(\mathbf{p}_F) + b(\eta, \mathbf{p}_F), \tag{82.14}$$

where $a(\mathbf{p}_F)$ is a function of $\mathbf{p}_F$ only, and $|a| \gg |b|$. The fact that a (for which the part L_0 of the collision integral is zero) is large expresses the rapidity of the energy relaxation. Substituting (82.14) in (82.12) and neglecting the relatively small term $L_1(b)$, we find

$$-\left(e\mathbf{E} + \frac{\eta}{T}\nabla T\right) \cdot \mathbf{v}\frac{\partial n_0}{\partial \epsilon} = L_0(b) + L_1(a). \tag{82.15}$$

The two terms on the right are in general of the same order of magnitude. However, in the calculation of electrical or thermal conductivity, only one of these terms is important, as can be seen by means of the fact that the linearized electron–phonon operator $I_{e,\mathrm{ph}}$ (and therefore L_0 and L_1) acting on the function $\varphi(\eta, p_F)$ does not change its parity with respect to η†. We therefore divide φ into

†This has been shown in §79 for $I^{(1)}_{e,\mathrm{ph}}$. We shall not pause to give the exactly similar proof for $I_{e,\mathrm{ph},e}$.

parts φ_g and φ_u even and odd with respect to η:

$$\varphi_g = a + b_g, \quad \varphi_u = b_u$$

(the function a independent of η is obviously even). Substitution of $\varphi = \varphi_g + \varphi_u$ in (82.15), followed by separation of the terms odd and even in η, gives two equations:

$$-\frac{\eta}{T}\frac{\partial n_0}{\partial \epsilon}\mathbf{v}_F \cdot \nabla T = L_0(b_u), \tag{82.16}$$

$$-\frac{\partial n_0}{\partial \epsilon} e\mathbf{E} \cdot \mathbf{v}_F = L_0(b_g) + L_1(a); \tag{82.17}$$

on the left-hand sides, the velocity $\mathbf{v}$ has been replaced, with sufficient accuracy, by the velocity $\mathbf{v}_F$ on the Fermi surface, which is independent of η. Integration of the second equation with respect to η gives

$$e\mathbf{E} \cdot \mathbf{v}_F = \int L_1(a)\, d\eta, \tag{82.18}$$

since by (82.13) the L_0 term disappears.

The heat flux (for $\mathbf{E} = 0$) is entirely determined by the solution of equation (82.16), which contains only the operator L_0: as we should expect, it depends on the electron energy relaxation processes. It is calculated from that solution as the integral

$$\mathbf{q}' = \int \mathbf{v}\eta\delta\tilde{n} \,.\, 2d^3p/(2\pi)^3 \approx -\int \mathbf{v}_F\eta \frac{\partial n_0}{\partial \eta} b_u \frac{2d^3p}{(2\pi)^3}; \tag{82.19}$$

the part of φ that is even in η makes no contribution, because the resulting integrand is an odd function.

The operator L_0 is the principal part of the electron–phonon collision integral. The corresponding effective collision frequency is therefore $\nu_{e,\mathrm{ph}}$ from (82.7); more precisely, this quantity is the effective collision frequency as regards energy exchange. The corresponding electron mean free path is $l \sim v_F/\nu_{e,\mathrm{ph}}$. The thermal conductivity can be estimated from the formula (7.10) in the kinetic theory of gases: $\kappa \sim c\bar{v}lN$. In the present case, N is the number density of electrons, c the electronic part of the specific heat (per conduction electron) and $\bar{v} \sim v_F$. The quantities N and v_F are independent of the temperature; the specific heat of an electron Fermi liquid is proportional to T, and from (82.7) the mean free path $l \propto T^{-3}$. Since the heat flux thus calculated refers to $\mathbf{E} = 0$, the coefficient in it is not the thermal conductivity κ itself but $\kappa' = \kappa + T\sigma\alpha^2$; see (78.3). Thus $\kappa' \propto T^{-2}$. The term $T\sigma\alpha^2$ is small in comparison with κ' (see the next-but-one footnote), and so $\kappa \propto T^{-2}$. Putting, for a rough estimate, $c \sim m^* p_F T/N\hbar^3$ in ordinary units (*SP* 2, (1.15)), we find

$$\kappa \sim (\epsilon_F p_F/\hbar^2)\Theta^2/T^2. \tag{82.20}$$

The electrical conductivity is obtained by solving the equation (82.18), which contains only the operator L_1: as is to be expected, the electric current depends on the processes of relaxation with respect to directions of the electron quasi-momentum. It has been noted at the beginning of § 81 that these processes are of the nature of diffusion along the Fermi surface. We shall show in § 83 how the transport equation (82.18) can in fact be put into the form of a diffusion equation. The temperature dependence of the electrical conductivity, however, can be ascertained from the following simple arguments.

The movement along the Fermi surface takes place in small jumps $k \sim T/u$; this acts as the "mean free path" l_p in momentum space, and the frequency of "scattering events" is the same as the electron–phonon collision frequency $\nu_{e,ph}$. The diffusion coefficient along the Fermi surface can be estimated by the formula $D \sim l\bar{v} \sim l^2\nu$ from the kinetic theory of gases, with l and ν replaced by l_p and $\nu_{e,ph}$. We thus have (in ordinary units)

$$D_p \sim (p_F^2\Theta/\hbar)(T/\Theta)^5. \tag{82.21}$$

From this we can find the relaxation time which is to appear in the estimate of the electrical conductivity according to (78.16): $\sigma \sim e^2Nv_F\tau/p_F$. It is the time in which the electron quasi-momentum changes by an amount of the order of itself. That is, in the time τ the electron must diffuse a distance $\sim p_F$ along the Fermi surface. In a diffusional motion, the mean square of the displacement is proportional to the time (and to the diffusion coefficient). We thus find $p_F^2 \sim D_p\tau$, and for the conductivity (in ordinary units)

$$\sigma \sim (\hbar e^2N/m^*\Theta)(\Theta/T)^5. \tag{82.22}$$

At low temperatures, therefore, the conductivity is proportional to T^{-5}.†

Let us now consider the thermoelectric coefficient. Here the position is similar to that at high temperatures. If the current **j** is calculated from the function b_u, the solution of (82.16), then, since this is an odd function of η, the integral is zero in the first approximation, and a non-zero result is obtained only when we include the next term in η/ϵ_F in the expansion of the integrand. As when $T \gg \Theta$, this gives the thermoelectric coefficient (in ordinary units)

$$\alpha^{\mathrm{I}} \sim T/e\epsilon_F, \tag{82.23}$$

instead of the "normal" order of magnitude $\alpha \sim 1/e$.†

Another contribution to the thermoelectric coefficient arises from the term χ_1, neglected in (82.5), in the phonon function χ: this contribution is due to the phonon drag acting on the electrons. If this term is retained, the collision integral (82.9)

†This result was first derived by F. Bloch (1929).

†From the estimates (82.20)–(82.23) we see that $T\alpha^2\sigma/\kappa \sim (\Theta/\epsilon_F)^2 \ll 1$, and this justifies the approximation used in deriving (82.21).

contains a further term

$$I^{(2)}_{e,\text{ph}}(\chi_1) \sim \nu_{e,\text{ph}}\chi_1 \partial N_0/\partial\omega \sim -\nu_{e,\text{ph}} u|\nabla T|/\nu_{\text{ph},e}T, \tag{82.24}$$

which may then be taken to the left-hand side of the transport equation (82.12), where it is to be compared with the term

$$-\frac{\partial n_0}{\partial T}\frac{\eta}{T}\mathbf{v}\cdot\nabla T. \tag{82.25}$$

The term (82.24) is small in comparison with (82.25), in the ratio T^2/Θ^2; the estimate is analogous to (82.8). The inclusion of it, however, gives a term proportional to ∇T in the solution φ of the transport equation, and this is not an odd function of η. Hence, in calculating the relevant contribution to the current, there is no further small factor, and the thermoelectric coefficient contains a term

$$\alpha^{\text{II}} \sim T^2/e\Theta^2 \tag{82.26}$$

(L. É. Gurevich, 1946).†

As the temperature decreases, so does the electron–phonon collision frequency, and ultimately the collisions between electrons and impurity atoms become predominant in causing the electrical and thermal resistance. Because of the different temperature dependence, the transition to "residual thermal resistance" takes place later than that to residual electrical resistance.

In very pure metals, there can exist a range of temperatures in which the transport properties of the metal are governed by collisions between electrons. The corresponding mean free path in the electron liquid in a metal, as in any Fermi liquid, varies with the temperature as T^{-2}, and the small expansion parameter is the ratio T/ϵ_F (see §75). When $T \sim \epsilon_F$, this mean free path must become $\sim d$, and so

$$l_{ee} \sim d(\epsilon_F/T)^2. \tag{82.27}$$

The temperature dependence of the electrical and thermal conductivities is then

$$\sigma \propto T^{-2}, \quad \kappa \propto T^{-1} \tag{82.28}$$

(L. D. Landau and I. Ya. Pomeranchuk 1936). When the temperature falls, the effective electron–electron collision frequency ν_{ee} decreases more slowly than the electron–phonon collision frequency $\nu_{e,\text{ph}}$. However, since the small parameter in ν_{ee}

†Here, the following comment is necessary. Since the phonon quasi-momentum is small, the law of conservation of energy gives $\epsilon(\mathbf{p}) - \epsilon(\mathbf{p}-\mathbf{k}) \approx \mathbf{v}_F\cdot\mathbf{k} \approx \pm\omega(\mathbf{k})$, from which we see that the angle θ between $\mathbf{v}_F$ and $\mathbf{k}$ is almost $\frac{1}{2}\pi$: $\cos\theta \sim \omega/v_F k \sim u/v_F \ll 1$. In the isotropic case the quasi-momentum $\mathbf{k}$ and the velocity $\mathbf{u}$ of the phonon are in the same direction, and so the product $\mathbf{u}\cdot\mathbf{v}_F$ is also small. A similar product occurs in the integral which gives the current in terms of the function φ proportional to $\mathbf{u}\cdot\nabla T$; this would cause, in the isotropic case, an additional small factor in α^{II}. In an anisotropic crystal, however, including those with cubic symmetry, there is in general no reason for such a small factor to occur.

is T/ϵ_F, and not T/Θ as in $\nu_{e,\text{ph}}$, electron–electron collisions can play a predominant role only at very low temperatures.

Note also that the laws (82.28) can in principle relate to cases with either open or closed Fermi surfaces. Since the electron quasi-momenta are large, the necessary existence of Umklapp processes does not in general give rise to any additional small factor for closed Fermi surfaces.

PROBLEM

Calculate the thermoelectric coefficient α for a metal with a closed Fermi surface at low temperatures, neglecting Umklapp processes.

SOLUTION. The electron transport equation is

$$-e\mathbf{E}\cdot\frac{\partial n_0}{\partial \mathbf{p}}-\frac{\epsilon-\mu}{T}\frac{\partial n_0}{\partial \epsilon}\mathbf{v}\,.\,\nabla T = C_{e,\text{ph}}(n). \tag{1}$$

The phonon transport equation may be written

$$-\frac{\omega}{T}\frac{\partial N_0}{\partial \mathbf{k}}\cdot\nabla T = C_{\text{ph},e}(N), \tag{2}$$

since

$$\mathbf{u}\frac{\partial N_0}{\partial T} = -\frac{\omega}{T}\frac{\partial N_0}{\partial \omega}\mathbf{u} = -\frac{\omega}{T}\frac{\partial N_0}{\partial \mathbf{k}}.$$

Multiplying equations (1) and (2) by $\mathbf{p}$ and $\mathbf{k}$ respectively, and integrating them over $2d^3p/(2\pi)^3$ and $d^3k/(2\pi)^3$ respectively, we add them term by term; the right-hand side is zero, by the conservation of the total quasi-momentum of electrons and phonons in the absence of Umklapp processes. The result is

$$\int e\mathbf{E}\cdot\frac{\partial n_0}{\partial \mathbf{p}}\mathbf{p}\frac{2d^3p}{(2\pi)^3}+\tfrac{1}{3}\nabla T\int\frac{\epsilon-\mu}{T}\frac{\partial n_0}{\partial \epsilon}\mathbf{v}\,.\,\mathbf{p}\frac{2d^3p}{(2\pi)^3}+\tfrac{1}{3}\nabla T\int\frac{\omega}{T}\left(\frac{\partial N_0}{\partial \mathbf{k}}\cdot\mathbf{k}\right)\frac{d^3k}{(2\pi)^3}=0; \tag{3}$$

the second and third integrals are written on the assumption that the crystal has cubic symmetry.

The first integral in (3) is transformed as in the derivation of (81.4), and gives $-e\mathbf{E}(N_e-N_h)$. The second integral is calculated as in the derivation of (78.12), and is $-AT\nabla T$, where

$$A=\frac{\pi^2}{9}\left[\frac{\partial}{\partial \epsilon}\int \mathbf{v}\,.\,\mathbf{p}\frac{dS}{v(2\pi)^3}\right]_{\epsilon=\epsilon_F};$$

the integral is taken over a surface of constant energy ϵ. The third integral, after integration by parts, becomes

$$-\frac{\nabla T}{3T}\int N_0(3\omega+\mathbf{k}\,.\,\mathbf{u})\frac{d^3k}{(2\pi)^3};$$

the integral over the faces of the reciprocal lattice cell is zero, because of the rapid decrease of N_0 with increasing ω at low temperatures. For long-wavelength acoustic phonons (the only ones that are important at low temperatures), the velocity $\mathbf{u}$ and the ratio $\mathbf{k}=\mathbf{k}/\omega$ depend only on the directions of $\mathbf{k}$, not on ω. Using for the integral with respect to ω the usual expression, we find that the third integral in (3) is $-BT^3\nabla T$, where

$$B=\frac{\pi^4}{15}\sum\int(1+\tfrac{1}{3}\boldsymbol{\kappa}\,.\,\mathbf{u})\kappa^2\frac{do_{\mathbf{k}}}{(2\pi)^3},$$

and the summation is over the three acoustic branches of the phonon spectrum.

Equation (3) thus becomes

$$-e\mathbf{E}(N_e-N_h)=\nabla T(AT+BT^3).$$

Comparison with (78.1) (for $\mathbf{j} = 0$) gives the thermoelectric coefficient

$$\alpha = (AT + BT^3)/(N_h - N_e). \tag{4}$$

The condition $\mathbf{j} = 0$ can be met by means of a suitably chosen term of the form (81.1) in the solution of the transport equation. In accordance with the discussion in §81, the expression (4) is finite for an uncompensated metal, but becomes infinite when $N_e = N_h$.

§83. Electron diffusion on the Fermi surface

In this section we shall show how the transport equation (82.17) for the problem of electrical conduction at low temperatures can be reduced to a diffusion equation.† Having only this problem in mind, we shall consider only the part of the function φ that is independent of $\eta = \epsilon - \mu$, and denote it by $\varphi(\mathbf{p}_F)$ instead of $a(\mathbf{p}_F)$ as in §82. We shall again have in mind the case of open Fermi surfaces.

The function

$$\frac{\delta \tilde{n}}{(2\pi)^3} = -\frac{\partial n_0}{\partial \epsilon}\frac{\varphi}{(2\pi)^3}$$

is the non-equilibrium change in the electron distribution in momentum space. From this we can go to the distribution over the Fermi surface by writing the volume element d^3p as $d\epsilon\, dS/v$ (74.19), integrating over $d\epsilon = d\eta$, and approximately replacing the area element dS on the constant-energy surface, and the speed v, which depend on ϵ, by their values dS_F and v_F on the Fermi surface. The function φ is, by hypothesis, independent of ϵ, and the integration of the factor $-\partial n_0/\partial \epsilon$ gives unity. The distribution density on the Fermi surface is then

$$\varphi(\mathbf{p}_F)/(2\pi)^3 v_F. \tag{83.1}$$

For clarity in the derivation, we first write the transport equation (82.17) with the partial derivative with respect to time on the left-hand side, as if the distribution were not stationary:

$$-\frac{\partial n_0}{\partial \epsilon}\frac{\partial \varphi}{\partial t} - e\mathbf{E}\cdot\mathbf{v}_F\frac{\partial n_0}{\partial \epsilon} = L_1(\varphi).$$

Here the term in L_0 is omitted which in any case disappears when the equation is integrated over $d\eta/v_F$:

$$\frac{\partial}{\partial t}\frac{\varphi}{v_F} - \int L_1(\varphi)\frac{d\eta}{v_F} = -e\mathbf{E}\cdot\mathbf{v}_F/v_F. \tag{83.2}$$

The first term on the left is the rate of change of the electron density on the Fermi surface. This equation must have the form of a continuity equation, i.e. the second

†The proof given below is that of R. N. Gurzhi and A. I. Kopeliovich (1971).

term on the left must be the divergence of the electron flux s on the Fermi surface, and the electric field term on the right acts as the source or sink density. Here we are concerned with a two-dimensional divergence on a curved surface, but it may be conveniently written in three-dimensional terms:

$$-\int L_1(\varphi)\, d\eta/v_F = \{\nabla_{\mathbf{p}} - \mathbf{n}_F(\mathbf{n}_F\,.\,\nabla_{\mathbf{p}})\}\,.\,\mathbf{s}, \tag{83.3}$$

where $\nabla_{\mathbf{p}}$ is the ordinary operator of differentiation with respect to Cartesian coordinates in **p**-space, and the operator in the braces is its projection on the tangent plane to the Fermi surface at any specified point ($\mathbf{n}_F$ being a unit vector along the normal).† The vector $\mathbf{s}(\mathbf{p}_F)$ is specified on the Fermi surface, but in (83.3) it is formally regarded as being specified in all space (though depending only on the direction of $\mathbf{p}_F$). The transport equation, in which we now omit the time derivative, becomes

$$\{\nabla_{\mathbf{p}} - \mathbf{n}_F(\mathbf{n}_F\,.\,\nabla_{\mathbf{p}})\}\mathbf{s} = -e\mathbf{E}\,.\,\mathbf{v}_F/v_F. \tag{83.4}$$

The problem is to find the flux s in terms of φ.

We use Cartesian coordinates in **p**-space, with the origin on the Fermi surface at the point where $\mathbf{s}(\mathbf{p}_F)$ is being calculated, and the z-axis along the normal there. By definition, the flux component s_x is the difference between the numbers of electrons per unit time crossing (as a result of collisions) a strip of unit width in the yz-plane from left to right (i.e. in the positive x-direction) and from right to left.

Let us consider the difference between the number of events in which a phonon with quasi-momentum **k** in a given range d^3k is emitted by an electron with quasi-momentum in a range d^3p, and the number of inverse events in which such a phonon is absorbed. It is minus the first term in the integrand in (79.9):

$$d^3p\,\frac{d^3k}{(2\pi)^3}\frac{\partial N_0}{\partial\omega}w(n_0' - n_0)\delta(\epsilon - \epsilon' - \omega_{\mathbf{k}})(\varphi_{\mathbf{p}'} - \varphi_{\mathbf{p}} + \chi_{\mathbf{k}}), \tag{83.5}$$

with $\mathbf{p} = \mathbf{p}' + \mathbf{k}$.‡ The phonon function $\chi_{\mathbf{k}}$ here is to be expressed in terms of φ by

†This operator appears in the two-dimensional analogue of Gauss's theorem,

$$\oint \mathbf{e}\,.\,\mathbf{s}\,dl = \int \{\nabla - \mathbf{n}(\mathbf{n}\,.\,\nabla)\}\,.\,\mathbf{s}\,dS.$$

The integral on the left is taken round a closed contour on the surface in question (**e** being a unit vector along the outward normal to the contour in the tangent plane to the surface at the point considered); the integral on the right is taken over the part of the surface that is enclosed by the contour.

‡In the foregoing arguments we have omitted a factor $(2\pi)^{-3}$ in the definition of the surface density (83.1). A corresponding factor is accordingly omitted from (83.5) also.

We have agreed, in the case of open Fermi surfaces, to include values of the electron quasi-momentum throughout the reciprocal lattice (see §81); the law of conservation of quasi-momentum is therefore written without the **b** term.

(82.5):

$$\chi_{\mathbf{k}} = -\frac{1}{\nu_{\text{ph},e}} \int w(n_0' - n_0)\delta(\epsilon - \epsilon' - \omega_{\mathbf{k}})(\varphi_{\mathbf{p}'} - \varphi_{\mathbf{p}}) \frac{2d^3p}{(2\pi)^3}, \tag{83.6}$$

with $\nu_{\text{ph},e}$ from (82.6).

If $k_x < 0$, the emission of the phonon will result in the passage through the strip (from left to right) of those electrons for which the x-component of the original quasi-momentum is in the range

$$k_x < p_x < 0; \tag{83.7a}$$

for such values of $\mathbf{p}$, (83.5) gives a positive contribution to the flux s_x. If $k_x > 0$, the emission of the phonon results in the passage through the strip (from right to left) of electrons with

$$0 < p_x < k_x; \tag{83.7b}$$

the corresponding contribution to s_x is negative.

It is now clear that to find s_x we must (1) integrate the expression (83.5) over a unit range of p_y and over the whole range of p_z (because of the rapid convergence, the latter integration can be extended from $-\infty$ to $+\infty$); (2) integrate over the range (83.7) of p_x (in view of the slow variation of all quantities with p_x on the Fermi surface, this reduces simply to multiplication by the length of the range, i.e. by $-k_x$ when we take account of the sign of the result in s_x); (3) integrate over d^3k.

The flux component s_y differs from s_x only in that k_x is replaced by k_y in the integrand. The flux may therefore be written in the vector form

$$\mathbf{s}(\mathbf{p}_F) = -\int \frac{d^3k}{(2\pi)^3} \int_{-\infty}^{\infty} \boldsymbol{\kappa} \left\{ \frac{\partial N_0}{\partial \omega} w(n_0' - n_0)\delta(\epsilon - \epsilon' - \omega)(\varphi_{\mathbf{p}'} - \varphi_{\mathbf{p}} + \chi_{\mathbf{k}}) \right\} dp_z, \tag{83.8}$$

where $\boldsymbol{\kappa}$ is the projection of $\mathbf{k}$ on the tangent plane at the point $\mathbf{p}_F$.

First of all, we write $d^3k = dk_z\, d^2\kappa$ and integrate with respect to k_z. Since $\mathbf{k}$ is small, we can transform the argument of the delta function in (83.8):

$$\delta(\epsilon_{\mathbf{p}} - \epsilon_{\mathbf{p}-\mathbf{k}} - \omega_{\mathbf{k}}) \approx \delta(\mathbf{k} \cdot \mathbf{v}_F - \omega) = \frac{1}{v_F}\delta(k_z - \omega/v_F);$$

$\mathbf{v}_F$ is along the normal to the Fermi surface. The integration with respect to k_z removes the delta function and replaces k_z everywhere by ω/v_F. Since $\omega/v_F \sim ku/v_F \ll k$, we can put simply $k_z = 0$, i.e. make the change

$$\mathbf{k} \to \boldsymbol{\kappa}. \tag{83.9}$$

The integration over $dp_z = d\epsilon/v_z$ can also be carried out in a general form, since the only rapidly varying function of ϵ in the integrand is the difference

$$n_0(\epsilon - \omega) - n_0(\epsilon) \approx -\omega \partial n_0/\partial \epsilon;$$

the integration with respect to ϵ converts this factor into ω. The expression (83.8) now becomes

$$\mathbf{s}(\mathbf{p}_F) = -\frac{1}{2\pi v_F^2}\int \boldsymbol{\kappa}\omega_\kappa \frac{\partial N_0(\omega_\kappa)}{\partial \omega_\kappa} w(\varphi_{\mathbf{p}'} - \varphi_{\mathbf{p}} + \chi_\kappa)\frac{d^2\kappa}{(2\pi)^2}. \tag{83.10}$$

To transform the integral further, we again use the smallness of $\mathbf{k}$ to write

$$\varphi(\mathbf{p}-\mathbf{k}) - \varphi(\mathbf{p}) \approx -\mathbf{k}\,.\,\partial\varphi/\partial\mathbf{p} \approx -\boldsymbol{\kappa}\,.\,\partial\varphi/\partial\mathbf{p} = -\kappa\mathbf{t}\,.\,\partial\varphi/\partial\mathbf{p},$$

where $\mathbf{t} = \boldsymbol{\kappa}/\kappa$ is a unit vector tangential to the Fermi surface, in the direction of $\boldsymbol{\kappa}$. Since a similar difference occurs in the integral (83.6), we can put $\chi(\mathbf{k})$ in the form

$$\chi(\mathbf{k}) = \boldsymbol{\kappa}\,.\,\mathbf{a}(\mathbf{t}). \tag{83.11}$$

Lastly, from (79.4),

$$w = \kappa M(\mathbf{p}_F, \mathbf{t}). \tag{83.12}$$

With this notation,

$$\mathbf{s} = -\frac{1}{2\pi v_F^2}\int \mathbf{t}\kappa^3\omega_\kappa \frac{\partial N_0}{\partial \omega_\kappa} M\left(\mathbf{t}\,.\,\mathbf{a} - \mathbf{t}\cdot\frac{\partial\varphi}{\partial\mathbf{p}}\right)\frac{\kappa\, d\kappa\, d\phi}{(2\pi)^2}, \tag{83.13}$$

where ϕ is the polar angle of directions of $\boldsymbol{\kappa}$ in the tangent plane.

The integration with respect to κ in (83.13) reduces to the calculation of the integral

$$J = \int_0^\infty \kappa^4\omega_\kappa \frac{\partial N_0}{\partial \omega_\kappa}\, d\kappa;$$

because of the rapid convergence, the integration may be extended to ∞. The energy of a phonon with a small quasi-momentum $\boldsymbol{\kappa} = \kappa\mathbf{t}$ is $\omega_\kappa = u(\mathbf{t})\kappa$. Hence

$$J = \frac{1}{u^5}\int_0^\infty \omega^5 \frac{\partial N_0}{\partial\omega}\, d\omega = -\frac{5}{u^5}\int_0^\infty N_0\omega^4\, d\omega = -\frac{5T^5}{u^5}\int_0^\infty \frac{x^4\, dx}{e^x - 1}$$
$$= -120\zeta(5)T^5/u^5;$$

the value of the zeta function is $\zeta(5) = 1.037$.

We thus arrive at the following expression for the electron flux along the Fermi surface:

$$\mathbf{s} = -\frac{30\zeta(5)T^5}{\pi^2 v_F^2}\left\langle \frac{M(\mathbf{t})}{u^5(\mathbf{t})}\mathbf{t}\left(\mathbf{t}\cdot\frac{\partial\varphi}{\partial\mathbf{p}} - \mathbf{t}\,.\,\mathbf{a}\right)\right\rangle, \tag{83.14}$$

where the angle brackets denote averaging over the directions of $\mathbf{t}$ in the tangent plane at a given point $\mathbf{p}_F$ on the Fermi surface. It remains for us to simplify as far as possible the expression for **a**.

With the definition (83.11), we have from (83.6)

$$\mathbf{a} = \frac{\int M(n_0' - n_0)\delta(\epsilon - \epsilon' - \omega)(\partial\varphi/\partial\mathbf{p})\, d^3p}{\int M(n_0' - n_0)\delta(\epsilon - \epsilon' - \omega)\, d^3p},$$

where common factors in the numerator and the denominator have been cancelled. The integration over d^3p is replaced by one over $dS_F\, d\epsilon/v_F$ (see the beginning of this section). Only the factor $n_0(\epsilon - \omega) - n_0(\epsilon)$, which is the same in both integrals, depends on ϵ; the result of the integration over $d\epsilon$ cancels in the numerator and denominator. The argument of the delta function may then be written in the form $\mathbf{k}\cdot\mathbf{v}_F - \omega \approx \boldsymbol{\kappa}\cdot\mathbf{v}_F$, quantities of relative order u/v_F being neglected. The final result is

$$\mathbf{a} = \frac{\int v_F^{-2} M\delta(\mathbf{n}\cdot\mathbf{t})(\partial\varphi/\partial\mathbf{p})\, dS_F}{\int v_F^{-2} M\delta(\mathbf{n}\cdot\mathbf{t})\, dS_F}, \tag{83.15}$$

M being a function of the position $\mathbf{p}_F$ on the Fermi surface and of the direction **t**, and **n** being the unit vector along the normal. As a consequence of the presence of the delta functions, the integrals are in fact taken only along a curve on the Fermi surface where the normal is perpendicular to the direction **t** of the phonon quasi-momentum.

Formulae (83.4), (83.14) and (83.15) solve the problem of bringing the transport equation into the form of a diffusion equation. The result is an integro-differential equation. The flux (83.14) may be written as

$$s_\alpha = -D_{\alpha\beta}(\partial\varphi/\partial p_\beta - a_\beta), \tag{83.16}$$

where

$$D_{\alpha\beta} = T^5 \frac{30\zeta(5)}{\pi^2 v_F^2}\left\langle \frac{M(\mathbf{t})}{u^5(\mathbf{t})} t_\alpha t_\beta \right\rangle \tag{83.17}$$

and α, β are two-dimensional vector suffixes. The first term has the usual differential form with the diffusion coefficient tensor $D_{\alpha\beta}$; it relates to electron scattering by equilibrium phonons. The second, integral, term is due to electron drag by non-equilibrium phonons.

The current density is calculated from the functions φ as the integral

$$\mathbf{j} = -\frac{2e}{(2\pi)^3}\int \varphi\mathbf{n}\, dS_F.$$

It is clear from (83.4), with **s** from (83.16) and (83.17), that φ (and therefore the conductivity of the metal) varies with temperature as T^{-5}, in accordance with the result in §82. Note that the electron drag by phonons does not affect this law, though it does affect the form of the transport equation.

§84. Galvanomagnetic phenomena in strong fields. General theory

The characteristic dimensionless parameter governing the effect of a magnetic field on the electrical conductivity of a metal is the ratio r_B/l, where r_B is the Larmor electron orbit radius and l the mean free path.

It is known (*SP* 2, §57) that the motion of a conduction electron in a magnetic field is almost always quasi-classical, because the ratio $\hbar\omega_B/\epsilon_F$ (where ω_B is the Larmor frequency) is very small. The trajectory in momentum space is then the circumference of a cross-section of a constant-energy surface $\epsilon(\mathbf{p}) = \text{constant}$ by a plane $p_z = \text{constant}$, the z-axis being parallel to the field. Since the energy of the electrons is close to the limiting energy ϵ_F, the constant-energy surfaces in question are close to the Fermi surface. Hence the size of the trajectory in momentum space is given by the linear dimension p_F of the appropriate cross-section of the Fermi surface. The size of the trajectory in ordinary space is

$$r_B \sim cp_F/eB,$$

and is inversely proportional to the magnetic field. In galvanomagnetic phenomena, therefore, fields are to be regarded as weak for which $r_B \gg l$, and as strong for which

$$r_B \ll l. \tag{84.1}$$

For weak magnetic fields, the transport treatment does not (for a general electron dispersion relation) lead to anything beyond the results of the purely phenomenological theory. The nature of the magnetic field dependence of the conductivity tensor components $\sigma_{\alpha\beta}$ in this case corresponds simply to an expansion in powers of B, taking account of the requirements imposed by the principle of symmetry of the kinetic coefficients (see *ECM*, §21).

In strong magnetic fields, however, the transport treatment is needed in order to find this dependence. The condition (84.1) for a strong field is in practice satisfied only at low temperatures, where the mean free path l is sufficiently long. The metal is then usually in the range of the residual resistance due to electron scattering by impurity atoms, and we shall have this case in mind. The interaction of the conduction electrons with an impurity atom takes place at distances of the order of the lattice constant d. If $r_B \ll l$ but $r_B \gg d$, the presence of the magnetic field does not affect this interaction, nor therefore the collision integral. Under these conditions, the magnetic field dependence of the conductivity tensor is not affected by the specific form of the collision integral. It does depend considerably, however, on the structure of the conduction electron energy spectrum, i.e. on the form of the Fermi surface.†

†The theory given below is due to I. M. Lifshitz, M. Ya. Azbel' and M. I. Kaganov (1956).

Let us now construct the transport equation describing galvanomagnetic phenomena.

The distribution function is here suitably expressed not in terms of the Cartesian components of the quasi-momentum **p** but in terms of other variables related to the electron trajectory: the energy ϵ, the quasi-momentum component p_z along the magnetic field (the z-axis) and the "time for the electron to move along the momentum trajectory" from some fixed point to the point in question. This latter variable, which we denote by τ, is brought in by means of the quasi-classical equation of motion of a conduction electron in a magnetic field,

$$d\mathbf{p}/d\tau = -e\mathbf{v}\times\mathbf{B}/c, \quad \mathbf{v} = \partial\epsilon/\partial\mathbf{p};$$

the x- and y-components of this are

$$dp_x/d\tau = -ev_yB/c, \quad dp_y/d\tau = ev_xB/c. \tag{84.2}$$

Taking the sum of the squares of these equations and using the element of length ds on the momentum trajectory in the xy-plane ($ds^2 = dp_x^2 + dp_y^2$), we obtain

$$d\tau = (c/eB)\, ds/v_\perp, \quad v_\perp^2 = v_x^2 + v_y^2; \tag{84.3}$$

integration of this equation gives the new variable τ in terms of the old variables p_x, p_y, p_z.

The left-hand side of the transport equation† is, in the new variables,

$$\frac{dn}{dt} = \frac{\partial n}{\partial \epsilon}\dot{\epsilon} + \frac{\partial n}{\partial p_z}\dot{p}_z + \frac{\partial n}{\partial \tau}\dot{\tau}. \tag{84.4}$$

The distribution function will, as usual, be sought in the form

$$n = n_0(\epsilon) + \delta\tilde{n}(\epsilon, p_z, \tau). \tag{84.5}$$

It has been shown at the end of §74 that, in static electric and magnetic fields, the transport equation, linearized with respect to $\delta\tilde{n}$, for quasi-particles in a Fermi liquid has the same form as for particles in a Fermi gas. The derivatives $\dot{\epsilon}$, $\dot{p}_z$ and $\dot{\tau}$ are to be expressed by means of the equation of motion of an individual electron in an electromagnetic field:

$$\dot{\mathbf{p}} = -e\mathbf{E} - e\mathbf{v}\times\mathbf{B}/c. \tag{84.6}$$

Hence we have

$$\dot{\epsilon} = (\partial\epsilon/\partial\mathbf{p})\,.\,\dot{\mathbf{p}} = -e\mathbf{v}\,.\,\mathbf{E};$$

†The use of the quasi-classical transport equation implies the neglect of effects due to quantization of energy levels in the magnetic field. These will be discussed in §90.

the magnetic field does not appear, since it does no work on the charge. For a field **B** in the z-direction we have $\dot{p}_z = -eE_z$. Lastly, a comparison of (84.2) and (84.6) shows that the derivative $d\tau/dt$ differs from unity only because of the field **E**; and this difference need not be taken into account.

Since the equilibrium distribution function n_0 depends only on ϵ, and ϵ, p_z and τ are independent variables, we have $\partial n_0/\partial p_z = 0$, $\partial n_0/\partial \tau = 0$. The electric field is regarded as being extremely weak, and in linearizing the transport equation the terms which contain both the small quantities $\delta\tilde{n}$ and **E** are to be omitted. The expression (84.4) then reduces to

$$\frac{dn}{dt} \approx -\frac{\partial n_0}{\partial \epsilon} e\mathbf{v} \,.\, \mathbf{E} + \frac{\partial \delta\tilde{n}}{\partial t}.$$

We write

$$\delta\tilde{n} = (\partial n_0/\partial \epsilon) e\mathbf{E} \,.\, \mathbf{g}, \quad \mathbf{g} = \mathbf{g}(\epsilon, p_z, \tau); \tag{84.7}$$

cf. (78.6). The left-hand side of the transport equation then becomes finally

$$\frac{dn}{dt} = \frac{\partial n_0}{\partial \epsilon} e\mathbf{E} \cdot \left(-\mathbf{v} + \frac{\partial \mathbf{g}}{\partial \tau}\right). \tag{84.8}$$

The collision integral on the right of the transport equation, after linearization, is written in the form

$$C(n) = (\partial n_0/\partial \epsilon) e\mathbf{E} \,.\, I(\mathbf{g}); \tag{84.9}$$

in the collision integral for elastic scattering by impurity atoms, any factor in $\delta\tilde{n}$ that depends only on ϵ can be taken outside the integral. The specific form of the linear integral operator $I(\mathbf{g})$ need not be stated.

Equating (84.8) and (84.9), we have finally the transport equation determining the function **g**:

$$\partial \mathbf{g}/\partial \tau - I(\mathbf{g}) = \mathbf{v}. \tag{84.10}$$

The conductivity tensor is given by the integral (78.9):

$$\sigma_{\alpha\beta} = -e^2 \int \frac{\partial n_0}{\partial \epsilon} v_\alpha g_\beta \frac{2d^3p}{(2\pi\hbar)^3}.$$

In this integral, the change to the new variables is made by the substitution $d^3p \to |J|\, d\epsilon\, dp_z\, d\tau$, where

$$J = \partial(p_x, p_y, p_z)/\partial(\tau, \epsilon, p_z)$$

is the Jacobian of the transformation, which is easily found directly from the equations (84.2) which define the variable τ. Writing both sides of the first equation

(84.2), for instance, as Jacobians,

$$\frac{\partial(p_x, \epsilon, p_z)}{\partial(\tau, \epsilon, p_z)} = -\frac{eB}{c}\frac{\partial(\epsilon, p_x, p_z)}{\partial(p_y, p_x, p_z)},$$

and multiplying both sides by $\partial(p_y, p_x, p_z)/\partial(\epsilon, p_x, p_z)$, we find $|J| = eB/c$. Neglecting the thermal blurring of the distribution n_0, we put as usual $\partial n_0/\partial\epsilon = -\delta(\epsilon - \epsilon_F)$, obtaining as the final expression

$$\sigma_{\alpha\beta} = \frac{2e^3B}{c(2\pi\hbar)^3}\int v_\alpha g_\beta \, d\tau \, dp_z, \tag{84.11}$$

the integration being taken over the Fermi surface.

According to the definition (84.3), τ is proportional to $1/B$. The term $\partial\mathbf{g}/\partial\tau$ in the linear equation (84.10) is therefore proportional to B, and is thus large compared with the other terms. This makes it possible to solve the equation by successive approximation as a series in powers of $1/B$:

$$\mathbf{g} = \mathbf{g}^{(0)} + \mathbf{g}^{(1)} + \cdots, \tag{84.12}$$

where $\mathbf{g}^{(n)} \propto B^{-n}$.† The terms in this series satisfy the equations

$$\left.\begin{aligned} \partial\mathbf{g}^{(0)}/\partial\tau &= 0, \\ \partial\mathbf{g}^{(1)}/\partial\tau &= I(\mathbf{g}^{(0)}) + \mathbf{v}, \\ \partial\mathbf{g}^{(2)}/\partial\tau &= I(\mathbf{g}^{(1)}), \ldots \end{aligned}\right\} \tag{84.13}$$

The solution of these equations is

$$\left.\begin{aligned} \mathbf{g}^{(0)} &= \mathbf{C}^{(0)}, \\ \mathbf{g}^{(1)} &= \int_0^\tau [I(\mathbf{C}^{(0)}) + \mathbf{v}(\tau)]\, d\tau + \mathbf{C}^{(1)}, \\ \mathbf{g}^{(2)} &= \int_0^\tau I(\mathbf{g}^{(1)})\, d\tau + \mathbf{C}^{(2)}, \ldots, \end{aligned}\right\} \tag{84.14}$$

where $\mathbf{C}^{(0)}, \mathbf{C}^{(1)}, \ldots$ are functions of ϵ and p_z only.

The function $\mathbf{g}$ must satisfy certain conditions. If the electron momentum trajectories (i.e. the perimeters of cross-sections of the Fermi surface by planes p_z = constant) are closed, the motion of the electrons is periodic; accordingly, the function $\mathbf{g}(\epsilon, p_z, \tau)$ must be periodic in τ (the period T depending on p_z). If the trajectory is open, however, the motion in momentum space is infinite and $\mathbf{g}$ need only satisfy the condition of being finite.

Let us now average equations (84.13) with respect to τ. If the functions $\mathbf{g}$ are

†As in §59 when calculating the transport coefficients of a plasma in a strong magnetic field.

periodic, the mean value over the period,

$$\overline{\frac{\partial \mathbf{g}}{\partial \tau}} = \frac{1}{T}\int_0^T \frac{\partial \mathbf{g}}{\partial \tau}\,d\tau = \frac{\mathbf{g}(T)-\mathbf{g}(0)}{T},$$

is zero, since $\mathbf{g}(T) = \mathbf{g}(0)$. If they are not periodic, the averaging is over an infinite range of τ, and the mean value is zero since $\mathbf{g}$ is finite. In all cases, therefore, averaging the equations gives

$$\overline{I(\mathbf{g}^{(0)})} \equiv \overline{I(\mathbf{C}^{(0)})} = -\bar{\mathbf{v}}, \quad \overline{I(\mathbf{g}^{(1)})} = 0, \ldots; \tag{84.15}$$

these relations determine in principle the functions $\mathbf{C}^{(0)}$, $\mathbf{C}^{(1)}, \ldots$

In going on to calculate the conductivity tensor, let us first recall some general properties of it which follow from the phenomenological theory (*ECM*, § 21).

The principle of the symmetry of the kinetic coefficients gives

$$\sigma_{\alpha\beta}(\mathbf{B}) = \sigma_{\beta\alpha}(-\mathbf{B}). \tag{84.16}$$

The tensor $\sigma_{\alpha\beta}$ may be separated into symmetric and antisymmetric parts:

$$\sigma_{\alpha\beta} = \sigma^{(s)}_{\alpha\beta} + \sigma^{(a)}_{\alpha\beta}. \tag{84.17}$$

For these we have, with (84.16),

$$\left.\begin{aligned} \sigma^{(s)}_{\alpha\beta}(\mathbf{B}) &= \sigma^{(s)}_{\beta\alpha}(\mathbf{B}) = \sigma^{(s)}_{\alpha\beta}(-\mathbf{B}), \\ \sigma^{(a)}_{\alpha\beta}(\mathbf{B}) &= -\sigma^{(a)}_{\beta\alpha}(\mathbf{B}) = -\sigma^{(a)}_{\alpha\beta}(-\mathbf{B}). \end{aligned}\right\} \tag{84.18}$$

The components $\sigma^{(s)}_{\alpha\beta}$ and $\sigma^{(a)}_{\alpha\beta}$ are therefore even and odd functions of $\mathbf{B}$ respectively. Instead of the antisymmetric tensor $\sigma^{(a)}_{\alpha\beta}$, we can use its dual axial vector $\mathbf{a}$, defined by

$$a_{xy} = a_z, \quad a_{zx} = a_y, \quad a_{yz} = a_x.$$

The components of the current density vector are then

$$j_\alpha = \sigma_{\alpha\beta}E_\beta = \sigma^{(s)}_{\alpha\beta}E_\beta + (\mathbf{E}\times\mathbf{a})_\alpha. \tag{84.19}$$

The dissipation of energy when the current flows is determined only by the symmetric part of the conductivity tensor: $\mathbf{j}\cdot\mathbf{E} = \sigma^{(s)}_{\alpha\beta}E_\alpha E_\beta$. The inverse tensor $\rho_{\alpha\beta} = \sigma^{-1}{}_{\alpha\beta}$ can thus also be separated into symmetric and antisymmetric parts, the latter having a dual axial vector $\mathbf{b}$. Then $\mathbf{E}$ is expressed in terms of $\mathbf{j}$ by

$$E_\alpha = \rho^{(s)}_{\alpha\beta}j_\beta + (\mathbf{j}\times\mathbf{b})_\alpha. \tag{84.20}$$

The terms $\mathbf{E}\times\mathbf{a}$ in the current and $\mathbf{j}\times\mathbf{b}$ in the field represent the *Hall effect*.

§85. Galvanomagnetic phenomena in strong fields. Particular cases

Closed trajectories

Let us begin with cases where all the momentum trajectories of electrons (i.e. for all p_z) with a given direction of **B** are closed. This is true for any direction of **B** if the Fermi surfaces are closed. With open Fermi surfaces, cases can occur where the trajectories are closed for any direction of **B**, and where the cross-sections are closed only for certain directions (or certain ranges of directions) of the field.

In movement in a closed trajectory (in the xy-plane), the mean values of the velocities in the plane are zero: $\overline{v_x} = \overline{v_y} = 0$, as is clear from the equations of motion (84.2) with allowance for the fact that p_x and p_y return to their initial values after a traversal of the whole trajectory. The value of $\overline{v_z}$ is always non-zero, because the motion in the direction of the field is infinite. The first equation (84.15) now gives

$$\overline{I(C_x^{(0)})} = \overline{I(C_y^{(0)})} = 0,$$

whence $C_x^{(0)} = C_y^{(0)} = 0$.† The solution (84.14) then becomes

$$\left.\begin{aligned} g_x &= (c/eB)p_y + C_x^{(1)} + g_x^{(2)} + \cdots, \\ g_y &= -(c/eB)p_x + C_y^{(1)} + g_y^{(2)} + \cdots, \\ g_z &= C_z^{(0)} + g_z^{(1)} + \cdots; \end{aligned}\right\} \tag{85.1}$$

the integration of $\mathbf{v}(\tau)$ is carried out by means of equations (84.2).

The components of the stress tensor are calculated from equation (84.11). For example,

$$\sigma_{xx} = \frac{2e^2}{(2\pi\hbar)^3}\int\oint \frac{dp_y}{d\tau}\left[\frac{c}{eB}p_y + C_x^{(1)} + g_x^{(2)}\right] d\tau\, dp_z,$$

with v_x again given by (84.2). Since $C_x^{(1)}$ is independent of τ, the integration with respect to τ in the first two terms amounts to that of the derivatives $dp_y^2/d\tau$ and $dp_y/d\tau$, and gives zero. Thus the only contribution to the integral comes from the $g_x^{(2)}$ term, so that $\sigma_{xx} \propto B^{-2}$.

Next we calculate

$$\sigma_{xy} = \frac{2e^2}{(2\pi\hbar)^3}\int\oint \frac{dp_y}{d\tau}\left[-\frac{c}{eB}p_x + C_y^{(1)}\right] d\tau\, dp_z.$$

Integration of the second term again gives zero, and in the first term

$$\oint p_x \frac{dp_y}{d\tau} d\tau = \int p_x\, dp_y = \pm S(p_z),$$

†There is no reason why the linear homogeneous equation $\overline{I(C)} = 0$ should have any solution other than the trivial $C = 0$.

where $S(p_z)$ is the area of the cross-section of the Fermi surface by the plane $p_z = \text{constant}$. The plus and minus signs relate to cases where the perimeter encloses a region of smaller and larger energies respectively, i.e. the closed trajectory is an electron and a hole trajectory respectively (see *SP* 2, §61); we denote the area S by S_e and S_h in these two cases. The difference in sign is due to a change in the direction of traversal of the trajectory. The integration of S with respect to p_z gives the volume Ω in momentum space within the Fermi surface; if the closed trajectories are on an open Fermi surface, then Ω is the volume between that surface and the faces of the reciprocal lattice cell. Thus

$$\sigma_{xy} = \frac{ec}{B}\frac{2(\Omega_h - \Omega_e)}{(2\pi\hbar)^3} = \frac{ec}{B}(N_h - N_e), \tag{85.2}$$

where Ω_e and Ω_h are the volumes of the electron and hole cavities of the Fermi surface. The quantities

$$N_e = 2\Omega_e/(2\pi\hbar)^3, \quad N_h = 2\Omega_h/(2\pi\hbar)^3$$

are respectively the numbers of electron-occupied states with energies $\epsilon < \epsilon_F$ and free states with $\epsilon > \epsilon_F$, per unit volume of the crystal. For closed Fermi surfaces these concepts have a quite definite meaning; N_e and N_h are characteristics of the electronic spectrum of the metal and independent of the direction of the field **B**. For open surfaces, their meaning becomes more conventional, as they may depend on the direction of **B**.

The expression (85.2) is an odd function of **B** and therefore belongs to the antisymmetric part of the tensor $\sigma_{\alpha\beta}$.† The component $\sigma_{xy}^{(s)}$ of the symmetric part is given by the next term in the expansion of σ_{xy}, which is proportional to B^{-2}.

The dependence on B of the remaining components $\sigma_{\alpha\beta}$ is found similarly. For example,

$$\sigma_{zz} = \frac{2e^3B}{(2\pi\hbar)^3c}\int\oint v_z C_z^{(0)}\,d\tau\,dp_z.$$

The integration with respect to τ brings in a factor B^{-1}, and $C_z^{(0)}$ is independent of B; hence σ_{zz} also is independent of B.

The result is

$$\sigma_{zz}^{(s)} = \text{constant}, \quad \text{other } \sigma_{\alpha\beta}^{(s)} \propto B^{-2}, \quad \mathbf{a} \propto B^{-1}. \tag{85.3}$$

All components $\sigma_{\alpha\beta}^{(s)}$ and **a** depend on the form of the collision integral, except

$$a_z = (ec/B)(N_h - N_e).$$

†It is clear from the derivation of the transport equation that B appears in it not as the magnitude of the vector **B** but as the component $B_z = B$. The change $\mathbf{B} \to -\mathbf{B}$ therefore requires also $B \to -B$ in the formulae given.

All the $\sigma_{\alpha\beta}$ except σ_{zz} tend to zero as $B \to \infty$. The physical reason for this behaviour is the localization of electrons on orbits small compared with the mean free path; σ_{zz} is finite because the motion of electrons along the magnetic field always remains infinite.

The small parameter in the expansion is the ratio r_B/l. Hence the components $\sigma^{(s)}_{\alpha\beta}$ that are proportional to B^{-2} may be estimated in order of magnitude as

$$\sigma^{(s)} \sim \sigma_0 (r_B/l)^2, \quad \sigma_0 \sim Ne^2 l/p_F.$$

Note that $\sigma^{(s)} \propto 1/l$; this means that, as the mean free path increases, the transverse conductivity in the magnetic field tends to zero, not to infinity as when the field is absent.

The components of the antisymmetric part of the tensor $\sigma_{\alpha\beta}$ are estimated as

$$\sigma^{(a)} \sim \sigma_0 r_B/l \sim ecN/B.$$

It must be emphasized, however, that the fact that this estimate is independent of l does not mean that the exact values of the $\sigma^{(a)}_{\alpha\beta}$ (apart from $\sigma^{(a)}_{xy}$) are independent of the specific form of the collision integral; an exact calculation of the tensor $\sigma_{\alpha\beta}$ would require a complete determination of the functions $\mathbf{C}^{(1)}$ and $\mathbf{g}^{(2)}$ by solving the specific transport equation.

From (85.3) we can find also the limiting dependences on B for the components of the inverse tensor $\rho_{\alpha\beta} = \sigma^{-1}{}_{\alpha\beta}$.† Retaining only the terms of the lowest order in $1/B$, we find

$$\rho^{(s)}_{\alpha\beta} = \text{constant}, \quad b_x,\ b_y = \text{constant}, \quad b_z \propto B, \tag{85.4}$$

and all these quantities depend on the form of the collision integral, except for

$$b_z \approx -1/a_z = B/ec(N_e - N_h). \tag{85.5}$$

All the components $\rho^{(s)}_{\alpha\beta}$ tend to constant limits as $B \to \infty$.

Compensated metals, in which $N_e = N_h$, need special treatment. The expression (85.2) is then zero, and the expansion of $\sigma^{(a)}_{xy}$ begins with the term proportional to B^{-3}. In this case, therefore,

$$a_x, \quad a_y \propto B^{-1}, \quad a_z \propto B^{-3}; \tag{85.6}$$

†The inverse tensor must, of course, be calculated from $\sigma_{\alpha\beta} = \sigma^{(s)}_{\alpha\beta} + \sigma^{(a)}_{\alpha\beta}$, and only then be separated into symmetric and antisymmetric parts. We thus find

$$\rho^{(s)}_{\alpha\beta} = \frac{1}{\sigma}\{\sigma^{(s)-1}{}_{\alpha\beta}\sigma^{(s)} + a_\alpha a_\beta\}, \quad b_\alpha = -\frac{1}{\sigma}\sigma^{(s)}_{\alpha\beta} a_\beta,$$

where $\sigma = \sigma^{(s)} + \sigma^{(s)}_{\alpha\beta} a_\alpha a_\beta$ is the determinant of $\sigma_{\alpha\beta}$, and $\sigma^{(s)}$ is the determinant of its symmetric part; see *ECM*, §21, Problem.

the dependence of $\sigma^{(s)}_{\alpha\beta}$ on **B** is as before. For the inverse tensor, we now have

$$\left.\begin{aligned}\rho^{(s)}_{zz} = \text{constant}, \quad \rho^{(s)}_{yz}, \rho^{(s)}_{xz} = \text{constant},\\ \rho^{(s)}_{xy}, \rho^{(s)}_{xx}, \rho^{(s)}_{yy} \propto B^2, \mathbf{b} \propto B.\end{aligned}\right\} \tag{85.7}$$

Open trajectories

For metals with open Fermi surfaces, which allow open trajectories, several cases are possible; we shall here consider only one of these, which illustrates the characteristic features of the situation.

Let us take a Fermi surface of the corrugated-cylinder type, passing continuously from one reciprocal lattice cell to the next (Fig. 30). If the magnetic field is not perpendicular to the cylinder axis, all cross-sections are closed, and the asymptotic dependence of $\sigma_{\alpha\beta}$ on **B** is again given by (85.3).

If, however, the magnetic field is perpendicular to the cylinder axis, there are open cross-sections. As usual, we take the z-axis parallel to the field and the x-axis here parallel to the cylinder axis; Fig. 31 shows a cut across the part of the Fermi surface in one cell. The trajectories are open when $|p_z| < |p_1|$, and are infinite in the direction of the p_x-axis. The mean velocity values are

$$\overline{v_x} = (c/eB)\,\overline{dp_y/d\tau} = 0, \quad \overline{v_y} = -(c/eB)\,\overline{dp_x/d\tau} \neq 0,$$

since p_x varies without limit; as always, $\overline{v_z} \neq 0$. The non-zero components of $\mathbf{C}^{(0)}$ in the solution of the transport equation are now $C^{(0)}_y$ and $C^{(0)}_z$; in (85.1), the second line is therefore replaced by

$$g_y = C^{(0)}_y + g^{(1)}_y + \cdots.$$

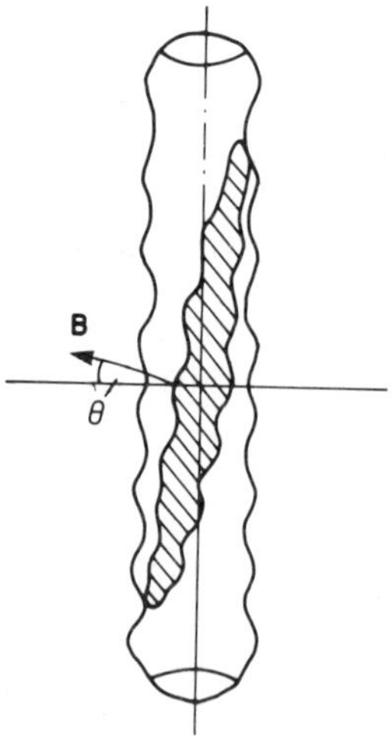

FIG. 30.

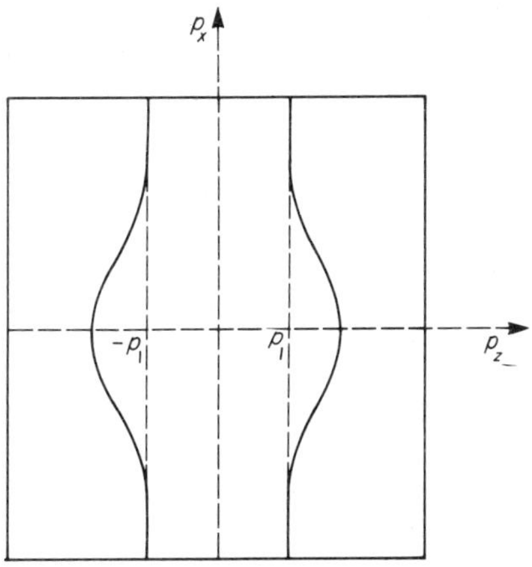

FIG. 31.

In the same manner as above, we now find

$$\sigma^{(s)}_{xx} \propto B^{-2}, \quad \text{other } \sigma^{(s)}_{\alpha\beta} \text{ constant}, \quad a_x \propto B^{-3}, \quad a_y, a_z \propto B^{-1}, \tag{85.8}$$

Hence, for the inverse tensor,

$$\rho^{(s)}_{xx} \propto B^2, \quad \text{other } \rho^{(s)}_{\alpha\beta} \text{ constant}, \quad b_x \propto B^{-1}, \quad b_y, b_z \propto B. \tag{85.9}$$

Thus there is a sharp anisotropy of the resistance in the plane perpendicular to the magnetic field: the resistance ρ_{yy} along the y-axis tends to a constant, whereas that along the x-axis increases as the square of the field.†

Another feature of the galvanomagnetic properties of metals with an open Fermi surface is their considerable dependence on the direction of a strong magnetic field. Here the change takes place as the direction of **B** approaches the plane perpendicular to the cylinder axis, and the relations (85.3) and (85.4) are replaced by (85.8) and (85.9). When **B** is at a small angle θ to that plane (Fig. 30), the momentum trajectory of the electron becomes large, with dimensions of the order of p_F/θ, where p_F is the transverse dimension of the cylindrical Fermi surface. Accordingly, the trajectories in actual space also become large, with dimensions of the order of r_B/θ, where r_B is the Larmor radius corresponding to the momentum p_F. For angles such that $r_B/\theta l \sim 1$, the expansion in powers of r_B/l used above becomes inapplicable, and it is in this range of angles that the field dependence of the resistance changes.

The whole of this discussion has related, of course, to single crystals. In polycrystalline materials there is an averaging of the anisotropic galvanomagnetic properties, depending on the directional distribution of the crystallites.

†The electron trajectory in the xy-plane of actual space differs from that in the p_xp_y-plane of momentum space only in scale and in a rotation through 90° (*SP* 2, §57). In the present case, therefore, the motion of the electron in actual space is infinite in the y-direction.

The thermomagnetic phenomena in a metal in a strong magnetic field could be discussed similarly. In particular, the components of the electronic thermal conductivity tensor would be found to tend to zero as $B \to \infty$. Under these conditions, the transfer of heat by phonons becomes important, it is necessary to take account of the electron–phonon interaction, and the whole situation is greatly complicated.

§86. Anomalous skin effect

It is known from macroscopic electrodynamics that an alternating electromagnetic field is damped within a conductor, and not only the field but also the resulting electric current is concentrated near the surface of the conductor. This is called the *skin effect*. The following formulae have been given in *ECM*, §§45 and 46.

The quasi-steady electromagnetic field in the metal satisfies Maxwell's equations

$$\text{curl}\,\mathbf{E} = -(1/c)\partial\mathbf{B}/\partial t, \tag{86.1}$$

$$\text{curl}\,\mathbf{B} = 4\pi\mathbf{j}/c, \quad \text{div}\,\mathbf{B} = 0; \tag{86.2}$$

the metal is assumed non-magnetic, so that $\mathbf{H} = \mathbf{B}$ in it. Here, of course, we assume that the general condition for validity of the macroscopic equations is satisfied: the distances δ over which the field varies significantly are large in comparison with atomic dimensions. If moreover these distances are large in comparison with the conduction electron mean free path l, then the relation between the current density $\mathbf{j}$ and the field $\mathbf{E}$ is given by linear expressions connecting their values at one point in space: $j_\alpha = \sigma_{\alpha\beta}E_\beta$, where $\sigma_{\alpha\beta}$ is the conductivity tensor. The skin effect is then said to be *normal*. In discussing this case, we shall assume the medium isotropic, or else a crystal with cubic symmetry; the tensor $\sigma_{\alpha\beta}$ then reduces to a scalar, and $\mathbf{j} = \sigma\mathbf{E}$.

Let us take a simple geometry in which the metal occupies the half-space $x > 0$ bounded by the plane $x = 0$. A uniform external electric field is applied, parallel to the surface of the metal and varying in time with frequency ω. Equations (86.1) and (86.2) become

$$\text{curl}\,\mathbf{E} = i\omega\mathbf{B}/c, \quad \text{curl}\,\mathbf{B} = 4\pi\sigma\mathbf{E}/c, \quad \text{div}\,\mathbf{B} = 0. \tag{86.3}$$

The symmetry of the problem shows that the distributions of all quantities in the metal are functions of the coordinate x only. The first equation (86.3) then shows that the magnetic field $\mathbf{B}$ is everywhere parallel to the boundary plane. We can satisfy all the equations by supposing that the electric field $\mathbf{E}$ is everywhere in that plane also. This also fulfils automatically the necessary boundary condition that the current component normal to the surface of the metal is zero: from $E_x = 0$, it follows that $j_x = 0$ everywhere.†

†The situation is different in an anisotropic medium. To satisfy this condition there, an electric field normal to the surface would have to be present, as well as the tangential one.

Eliminating **B** from the first two equations (86.3), we find

$$\text{curl curl } \mathbf{E} = \text{grad div } \mathbf{E} - \triangle\mathbf{E} = 4\pi i\omega\sigma\mathbf{E}/c^2.$$

For the tangential field, which depends only on x, div $\mathbf{E} = 0$, and this equation becomes

$$\mathbf{E}'' = -4\pi i\omega\sigma\mathbf{E}/c^2, \tag{86.4}$$

the prime denoting differentiation with respect to x. The solution which tends to zero as $x \to \infty$ is

$$\mathbf{E} = \mathbf{E}_0 e^{-i\omega t e(u^i - 1)x/\delta}, \tag{86.5}$$

where $\mathbf{E}_0$ is the field amplitude at the surface of the metal, and

$$\delta = c/\sqrt{(2\pi\sigma\omega)}. \tag{86.6}$$

The quantity δ is called the *field penetration depth*; it decreases with increasing frequency of the field. The magnetic field in the metal decreases according to the same law; it follows from (86.3) that **E** and **B** are everywhere related by $\mathbf{E} = \zeta\mathbf{B} \times \mathbf{n}$, where **n** is a unit vector normal to the surface and into the metal, i.e. in the positive x-direction, with

$$\zeta = (1 - i)\omega\delta/2c = (1 - i)\sqrt{(\omega/8\pi\sigma)}. \tag{86.7}$$

In particular, this relation exists between the values of the fields at the surface of the metal:

$$\mathbf{E}_0 = \zeta\mathbf{B}_0 \times \mathbf{n}. \tag{86.8}$$

The quantity ζ is called the *surface impedance* of the metal. Its real part governs the field energy dissipation in the metal (*ECM*, §67).

In order that the relation $\mathbf{j} = \sigma\mathbf{E}$ should hold between the current and the field at the same point and the same time, the electron mean free path l and mean free time $\tau \sim l/v_F$ must satisfy the conditions $l \ll \delta$ and $\tau\omega \ll 1$, i.e. l must be small in comparison with the characteristic distance δ over which the field varies, and τ small in comparison with the field period. When the first of these conditions is not satisfied, the relation between the field and the current is no longer local and there is a spatial dispersion of the conductivity. When the second condition is not satisfied, there is a frequency dispersion of the conductivity. The transport equation is then needed to determine the relation between the current and the field.

The nature of the skin effect thus depends on the relative magnitude of the three characteristic dimensions δ, l, and v_F/ω. The normal skin effect described by (86.5)–(86.8) corresponds to the lowest range of frequencies, such that

$$l \ll \delta, \quad l \ll v_F/\omega. \tag{86.9}$$

As the field frequency increases, or the mean free path increases (with decreasing temperature of the metal), the penetration depth decreases. In metals, the condition $l \ll \delta$ is usually the first to be violated, and the current–field relation becomes non-local; the skin effect is then said to be *anomalous*. In this section, we shall consider the limiting case where

$$\delta \ll l, \quad \delta \ll v_F/\omega. \tag{86.10}$$

There may be any relation between l and v_F/ω.†

The solution of the boundary problem of the skin effect begins from the auxiliary problem of the relation in an infinite metal between the current and the electric field

$$\mathbf{E} = \mathbf{E}_0 e^{i(\mathbf{k}\,.\,\mathbf{r}-\omega t)}$$

variable in time and space. The wave vector of the field is assumed to satisfy the inequalities

$$1/k \ll l, \quad 1/k \ll v_F/\omega, \tag{86.11}$$

corresponding to (86.10). The change δn in the electron distribution varies in the same manner as the field.

Since $v_F k \gg v_F/l \sim 1/\tau$, the collision integral $C(n) \sim \delta n/\tau$ in the transport equation may be neglected in comparison with the spatial derivative term $\mathbf{v}\,.\,\partial n/\partial \mathbf{r} \sim v_F k \delta n$. Since $k v_F \gg \omega$, we can also neglect the time derivative $\partial n/\partial t \sim \omega \delta n$.

With the latter approximation, the transport equation for the quasi-particles in an electron Fermi liquid again reduces to the equation for a gas when the distribution function is redefined, δn being replaced by $\delta\tilde{n}$ from (74.13). In the present case, these approximations bring the transport equation to the simple form

$$\mathbf{v}\,.\,\partial\delta\tilde{n}/\partial\mathbf{r} - e\mathbf{E}\,.\,\partial n_0/\partial\mathbf{p} = 0.$$

Putting

$$\partial\delta\tilde{n}/\partial\mathbf{r} = i\mathbf{k}\delta\tilde{n}, \quad \partial n_0/\partial\mathbf{p} = \mathbf{v}\,\partial n_0/\partial\epsilon,$$

we then find

$$\delta\tilde{n} = -i(e\mathbf{E}\,.\,\mathbf{v}/\mathbf{k}\,.\,\mathbf{v})\partial n_0/\partial\epsilon. \tag{86.12}$$

This expression has a pole at $\mathbf{k}\,.\,\mathbf{v} = 0$. To calculate the current

$$\mathbf{j} = -e\int \mathbf{v}\delta\tilde{n}\,.\,2d^3p/(2\pi\hbar)^3,$$

†The equality $\delta \sim l$ is reached when $\omega \sim c^2/\sigma l^2$, i.e. (with the estimate $\sigma \sim le^2N/p_F$) when $\omega \sim c^2 p_F/e^2 l^3 N$. This is compatible with the inequality $\delta \sim l \ll v_F/\omega$ if $l \gg c/\Omega$, where $\Omega \sim (Ne^2/m^*)^{1/2}$ is the plasma frequency of the metal ($m^* \sim p_F/v_F$ being the effective mass of the conduction electrons). For ordinary metals, $\Omega \sim 10^{15}$–10^{16} s^{-1}.

this pole must be avoided by writing $\mathbf{k}\cdot\mathbf{v}\to\mathbf{k}\cdot\mathbf{v}-i0$:†

$$\mathbf{j} = ie^2\int\frac{\mathbf{v}(\mathbf{E}\cdot\mathbf{v})}{\mathbf{k}\cdot\mathbf{v}-i0}\frac{\partial n_0}{\partial\epsilon}\frac{2d^3p}{(2\pi\hbar)^3}. \tag{86.13}$$

Neglecting as usual the thermal blurring of the equilibrium distribution function, we write $\partial n_0/\partial\epsilon = -\delta(\epsilon-\epsilon_F)$ and transform the integral over d^3p into one over the Fermi surface by (74.20). According to the usual formula of differential geometry, the area element is $dS = do_\nu/K$, where do_ν is a solid-angle element for the direction of the normal $\boldsymbol{\nu}$ to the surface and K is the Gaussian curvature of the surface, i.e. $K = 1/R_1R_2$, the reciprocal of the product of its principal radii of curvature at the point concerned. Noting also that the direction of the normal at any point on the Fermi surface is the same as that of the velocity $\mathbf{v} = \partial\epsilon/\partial\mathbf{p}$, we find

$$\mathbf{j} = -\frac{2ie^2}{(2\pi\hbar)^3}\int\frac{\boldsymbol{\nu}(\mathbf{E}\cdot\boldsymbol{\nu})}{K(\boldsymbol{\nu})}\frac{do_\nu}{\mathbf{k}\cdot\boldsymbol{\nu}-i0}. \tag{86.14}$$

If the direction of $\boldsymbol{\nu}$ is specified by the azimuthal and polar angles φ and θ relative to the direction of $\mathbf{k}$ as polar axis, $\mathbf{k}\cdot\boldsymbol{\nu} = k\cos\theta$ and $do_\nu = \sin\theta\, d\varphi\, d\theta$.

The integration in (86.14) with respect to the variable $\mu = \cos\theta$ is taken over the segment $-1\le\mu\le 1$ of the real axis, passing along a semicircle below the pole $\mu = 0$. It is easy to see that the integral along the straight segments (i.e. the principal value) is then zero, leaving only the contribution from the semicircle. To prove this, we note that the function $\epsilon(\mathbf{p})$ is even, and therefore the Fermi surface $\epsilon(\mathbf{p}) = \epsilon_F$ is unchanged when $\mathbf{p}\to-\mathbf{p}$; since a change in the sign of $\mathbf{p}$ changes the sign of the normal vector $\boldsymbol{\nu}$, it follows that $K(-\boldsymbol{\nu}) = K(\boldsymbol{\nu})$. The integral in (86.14) can therefore be written as

$$\frac{1}{2}\left\{\int\frac{\boldsymbol{\nu}(\mathbf{E}\cdot\boldsymbol{\nu})\,do_\nu}{K(\boldsymbol{\nu})(\mathbf{k}\cdot\boldsymbol{\nu}-i0)} - \int\frac{\boldsymbol{\nu}(\mathbf{E}\cdot\boldsymbol{\nu})\,do_\nu}{K(\boldsymbol{\nu})(\mathbf{k}\cdot\boldsymbol{\nu}+i0)}\right\},$$

where the braces contain the sum of the integrals obtained from each other by the change of variable of integration $\boldsymbol{\nu}\to-\boldsymbol{\nu}$; the statement made above is then obvious.

At the pole of the integrand, $\mathbf{k}\cdot\boldsymbol{\nu} = k\cos\theta = 0$, i.e. the normal $\boldsymbol{\nu}$ is perpendicular to the given direction of the wave vector $\mathbf{k}$. The residue with respect to the variable $\cos\theta$ is therefore the integral

$$\int\frac{\boldsymbol{\nu}(\mathbf{E}\cdot\boldsymbol{\nu})}{kK(\boldsymbol{\nu})}d\varphi$$

taken along the geometric locus of points on the Fermi surface for which $\boldsymbol{\nu}\perp\mathbf{k}$.

Thus we have finally the relation between the current and the field in the form

$$j_\alpha = \sigma_{\alpha\beta}(\mathbf{k})E_\beta, \tag{86.15}$$

†This corresponds to the usual $\omega\to\omega+i0$ in $\omega-\mathbf{k}\cdot\mathbf{v}$.

where

$$\sigma_{\alpha\beta}(\mathbf{k}) = 2\pi e^2 A_{\alpha\beta}/(2\pi\hbar)^3 k, \quad A_{\alpha\beta} = \int_0^{2\pi} [\nu_\alpha \nu_\beta / K(\varphi)]\, d\varphi \tag{86.16}$$

is a real tensor in the plane perpendicular to **k**; if the direction of **k** is taken as the x-axis, α and β take the values y and z. The vector **j** lies in this plane, and is therefore transverse to **k**.

The contribution to the current comes only from electrons with $\mathbf{v}\cdot\mathbf{k} = 0$, i.e. moving perpendicular to the wave vector. This is a natural consequence of an approximation in which the mean free path is regarded as indefinitely long: in motion at an angle to **k**, the electron passes through a field oscillating in space, and these oscillations reduce to zero the overall action of the field on the electron. In the next approximation, when the finite value of kl is taken into account, there is a contribution to the current from electrons moving in a small range of angles $\sim 1/kl$ to the plane perpendicular to **k**.

Let us now go on to the subject of field penetration in the anomalous skin effect. This is a half-space problem, to be solved with the boundary conditions at the surface of the metal. These conditions for the distribution function depend on the physical properties of the surface with respect to electrons incident on it. It is important, however, that in this case the current is due essentially only to electrons moving almost parallel to the metal surface. For these, the law of reflection is largely independent of the degree of perfection of the metal surface, and approximates to specular reflection, the electrons being reflected with reversed components of the velocity **v** normal to the surface but unchanged tangential components; in order not to interrupt the discussion, a more detailed treatment of this will be postponed to the end of the section.

Specular reflection corresponds to the following boundary condition on the distribution function:

$$\delta\tilde{n}(v_x, v_y, v_z) = \delta\tilde{n}(-v_x, v_y, v_z) \quad \text{for} \quad x = 0. \tag{86.17}$$

With this condition, the half-space problem is equivalent to that of an infinite medium in which the field distribution is symmetrical about the plane $x = 0$: $\mathbf{E}(t, x) = \mathbf{E}(t, -x)$. Electrons reflected from the boundary in the half-space ($x > 0$) problem correspond to electrons passing freely through the plane $x = 0$ from the side $x = 0$ in the infinite-medium problem.

In the problem of the extreme anomalous skin effect, we can suppose that the field **E** (which depends only on the one coordinate x) is everywhere parallel to the plane $x = 0$. According to (86.15), the current vector **j** lies in the same plane, and so the condition of zero current component normal to the metal surface at all points on the surface is necessarily satisfied.†

†In subsequent approximations, when the finiteness of the ratio δ/l is taken into account, there are components $\sigma_{\alpha x}$ and σ_{xx} of the conductivity tensor as well as $\sigma_{\alpha\beta}$. To satisfy the boundary condition $j_x = 0$, we must then include the field E_x normal to the surface, as already noted in the first footnote to this section.

Without the assumption that $\mathbf{j} = \sigma\mathbf{E}$, we have for the two-dimensional vector $\mathbf{E}$, in place of (86.4),

$$\mathbf{E}'' = -4\pi i\omega\mathbf{j}/c^2. \tag{86.18}$$

We shall leave the time factor $e^{-i\omega t}$ implicit in the subsequent formulae, so that $\mathbf{E}$, $\mathbf{j}$, etc., will be functions of x only.

The function $\mathbf{E}(x)$, continued symmetrically into the range $x < 0$, is continuous at $x = 0$, but the derivative $\mathbf{E}'(x)$, being an odd function of x, has a discontinuity there, changing sign as x passes through zero. According to (86.1), these derivatives are related to the magnetic field by

$$\mathbf{E}' = i\omega\mathbf{B}\times\mathbf{n}/c,$$

where $\mathbf{n}$ is again a unit vector in the x-direction. In the half-space problem, the condition at $x = 0$ would therefore be $\mathbf{E}' = i\omega\mathbf{B}_0\times\mathbf{n}/c$, where $\mathbf{B}_0$ is the field at the boundary of the metal. In the infinite-medium problem, this corresponds to

$$\mathbf{E}'(+0) - \mathbf{E}'(-0) = 2i\omega\mathbf{B}_0\times\mathbf{n}/c.$$

We multiply both sides of (86.18) by e^{-ikx} and integrate with respect to x from $-\infty$ to ∞.† On the left-hand side, we write

$$\int_{-\infty}^{\infty}\mathbf{E}''e^{-ikx}\,dx = \int_{-\infty}^{0}(\mathbf{E}'e^{-ikx})'\,dx + \int_{0}^{\infty}(\mathbf{E}'e^{-ikx})'\,dx + ik\int_{-\infty}^{\infty}\mathbf{E}'e^{-ikx}\,dx.$$

Since the field $\mathbf{E}(x)$ is zero at infinity, the first two integrals give just the difference $\mathbf{E}'(-0) - \mathbf{E}'(+0)$. In the last term we can simply integrate by parts, since $\mathbf{E}(x)$ itself is continuous. The result is

$$2i\omega\mathbf{B}_0\times\mathbf{n}/c + k^2\mathbf{E}(k) = 4\pi i\omega\mathbf{j}(k)/c^2,$$

where $\mathbf{E}(k)$ and $\mathbf{j}(k)$ are the Fourier transforms of $\mathbf{E}(x)$ and $\mathbf{j}(x)$.

According to (86.15), these transforms are related by $j_\alpha(k) = \sigma_{\alpha\beta}(k)E_\beta(k)$. We then find the expression

$$E_\alpha(k) = \zeta_{\alpha\beta}(k)(\mathbf{B}_0\times\mathbf{n})_\beta, \tag{86.19}$$

where $\zeta_{\alpha\beta}(k)$ is a two-dimensional tensor specified by means of its inverse:

$$\zeta^{-1}{}_{\alpha\beta}(k) = -\frac{c}{2i\omega}\left[k^2\delta_{\alpha\beta} - \frac{4\pi i\omega}{c^2}\sigma_{\alpha\beta}(|k|)\right]. \tag{86.20}$$

The argument of $\sigma_{\alpha\beta}$ is written as $|k|$, as a reminder that it is the magnitude of the vector $\mathbf{k}$.

†The subsequent calculations are formally identical with those in the problem of magnetic field penetration into a superconductor (*SP* 2, §52).

The function $\mathbf{E}(x)$ itself is obtained from (86.19) on multiplying by e^{ikx} and integrating over $dk/2\pi$. Since $\zeta_{\alpha\beta}(k)$ is even, we have

$$E_\alpha(x) = \frac{1}{\pi}\int_0^\infty \zeta_{\alpha\beta}(k)\cos kx\, dk\,.\,(\mathbf{B}_0\times\mathbf{n})_\beta. \tag{86.21}$$

In particular, the field at the boundary of the metal is

$$E_{0\alpha} = \zeta_{\alpha\beta}(\mathbf{B}_0\times\mathbf{n})_\beta, \qquad \zeta_{\alpha\beta} = \frac{1}{\pi}\int_0^\infty \zeta_{\alpha\beta}(k)\, dk. \tag{86.22}$$

For an actual calculation of the surface impedance, we take the principal axes of the symmetrical tensor $\sigma_{\alpha\beta}(k)$ as the y and z axes. The tensor $\zeta_{\alpha\beta}$ is brought to principal axes along with $\sigma_{\alpha\beta}$, and its principal values are

$$\zeta^{(\alpha)} = -\frac{2i\omega}{\pi c}\int_0^\infty \frac{dk}{k^2 - ib^{(\alpha)}/k}, \qquad b^{(\alpha)} = \frac{\omega e^2 A^{(\alpha)}}{\pi c^2\hbar^3},$$

where the $A^{(\alpha)}$ are the principal values of the tensor $A_{\alpha\beta}$. The integration gives†

$$\zeta^{(\alpha)} = (1 - i\surd 3)\frac{2\pi^{1/3}\hbar}{3^{3/2}}\left(\frac{\omega^2}{ce^2A^{(\alpha)}}\right)^{1/3}. \tag{86.23}$$

The quantities $A^{(\alpha)}$ depend only on the shape and size of the Fermi surface. The impedance (86.23) does not depend at all on the electron mean free path. For an order-of-magnitude estimate, we can assume that the radii of curvature of the Fermi surface are $\sim p_F$; then $A \sim p_F^2$, and

$$\zeta \sim (\hbar^3\omega^2/ce^2p_F^2)^{1/3}. \tag{86.24}$$

The real part of the impedance determines the dissipation of the field energy in the metal. In the approximation concerned (where electron collisions are ignored) this dissipation is of the Landau-damping type.‡

†The contour of integration (the right-hand half of the real axis) can be turned through $-\pi/6$ in the complex k-plane without crossing any poles of the integrand. Integration along the line $k = ue^{-i\pi/6}$ gives

$$I \equiv \int_0^\infty \frac{k\,dk}{k^3 - ib} = e^{i\pi/6}\int_0^\infty \frac{u\,du}{u^3 + b},$$

and on substituting $u^3 + b = b/\xi$,

$$I = \frac{e^{i\pi/6}}{3b^{1/3}}\int_0^1 \xi^{-2/3}(1-\xi)^{-1/3}\, d\xi$$
$$= \frac{\Gamma(1/3)\Gamma(2/3)}{3b^{1/3}\Gamma(1)}e^{i\pi/6}$$
$$= \frac{\pi(\surd 3 + i)}{3^{3/2}b^{1/3}}.$$

‡The phenomena which are the essence of the anomalous skin effect were first noted by H. London (1940). The qualitative theory of the effect is due to A. B. Pippard (1947), and the quantitative theory given here is due to G. E. H. Reuter and E. H. Sondheimer (1948).

The law of damping of the electric field within the metal in the anomalous skin effect is not exponential, and so the concept of the penetration depth has not the same literal significance as in (86.5). Since the integrand in (86.21) contains the oscillating factor $\cos kx$, the integral for any given x is mainly governed by the range $k \sim 1/x$. A considerable decrease of the function $\mathbf{E}(x)$ occurs when these $k \gg b^{1/3}$.† The penetration depth δ is therefore of the order of $b^{-1/3}$, or

$$\delta \sim (c^2\hbar^3/\omega e^2 A)^{1/3} \sim (c^2\hbar^3/\omega e^2 p_F^2)^{1/3}. \qquad (86.25)$$

As the frequency increases, this depth continues to decrease, but more slowly than for the normal effect. The values given by (86.6) and (86.25), which we denote by δ_n and δ_a, are comparable in order of magnitude when $\delta \sim l$. Since one of them decreases as $\omega^{-1/2}$ and the other as $\omega^{-1/3}$, it is clear that for a given value of ω we have $\delta_a^3 \sim \delta_n^2 l$.

Lastly, some comments about the nature of electron reflection from the metal boundary. If the surface is ideal (free from defects) and coincides with a crystallographic plane, the configuration of atoms in it has the periodicity corresponding to the translational symmetry of the crystal lattice. In that case, the reflection of an electron conserves not only the energy but also the tangential components p_y and p_z of the electron quasi-momentum. The normal component p_x' of the quasi-momentum of the reflected electron is determined from the value p_x for the incident electron by means of the equation

$$\epsilon(p_x', p_y, p_z) = \epsilon(p_x, p_y, p_z), \qquad (86.26)$$

and we must have $v_x' = \partial\epsilon/\partial p_x' > 0$, the reflected electron moving away from the boundary (the velocity of the incident electron is $v_x = \partial\epsilon/\partial p_x < 0$). Equation (86.26) may have several such solutions, and in general $v_x' \neq -v_x$.

For electrons at glancing incidence, however, these solutions always include one that corresponds to a small change in the quasi-momentum, with $v_x' = -v_x$ (i.e. the reflection is literally specular): an electron moving almost parallel to the boundary has a small $v_x = \partial\epsilon/\partial p_x$, and this means that the electron corresponds to a point P on the constant-energy surface in **p**-space that is near the extremum of ϵ as a function of p_x (where $\partial\epsilon/\partial p_x = 0$). Near such a point, on the other side of the extremum, there is always a point P' where the derivative $\partial\epsilon/\partial p_x$ differs only in sign from its value at P.

It can be shown that the reflection of such an electron takes place, with very high probability, with this change of quasi-momentum. Moreover, the statement remains true for reflection from an imperfect surface having roughness of atomic dimensions, when there is, strictly speaking, no conservation of the tangential components of the quasi-momentum. An intuitive explanation is that the wave function

†When $x \gg \delta$, the integral (86.21) is governed by $k \ll b^{1/3}$. Then $\zeta(k) \sim k$, and the field $\mathbf{E}(x)$ decreases as x^{-2}.

of the electron varies slowly in the x-direction and so does not "notice" the atomic roughnesses of the surface.†

It is noteworthy that the value of the surface impedance in the highly anomalous skin effect is in fact fairly insensitive to the nature of the electron reflection. For example, in diffuse reflection (when all directions of the reflected electron are equally probable, whatever the angle of incidence), the value of the impedance differs from (86.23) only by a factor 9/8. The boundary condition for diffuse reflection from a plane surface is $\delta\tilde{n}(v_x > 0, v_y, v_z) = 0$ for $x = 0$. Here, however, the Fourier method is unsuitable, and the problem has to be solved by what is called the Wiener–Hopf method.‡

§87. Skin effect in the infra-red range

We have now considered two limiting cases of the skin effect: the normal effect, when the mean free path l is the smallest of the three characteristic distances δ, l and v_F/ω; and the anomalous effect, when the penetration depth δ is the smallest. Let us now take a third case, in which the smallest is

$$v_F/\omega \ll \delta, \quad v_F/\omega \ll l. \tag{87.1}$$

This is reached in a natural way from the anomalous effect by further increasing the frequency; although the penetration depth then decreases, the product $\omega\delta$ increases as $\omega^{2/3}$. For ordinary metals, the conditions (87.1) are satisfied in the infra-red range.

These conditions put a lower limit on the frequency. However, the validity of the results below, which are based on the theory of Fermi liquids, is also subject to an upper limit of frequency: $\hbar\omega \ll \epsilon_F$. If this is not satisfied, quasi-particles are excited from the depth of the Fermi distribution, and these have no meaning in the Fermi liquid theory.

To determine the relation between the current and the electric field, we must again go back to the transport equation. Now, however, because of the condition $\omega \gg v_F/\delta$, the term containing the time derivative is large compared with the term containing spatial derivatives, and because of the condition $\omega \gg v_F/l$ it is also large compared with the collision integral. Neglecting these terms, we have the transport equation in the form

$$\frac{\partial \delta n}{\partial t} - e\mathbf{E}\,.\,\mathbf{v}\frac{\partial n_0}{\partial \epsilon} = 0.$$

Putting $\partial\delta n/\partial t = -i\omega\delta n$, we obtain from this

$$\delta n = -(\partial n_0/\partial \epsilon)\psi, \quad \psi = e\mathbf{E}\,.\,\mathbf{v}/i\omega. \tag{87.2}$$

†These statements are proved in the review article by A. F. Andreev, *Soviet Physics Uspekhi* **14**, 609, 1972.

‡See G. E. H. Reuter and E. H. Sondheimer, *Proceedings of the Royal Society* A, **195**, 336, 1948.

The absence of the term containing the coordinate derivatives means that there is no spatial dispersion. In this sense the skin effect is normal again. The presence of the time-derivative term, however, causes frequency dispersion of the conductivity. The situation here is similar to that in calculating the permittivity of a collisionless plasma. The only difference lies in the anisotropy of the metal and in the Fermi-liquid effects. The latter have the result that the current density is given by an integral that depends not only on the distribution function δn but also on the interaction function $f(\mathbf{p}, \mathbf{p}')$ of the quasi-particles (the conduction electrons). It should be noted that, because of the presence of the term $\partial \delta n/\partial t$ in the transport equation, it is not possible here to eliminate the interaction of the quasi-particles by using the effective distribution function $\delta \tilde{n}$.

According to (74.21) and (74.22), the current density is expressed in terms of the correction to the electron distribution function by

$$\mathbf{j} = -e \int \boldsymbol{\nu} \left[\psi(\mathbf{p}_F) + \int f(\mathbf{p}_F, \mathbf{p}_F') \psi(\mathbf{p}_F') \frac{dS_F'}{v_F'(2\pi\hbar)^3} \right] \frac{2dS_F}{(2\pi\hbar)^3},$$

where $\boldsymbol{\nu}$ is a unit vector in the direction of the velocity $\mathbf{v}_F$, which is the same as the normal vector to the Fermi surface. Substituting ψ from (87.2), we find the relation between the current and the field in the form $j_\alpha = \sigma_{\alpha\beta}(\omega) E_\beta$, with the conductivity tensor

$$\left. \begin{aligned} \sigma_{\alpha\beta} &= -(e^2/i\omega m) N_{\alpha\beta}^{(\text{eff})}, \\ N_{\alpha\beta}^{(\text{eff})} &= \int \nu_\alpha \left[v_F \nu_\beta + \int f(\mathbf{p}_F, \mathbf{p}_F') \nu_\beta' \frac{dS_F'}{(2\pi\hbar)^3} \right] \frac{2dS_F}{(2\pi\hbar)^3}. \end{aligned} \right\} \tag{87.3}$$

The symmetry of the tensor $N_{\alpha\beta}^{(\text{eff})}$ is determined by that of the crystal, and does not depend on the direction of the field as in (86.15). In a crystal with cubic symmetry, which we shall assume for the sake of simplicity, this tensor, and therefore $\sigma_{\alpha\beta}$, reduce to a scalar: $N_{\alpha\beta}^{(\text{eff})} = N^{(\text{eff})} \delta_{\alpha\beta}$, and

$$\sigma(\omega) = -(e^2/im\omega) N^{(\text{eff})}. \tag{87.4}$$

The description of the properties of the metal by means of this conductivity can be replaced, in the usual manner, by a description in terms of the permittivity

$$\epsilon(\omega) = 1 + i\,.\,4\pi\sigma(\omega)/\omega = 1 - 4\pi e^2 N^{(\text{eff})}/m\omega^2. \tag{87.5}$$

The notation $N^{(\text{eff})}$ is used by analogy with the limiting expression (*ECM*, §59) for the permittivity at very high frequencies: $\epsilon = 1 - 4\pi e^2 N/m\omega^2$, where N is the total number of electrons per unit volume of the substance. Thus $N^{(\text{eff})}$ in the infra-red optics of metals represents the effective number of electrons; it depends on the interaction function of the conduction electrons.

Together with $N^{(\text{eff})}$, it is useful to define also the effective plasma frequency

$$\Omega = (4\pi e^2 N^{(\text{eff})}/m)^{1/2}. \tag{87.6}$$

The conductivity is then

$$\sigma = i\Omega^2/4\pi\omega. \tag{87.7}$$

The value of Ω is determined only by the parameters of the electron spectrum of the metal; as a rough estimate, therefore, it is equal to $\epsilon_F/\hbar$, the Fermi limiting energy. Since the present theory is restricted by the condition $\hbar\omega \ll \epsilon_F$, we have $\Omega \gg \omega$.

The penetration of the field into the metal is described by equation (86.4), which, on substitution of σ from (87.7), becomes

$$\mathbf{E}'' - (\Omega^2/c^2)\mathbf{E} = 0.$$

The solution which tends to zero as $x \to \infty$ is

$$\mathbf{E} = \mathbf{E}_0 e^{-x/\delta}, \quad \delta = c/\Omega; \tag{87.8}$$

for typical metals, $c/\Omega \sim 10^{-5}$ cm. The field is therefore damped exponentially, the penetration depth being independent of the frequency. The relation between the electric and magnetic fields is now given (as is easily seen by means of the first equation (86.3)) by (86.8) with impedance

$$\zeta = -i\omega\delta/c = -i\omega/\Omega. \tag{87.9}$$

The purely imaginary impedance denotes total reflection of the electromagnetic wave from the surface of the metal, without dissipation. This is to be expected, since the approximation used has taken no account of collisions of electrons, which are the cause of dissipation.

With (87.7), the basic conditions for this theory to be valid may be written

$$\Omega \gg \omega \gg \Omega v_F/c. \tag{87.10}$$

The left-hand inequality is usually compatible with $\hbar\omega \gg \Theta$, where Θ is the Debye temperature. The Fermi parameter v_F and the function f in (87.3) must then be taken not on the Fermi surface itself but for $|\epsilon - \epsilon_F| \gg \Theta$. It has been shown in *SP* 2, §65, that the electron–phonon interaction has the result that v_F in this range differs from that in the range $|\epsilon - \epsilon_F| \ll \Theta$ (which is important, for instance, as regards the static properties of the metal at low temperatures); the same is true of the quasi-particle interaction function f.

§88. **Helicon waves in metals**

The fact that an external alternating magnetic field does not penetrate into a metal means that undamped electromagnetic waves with frequencies up to the plasma frequency ($\omega \sim \Omega$) cannot propagate in a metal.

The position is quite different, however, in the presence of a static magnetic field **B**. This field alters the motion of the electrons and therefore greatly affects the electromagnetic properties of the metal. It is important that the motion becomes finite in the plane perpendicular to the field. In strong fields, when the Larmor radius $r_B \sim cp_F/eB$ of the orbit becomes small in comparison with the mean free path,

$$r_B \ll l, \tag{88.1}$$

or (equivalently) $\omega_B \tau \gg 1$, where $\omega_B \sim v_F/r_B \sim eB/m^*c$ is the Larmor frequency and $\tau \sim l/v_F$ the mean free time, the electrical conductivity in directions transverse to the field is greatly reduced, tending to zero as $B \to \infty$. We may say that in these directions the metal behaves like an insulator, and so the energy dissipation is reduced in waves whose electric field is polarized in the plane perpendicular to **B**. The propagation of such waves without damping (in the first approximation) thus becomes possible. The permissible wave frequencies are limited by the condition

$$\omega \ll \omega_B; \tag{88.2}$$

only if this is satisfied can the electron trajectories undergo appreciable curvature during the period of the field, thus altering the electromagnetic properties of the metal with regard to these frequencies.

The finite motion of the electron (in the plane perpendicular to **B**) presupposes that its momentum trajectory, a cross-section of the Fermi surface, is also finite. The above discussion therefore applies to metals with closed Fermi surfaces for any direction of **B**, but to those with open surfaces only for directions of **B** such that the cross-sections are closed. With open cross-sections, the electron motion remains infinite in the magnetic field, the conductivity does not decrease, and the propagation of electromagnetic waves in the directions concerned cannot occur.

Undamped electromagnetic waves in a metal may be regarded as Bose branches of the energy spectrum of the electron Fermi liquid. The macroscopic nature of these waves is shown by the wavelength, which is large compared with the lattice constant. For this reason, the excitations correspond to only a very small relative phase volume, and their contribution to the thermodynamic quantities in the metal is negligible.

We again write Maxwell's equations:

$$\operatorname{curl} \tilde{\mathbf{B}} = 4\pi \mathbf{j}/c, \quad \operatorname{curl} \mathbf{E} = -(1/c)\partial \tilde{\mathbf{B}}/\partial t, \tag{88.3}$$

where $\tilde{\mathbf{B}}$ denotes the weak alternating magnetic field of the wave, in contrast to the constant **B**. Elimination of $\tilde{\mathbf{B}}$ from these equations gives

$$\begin{aligned}\operatorname{curl}\operatorname{curl} \mathbf{E} &= \operatorname{grad}\operatorname{div} \mathbf{E} - \triangle \mathbf{E} \\ &= -(4\pi/c^2)\partial \mathbf{j}/\partial t.\end{aligned}$$

For a monochromatic plane wave, this becomes

$$(-k_\alpha k_\gamma + k^2 \delta_{\alpha\gamma})E_\gamma = 4\pi i \omega j_\alpha / c^2. \tag{88.4}$$

The field $\mathbf{E}$ is expressed in terms of the current by $E_\alpha = \rho_{\alpha\beta} j_\beta$, where $\rho_{\alpha\beta} = \sigma^{-1}{}_{\alpha\beta}$ is the resistivity tensor. We then obtain a set of linear homogeneous equations,

$$[k^2\rho_{\alpha\beta} - k_\alpha k_\gamma \rho_{\gamma\beta} - 4\pi i\omega\delta_{\alpha\beta}/c^2]j_\beta = 0. \tag{88.5}$$

The determinant of this yields the dispersion relation for the waves.

In §§84 and 85 we have derived the conductivity tensor for a metal (in the residual resistance range) in a strong magnetic field, for the stationary case. Let us now see how the results must be modified in the non-stationary case.

The periodicity of the electric field in time and space (and therefore that of the alternating part of the electron distribution function) causes the presence of terms

$$\frac{\partial \delta n}{\partial t} + \mathbf{v}\cdot\frac{\partial \delta \tilde{n}}{\partial \mathbf{r}} = -i\omega\delta n + i\mathbf{k}\,.\,\mathbf{v}\delta\tilde{n}$$

on the left of the transport equation; cf. (74.25). Similarly to (84.7), we put δn and $\delta\tilde{n}$ in the form

$$\delta n = (\partial n_0/\partial\epsilon)e\mathbf{E}\,.\,\mathbf{h}, \quad \delta\tilde{n} = (\partial n_0/\partial\epsilon)e\mathbf{E}\,.\,\mathbf{g}.$$

According to (74.21), the functions $\mathbf{h}$ and $\mathbf{g}$ are related by the linear integral expression

$$\mathbf{g} = \mathbf{h} + \int f(\mathbf{p}, \mathbf{p}_F')\mathbf{h}'\, dS_F'/v_F'(2\pi\hbar)^3 \equiv \hat{L}\mathbf{h}.$$

The transport equation then becomes

$$\partial\mathbf{g}/\partial t - [I(\mathbf{g}) + i\omega\hat{L}^{-1}\mathbf{g} - i(\mathbf{k}\,.\,\mathbf{v})\mathbf{g}] = \mathbf{v}. \tag{88.6}$$

It differs from (84.10) in that $I(\mathbf{g})$ is replaced by the expression in the square brackets, which depends not only on the nature of the electron scattering by impurity atoms but also on the interaction between the electrons.

Because of the condition $r_B \ll l$, the term $I(\mathbf{g})$ in (88.6) is much less than $\partial\mathbf{g}/\partial\tau$, as it was in (84.10). Because of the condition $\omega \ll \omega_B$, the term $i\omega\hat{L}^{-1}g \sim i\omega g$ is also small. We shall impose also a condition on the wave number, $kv_F \ll \omega_B$, i.e.,

$$kr_B \ll 1: \tag{88.7}$$

the wavelength must be much greater than the Larmor radius. Then the last term in the square brackets in (88.6) is small also. The method of solving the transport equation by successive approximation (§84) then remains valid, and therefore so do the results found there for the leading terms in the expansion of the conductivity tensor in powers of $1/B$. However, because of the presence of ω and $\mathbf{k}$ in (88.6), there will in general be frequency and spatial dispersion of the conductivity.

The presence of several characteristic parameters of length and time, and the variety of geometrical properties of the Fermi surfaces, causes a multiplicity of phenomena relating to the propagation of electromagnetic waves in metals. We shall consider in §§88 and 89 only some typical cases.

Let us take the case of an uncompensated metal with a closed Fermi surface. According to (85.4) and (85.5), the largest component of the resistivity tensor is

$$\rho_{xy} = -\rho_{yx} = B/ec(N_e - N_h); \tag{88.8}$$

this belongs to the non-dissipative (anti-Hermitian) part of the tensor. This component is entirely independent of the form of the collision integral, and therefore of the form of the expression in square brackets in (88.6). The formula (88.8) remains valid in the wave field, therefore.

The description of the medium by means of the resistivity tensor $\rho_{\alpha\beta}$ (or the conductivity tensor $\sigma_{\alpha\beta}$) is equivalent to the use of the permittivity tensor

$$\epsilon_{\alpha\beta} = 4\pi i\sigma_{\alpha\beta}/\omega, \quad \epsilon^{-1}{}_{\alpha\beta} = \omega\rho_{\alpha\beta}/4\pi i.$$

The tensor $\epsilon^{-1}{}_{\alpha\beta}$ here has only the components

$$\epsilon^{-1}{}_{xy} = -\epsilon^{-1}{}_{yx} = \omega B/4\pi ice(N_e - N_h).$$

This is the same as that found in §56 for waves in plasmas, except that the electron density N_e is replaced by the difference $N_e - N_h$. The results of §56 may therefore be applied immediately to these waves in metals, which are likewise called *helicon waves*.†

The dispersion relation for them is

$$\omega = cB|\cos\theta|/4\pi e|N_e - N_h|, \tag{88.9}$$

where θ is the angle between **k** and **B**. The electric field of the wave is elliptically polarized in the plane perpendicular to the magnetic field **B**. Taking the direction of **B** as the z-axis (as in §56), and the plane through **k** and **B** as the xz-plane, we find the electric field

$$E_y = \pm i|\cos\theta|E_x, \tag{88.10}$$

the upper and lower signs relating to the cases $N_e > N_h$ and $N_e < N_h$ respectively.

§89. Magnetoplasma waves in metals

Let us now consider waves in a compensated ($N_e = N_h$) metal having a closed Fermi surface. In addition to the obligatory conditions (88.1) and (88.2), we shall

†The possibility of the propagation of such waves in metals was noted by O. V. Konstantinov and V. I. Perel' (1960).

assume that the inequalities

$$\omega \gg v_F/l, \quad \omega \gg kv_F \tag{89.1}$$

are also satisfied. The first of these implies that the collision integral $I(\mathbf{g})$ in the transport equation (88.6) is much less than the term $i\omega\hat{L}^{-1}\mathbf{g}$, and the second condition implies that the term $i(\mathbf{k}\cdot\mathbf{v})\mathbf{g}$ is also small. Neglecting these terms, we have the equation

$$\partial \mathbf{g}/\partial \tau - i\omega\hat{L}^{-1}\mathbf{g} = \mathbf{v}, \tag{89.2}$$

which is obtained from (84.10) on replacing the term $I(\mathbf{g})$ by $i\omega\hat{L}^{-1}\mathbf{g}$.

The results derived in §85 for the resistivity tensor in the stationary case therefore remain valid, except that the small parameter in the expansion in powers of $1/B$ is not r_B/l but $-i\omega/\omega_B$. There is no spatial dispersion of the conductivity, but there is frequency dispersion.

According to (85.7), in the stationary case the leading terms in the expansion of the resistivity tensor for a compensated metal are

$$\rho_{zz} = \text{constant}; \quad \rho_{xx}, \rho_{yy}, \rho_{xy} \propto B^2; \quad \rho_{xz}, \rho_{yz} \propto B. \tag{89.3}$$

To bring out the parameter r_B/l in this tensor, however, we must ascertain how not only B but also l occurs in its components. To do so, we write, for example, the estimate

$$\rho_{xx} \sim \rho_0(l/r_B)^2 \sim (B/ecN)l/r_B.$$

where $\rho_0 \sim p_F/Ne^2l$. Similarly

$$\rho_{yz} \sim \rho_0 l/r_B \sim B/ecN, \quad \rho_{zz} \sim \rho_0 \sim (B/ecN)r_B/l.$$

Now, with the above-mentioned change of expansion parameter, we find the tensor $\rho_{\alpha\beta}(\omega)$ as

$$\rho_{\alpha\beta} = (B/ecN)\begin{bmatrix} (\omega_B/-i\omega)a_{xx} & (\omega_B/-i\omega)a_{xy} & a_{xz} \\ (\omega_B/-i\omega)a_{xy} & (\omega_B/-i\omega)a_{yy} & a_{yz} \\ -a_{xz} & -a_{yz} & (-i\omega/\omega_B)a_{zz} \end{bmatrix}, \tag{89.4}$$

where all the $a_{\alpha\beta} \sim 1$ are dimensionless real coefficients; the quantities N and m^* (in $\omega_B = eB/m^*c$) are to be regarded here as parameters chosen in some manner and having the correct order of magnitude. All the terms in (89.4) belong to the anti-Hermitian (i.e. non-dissipative) part of the tensor. It is therefore evident that including only these terms will give undamped waves. In the general case where the directions of $\mathbf{B}$ and $\mathbf{k}$ are arbitrary, the wave dispersion relation is expressed by fairly lengthy formulae. We shall consider only a particular case which illustrates the fundamental properties of these waves.

We shall suppose that the metal crystal lattice has an axis of symmetry with order higher than 2, and that the field **B** is along this axis (the z-axis). The quantities a_{xx}, a_{yy} and $a_{xy} = a_{yx}$ form a two-dimensional symmetrical tensor in the xy-plane, which with the symmetry stated reduces to a scalar: $a_{xx} = a_{yy} \equiv a_1$, $a_{xy} = 0$. The quantities a_{xz} and a_{yz} form a two-dimensional vector in the same plane, and with the symmetry stated are zero. Thus there remain only the components

$$\rho_{xx} = \rho_{yy} = (B/ecN)(\omega_B/-i\omega)a_1, \quad \rho_{zz} = (B/ecN)(-i\omega/\omega_B)a_2. \tag{89.5}$$

We again take the xz-plane to contain the directions of **k** and **B**. If ρ_{zz}, which is small compared with ρ_{xx}, is neglected, the dispersion relation separates into the two equations

$$4\pi i\omega/c^2 - k^2\rho_{yy} = 0, \quad 4\pi i\omega/c^2 - k_z^2\rho_{xx} = 0;$$

here we assume that the angle θ between **k** and **B** is not too close to $\frac{1}{2}\pi$, so that k_z^2 is not too small ($\cos\theta \gg \omega/\omega_B$). Hence we find as the dispersion relations for the two types of wave

$$\left.\begin{aligned} \omega^{(1)} &= ku_A\sqrt{a_1}, \\ \omega^{(2)} &= ku_A|\cos\theta|\sqrt{a_1}, \end{aligned}\right\} \tag{89.6}$$

where†

$$u_A = B/(4\pi Nm^*)^{1/2}. \tag{89.7}$$

These are called *magnetoplasma waves*. The two types are respectively analogous to the fast magnetosonic waves and the Alfvén waves in plasmas.‡ Oscillations corresponding to the slow magnetosonic waves cannot have a speed ω/k that satisfies the second condition (89.1), and therefore cannot occur in metals.

§90. Quantum oscillations of the conductivity of metals in a magnetic field

The theory of galvanomagnetic effects given in §§84 and 85 was quasi-classical, in the sense that quantum behaviour appeared only in the electron distribution function; the discreteness of the energy levels in the magnetic field (with closed electron trajectories) was not taken into account. This discreteness causes, however, a qualitatively new effect, namely oscillations of the conductivity as a function of the magnetic field, called the *Shubnikov–de Haas effect*. This is analogous to the oscillations of the magnetic moment (the de Haas–van Alphen

†With the dispersion relations (89.6), (89.7), the condition $kv_F \ll \omega$ signifies that we must have $u_A \gg v_F$. In attainable fields B, this can in practice be satisfied only in semi-metals (such as bismuth) with a low carrier density.

‡The possibility of such waves was noted by S. J. Buchsbaum and J. Golt (1961). The theory given here is due to É. A. Kaner and V. G. Skobov (1963).

effect), but the theory of it is more complex because it is a transport effect and not a thermodynamic effect. We shall consider it in terms of a model of non-interacting electrons, leaving aside the question (which does not appear to have been investigated) of Fermi-liquid effects.

As in § 84, the magnetic field will be assumed strong in the sense of the condition (84.1), which we write in the form

$$\omega_B \tau \gg 1, \tag{90.1}$$

where τ is the electron mean free time and

$$\omega_B = eB/m^*c \tag{90.2}$$

the Larmor frequency; m^* is the cyclotron mass of the electrons.†

At the same time, of course, the field must not be so strong as to violate the quasi-classicality condition

$$\hbar\omega_B \ll \epsilon_F. \tag{90.3}$$

There may be any relationship between $\hbar\omega_B$ and T.

We shall examine only the quantum oscillations of the transverse (relative to the magnetic field, which is in the z-direction) conductivity, and assume in order to simplify the formulae, that the crystal has a symmetry axis (of order >2) parallel to the magnetic field. In such a crystal the symmetric (dissipative) part of the conductivity tensor has only the components $\sigma_{xx} = \sigma_{yy}$ and σ_{zz}. The comparative simplicity of the problem for the transverse components arises because for them the influence of collisions may (as we have seen in § 84) be regarded as a small perturbation in comparison with the influence of the magnetic field; this is not so for the longitudinal conductivity σ_{zz}.‡

As in § 84, we consider a metal in the residual resistance range, so that we are concerned with collisions between electrons and impurity atoms. Since these collisions are elastic, electrons with different energies take part independently in producing the electric current.

Let $g(\epsilon)$ be the number of quantum states of an electron per unit energy range. Then the spatial number density of electrons with energies in the range $d\epsilon$ is $n(\epsilon)g(\epsilon)\,d\epsilon$, where $n(\epsilon)$ denotes the state occupation numbers. Let $j_y(\epsilon)$ be the transverse current density generated by these electrons. When both an electric field and an electron density gradient are present, the current density is the sum

$$j_y(\epsilon) = eD(\epsilon)g(\epsilon)\partial n/\partial y + \sigma_{yy}(\epsilon)E_y. \tag{90.4}$$

†The definition (*SP* 2 (57.6)) is $m^* = (1/2\pi)\partial S/\partial\epsilon$, where $S(\epsilon, p_z)$ is the area of the cross-section of the constant-energy surface by a plane p_z = constant; this surface is here defined in **p**-space, not in $\mathbf{p}/\hbar$-space as in *SP* 2.

‡For the antisymmetric part of the conductivity tensor, the quantum oscillations occur only in the second approximation with respect to $1/\omega_B\tau$.

The first term is the diffusive charge transfer; $D(\epsilon)$ is the diffusion coefficient (in actual space) for electrons with energy ϵ. The current (90.4) must be zero for the distribution

$$n_0(\epsilon - e\varphi) \approx n_0(\epsilon) - e\varphi \partial n_0/\partial \epsilon,$$

which corresponds to statistical equilibrium of the electron gas in a weak static electric field with potential $\varphi(\mathbf{r})$ (n_0 being the Fermi distribution). Hence we have as the relation between $\sigma_{yy}(\epsilon)$ and $D(\epsilon)$

$$\sigma_{yy}(\epsilon) = -e^2 g(\epsilon) D(\epsilon) \partial n_0/\partial \epsilon.$$

The total electrical conductivity, including the contributions from electrons of all energies, is

$$\sigma_{yy} = -e^2 \int g(\epsilon) D(\epsilon) \frac{\partial n_0}{\partial \epsilon} d\epsilon = -e^2 \sum_s D(\epsilon_s) \frac{\partial n_0(\epsilon_s)}{\partial \epsilon}. \tag{90.5}$$

The summation in the last expression is over all quantum states of the electron; s conventionally denotes the set of all the state quantum numbers. This formula reduces the calculation of the conductivity to that of the electron diffusion coefficient in the absence of the electric field.

The diffusion coefficient is in turn expressed in terms of the properties of the microscopic scattering events by a formula of the type (21.4):

$$D = \sum (\Delta y)^2 / 2\delta t,$$

where the summation is over the collisions undergone by an electron in the time δt, and Δy is the change in the mean value of the electron coordinate y in the collision (the electron motion is finite in the plane perpendicular to the field, and in the intuitive picture of quasi-classical orbits Δy is the displacement of the orbit centre). Let

$$N_{\text{imp}} W_{s's} \delta(\epsilon_s - \epsilon_{s'})$$

denote the probability of an electron transition from state s to state s' in scattering; the delta function expresses the fact that the scattering is elastic, and the factor N_{imp}, the concentration of impurity atoms, expresses the fact that scattering by randomly distributed atoms takes place independently. The diffusion coefficient is then

$$D(\epsilon_s) = \tfrac{1}{2} N_{\text{imp}} \sum_{s'} (y_s - y_{s'})^2 W_{ss'} \delta(\epsilon_s - \epsilon_{s'}),$$

where y_s is the mean value of the coordinate in state s. Substituting this expression in (90.5), we find the conductivity

$$\sigma_{yy} = -\tfrac{1}{2}e^2 N_{\text{imp}} \sum_{s,s'} (y_s - y_{s'})^2 \frac{\partial n_0(\epsilon_s)}{\partial \epsilon} W_{s's}\delta(\epsilon_s - \epsilon_{s'}) \tag{90.6}$$

(Ş. Ţiţeica 1935, B. I. Davydov and I. Ya. Pomeranchuk 1939).†

In practical applications of this formula the significance of s must be made explicit. Discrete quantization of conduction electron energy levels in a magnetic field occurs when there are closed quasi-classical trajectories in **p**-space (i.e. closed cross-sections of the constant-energy surfaces), and we shall assume that this is so. The quantum states are defined by four numbers:

$$s = (n, P_x, P_z = p_z, \sigma), \tag{90.7}$$

where n is a (large) positive integer; $\sigma = \pm 1$ denotes the value of the electron spin component; P_x and P_z are components of the generalized quasi-momentum $\mathbf{P} = \mathbf{p} - e\mathbf{A}/c$. The vector potential of the magnetic field is chosen in the gauge $A_x = -By$, $A_y = A_z = 0$. Because the coordinates x and z are cyclic, the generalized quasi-momentum components P_x and P_z are conserved; see *SP* 2, §58. The energy levels depend on only the three quantum numbers n, p_z and σ; they are given by

$$\epsilon_{n\sigma}(p_z) = \epsilon(n, p_z) + \sigma\beta B\xi_n(p_z), \tag{90.8}$$

the function $\epsilon(n, p_z)$ being the solution of the equation

$$S(\epsilon, p_z) = 2\pi(e\hbar B/c)(n + \tfrac{1}{2}). \tag{90.9}$$

In the second term in (90.8), $\beta = e\hbar/2mc$ is the Bohr magneton, and the factor $\xi_n(p_z)$ represents the change in the electron magnetic moment due to spin–orbit interaction in the lattice.

The conductivity tensor considered in §§84 and 85 is actually the result of averaging the exact functions $\sigma_{\alpha\beta}(B)$ over small quantum oscillations. In particular, from (85.3), the transverse conductivity thus averaged is $\bar{\sigma}_{yy} \propto B^{-2}$. We shall show, first of all, that this result follows from (90.6), and ascertain the relation between the quantities $W_{s's}$ in this formula and the function $w(\mathbf{p}', \mathbf{p})$ in the quasi-classical collision integral (78.14) for electrons and impurities.

It has been noted in §84 that the condition for quasi-classical motion of the electrons ensures also that the scattering process is independent of the magnetic field. The scattering probability in the absence of the field, with change of quasi-momentum from **p** to **p**$'$, has been expressed in the collision integral (78.14) as

$$w(\mathbf{p}', \mathbf{p})\delta(\epsilon - \epsilon')\, d^3p'/(2\pi\hbar)^3. \tag{90.10}$$

In order to put this in a form suitable also for scattering in a magnetic field, it need

†In scattering by impurities, the Pauli principle does not affect the formulae; cf. the collision integral (78.14), in which the products nn' associated with this principle cancel.

only be changed to variables which remain meaningful for motion in the field:

$$w(P'_x, p'_z, \epsilon'; P_x, p_z, \epsilon)\delta(\epsilon - \epsilon')\, dP_x\, dp_z\, d\epsilon/(2\pi\hbar)^3 v_y; \tag{90.11}$$

the derivative $v_y = \partial\epsilon/\partial p_y$ also is understood to be expressed in terms of the new variables. The y-coordinate in motion in a quasi-classical trajectory is related to the generalized quasi-momentum by $P_x = p_x + eBy/c$; hence the mean value over the trajectory is

$$\bar{y} = (c/eB)[P_x - \bar{p}_x(\epsilon, p_z)] \equiv \kappa/B. \tag{90.12}$$

The conductivity $\bar{\sigma}_{yy}$ averaged over the oscillations is found from (90.6) by replacing the summation over the discrete variable s by integration over the continuous variable ϵ. Using for brevity the notation

$$a(\epsilon, p'_z, p_z) = \frac{1}{2}\int (\kappa - \kappa')^2 w\, dP_x\, dP'_x/v_y v'_y(2\pi\hbar)^4, \tag{90.13}$$

we find

$$\bar{\sigma}_{yy} = -\frac{e^2 N_{\text{imp}}}{B^2}\int a\frac{\partial n_0}{\partial\epsilon}\delta(\epsilon - \epsilon')\, d\epsilon\, d\epsilon' \frac{2\, dp_z\, dp'_z}{(2\pi\hbar)^2}; \tag{90.14}$$

the factor 2 comes from the two directions of the electron spin, and the scattering probability is assumed independent of the spin, so that the spin component is unchanged. The delta function is removed by the integration over ϵ'; in that over ϵ, we can regard the slowly varying factor a as constant, and equal to its value for $\epsilon = \mu$, and integrate only the derivative $\partial n_0/\partial\epsilon$. The result is

$$\bar{\sigma}_{yy} = \frac{e^2 N_{\text{imp}}}{B}\int a\frac{2\, dp_z\, dp'_z}{(2\pi\hbar)^2} \equiv \frac{1}{B^2}\int b(p_z)\,.\,2\, dp_z. \tag{90.15}$$

Let us now take into account the discreteness of the levels. The integration in (90.14) over the continuous variable ϵ (with fixed P_x and p_z) must then be replaced by a summation over n, with

$$\int \cdots d\epsilon \to \hbar\omega_B \sum_n \cdots,$$

where

$$\hbar\omega_B = \partial\epsilon(n, p_z)/\partial n,$$

as is clear from (90.9) and the definition of the cyclotron mass m^*. With the notation given above, we have

$$\sigma_{yy} = -\frac{e^2 N_{\text{imp}}}{B^2}\int \sum_{n,n',\sigma} a(\epsilon_{n\sigma}, p'_z, p_z)\frac{\partial n_0(\epsilon_{n\sigma})}{\partial\epsilon} \times$$

$$\times\, \delta(\epsilon_{n\sigma} - \epsilon_{n'\sigma})\hbar\omega_B\hbar\omega'_B \frac{dp_z\, dp'_z}{(2\pi\hbar)^3}; \tag{90.16}$$

because of the integration over both variables p_z and p_z', the function a may be regarded as symmetrical with respect to them.

The oscillatory part $\tilde{\sigma}_{yy}$ of this expression is separated by means of the Poisson summation formula (cf. *SP* 2, § 63):

$$\tfrac{1}{2}F(0) + \sum_{n=1}^{\infty} F(n) = \int_0^{\infty} F(x)\,dx + 2\,\mathrm{re}\sum_{l=1}^{\infty}\int_0^{\infty} F(x)e^{2\pi ilx}\,dx, \tag{90.17}$$

and arises from the sum over l in this formula, whereas the averaged $\bar{\sigma}_{yy}$ comes from the first (integral) term.

We shall suppose that the oscillation amplitude is small in comparison with the averaged $\bar{\sigma}_{yy}$; this imposes a certain condition on the magnetic field strength (see (90.26)). It is then sufficient to take account of the oscillatory part in only one of the sums over n and n' in (90.16). Using the symmetry of a with respect to p_z and p_z', and defining b by analogy with (90.15), we have

$$\tilde{\sigma}_{yy} = \frac{4}{B^2}\,\mathrm{re}\sum_{l=1}^{\infty}\sum_{\sigma=\pm 1}\tilde{J}_{l\sigma}, \tag{90.18}$$

where $\tilde{J}_{l\sigma}$ is the oscillatory part of the integral

$$J_{l\sigma} = -\int_0^{\infty} dn \int b(\epsilon_{n\sigma}, p_z)\frac{\partial n_0(\epsilon_{n\sigma})}{\partial\epsilon}\frac{\partial\epsilon_{n\sigma}}{\partial n}e^{2\pi iln}\,dp_z.$$

With $\epsilon(n, p_z)$ from (90.8) as a variable of integration in place of n, we integrate by parts with respect to ϵ (and can regard the slowly varying factor b as a constant). The integrated term does not give an oscillatory dependence on the field (and is only a small correction to $\bar{\sigma}_{yy}$); omitting it, we have

$$\tilde{J}_{l\sigma} = 2\pi il\int_0^{\infty}\int \frac{b(\epsilon, p_z)}{\exp[(\epsilon-\mu_\sigma)/T]+1}\frac{\partial n}{\partial\epsilon}\,dp_z\,d\epsilon. \tag{90.19}$$

Here $\mu_\sigma = \mu - \sigma\beta\xi B$, and

$$n(\epsilon, p_z) = \frac{cS(\epsilon, p_z)}{2\pi e\hbar B} - \frac{1}{2} \tag{90.20}$$

(cf. (90.9)); in the argument of the function $b(\epsilon_{n\sigma}, p_z)$, the term $\beta\xi B$ has been neglected in comparison with the large quantity ϵ.

The integration over p_z in (90.19) is carried out in exactly the same way as that which occurred in *SP* 2 (63.8) in the study of the de Haas–van Alphen effect. The integral is governed by the ranges near the points $p_z = p_{z,\mathrm{ex}}(\epsilon)$ at which $n(\epsilon, p_z)$ (i.e. the cross-sectional area S) has an extremum as a function of p_z. The result is

$$\tilde{J}_{l\sigma} = \sum_{\mathrm{ex}}\int_0^{\infty}\frac{2\pi i\sqrt{l}\,\exp\{2\pi iln_{\mathrm{ex}} \pm (1/4)i\pi\}b_{\mathrm{ex}}(\epsilon)}{\{\exp[(\epsilon-\mu_\sigma)/T]+1\}\,.\,|\partial^2 n/\partial p_z^2|_{\mathrm{ex}}^{1/2}}\frac{dn_{\mathrm{ex}}}{d\epsilon}\,d\epsilon, \tag{90.21}$$

where

$$n_{ex}(\epsilon) = n(\epsilon, p_{z,ex}(\epsilon)), \quad b_{ex}(\epsilon) = b(\epsilon, p_{z,ex}(\epsilon)),$$

and the $\pm$ signs in the exponent refer to the cases in which $p_{z,ex}$ is respectively a maximum or a minimum of $n(\epsilon, p_z)$; the summation is over all extrema.

The integral (90.21) in turn is exactly analogous to the integral *SP* 2 (63.9), differing only in the slowly varying factors b and $dn_{ex}/d\epsilon = cm^*_{ex}/e\hbar B$ in the integrand; these factors, and $|\partial^2 n/\partial p_z^2|_{ex}^{-1/2}$, may be replaced by their values at $\epsilon = \mu$, i.e. on the Fermi surface. The integration over ϵ and the summation over σ then lead to the final result

$$\left.\begin{aligned} \tilde{\sigma}_{yy} &= \sum_{ex}\sum_{l=1}^{\infty}(-1)^l \sigma_{yy}^{(l)} \cos\left\{l\frac{cS_{ex}}{e\hbar B} \pm \frac{1}{4}\pi\right\}, \\ \sigma_{yy}^{(l)} &= \frac{2^{5/2}\pi^{1/2}(e\hbar)^{1/2} b_{ex}}{c^{1/2}B^{3/2}l^{1/2}}\left|\frac{\partial^2 S}{\partial p_z^2}\right|_{ex}^{-1/2}\frac{\lambda_l}{\sinh\lambda_l}\cos(\pi l\xi_{ex}m^*_{ex}/m), \\ \lambda_l &= 2\pi^2 lT/\hbar\omega_B, \quad \omega_B = eB/m^*_{ex}c, \end{aligned}\right\} \tag{90.22}$$

where S_{ex}, ξ_{ex}, m^*_{ex} and b_{ex} are taken at $\epsilon = \mu$ on the Fermi surface.†

If for a given direction of **B** there is only one extremal cross-section of the Fermi surface, there is proportionality between the oscillatory parts of the conductivity σ_{yy} and the longitudinal magnetic susceptibility. Comparison of (90.22) with *SP* 2 (63.13) gives

$$\tilde{\sigma}_{yy} = \frac{(2\pi)^4\hbar^3 m^*_{ex} b_{ex}}{S_{ex}^2}\frac{\partial \tilde{M}_z}{\partial B}. \tag{90.23}$$

The foregoing calculations presuppose that the oscillation amplitude of the conductivity is small in comparison with its averaged value. This requirement is indeed the condition for the whole theory in §§84 and 85 to be applicable: it is clear that the averaged values have real significance only if they form the major part of the conductivity tensor.

When $\hbar\omega_B \sim T$, the oscillation amplitude is determined by the leading terms in the sum in (90.22), with $l \sim 1$, $\lambda_l \sim 1$. According to the definition in (90.15), the value of b_{ex} can be estimated as $b_{ex} \sim \bar{\sigma}B^2/p_F$. The derivative $\partial^2 S/\partial p_z^2 \sim 1$. Hence we have the following estimate of the oscillation amplitude:

$$\tilde{\sigma}/\bar{\sigma} \sim (\hbar\omega_B/\epsilon_F)^{1/2}, \quad \hbar\omega_B \sim T. \tag{90.24}$$

This ratio is small, because of the obligatory condition (90.3).

If $T \ll \hbar\omega_B$, however, the estimate is changed. The oscillation amplitude is then determined by the sum of a large number of terms in (90.22), having $\lambda_l \sim 1$, i.e. $l \sim \hbar\omega_B/T \gg 1$. The number of such terms is of the order of l itself. In comparison with the previous estimate, we now have an additional factor $l^{-1/2}l \sim (\hbar\omega_B/T)^{1/2}$, so

†The oscillations of the conductivity were discussed by A. I. Akhiezer (1939) and by B. I. Davydov and I. Ya. Pomeranchuk (1939) for a quadratic electron dispersion relation, and by A. M. Kosevich and V. V. Andreev (1960) for any dispersion relation.

that

$$\tilde{\sigma}/\bar{\sigma} \sim (\hbar\omega_B/\epsilon_F)^{1/2}(\hbar\omega_B/T)^{1/2}. \tag{90.25}$$

The requirement that this ratio be small leads to the condition

$$\hbar\omega_B \ll (\epsilon_F T)^{1/2}. \tag{90.26}$$

PROBLEM

Determine the transverse conductivity of an electron gas with a quadratic dispersion relation ($\epsilon = p^2/2m$). The electrons are scattered isotropically by impurity atoms, with a cross-section that is independent of the energy.

SOLUTION. The problem amounts to the calculation of $b(p_z)$ in (90.15) and (90.23). With a quadratic dispersion relation, $\mathbf{p} = m\mathbf{v}$, and $\bar{\mathbf{p}} = 0$ since the mean velocity along a closed trajectory is $\bar{\mathbf{v}} = 0$; hence, by (90.12), $\kappa = cP_x/e$. According to the discussion in the text, when calculating the mean value of $(\kappa - \kappa')^2$ we can regard the scattering process as independent of the magnetic field. The difference between $\mathbf{P}$ and $\mathbf{p}$ is then unimportant: if the position of the scattering atom is taken as $\mathbf{r} = 0$, we have $\mathbf{P} = \mathbf{p}$.

In the case under consideration, the scattering probability has the form $v\sigma_0 do'/4\pi$, where do' is the solid-angle element for directions of the momentum $\mathbf{p}'$ after scattering, and σ_0 is the constant total scattering cross-section. This expression may be put in the equivalent form

$$(\sigma_0/4\pi m)\, dp_z'\, d\varphi'\, \delta(\epsilon - \epsilon')\, d\epsilon',$$

where φ' is the azimuthal angle for the direction of $\mathbf{p}'$ in the xy-plane, and it here replaces (90.11). Similarly, we write the volume element in $\mathbf{p}$-space as $d^3p \to m\, dp_z\, d\varphi\, d\epsilon$, and

$$p_x = (2m\epsilon - p_z^2)^{1/2} \cos\varphi.$$

We then find

$$a(\epsilon, p_z, p_z') = \frac{c^2\sigma_0}{8\pi e^2}\int (p_x - p_x')^2 \frac{d\varphi\, d\varphi'}{2\pi\hbar} = \frac{\sigma_0 c^2}{8e^2\hbar}(4m\epsilon - p_z^2 - p_z'^2)$$

and

$$b(\epsilon, p_z) = e^2 N_{\text{imp}} \int_{-\surd(2m\epsilon)}^{\surd(2m\epsilon)} a\, dp_z'/(2\pi\hbar)^2$$

$$= \frac{c^2\surd(2m\epsilon)}{16\pi^2\hbar^3 l}\left(\frac{10}{3}m\epsilon - p_z^2\right),$$

where $l = 1/\sigma_0 N_{\text{imp}}$ is the mean free path.

The averaged conductivity is calculated from (90.15):

$$\bar{\sigma}_{yy} = c^2 p_F N/B^2 l,$$

where $N = p_F^3/3\pi^2\hbar^3$ is the number density of electrons. The cross-sectional area of the Fermi sphere is greatest at $p_z = 0$, and $S_{\text{ex}} = \pi p_F^2$. Hence

$$b_{\text{ex}} = 5c^2 N/16l.$$

The oscillatory part of the conductivity is, form (90.23),

$$\tilde{\sigma}_{yy} = B^2 \bar{\sigma}_{yy} \frac{5}{6N\epsilon_F}\frac{\partial \tilde{M}_z}{\partial B}.$$

That of the magnetization, $\tilde{M}_z$, for the model in question is given by *SP* 1 (60.6).

CHAPTER X

THE DIAGRAM TECHNIQUE FOR NON-EQUILIBRIUM SYSTEMS

§91. The Matsubara susceptibility

THE study of the behaviour of various systems in a weak alternating external field usually amounts to the calculation of the appropriate generalized susceptibilities. In this section we shall derive expressions relating the generalized susceptibility to an auxiliary quantity which may be calculated by means of the Matsubara diagram technique; this opens the way to the use of such a technique to study the transport properties of various systems (A. A. Abrikosov, I. E. Dzyaloshinskiĭ and L. P. Gor'kov, 1962).

The generalized susceptibility $\alpha(\omega)$ is defined as follows (*SP* 1, § 123). Let the external action on the system be described by the inclusion in the Hamiltonian of a perturbing operator

$$\hat{V}(t) = -\hat{x}f(t), \tag{91.1}$$

where $\hat{x}$ is the Schrödinger (time-independent) operator of some physical quantity describing the system, and the perturbing generalized force $f(t)$ is a given function of time; we assume that the mean value of x is zero in the absence of the external action. Then, in the first approximation with respect to f, there is a linear relation between the Fourier components of the mean value $\bar{x}(t)$ and the force $f(t)$, and the generalized susceptibility is the coefficient in this relation:

$$\bar{x}_\omega = \alpha(\omega)f_\omega. \tag{91.2}$$

According to the Kubo formula (*SP* 1, § 126), the function $\alpha(\omega)$ can be expressed in the operator form

$$\alpha(\omega) = i\int_0^\infty e^{i\omega t}\langle \hat{x}_0(t)\hat{x}_0(0) - \hat{x}_0(0)\hat{x}_0(t)\rangle\, dt, \tag{91.3}$$

where $\hat{x}_0(t)$ is the Heisenberg operator defined in terms of the unperturbed Hamiltonian of the system (indicated by the subscript 0), and the averaging is over the specified unperturbed stationary state of the system, or over the Gibbs distribution with the unperturbed Hamiltonian.†

†Throughout this chapter we take $\hbar = 1$.

Let us now consider, purely formally, a system obeying the "Matsubara" equations of motion, which differ from the actual equations by the change of the time $t \to -i\tau$; the new variable τ takes values in the finite range

$$-1/T \leqslant \tau \leqslant 1/T. \tag{91.4}$$

Let this system be subjected to a perturbation

$$\hat{V}(\tau) = -\hat{x}f(\tau). \tag{91.5}$$

The mean value $\bar{x}$ will then also be a function of τ. We expand the function $f(\tau)$ in Fourier series on the range (91.4):

$$f(\tau) = \sum_{s=-\infty}^{\infty} f_s e^{-i\zeta_s\tau}, \quad \zeta_s = 2\pi sT, \tag{91.6}$$

and the function $\bar{x}(\tau)$ similarly.† The *Matsubara susceptibility* is defined as the coefficient of proportionality between the components of the two expansions:

$$\bar{x}_x = \alpha_M(\zeta_s)f_s. \tag{91.7}$$

Our aim is now, firstly, to obtain for $\alpha_M(\zeta_s)$ a formula analogous to (91.3) and, secondly, to find a relation between $\alpha_M(\zeta_s)$ and the function $\alpha(\omega)$ which is sought.

For the first task, let $\hat{H}$ be the unperturbed Hamiltonian of the system. The "exact" Matsubara operator of x is calculated as‡

$$\hat{x}^M(\tau) = \hat{\sigma}^{-1}(\tau, 0)\hat{x}_0{}^M(\tau)\hat{\sigma}(\tau, 0), \tag{91.8}$$

where $\hat{\sigma}$ is the Matsubara S-matrix:

$$\hat{\sigma}(\tau, 0) = \mathrm{T}_\tau \exp\left\{-\int_0^\tau \hat{V}_0{}^M(\tau')\,d\tau'\right\}, \tag{91.9}$$

and the subscript 0 denotes operators in the Matsubara "interaction representation":§

$$\hat{x}_0{}^M(\tau) = \exp(\tau\hat{H}_0)\hat{x}\exp(-\tau\hat{H}_0) \tag{91.10}$$

and similarly for $\hat{V}_0{}^M(\tau)$. In first-order perturbation theory, the expression (91.9) reduces to

$$\hat{\sigma}(\tau, 0) \approx 1 - \int_0^\tau \hat{V}_0{}^M(\tau')\,d\tau'. \tag{91.11}$$

†For x, which has a classical limit, we must use the technique corresponding to Bose statistics; the expansion (91.6) is then made in terms of "even frequencies" ζ_s.

‡All the concepts and formulae used below may be found in *SP* 2, §38.

§ Formula (91.8) is valid even when the initial operator $\hat{V}(\tau)$ depends explicitly on τ, although this has not been implied in the derivation given in *SP* 2, §38.

The value averaged over the Gibbs distribution is

$$\bar{x}(\tau) = \mathrm{tr}\{e^{-\hat{H}/T}\hat{x}^M(\tau)\}. \tag{91.12}$$

According to *SP* 2 (38.6), we have

$$\begin{aligned} e^{-\hat{H}/T} &= e^{-\hat{H}_0/T}\hat{\sigma}(1/T, 0) \\ &\approx e^{-\hat{H}_0/T}\left(1 - \int_0^{1/T} \hat{V}_0{}^M(\tau')\, d\tau'\right), \end{aligned}$$

and by (91.8) and (91.11)

$$\hat{x}^M(\tau) \approx \hat{x}_0{}^M(\tau) - \int_0^\tau \{\hat{x}_0{}^M(\tau)\hat{V}_0{}^M(\tau') - \hat{V}_0{}^M(\tau')\hat{x}_0{}^M(\tau)\}\, d\tau'.$$

Substitution of these expressions in (91.12) gives with the same accuracy

$$\begin{aligned} \bar{x}(\tau) = \mathrm{tr}\Big\{e^{-\hat{H}_0/T}\Big[\int_0^\tau [\hat{V}_0{}^M(\tau')\hat{x}_0{}^M(\tau) - \hat{x}_0{}^M(\tau)\hat{V}_0{}^M(\tau')]\, d\tau' \\ - \int_0^{1/T} \hat{V}_0{}^M(\tau')\hat{x}_0(\tau)\, d\tau'\Big]\Big\}. \end{aligned}$$

In the first integral, the variable $\tau' < \tau$; in the second, we divide the range of integration into those from 0 to τ and from τ to $1/T$. After cancelling, and substituting for $\hat{V}_0(\tau)$ from (91.5), we see that the result may be written

$$\bar{x}(\tau) = \int_0^{1/T} f(\tau')\langle \mathrm{T}_\tau \hat{x}_0{}^M(\tau)\hat{x}_0{}^M(\tau')\rangle\, d\tau'; \tag{91.13}$$

the operator T_τ of chronological ordering with respect to τ places the factors in order of increasing τ from right to left, without changing the sign of the product. The averaging in (91.13) is over the Gibbs distribution with the Hamiltonian $\hat{H}_0$. The result of the averaging depends only on the difference $\tau - \tau'$. Finally, putting $f(\tau')$ in the form of the Fourier expansion (91.6), we obtain the required formula for the Matsubara susceptibility:

$$\alpha_M(\zeta_s) = \int_0^{1/T} e^{i\zeta_s\tau}\langle \mathrm{T}_\tau \hat{x}_0{}^M(\tau)\hat{x}_0{}^M(0)\rangle\, d\tau. \tag{91.14}$$

We see that $\alpha_M(\zeta_s)$ is expressed in terms of the Fourier component of the Matsubara Green's function constructed from the operators $\hat{x}$; cf. the definition in *SP* 2 (37.2). Note the difference from formula (91.3) for $\alpha(\omega)$, which contains the retarded (with respect to the time t) commutator, not the chronological product.

To deal with the second part of the problem, that is, to find the relation between the functions $\alpha(\omega)$ and $\alpha_M(\zeta_s)$, we must start from (91.3) and (91.1), and express these functions in terms of the matrix elements of the operator $\hat{x}$. We shall not give the relevant calculations here, since they are practically identical with earlier similar ones (*SP* 1, § 126; *SP* 2, §§ 36, 37), but simply give the result:

$$\alpha(\omega) = \sum_{m,n} e^{-E_n/T} \frac{|x_{mn}|^2}{\omega - \omega_{mn} + i0}(1 - e^{-\omega_{mn}/T}), \tag{91.15}$$

$$\alpha_M(\zeta_s) = \sum_{m,n} e^{-E_n/T} \frac{|x_{mn}|^2}{i\zeta_s - \omega_{mn}}(1 - e^{-\omega_{mn}/T}). \tag{91.16}$$

Here the x_{mn} are the matrix elements of the Schrödinger operator $\hat{x}$ with respect to the stationary states of the system, and $\omega_{mn} = E_m - E_n$. Comparison of the two expressions shows that

$$\alpha_M(\zeta_s) = \alpha(i\zeta_s), \quad \zeta_s > 0. \tag{91.17}$$

Since the generalized susceptibility $\alpha(\omega)$ is real on the positive half of the imaginary ω-axis, the function $\alpha_M(\zeta_s)$ is real when $\zeta_s > 0$. It is seen from (91.16) that $\alpha_M(-\zeta_s) = \alpha_M^*(\zeta_s)$. Thus $\alpha_M(\zeta_s)$ is a real even function of ζ_s, expressed in terms of $\alpha(\omega)$ by

$$\alpha_M(\zeta_s) = \alpha(i|\zeta_s|). \tag{91.18}$$

This gives the required relation. To determine $\alpha(\omega)$, we must construct a function that is analytic in the upper half of the ω-plane, whose values at discrete points $\omega = i\zeta_s$ on the positive half of the imaginary axis are equal to $\alpha_M(\zeta_s)$; this will give the required generalized susceptibility.

In the next chapter, the above method will be applied to the transport properties of superconductors.

To conclude, we shall show that from $\alpha(\omega)$ we can find the relaxation formula for the quantity x reaching its equilibrium value $x = 0$. To do so, we shall suppose that the initial non-equilibrium value of x is produced by a generalized force $f(t)$ acting when $t < 0$ but not thereafter. The value of $x(t)$ at a time t is determined by the values of f throughout the preceding time:

$$x(t) = \int_{-\infty}^{t} \alpha(t - t')f(t')\,dt',$$

the function $\alpha(t)$ being related to the generalized susceptibility by the inverse Fourier transformation

$$\alpha(t) = \int_{-\infty}^{\infty} \alpha(\omega)e^{-i\omega t}\,d\omega/2\pi;$$

cf. *SP* 1, § 123. If $f = 0$ when $t > 0$, then

$$x(t) = \int_{-\infty}^{0} \alpha(t - t')f(t')\,dt'.$$

The behaviour of $x(t)$ for large t is determined by the asymptotic form of $\alpha(t)$ as $t \to \infty$. The latter in turn is determined by the singularity of $\alpha(\omega)$ that is in the lower half-plane and closest to the real axis. In particular, the relaxation of x by a simple exponential law $x \propto e^{-t/\tau}$ with relaxation time τ corresponds to a simple pole of $\alpha(\omega)$ at $\omega = -i/\tau$.

§92. Green's functions for a non-equilibrium system

Problems in physical kinetics always involve the consideration of non-equilibrium states. Nevertheless, the application of the method described in §91 allows, in some cases, the calculation of kinetic quantities to be reduced to that of Green's functions for systems in thermodynamic equilibrium, and this shows the possibility of using a diagram technique (e.g. the Matsubara) which is essentially applicable to equilibrium states. Of course, this possibility is always limited to physical problems relating to states not far from equilibrium.

We will now proceed to set up a diagram technique that is in principle suitable for calculating the Green's functions of systems in any non-equilibrium states. The equations then obtained for the Green's functions are similar to the transport equations as regards their significance. As applied to equilibrium systems, however, the same technique makes it possible to obtain the Green's functions and the generalized susceptibilities (at non-zero temperatures) as functions of continuous real frequencies directly, without any need for analytical continuation; for this reason it may prove, in complex cases, more useful than the Matsubara technique.†

Green's function for a non-equilibrium system is defined in the same way as in the equilibrium case:

$$iG_{\sigma_1\sigma_2}(X_1, X_2) = \langle n|\mathrm{T}\hat{\Psi}_{\sigma_1}(X_1)\hat{\Psi}^+_{\sigma_2}(X_2)|n\rangle$$
$$= \begin{cases} \langle n|\hat{\Psi}_{\sigma_1}(X_1)\hat{\Psi}^+_{\sigma_2}(X_2)|n\rangle, & t_1 > t_2; \\ \mp\langle n|\hat{\Psi}^+_{\sigma_2}(X_2)\hat{\Psi}_{\sigma_1}(X_1)|n\rangle, & t_1 < t_2. \end{cases} \tag{92.1}$$

The only difference is that the averaging (denoted by $\langle n|\ldots|n\rangle$) is now taken over any quantum state of the system, and not necessarily over the stationary state as in the equilibrium case.‡ The upper sign here and below refers to Fermi statistics, and the lower sign to Bose statistics; in the latter case (for a system of spinless particles) the spin suffixes σ_1, σ_2 must of course be omitted. In the case of Bose statistics it is assumed that there is no condensation, i.e. that the systems concerned either do not have a conserved number of particles (phonons or photons), or are at temperatures above the point at which condensation begins. In an inhomogeneous non-equilibrium system, the function (92.1) depends on the pairs of

†This technique is due to L. V. Keldysh (1964). In some respects it resembles that developed by R. Mills (1962) for equilibrium states.

‡The definition of G in *SP* 2, §36, for an equilibrium system with $T \neq 0$ also involved averaging over the Gibbs distribution. Let us here mention once more that, according to the fundamental principles of statistical physics, the result of a statistical averaging for an equilibrium system is independent of whether it is carried out with respect to the exact wave function of the stationary state of a closed system or by means of the Gibbs distribution for a system in a "thermostat". The only difference is that in the first case the result of the averaging is expressed in terms of the energy and number of particles in the system, and in the second case in terms of the temperature and the chemical potential.

variables $X_1 = (t_1, \mathbf{r}_1)$ and $X_2 = (t_2, \mathbf{r}_2)$ separately, and not only on their difference $X_1 - X_2$ as in the equilibrium case.

The diagram technique should enable us to express the Green's function of a system of interacting particles in terms of the functions for an ideal gas. There is, however, a necessity to introduce other functions besides G. In order not to interrupt the subsequent analysis, the definitions and some properties of these functions will now be given.

For reasons which will appear in §93, it is appropriate to denote the function (92.1) by G^{--}: thus we write this definition in the form†

$$iG_{12}^{--} = \langle \mathrm{T}\hat{\Psi}_1\hat{\Psi}_2^+\rangle = \begin{cases} \langle \hat{\Psi}_1\hat{\Psi}_2^+\rangle, & t_1 > t_2; \\ \mp\langle \hat{\Psi}_2^+\hat{\Psi}_1\rangle, & t_1 < t_2. \end{cases} \tag{92.2}$$

The definition

$$iG_{12}^{++} = \langle \tilde{\mathrm{T}}\hat{\Psi}_1\hat{\Psi}_2^+\rangle = \begin{cases} \mp\langle \hat{\Psi}_2^+\hat{\Psi}_1\rangle, & t_1 > t_2; \\ \langle \hat{\Psi}_1\hat{\Psi}_2^+\rangle, & t_1 < t_2 \end{cases} \tag{92.3}$$

differs from (92.2) in that T is replaced by $\tilde{\mathrm{T}}$, which signifies that the operator factors are arranged in the reverse of chronological order, with decreasing time from right to left.

Two further functions are defined as the mean values of the products of Ψ-operators not in chronological order:

$$iG_{12}^{+-} = \langle \hat{\Psi}_1\hat{\Psi}_2^+\rangle, \quad iG_{12}^{-+} = \mp\langle \hat{\Psi}_2^+\hat{\Psi}_1\rangle. \tag{92.4}$$

The difference in the signs in these definitions for Fermi systems is due to the general rule that there must be a change of sign when Ψ-operators are interchanged.

The second function (92.4) with $t_1 = t_2 \equiv t$ is the same as the one-particle density matrix; written in full,

$$\mp iG^{-+}(t, \mathbf{r}_1; t, \mathbf{r}_2) = \mathcal{N}\rho(t, \mathbf{r}_1, \mathbf{r}_2) \tag{92.5}$$

(cf. *SP* 2 (7.17), (31.4)). Here it does not matter from which side t_2 tends to the limit t_1, since G^{-+} is continuous when $t_2 = t_1$. The value of iG^{+-} with $t_1 = t_2$ is related to that of iG^{-+} by

$$i\{G^{+-}(t, \mathbf{r}_1; t, \mathbf{r}_2) - G^{-+}(t, \mathbf{r}_1; t, \mathbf{r}_2)\} = \delta(\mathbf{r}_1 - \mathbf{r}_2), \tag{92.6}$$

which follows from the commutation rule for Fermi or Bose Ψ-operators.

†To lighten the notation, we shall regard the spin suffixes as included in the variables $X = (t, \mathbf{r}, \sigma)$. Where no misunderstanding is possible, we shall also simplify by using suffixes to represent the arguments X: $\Psi_1 \equiv \Psi(X_1)$, $G_{12} \equiv G(X_1, X_2)$, etc. Lastly, the averaging will be denoted by $\langle \ldots \rangle$ simply, not by $\langle n | \ldots | n \rangle$.

The four G functions thus defined are not independent. They are linearly related in a way that is obvious from their definitions:

$$G^{--}+G^{++}=G^{-+}+G^{+-}. \tag{92.7}$$

The functions G^{--} and G^{++} are also connected by the relation of "anti-Hermitian conjugacy" when their arguments are interchanged:

$$G_{12}^{--}=-G_{21}^{++*}. \tag{92.8}$$

The functions G^{-+} and G^{+-} are themselves anti-Hermitian:

$$G_{12}^{-+}=-G_{21}^{-+*},\quad G_{12}^{+-}=-G_{21}^{+-*}. \tag{92.9}$$

The relation between these functions and the retarded or advanced Green's functions will be important in the following discussions. These latter functions are defined similarly to those in the equilibrium case (cf. *SP* 2, § 36):

$$\left.\begin{aligned} iG_{12}^{R}&=\begin{cases}\langle\hat{\Psi}_1\hat{\Psi}_2^{+}\pm\hat{\Psi}_2^{+}\hat{\Psi}_1\rangle, & t_1>t_2,\\ 0, & t_1<t_2;\end{cases}\\ iG_{12}^{A}&=\begin{cases}0, & t_1>t_2,\\ -\langle\hat{\Psi}_1\hat{\Psi}_2^{+}\pm\hat{\Psi}_2^{+}\hat{\Psi}_1\rangle, & t_1<t_2.\end{cases}\end{aligned}\right\} \tag{92.10}$$

These two are Hermitian conjugates:

$$G_{12}^{A}=G_{21}^{R*}. \tag{92.11}$$

Direct comparison of the definitions (92.2)–(92.4) and (92.10) gives

$$\left.\begin{aligned} G^{R}&=G^{--}-G^{-+}=G^{+-}-G^{++},\\ G^{A}&=G^{--}-G^{+-}=G^{-+}-G^{++}.\end{aligned}\right\} \tag{92.12}$$

In the steady state with spatial homogeneity, when all functions depend only on the differences $t=t_1-t_2$ and $\mathbf{r}=\mathbf{r}_1-\mathbf{r}_2$, they can be Fourier-expanded with respect to these variables. From (92.8) and (92.11) the Fourier components satisfy the equations

$$G^{--}(\omega,\mathbf{p})=-[G^{++}(\omega,\mathbf{p})]^*,\quad G^{A}(\omega,\mathbf{p})=[G^{R}(\omega,\mathbf{p})]^*, \tag{92.13}$$

and it follows from (92.9) that the Fourier components $G^{+-}(\omega,\mathbf{p})$ and $G^{-+}(\omega,\mathbf{p})$ are imaginary.

For a system of non-interacting particles the function G^{--} satisfies the equation

$$\hat{G}_{01}^{-1}G_{12}^{(0)--}=\delta(X_1-X_2), \tag{92.14}$$

where $\hat{G}_0^{-1}$ denotes the differential operator

$$\hat{G}_0^{-1} = i\frac{\partial}{\partial t} - \epsilon(-i\nabla) + \mu = i\frac{\partial}{\partial t} + \frac{\triangle}{2m} + \mu, \tag{92.15}$$

$\epsilon(\mathbf{p}) = \mathbf{p}^2/2m$, and

$$\delta(X_1 - X_2) = \delta_{\sigma_1\sigma_2}\delta(t_1 - t_2)\delta(\mathbf{r}_1 - \mathbf{r}_2); \tag{92.16}$$

the superscript (0) to G indicates that this function pertains to an ideal gas, and the suffix 1 to $\hat{G}_0^{-1}$ indicates that the differentiation is with respect to the variables t_1 and $\mathbf{r}_1$. The delta function on the right of (92.14) arises from the discontinuity of the function G^{--} at $t_1 = t_2$.† The functions G^R and G^A have a similar discontinuity, and therefore $G^{(0)R}$ and $G^{(0)A}$ satisfy a similar equation. The function G^{++} has a discontinuity of the opposite sign at $t_1 = t_2$, and therefore

$$\hat{G}_{01}^{-1} G_{12}^{(0)++} = -\delta(X_1 - X_2). \tag{92.17}$$

Lastly, the functions G^{+-} and G^{-+} are continuous at $t_1 = t_2$, and so, for an ideal gas, they satisfy the equations‡

$$\hat{G}_{01}^{-1} G_{12}^{(0)+-} = 0, \quad \hat{G}_{01}^{-1} G_{12}^{(0)-+} = 0. \tag{92.18}$$

We shall calculate all G functions for a stationary homogeneous state of an ideal gas, with some (not necessarily the equilibrium) momentum distribution $n_\mathbf{p}$ of the particles. To simplify the formulae, we shall suppose that this distribution is independent of the spin. The spin dependence of the G functions (in Fermi statistics) then separates as a factor $\delta_{\sigma_1\sigma_2}$, and we shall omit this factor together with the spin suffixes.

The Ψ-operators for an ideal gas are written as ordinary expansions

$$\hat{\Psi}_0(t, \mathbf{r}) = \frac{1}{\sqrt{\mathcal{V}}}\sum_{\mathbf{p}} \hat{a}_\mathbf{p} \exp\{i[\mathbf{p}\cdot\mathbf{r} - \epsilon(\mathbf{p})t + \mu t]\}, \tag{92.19}$$

and similarly for $\hat{\Psi}_0^+$; see *SP* 2 (9.3). When these expressions are substituted in the definitions of the G functions, it must be remembered that the only non-zero diagonal matrix elements are those of products of particle annihilation and creation operators with the same $\mathbf{p}$:

$$\langle \hat{a}_\mathbf{p}^+ \hat{a}_\mathbf{p}\rangle = n_\mathbf{p}, \quad \langle \hat{a}_\mathbf{p}\hat{a}_\mathbf{p}^+\rangle = 1 \mp n_\mathbf{p}.$$

†See *SP* 2, §9. The derivation given there does not depend on the assumed averaging over the ground state of the system, and remains valid for averaging over any quantum state.

‡If the differentiation is with respect to the second, not the first, variables in the G functions, the sign of $i\partial/\partial t$ must be changed, i.e. the operator $\hat{G}_{01}^{-1}$ must be replaced by $\hat{G}_{02}^{-1}{}^*$:

$$\hat{G}_{02}^{-1}{}^* G_{12}^{(0)--} = \delta(X_1 - X_2), \tag{92.14a}$$

and so on.

Thus we find, for example,

$$G^{(0)-+}(t, \mathbf{r}) = \pm \frac{i}{\mathcal{V}} \int n_{\mathbf{p}} \exp\{i\mathbf{p}\,.\,\mathbf{r} - i\epsilon(\mathbf{p})t + i\mu t\} \frac{\mathcal{V} d^3p}{(2\pi)^3},$$

where $t = t_1 - t_2$, $\mathbf{r} = \mathbf{r}_1 - \mathbf{r}_2$. With an identical transformation of this expression into

$$G^{(0)-+}(t, \mathbf{r}) = \pm 2\pi i \int n_{\mathbf{p}} \exp(i\mathbf{p}\,.\,\mathbf{r} - i\omega t)\delta(\omega - \epsilon + \mu)\, d\omega\, d^3p/(2\pi)^4,$$

we see that

$$G^{(0)-+}(\omega, \mathbf{p}) = \pm 2\pi i n_{\mathbf{p}} \delta(\omega - \epsilon + \mu). \tag{92.20}$$

Similarly,

$$G^{(0)+-}(\omega, \mathbf{p}) = -2\pi i(1 \mp n_{\mathbf{p}})\delta(\omega - \epsilon + \mu). \tag{92.21}$$

To calculate G^R, it is most convenient to start directly from the equation

$$\left[i\frac{\partial}{\partial t} - \epsilon(-i\nabla) + \mu\right] G^{(0)R}(t, \mathbf{r}) = \delta(t)\delta(\mathbf{r}),$$

solve it by the Fourier method, and use the fact that $G^R(\omega, \mathbf{p})$ cannot have a singularity in the upper half of the ω-plane. This gives immediately

$$G^{(0)R}(\omega, \mathbf{p}) = [\omega - \epsilon(\mathbf{p}) + \mu + i0]^{-1}; \tag{92.22}$$

the function $G^{(0)A}(\omega, \mathbf{p})$ is found from this, in accordance with (92.13), by simply taking the complex conjugate.

Lastly, (92.12) then gives

$$\begin{aligned} G^{(0)--}(\omega, \mathbf{p}) &= [\omega - \epsilon(\mathbf{p}) + \mu + i0]^{-1} \pm 2\pi i n_{\mathbf{p}} \delta(\omega - \epsilon + \mu) \\ &= \mathrm{P}\frac{1}{\omega - \epsilon + \mu} + i\pi(\pm 2n_{\mathbf{p}} - 1)\delta(\omega - \epsilon + \mu). \end{aligned} \tag{92.23}$$

Note that (92.22) is independent of the properties of the state (i.e. of the distribution $n_{\mathbf{p}}$) over which the averaging is carried out. This property of $G^{(0)R}$ (and of $G^{(0)A}$) is not in fact dependent on the homogeneity and stationarity of the state of the system, which were assumed in the derivation of (92.22): the function $G^{(0)R}(X_1, X_2)$ is necessarily dependent only on the difference $X_1 - X_2$.

For an equilibrium system, $n_{\mathbf{p}}$ in (92.21)–(92.23) is to be taken as the Fermi or Bose distribution function. The G functions are then expressed in terms of T and μ; this achieves the change from averaging over a given stationary quantum state to averaging over the Gibbs distribution.

PROBLEM

Find Green's functions for a homogeneous stationary state of a phonon gas in a liquid.

SOLUTION. Similarly to the definitions (92.4), we have for the phonon field

$$iD_{12}^{+-} = \langle \hat{\rho}_1' \hat{\rho}_2' \rangle, \; iD_{12}^{-+} = \langle \hat{\rho}_2' \hat{\rho}_1' \rangle, \tag{1}$$

where $\hat{\rho}' = \hat{\rho}'^{+}$ is the operator of the variable part of the density of the medium. Since this operator is self-conjugate, the functions (1) are related by

$$D_{12}^{+-} = D_{21}^{-+}, \tag{2}$$

and they of course again have the property (92.9).

For a gas of non-interacting phonons (see *SP* 2 (24.10)),

$$\hat{\rho}' = \hat{\rho}'^{+} = \sum_{\mathbf{k}} i\left(\frac{\rho_0 k}{2u\mathcal{V}}\right)^{1/2} (\hat{c}_{\mathbf{k}} e^{i(\mathbf{k}\cdot\mathbf{r}-ukt)} - \hat{c}_{\mathbf{k}}^{+} e^{-i(\mathbf{k}\cdot\mathbf{r}-ukt)}), \tag{3}$$

where ρ_0 is the unperturbed density and u the speed of sound. Substituting (3) in (1) and changing from summation to integration, we have

$$iD^{(0)-+}(t, \mathbf{r}) = \frac{\rho_0}{2u} \int \{\langle \hat{c}_{\mathbf{k}}^{+} \hat{c}_{\mathbf{k}} \rangle e^{i(\mathbf{k}\cdot\mathbf{r}-ukt)}$$

$$+ \langle \hat{c}_{\mathbf{k}} \hat{c}_{\mathbf{k}}^{+} \rangle e^{-i(\mathbf{k}\cdot\mathbf{r}-ukt)}\} \frac{k\,d^3k}{(2\pi)^3},$$

or, replacing the variable of integration $\mathbf{k}$ by $-\mathbf{k}$ in the second term and expressing the mean values in terms of the occupation numbers $N_{\mathbf{k}}$ of the phonon states,

$$iD^{(0)-+}(t, \mathbf{r}) = \int \frac{\rho_0 k}{2u} \{N_{\mathbf{k}} e^{-iukt} + (1 + N_{-\mathbf{k}}) e^{iukt}\} e^{i\mathbf{k}\cdot\mathbf{r}} \frac{d^3k}{(2\pi)^3}.$$

The integrand (without the factor $e^{i\mathbf{k}\cdot\mathbf{r}}$) is the Fourier component with respect to the coordinates. Expansion with respect to time, also, gives

$$iD^{(0)+-}(\omega, \mathbf{k}) = \frac{1}{u} \pi \rho_0 k \{N_{\mathbf{k}} \delta(\omega - uk) + (1 + N_{-\mathbf{k}}) \delta(\omega + uk)\}. \tag{4}$$

For the function $D^{(0)+-}$ we have, according to (2),

$$D^{(0)+-}(\omega, \mathbf{k}) = D^{(0)-+}(-\omega, -\mathbf{k}). \tag{5}$$

Two further Green's functions are defined by

$$iD_{12}^{--} = \langle \mathrm{T} \hat{\rho}_1' \hat{\rho}_2' \rangle, \quad iD_{12}^{++} = \langle \tilde{\mathrm{T}} \hat{\rho}_1' \hat{\rho}_2' \rangle, \tag{6}$$

with

$$D_{12}^{--} = D_{21}^{--}, \quad D_{12}^{++} = D_{21}^{++}. \tag{7}$$

For non-interacting phonons, a similar calculation gives (cf. *SP* 2, §31, Problem)

$$D^{(0)--}(\omega, \mathbf{k}) = -[D^{(0)++}(\omega, \mathbf{k})]^*$$

$$= \frac{\rho_0 k}{2u} \left\{ \left[\frac{1}{\omega - uk + i0} - \frac{1}{\omega + uk - i0} \right] - 2\pi i [N_{\mathbf{k}} \delta(\omega - uk) + N_{-\mathbf{k}} \delta(\omega + uk)] \right\}. \tag{8}$$

In accordance with (7), $D^{(0)--}(\omega, \mathbf{k}) = D^{(0)--}(-\omega, -\mathbf{k})$.

It follows from (8) that in the coordinate representation the function $D^{(0)--}(t, \mathbf{r})$ satisfies the equation

$$\left(\frac{\partial^2}{\partial t^2} - u^2 \triangle\right) D^{(0)--}(t, \mathbf{r}) = \rho_0 \delta(t) \triangle \delta(\mathbf{r}), \tag{9}$$

which replaces (92.14) for the Green's functions of ordinary particles.

§93. The diagram technique for non-equilibrium systems

The whole of the diagram technique is based on separating in the Hamiltonian of the system the interaction operator: $\hat{H} = \hat{H}_0 + \hat{V}$, where $\hat{H}_0$ is the Hamiltonian for a system of non-interacting particles. The diagram technique is a perturbation theory with respect to $\hat{V}$.

For a non-equilibrium system, the technique is constructed in the same way as in the equilibrium case with $T = 0$.† The Green's function $G \equiv G^{--}$ is expressed in terms of the Ψ-operators in the interaction representation (i.e. for an ideal gas) by

$$iG^{--}_{12} = \langle \hat{S}^{-1}\mathrm{T}[\hat{\Psi}_{01}\hat{\Psi}^{+}_{02}\hat{S}]\rangle, \tag{93.1}$$

where

$$\hat{S} \equiv \hat{S}(\infty, -\infty) = \mathrm{T}\exp\left(-i\int_{-\infty}^{\infty} \hat{V}_0(t)\,dt\right), \tag{93.2}$$

and $\hat{V}_0(t)$ is the operator $\hat{V}$ in the interaction representation. The averaging in (93.1) is over some state of the system of non-interacting particles. It will be convenient to assume that this is a stationary homogeneous state but not the ground state; we shall see later that the initial state can be eliminated, and the theory formulated so that the equations are independent of it. There is a difference here from the case $T = 0$, where the averaging is over the ground state. This difference is very important: the averaging is over the ground state. This difference is very important: the averaging of the operator $\hat{S}^{-1}$ cannot be separated from that of the other factors as in the derivation of *SP* 2 (12.14) from (12.12), because a non-ground state is not transformed into itself by the operator $\hat{S}^{-1}$, but into a superposition of other excited states, which may be intuitively regarded as the result of all possible processes of mutual scattering of quasi-particles.‡

The expression (93.1) is to be expanded in powers of $\hat{V}$. It is convenient first to transform $\hat{S}^{-1}$, using the unitarity of $\hat{S}$ and the fact that the operator $\hat{V}$ is Hermitian:

$$\hat{S}^{-1} = \hat{S}^{+} = \tilde{\mathrm{T}}\exp\left(i\int_{-\infty}^{\infty} \hat{V}(t)\,dt\right); \tag{93.3}$$

the symbol $\tilde{\mathrm{T}}$, which denotes anti-chronological ordering, has been defined in § 92.

Expanding $\hat{S}$ and $\hat{S}^{-1}$ in series and substituting in (93.1), we obtain a sum of various terms, in each of which an averaging is to be performed by means of Wick's theorem, and a diagram is associated with each way of contracting the Ψ-operators in pairs.§

†The following discussion is essentially based on that in *SP* 2, §§ 12 and 13.

‡For the same reason, the diagram technique given in *SP* 2, §§ 12 and 13, is not in general applicable, even when $T = 0$, for the case where alternating external fields are present (i.e. when $\hat{V}$ depends explicitly on the time even in the Schrödinger representation): the alternating fields excite the ground state of the system. It must be emphasized, however, that the technique described here is valid even when an alternating field is present.

§In the macroscopic limit, the validity of Wick's theorem does not depend on the homogeneous stationary state over which the averaging is carried out; see *SP* 2, § 13, end.

First of all, as in the "ordinary" diagram technique with $T=0$, only the connected diagrams (not containing separate vacuum loops) need be considered. The vacuum loops cancel out, as is easily seen by examining the first few diagrams, which show the general principle involved.

If all the contractions that yield a connected diagram are made in the factor $\mathrm{T}\hat{\Psi}_1\hat{\Psi}_2^+\hat{S}$ in (93.1), we obtain terms represented by the ordinary diagrams described in *SP* 2, § 13, though of course with a different specific form for the functions corresponding to the continuous lines. These are diagrams in the coordinate representation; the change to the momentum representation is unsuitable for non-equilibrium states (when the G functions depend on the variables X_1 and X_2 separately). Other terms arise from contractions involving also Ψ-operators from $\hat{S}^{-1}=\hat{S}^+$. In each order of perturbation theory, they are obtained from the ordinary terms on replacing any factor $\hat{V}$ from $\hat{S}$ by a factor from $\hat{S}^+$. Such terms are represented by diagrams of the same graphical form but with a somewhat different rule for reading them. The changes result from three causes: (1) in $\hat{S}^+$ the interaction operators appear as $+i\hat{V}$, not $-i\hat{V}$ as in $\hat{S}$; (2) all the Ψ-operators in $\hat{S}^+$ are always to the left of the operators in the product $\mathrm{T}\hat{\Psi}_1\hat{\Psi}_2^+\hat{S}$; (3) within the factor $\hat{S}^+$, the operators are ordered as a $\tilde{\mathrm{T}}$ (not T) product.

Let us now consider how these changes affect the construction of the diagram technique in the simple case of a system of particles (fermions, say) in an external field $U(t,\mathbf{r})\equiv U(X)$.

The first-order terms in the expansion of the expression (93.1) are

$$\left\langle \mathrm{T}\hat{\Psi}_1\hat{\Psi}_2^+\left(-i\int\hat{\Psi}_3^+U_3\hat{\Psi}_3\,d^4X_3\right)\right\rangle+\left\langle \tilde{\mathrm{T}}i\int\hat{\Psi}_3^+U_3\hat{\Psi}_3\,d^4X_3\,.\,\mathrm{T}\hat{\Psi}_1\hat{\Psi}_2^+\right\rangle.$$

The second term in this sum is characteristic of the situation in question; on averaging over the ground state, only the first term would have to be considered. In the first term, all four Ψ-operators are in the T product; their contractions in pairs according to

$$\mathrm{T}\Psi_1\Psi_2^+(-i\Psi_3^+U_3\Psi_3) \tag{93.4}$$

give factors $G_{32}^{(0)--}$ and $G_{13}^{(0)--}$. In the second term, the Ψ-operators contracted are not mutually ordered by T or $\tilde{\mathrm{T}}$:

$$\tilde{\mathrm{T}}(i\Psi_3^+U_3\Psi_3)\mathrm{T}(\Psi_1\Psi_2^+); \tag{93.5}$$

their contractions give factors $G_{32}^{(0)+-}$ and $G_{13}^{(0)--}$; $+iU_3$ replaces $-iU_3$.

The graphical elements differ from those occurring in the ordinary diagram technique by having additional symbols + or − at the ends of the lines. Broken lines with + or − at one end (a vertex of the diagram) denote factors $+iU(X)$ or $-iU(X)$:

$$\underset{+}{-----\cdot}=+iU(X),\qquad \underset{-}{-----\cdot}=-iU(X) \tag{93.6}$$

(cf. *SP* 2, § 19). Continuous lines with + or − at each end are associated with the various G functions:

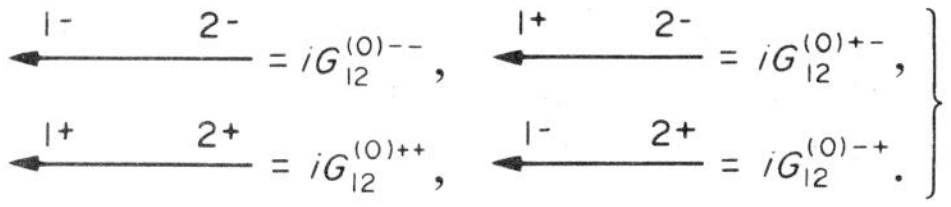

(93.7)

The numbers at the ends of the lines show the arguments of the functions (the variables X_1 and X_2).

The two terms (93.4) and (93.5) are then represented by the diagrams

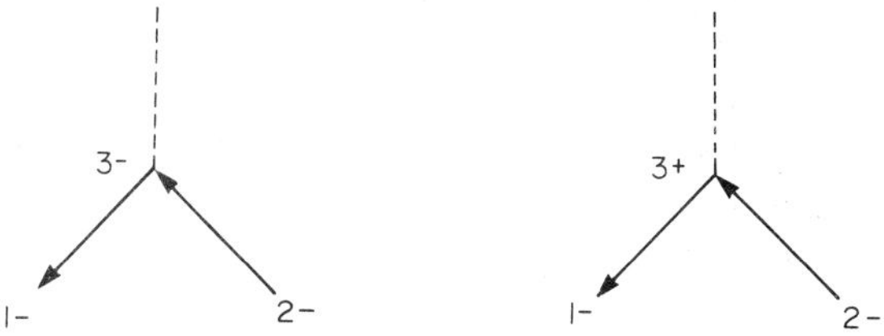

(93.8)

The two outer ends of the continuous lines are marked −, these being corrections to the functions G^{--}. Integration is implied with respect to the variables corresponding to the vertex of the diagram.† In analytical form,

$$iG_{12}^{(1)--} = \int \{iG_{13}^{(0)--} iG_{32}^{(0)--}(-iU_3) + iG_{13}^{(0)-+} iG_{32}^{(0)+-} iU_3\}\, d^4X_3. \tag{93.9}$$

In the next (second) order of perturbation theory, the correction to the functions G^{--} is given by four diagrams:

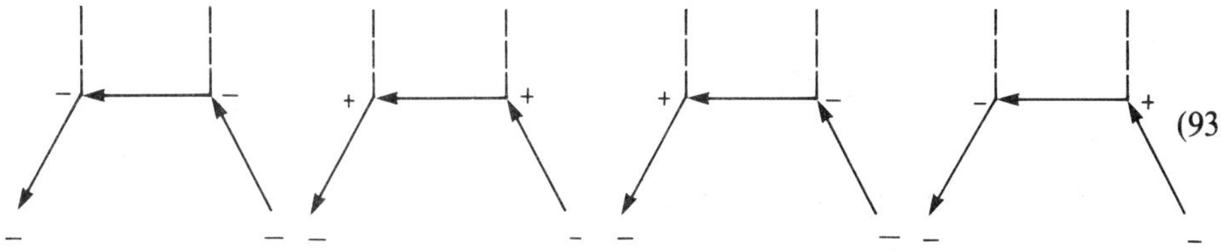

(93.10)

(the numerals are omitted). The signs + or − at each vertex of the diagram relate to the ends of all three of the lines that meet there.

Similarly, the correction terms in the other G functions are represented by diagrams with other signs at the two outer ends of the continuous lines. For example, in the first order the function G^{-+} has two diagrams:

(93.11)

†More precisely, integration over $dt\, d^3x$ and summation over a pair of like spin suffixes. The latter will be regarded here as included in the integration over d^4X.

Thus the diagrams in the Keldysh technique are obtained from those in the ordinary technique by assigning additional indices + or − in all possible ways to their vertices and free ends. This rule remains valid in the diagram technique for other types of interaction.

For a system with a pair interaction between particles, in the ordinary diagram technique the interaction potential of two particles is associated with an internal broken line. We now assign to the ends of such a line a further pair of like signs + or −:

$$\left.\begin{aligned} &\underset{-\,-\,-\,-\,-}{1+\qquad 2+} = iU(X_1 - X_2) \equiv i\delta(t_1 - t_2)U(\mathbf{r}_1 - \mathbf{r}_2),\\ &\underset{-\,-\,-\,-\,-}{1-\qquad 2-} = -iU(X_1 - X_2). \end{aligned}\right\} \qquad (93.12)$$

For example, the first-order correction to the functions G^{--} for a system with pair interaction is represented by a sum of four diagrams:

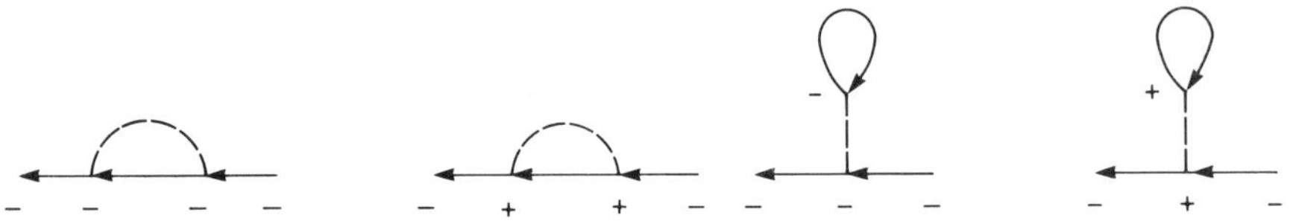

(93.13)

instead of the two diagrams *SP* 2 (13.13) in the ordinary technique. The continuous line forming a closed loop is again associated with a factor $N_0(\mu, T)$ (the ideal gas density) for either sign of the vertex.

It has already been mentioned that the Keldysh diagram technique is applicable also to equilibrium systems with $T \neq 0$. Let us suppose that there is no external field, and change from the coordinate to the momentum representation, expanding all G functions as Fourier integrals. Then each line in the diagrams is, as usual, assigned a definite 4-momentum, and the functions $U(Q)$ and $G^{(0)}(P)$ in the momentum representation are associated with these lines by the same rules.

When $T = 0$, the Fermi distribution function is

$$\begin{aligned} n_{\mathbf{p}} &= 1 \quad \text{for} \quad p < p_F,\\ &= 0 \quad \text{for} \quad p > p_F. \end{aligned}$$

Hence, from (92.20) and (92.21), we have for a Fermi system with $T = 0$

$$G^{(0)-+}(P) = 0 \quad \text{for} \quad p > p_F, \qquad G^{(0)+-}(P) = 0 \quad \text{for} \quad p < p_F,$$

and all the diagrams for G^{--} that contain "plus" vertices are identically zero. Thus the Keldysh diagram technique, as applied to equilibrium systems, becomes the ordinary diagram technique directly when $T = 0$, unlike the Matsubara technique.

§94. Self-energy functions

Like any "reasonable" diagram technique, the Keldysh technique allows the diagrams to be summed in "blocks". The most important of these are the self-energy functions.

This concept arises (*SP* 2, § 14) in considering Green's function diagrams that cannot be divided into two parts joined only by one continuous line. We can separate the factors $iG^{(0)}$ corresponding to the two end lines of such a diagram and express it (in the coordinate representation as a function of the two arguments X_1 and X_2) in the form

$$\int iG^{(0)}_{13}(-i\Sigma_{34})iG^{(0)}_{42}\, d^4X_3\, d^4X_4.$$

The function $-i\Sigma_{34}$, which stands for the whole of the inner part of the diagram, is called a *self-energy function*. The exact self-energy function, denoted by $-i\Sigma$, is the sum of all possible diagrams of this type. In accordance with the fact that in this technique each vertex in the diagram has to be given the sign + or −, there are four exact self-energy functions, corresponding to the signs of their "exit" and "entry" vertices; they are denoted by Σ^{--}, Σ^{++}, Σ^{-+} and Σ^{+-}.

The exact G functions are expressed in terms of the exact Σ functions by identities which may be written graphically for G^{--} as

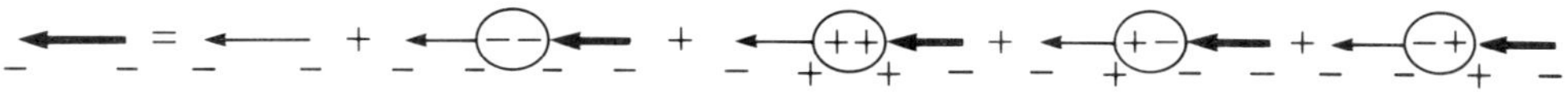

(94.1)

and similarly for the other functions; the thick lines are exact G functions and the ovals Σ functions (cf. *SP* 2 (14.4)). In analytical form,

$$G^{--}_{12} = G^{(0)--}_{12} + \int \{G^{(0)--}_{14}\Sigma^{--}_{43}G^{--}_{32} + G^{(0)-+}_{14}\Sigma^{++}_{43}G^{+-}_{32} + G^{(0)-+}_{14}\Sigma^{+-}_{43}G^{--}_{32} + G^{(0)--}_{14}\Sigma^{-+}_{43}G^{+-}_{32}\}\, d^4X_3\, d^4X_4 \qquad (94.2)$$

with three more equations for the other G functions.

These equations may be compactly written by means of the matrices

$$G = \begin{pmatrix} G^{--} & G^{-+} \\ G^{+-} & G^{++} \end{pmatrix}, \quad \Sigma = \begin{pmatrix} \Sigma^{--} & \Sigma^{-+} \\ \Sigma^{+-} & \Sigma^{++} \end{pmatrix}. \qquad (94.3)$$

Then the four equations such as (94.2) can be written jointly as one matrix equation

$$G_{12} = G^{(0)}_{12} + \int G^{(0)}_{14}\Sigma_{43}G_{32}\, d^4X_3\, d^4X_4, \qquad (94.4)$$

the factors in the integrand being combined by the rule of matrix multiplication.

The equations (92.14)–(92.18) satisfied by the ideal-gas G functions are similarly written jointly as

$$\hat{G}^{-1}_{01}G^{(0)}_{12} = \sigma_z\delta(X_1 - X_2), \qquad (94.5)$$

where†

$$\sigma_z = \begin{pmatrix} 1 & 0 \\ 0 & -1 \end{pmatrix}.$$

Let us now return to (94.4) and apply the operator $\hat{G}_{01}^{-1}$ to each side. Using (94.5), we obtain a set of four integro-differential equations written jointly as one matrix equation:

$$\hat{G}_{01}^{-1} G_{12} = \sigma_z \delta(X_1 - X_2) + \int \sigma_z \Sigma_{13} G_{32}\, d^4X_3. \tag{94.6}$$

This equation may be written in an equivalent alternative way, by noting that in the diagram form (94.1) the thick lines may just as well be on the left instead of on the right. In (94.2), therefore, the factors in each term of the integrand may be written in the order $G_{14}\Sigma_{43}G_{32}^{(0)}$. By applying the operator $\hat{G}_{02}^{-1*}$ (see the last footnote to § 92) to the resulting equations, we find

$$\hat{G}_{02}^{-1*} G_{12} = \sigma_z \delta(X_1 - X_2) + \int G_{13} \Sigma_{32} \sigma_z\, d^4X_3. \tag{94.7}$$

The self-energy functions themselves can be expressed as a series of skeleton diagrams whose graphical elements are thick continuous lines corresponding to exact G functions. For example, in a system of particles with pair interaction,

$-i\Sigma^{--} =$ 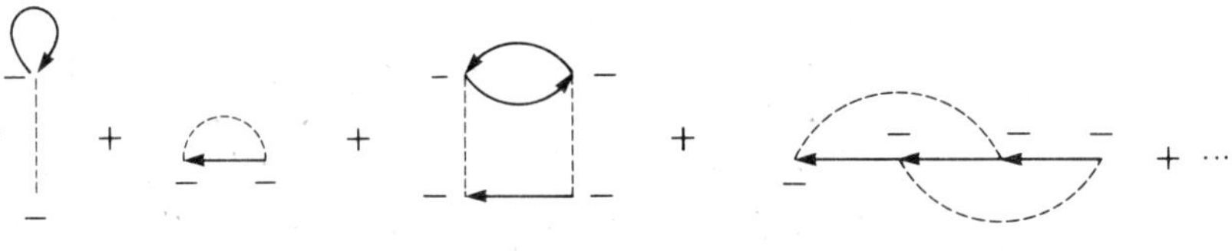 (94.8)

$-i\Sigma^{-+} =$ 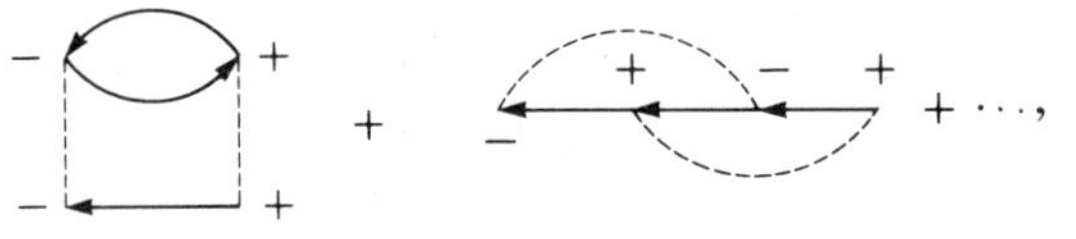 (94.9)

and similarly for Σ^{++} and Σ^{+-}; the further terms of the series contain diagrams with a larger number of broken lines.‡ Thus the equation (94.4) or (94.7) constitutes a complete (though very complicated) set of equations for the exact G functions.

Equations (94.6) do not involve the functions $G^{(0)}$ which depend on the choice of the "zero" state of a system of non-interacting particles. Thus there is no depend-

†The symbol σ_z, taken from the standard notation for the Pauli matrices, has of course no reference to spin here.

‡Cf. *SP* 2 (14.9), (14.10); all the diagrams of the first or second order listed there are among the skeleton diagrams (94.8).

ence on that choice.† But the occurrence of differential operators in the equations makes their solutions indefinite. This is manifested by the presence of the functions $G^{(0)}$ in the integral equations (94.4).

The set of equations (94.6) has, however, the disadvantage that it does not explicitly take account of the linear dependence of the G functions shown by (92.7). To avoid this disadvantage we must make a linear transformation of the matrix G in such a way that we can use (92.7) to reduce one of its elements to zero. This is done by means of the formula

$$G' = R^{-1}GR, \tag{94.10}$$

where

$$R = \frac{1}{\sqrt{2}}\begin{pmatrix}1 & 1\\ -1 & 1\end{pmatrix}, \quad R^{-1} = \frac{1}{\sqrt{2}}\begin{pmatrix}1 & -1\\ 1 & 1\end{pmatrix}.$$

It is easily seen that the transformed matrix is

$$G' = \begin{pmatrix}0 & G^A\\ G^R & F\end{pmatrix}, \tag{94.11}$$

where

$$F = G^{++} + G^{--} = G^{+-} + G^{-+}. \tag{94.12}$$

When the matrices $G^{(0)}$ and Σ are transformed in this way, equation (94.4) remains invariant.

The transformed matrix Σ is

$$\Sigma' = R^{-1}\Sigma R = \begin{pmatrix}\Omega & \Sigma^R\\ \Sigma^A & 0\end{pmatrix}, \tag{94.13}$$

with the notation

$$\Omega = \Sigma^{--} + \Sigma^{++}, \quad \Sigma^R = \Sigma^{--} + \Sigma^{-+}, \quad \Sigma^A = \Sigma^{--} + \Sigma^{+-}. \tag{94.14}$$

This may be proved by direct calculation, using the equation

$$\Sigma^{++} + \Sigma^{--} = -(\Sigma^{+-} + \Sigma^{-+}), \tag{94.15}$$

which follows from (92.7) and is easily derived by equating to zero the expression $\hat{G}_{01}^{-1}(G^{--} + G^{++} - G^{-+} - G^{+-})$ formed by means of equations (94.6).

†An important comment is needed here. When there is no external field, the functions G_0 depend only on the difference $X_1 - X_2$, and the functions G given by a series expansion in terms of the G_0 would have this property also. After the elimination of G_0, however, we can also consider solutions of (94.6) that depend on X_1 and X_2 separately.

Now expanding the transformed matrix equation (94.4), we obtain three equations. One of them is

$$G^A_{12} = G^{(0)A}_{12} + \int G^{(0)A}_{14}\Sigma^A_{43}G^A_{32}\, d^4X_3\, d^4X_4. \qquad (94.16)$$

The corresponding equation for G^R gives nothing new, since it is simply the Hermitian conjugate of (94.16). That equation, although it contains the function $G^{(0)A}$ pertaining to an ideal gas, does not depend on the "zero" state, since $G^{(0)A}$ does not do so, as noted in § 92.

Lastly, the third equation derived from (94.4), for the function F, contains terms involving the function $F^{(0)}$, which does depend on the "zero" state. These, however, are reduced to zero by the differential operator $\hat{G}_{01}^{-1}$, since $\hat{G}_{01}^{-1}F^{(0)} = 0$. The resulting equation is

$$\hat{G}_{01}^{-1}F_{12} = \int \{\Omega_{13}G^A_{02} + \Sigma^R_{13}F_{32}\}\, d^4X_3. \qquad (94.17)$$

Equations (94.16) and (94.17) constitute a complete description in principle of the behaviour of a non-equilibrium system. The second of them is an integro-differential equation, and forms a generalization of the Boltzmann equation; here it should be remembered that by (92.5) and (92.6) the functions G^{-+} and G^{+-}, and therefore F, are directly related to the particle distribution function in the system. The solution of (94.17) is arbitrary to the same extent as that of the transport equation. However, (94.16) is a purely integral equation and therefore brings no further arbitrariness into the solution.

There is nevertheless a fundamental feature of equations (94.16) and (94.17) by which they differ in general from the ordinary transport equation: they contain not one but two time variables, t_1 and t_2. We shall show in § 95 how this difference is removed in the quasi-classical case.

§ 95. The transport equation in the diagram technique

We shall use a simple example to show how the passage is made from equations of the type (94.16), (94.17) to the ordinary quasi-classical transport equation. Let us consider a slightly non-ideal Fermi gas at temperatures $T \sim \epsilon_F$, assuming the quasi-classicality conditions to be satisfied: the time intervals τ and distances L over which all quantities vary significantly satisfy the inequalities

$$\tau\epsilon_F \gg 1, \quad Lp_F \gg 1; \qquad (95.1)$$

cf. § 40. Although no new result is obtained in this case, of course, the analysis has some instructive features that will be useful in more complicated cases.

The quantized transport equation must determine the one-particle density matrix $\rho(t, \mathbf{r}_1, \mathbf{r}_2)$.† To go to the quasi-classical case, it is appropriate to use the mixed

†As in § 40, we assume that the electron distribution is independent of the spin, and omit the spin factor $\delta_{\sigma_1\sigma_2}$ from ρ.

coordinate–momentum representation, taking a Fourier expansion with respect to $\boldsymbol{\xi} = \mathbf{r}_1 - \mathbf{r}_2$ but retaining the coordinate dependence on $\mathbf{r} = \frac{1}{2}(\mathbf{r}_1 + \mathbf{r}_2)$. Here $\mathbf{r}_1 = \mathbf{r} + \frac{1}{2}\boldsymbol{\xi}$, $\mathbf{r}_2 = \mathbf{r} - \frac{1}{2}\boldsymbol{\xi}$, so that the Fourier transform is

$$\frac{1}{\mathcal{N}} n(t, \mathbf{r}, \mathbf{p}) = \int e^{-i\mathbf{p}\cdot\boldsymbol{\xi}} \rho(t, \mathbf{r} + \tfrac{1}{2}\boldsymbol{\xi}, \mathbf{r} - \tfrac{1}{2}\boldsymbol{\xi})\, d^3\xi. \tag{95.2}$$

The inverse transform is

$$\rho(t, \mathbf{r}_1, \mathbf{r}_2) = \frac{1}{\mathcal{N}} \int e^{i\mathbf{p}\cdot(\mathbf{r}_1 - \mathbf{r}_2)} n(t, \tfrac{1}{2}(\mathbf{r}_1 + \mathbf{r}_2), \mathbf{p}) \frac{d^3p}{(2\pi)^3}. \tag{95.3}$$

The integration of $n(t, \mathbf{r}, \mathbf{p})$ with respect to the coordinates gives the particle momentum distribution function, as is seen from the expression for this integral in terms of the original density matrix:

$$N_{\mathbf{p}} = \int n(t, \mathbf{r}, \mathbf{p})\, d^3x = \mathcal{N} \int e^{-i\mathbf{p}\cdot(\mathbf{r}_1 - \mathbf{r}_2)} \rho(t, \mathbf{r}_1, \mathbf{r}_2)\, d^3x_1\, d^3x_2. \tag{95.4}$$

The integration with respect to momenta gives the coordinate distribution, i.e. the spatial number density of particles, as we again see from the expression in terms of the density matrix:

$$N(t, \mathbf{r}) = \int n(t, \mathbf{r}, \mathbf{p})\, d^3p = \mathcal{N} \rho(t, \mathbf{r}, \mathbf{r}). \tag{95.5}$$

The function $n(t, \mathbf{r}, \mathbf{p})$ itself, in the general quantum case, cannot be regarded as the coordinate and momentum distribution function simultaneously; this would contradict the fundamental principles of quantum mechanics, and in any case the function $n(t, \mathbf{r}, \mathbf{p})$ defined by (95.2) is in general not even positive.

The function $n(t, \mathbf{r}, \mathbf{p})$ does, however, have the literal sense of a distribution function in the quasi-classical approximation. To see this, let us consider the operator of some physical quantity pertaining to an individual particle and depending on $\mathbf{r}$ and $\mathbf{p}$: $\hat{f} = f(\mathbf{r}, \hat{\mathbf{p}}) = f(\mathbf{r}, -i\nabla)$.† By definition of the density matrix, the mean value of f is

$$\bar{f} = \int [\hat{f}_1 \rho(t, \mathbf{r}_1, \mathbf{r}_2)]_{\mathbf{r}_1 = \mathbf{r}_2 = \mathbf{r}}\, d^3x,$$

where $\hat{f}_1$ acts on the variable $\mathbf{r}_1$. We substitute ρ in the form (95.3), and use the fact that with the conditions (95.1) n is a more slowly varying function of $\mathbf{r}_1$ than the factor $\exp(i\mathbf{p}\cdot\mathbf{r}_1)$. It is therefore sufficient to differentiate only the latter, which amounts to the change $-i\nabla_1 \to \mathbf{p}$. The expression for $\bar{f}$ then becomes

$$\bar{f} = \frac{1}{\mathcal{N}} \int f(\mathbf{r}, \mathbf{p}) n(t, \mathbf{r}, \mathbf{p})\, d^3x \frac{d^3p}{(2\pi)^3}, \tag{95.6}$$

†We shall take the particular case where all the operators ∇ are to the right of $\mathbf{r}$. In the quasi-classical approximation, this is not an important point.

which (since f is arbitrary) corresponds exactly to the definition of the classical distribution function.

We will now obtain the equations for the Green's function $G^{-+}(X_1, X_2)$ which is, by (92.5), most closely related to the density matrix. We use for this function the "four-dimensional" mixed representation

$$G^{-+}(X, P) = \int e^{iP\Xi} G^{-+}(X + \tfrac{1}{2}\Xi, X - \tfrac{1}{2}\Xi)\, d^4\Xi, \tag{95.7}$$

where $P = (\omega, \mathbf{p})$, $X = (t, \mathbf{r})$, $\Xi = (\xi_0, \boldsymbol{\xi})$, $t = \frac{1}{2}(t_1 + t_2)$, $\xi_0 = t_1 - t_2$. Then

$$n(t, \mathbf{r}, \mathbf{p}) = -i \int G^{-+}(X, P)\, d\omega/2\pi; \tag{95.8}$$

the integration over $d\omega/2\pi$ is equivalent to putting $t_1 = t_2$.

After these preliminary definitions, let us derive the transport equation. We take the $-+$ component of equations (94.6) and (94.7), and subtract term by term:

$$(\hat{G}_{02}^{-1*} - \hat{G}_{01}^{-1})G_{12}^{-+} = -\int (\Sigma_{13}^{--} G_{32}^{-+} + \Sigma_{13}^{-+} G_{32}^{++} + G_{13}^{-+}\Sigma_{32}^{++} + G_{13}^{--}\Sigma_{32}^{-+})\, d^4X_3. \tag{95.9}$$

The operator acting on the function G_{12}^{-+} on the left is

$$\begin{aligned} \hat{G}_{02}^{-1*} - \hat{G}_{01}^{-1} &= -i\left(\frac{\partial}{\partial t_1} + \frac{\partial}{\partial t_2}\right) - \frac{1}{2m}(\Delta_1 - \Delta_2) \\ &= -i\left(\frac{\partial}{\partial t} - \frac{i}{m}\nabla_{\mathbf{r}} \cdot \nabla_{\boldsymbol{\xi}}\right). \end{aligned}$$

We now take Fourier components (95.7) on each side of (95.9) and put $t_1 = t_2$ (or, equivalently, integrate over $d\omega/2\pi$). Using (95.8), we find that the left-hand side of (95.9) becomes

$$\frac{\partial n}{\partial t} + \frac{\mathbf{p}}{m} \cdot \frac{\partial n}{\partial \mathbf{r}},$$

which is the required form of the left-hand side of the transport equation for the distribution function $n(t, \mathbf{r}, \mathbf{p})$. The right-hand side of (95.9), after Fourier transformation, must therefore give the collision integral $C(n)$.

The change to Fourier components on the right-hand side must take account of the quasi-classicality conditions. The integral in (95.9) is a sum of terms of the form

$$\int \Sigma(X_1, X_3) G(X_3, X_2)\, d^4X_3.$$

We express the factors Σ and G as functions of the differences and averages of the "4-coordinates":

$$\int \Sigma(X_1 - X_3, \tfrac{1}{2}(X_1 + X_3)) G(X_3 - X_2, \tfrac{1}{2}(X_3 + X_2))\, d^4X_3.$$

In the change to Fourier components with respect to the first arguments, the important ranges are $|\mathbf{r}_1 - \mathbf{r}_3|$, $|\mathbf{r}_3 - \mathbf{r}_2| \sim 1/p$ for the coordinate differences, and $|t_1 - t_3|$, $|t_3 - t_2| \sim 1/\epsilon$ for the time differences. According to the conditions (95.1), Σ and G as functions of their second arguments vary only slightly in these ranges. We can therefore approximately replace those arguments by $X = \frac{1}{2}(X_1 + X_2)$:

$$\int \Sigma(X_1 - X_3, X) G(X_3 - X_2, X)\, d^4X_3,$$

and can then take the Fourier representation for a given value of X. The right-hand side of (95.9) then becomes

$$C(n) = -\int \{\Sigma^{-+}(G^{--} + G^{++}) + (\Sigma^{--} + \Sigma^{++})G^{-+}\}\, d\omega/2\pi$$

$$= \int \{-\Sigma^{-+}G^{+-} + \Sigma^{+-}G^{-+}\}\, d\omega/2\pi, \qquad (95.10)$$

where all the functions in the integrand have the same arguments $(X, P) \equiv (t, \mathbf{r}; \omega, \mathbf{p})$; in the second expression, we have used the relations (92.7) and (94.15).

Let us apply formula (95.10) to the model of an almost ideal Fermi gas discussed previously (*SP* 2, §§6 and 21). As there, we shall arbitrarily suppose that the potential $U(\mathbf{r}_1 - \mathbf{r}_2)$ of the interaction between particles satisfies the condition for perturbation theory to be applicable; in the change to the real interaction (which does not satisfy this condition) it is sufficient to express the result in terms of the scattering amplitude.

Having in view the determination of the collision integral in the first non-vanishing approximation of perturbation theory with regard to the particle interaction, we may suppose that the exact G functions in (95.10) are related to the distribution function n by the same formulae (92.20), (92.21) as in an ideal gas; this implies the neglect of small corrections, due to the interaction, in the gas particle energy $\epsilon = p^2/2m$.† The expressions (92.20) and (92.21) relate, strictly speaking, to the homogeneous steady state of the gas, but, in the quasi-classical case, because of the slow variation of n with coordinates and time, we can use the same expressions with $n_\mathbf{p}$ regarded as a function $n(t, \mathbf{r}, \mathbf{p})$, with t and $\mathbf{r}$ as parameters. The integration with respect to ω removes the delta functions, leaving

$$C(n) = i\Sigma^{-+}(\epsilon - \mu, \mathbf{p}; t, \mathbf{r})[1 - n(t, \mathbf{r}, \mathbf{p})]$$

$$+ i\Sigma^{+-}(\epsilon - \mu, \mathbf{p}; t, \mathbf{r}) n(t, \mathbf{r}, \mathbf{p}). \qquad (95.11)$$

It is clear from the form of this expression that the first term describes the "gain" of particles, possible only when $1 - n \neq 0$; the second describes the "loss", which is proportional to n. It remains to calculate the self-energy functions Σ^{-+} and Σ^{+-}.

†This approximation enables us to neglect the remaining components of (94.6), i.e. to regard them as satisfied identically in the relevant approximation.

The first non-vanhshing contribution to these comes from the second-order diagrams (cf. (94.9)); for example,

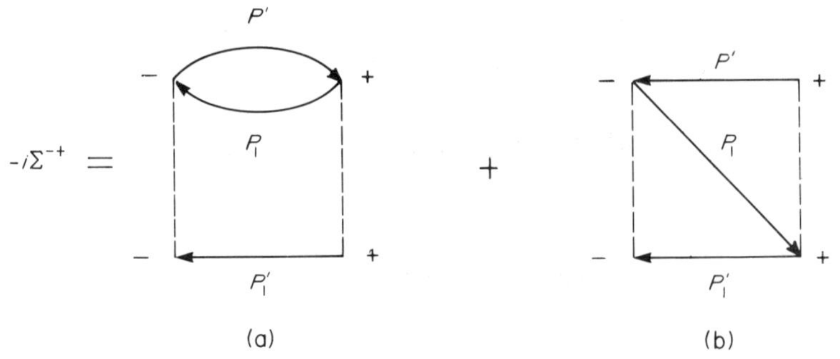

(95.12)

where $P_1' = P + P_1 - P'$. When U is replaced by U_0 (see below), the contributions to Σ from these two diagrams are related by $\Sigma_a = -2\Sigma_b$ (the minus sign comes from the closed loop in diagram (a), and the factor 2 from the spin summation in that loop; compare the analogous calculations in *SP* 2, §21). Expanding diagram (b) in analytical form, we find

$$i\Sigma^{-+}(P) = \int G^{-+}(P')G^{+-}(P_1)G^{-+}(P_1')U^2(\mathbf{p}_1 - \mathbf{p}')\, d^4P_1\, d^4P'/(2\pi)^8.$$

In a degenerate gas, the particle wavelength ($\sim 1/p$) is necessarily large in comparison with the range of the interaction forces, because of the condition for the gas to be rarefied (see *SP* 2, §6); this enables us to replace $U(\mathbf{p}_1 - \mathbf{p}')$ by its value for $\mathbf{p}_1 - \mathbf{p}' = 0$:

$$U_0 \equiv \int U(r)\, d^3x.$$

Substituting the expressions (92.20) and (92.21) for G^{-+} and G^{+-}, and eliminating the two delta functions by integration with respect to the "time" components of the 4-vectors P_1 and P', we see that the first term in (95.11) in fact coincides with the "gain" term in the collision integral (74.5) ($w = 2\pi U_0^2$). The calculation of Σ^{+-} is similar, and the second term in (95.11) coincides with the "loss" term in the same collision integral.

CHAPTER XI

SUPERCONDUCTORS

§96. High-frequency properties of superconductors. General formula

FORMULAE have been derived in *SP* 2, §51, which relate the current in a superconductor to the vector potential of the electromagnetic field there. Here, these formulae will be generalized to case of a field varying in time. As in *SP* 2, the investigation will be based on the BCS model, the electrons in a metal being regarded as an isotropic gas with a weak attraction between the particles.†

As always in metals (and the more so in superconductors), the displacement current may be neglected in Maxwell's equations:

$$\operatorname{curl} \mathbf{H} = 4\pi \mathbf{j}/c. \tag{96.1}$$

Hence, in this approximation,

$$\operatorname{div} \mathbf{j} = 0. \tag{96.2}$$

To describe the field, we choose the gauge in which the scalar potential $\varphi = 0$. The linear relation between the Fourier components (with respect to time and coordinates) of the current density and the vector potential of the field is written

$$j_\alpha(\omega, \mathbf{k}) = -Q(\omega, \mathbf{k})(\delta_{\alpha\beta} - k_\alpha k_\beta/k^2)A_\beta(\omega, \mathbf{k}), \tag{96.3}$$

which satisfies identically the equation (96.2), i.e. the condition $\mathbf{k} \cdot \mathbf{j}(\omega, \mathbf{k}) = 0$. The longitudinal part (parallel to $\mathbf{k}$) of the vector $\mathbf{A}$ does not appear in (96.3), nor therefore in the equations at all, so that it can be taken as zero, with the assumption that $\mathbf{k} \cdot \mathbf{A}(\omega, \mathbf{k}) = 0$. With this choice of $\mathbf{A}$, the relation between the current and the field reduces to

$$\mathbf{j}(\omega, \mathbf{k}) = -Q(\omega, \mathbf{k})\mathbf{A}(\omega, \mathbf{k}). \tag{96.4}$$

Our object is to calculate the function $Q(\omega, \mathbf{k})$. This is a generalized susceptibility, and to solve the problem we use the method described in §91. We formally include in the Hamiltonian of the superconductor a "vector potential" that depends on the Matsubara variable τ (and on the coordinates):‡

$$\mathbf{A}(\tau, \mathbf{r}) = \mathbf{A}(\zeta_s, \mathbf{k})e^{i(\mathbf{k}\cdot\mathbf{r} - \zeta_s\tau)}, \quad \zeta_s = 2\pi sT. \tag{96.5}$$

†The results in §§96 and 97 are due to J. Bardeen and D. C. Mattis (1958) and to A. A. Abrikosov, L. P. Gor'kov and I. M. Khalatnikov (1958).

‡In this section we put $\hbar = 1$.

Using Gor'kov's equations, we calculate the correction linear in **A** to the Matsubara Green's function:

$$\mathcal{G}(\tau_1, \mathbf{r}_1 : \tau_2, \mathbf{r}_2) = \mathcal{G}^{(0)}(\tau_1 - \tau_2, \mathbf{r}_1 - \mathbf{r}_2) + \mathcal{G}^{(1)}(\tau_1, \mathbf{r}_1; \tau_2, \mathbf{r}_2); \tag{96.6}$$

because of the "homogeneity in τ" and the spatial homogeneity of the unperturbed superconductor, $\mathcal{G}^{(0)}$ depends only on the differences of its arguments. The current density $\mathbf{j}(\tau, \mathbf{r})$ is expressed in terms of the Green's function by

$$\mathbf{j}(\tau, \mathbf{r}) = -\frac{ie}{m}[(\nabla' - \nabla)\mathcal{G}^{(1)}(\tau, \mathbf{r}; \tau', \mathbf{r}')]_{\mathbf{r}'=\mathbf{r}, \tau'=\tau+0} - \frac{e^2 N}{mc}\mathbf{A}(\tau, \mathbf{r}), \tag{96.7}$$

where N is the number density of particles.† With the field (96.5), this relation has in practice the form

$$\mathbf{j}(\tau, \mathbf{r}) = -Q_M(\zeta_s, \mathbf{k})\mathbf{A}(\tau, \mathbf{r}). \tag{96.8}$$

The coefficient Q_M is the Matsubara susceptibility, and by (91.18)

$$Q(i|\zeta_s|, \mathbf{k}) = Q_M(\zeta_s, \mathbf{k}). \tag{96.9}$$

To determine the required function $Q(\omega, \mathbf{k})$, it is necessary to make an analytical continuation from the points $\omega = i|\zeta_s|$ to the whole of the upper half-plane.

The calculation of Q_M is similar to the calculations in *SP* 2, § 51. In the potential gauge with div $\mathbf{A} = 0$, there is no correction to the gap Δ in the energy spectrum, and the linearized Gor'kov equations for the Green's functions $\mathcal{G}$ and $\bar{\mathcal{F}}$ are

$$\left.\begin{aligned}
&\left[-\frac{\partial}{\partial\tau} + \frac{\triangle}{2m} + \mu\right]\mathcal{G}^{(1)}(\tau, \mathbf{r}; \tau', \mathbf{r}') + \Delta\bar{\mathcal{F}}^{(1)}(\tau, \mathbf{r}; \tau', \mathbf{r}') \\
&\qquad = -\frac{ie}{mc}\mathbf{A}(\tau, \mathbf{r}) \cdot \nabla\mathcal{G}^{(0)}(\tau - \tau', \mathbf{r} - \mathbf{r}'), \\
&\left[\frac{\partial}{\partial\tau} + \frac{\triangle}{2m} + \mu\right]\bar{\mathcal{F}}^{(1)}(\tau, \mathbf{r}; \tau', \mathbf{r}') - \Delta\mathcal{G}^{(1)}(\tau, \mathbf{r}; \tau', \mathbf{r}') \\
&\qquad = \frac{ie}{mc}\mathbf{A}(\tau, \mathbf{r}) \cdot \nabla\bar{\mathcal{F}}^{(0)}(\tau - \tau', \mathbf{r} - \mathbf{r}').
\end{aligned}\right\} \tag{96.10}$$

With a field of the form (96.5), we can at once separate the dependence of $\mathcal{G}^{(1)}$ and $\bar{\mathcal{F}}^{(1)}$ on the sums $\tau + \tau'$ and $\mathbf{r} + \mathbf{r}'$, putting

$$\mathcal{G}^{(1)} = g(\tau - \tau', \mathbf{r} - \mathbf{r}')\exp[\tfrac{1}{2}i\mathbf{k} \cdot (\mathbf{r} + \mathbf{r}') - \tfrac{1}{2}i\zeta_s(\tau + \tau')] \tag{96.11}$$

†Cf. *SP* 2 (51.17). In making comparisons with the formulae in *SP* 2, § 51, it must be remembered that e is now a positive quantity, the unit charge.

and similarly for $\bar{\mathscr{F}}^{(1)}$ with f in place of g. After this change, the first equation (96.10), for example, becomes

$$\left[-\left(\frac{\partial}{\partial\tau}-\tfrac{1}{2}i\zeta_s\right)+\frac{1}{2m}(\nabla+\tfrac{1}{2}\mathbf{k})^2+\mu\right]g+\Delta f$$
$$=-\frac{ie}{mc}\mathbf{A}(\zeta_s,\mathbf{k})\exp[\tfrac{1}{2}i\mathbf{k}\,.\,(\mathbf{r}-\mathbf{r}')-\tfrac{1}{2}i\zeta_s(\tau-\tau')]\,.\,\nabla\mathscr{G}^{(0)}.$$

We now expand all quantities in Fourier series with respect to $\tau-\tau'$ and Fourier integrals with respect to $\mathbf{r}-\mathbf{r}'$:

$$g(\tau,\mathbf{r})=T\sum_{s'=-\infty}^{\infty}\int g(\zeta'_{s'},\mathbf{p})\exp[i\mathbf{p}\,.\,\mathbf{r}-i\zeta'_{s'}\tau]\,d^3p/(2\pi)^3 \tag{96.12}$$

and so on. We then obtain for the Fourier components a pair of algebraic equations:

$$\left.\begin{aligned}&\left[i(\zeta'_{s'}+\tfrac{1}{2}\zeta_s)-\frac{1}{2m}(\mathbf{p}+\tfrac{1}{2}\mathbf{k})^2+\mu\right]g(\zeta'_{s'},\mathbf{p})+\Delta f(\zeta'_{s'},\mathbf{p})\\&\qquad=\frac{e}{mc}\mathbf{p}\,.\,\mathbf{A}(\zeta_s,\mathbf{k})\mathscr{G}^{(0)}(\zeta'_{s'}-\tfrac{1}{2}\zeta_s,\mathbf{p}-\tfrac{1}{2}\mathbf{k}),\\&\left[-i(\zeta'_{s'}+\tfrac{1}{2}\zeta_s)-\frac{1}{2m}(\mathbf{p}+\tfrac{1}{2}\mathbf{k})^2+\mu\right]f(\zeta'_{s'},\mathbf{p})-\Delta f(\zeta'_{s'},\mathbf{p})\\&\qquad=-\frac{e}{mc}\mathbf{p}\,.\,\mathbf{A}(\zeta_s,\mathbf{k})\bar{\mathscr{F}}^{(0)}(\zeta'_{s'}-\tfrac{1}{2}\zeta_s,\mathbf{p}-\tfrac{1}{2}\mathbf{k}).\end{aligned}\right\} \tag{96.13}$$

The "unperturbed" Green's functions $\mathscr{G}^{(0)}$ and $\bar{\mathscr{F}}^{(0)}$ are expanded in Fourier series with "odd frequencies" $(2s'+1)\pi T$. It therefore follows from (96.13) that the "frequencies" $\zeta'_{s'}$ take the values

$$\zeta'_{s'}=(2s'+1-s)\pi T.$$

The functions $\mathscr{G}^{(0)}$ and $\bar{\mathscr{F}}^{(0)}$ are (see *SP* 2 (42.7), (42.8))

$$\left.\begin{aligned}\mathscr{G}^{(0)}(\zeta_s,\mathbf{p})&=-(i\zeta_s+\eta)/(\zeta_s^2+\epsilon^2),\\\bar{\mathscr{F}}^{(0)}(\zeta_s,\mathbf{p})&=\Delta/(\zeta_s^2+\epsilon^2),\end{aligned}\right\} \tag{96.14}$$

where

$$\eta=\frac{p^2}{2m}-\mu\approx v_F(p-p_F),\quad \epsilon^2=\Delta^2+\eta^2; \tag{96.15}$$

the constant Δ is assumed real. By using these formulae, we can easily put the solution of equations (96.13) in the form

$$g(\zeta'_{s'},\mathbf{p})=\frac{e}{mc}\mathbf{p}\,.\,\mathbf{A}(\zeta_s,\mathbf{p})\{\mathscr{G}^{(0)}(P_+)\mathscr{G}^{(0)}(P_-)+\bar{\mathscr{F}}^{(0)}(P_+)\bar{\mathscr{F}}^{(0)}(P_-)\}, \tag{96.16}$$

where

$$P_{\pm} = (\zeta'_{s'} \pm \tfrac{1}{2}\zeta_s, \mathbf{p} \pm \tfrac{1}{2}\mathbf{k}). \tag{96.17}$$

With (96.7), (96.11) and (96.12), we obtain for the current density

$$\mathbf{j}(\zeta_s, \mathbf{k}) = -\frac{2eT}{m} \sum_{s'=-\infty}^{\infty} \int \mathbf{p} g(\zeta'_{s'}, \mathbf{p}) \frac{d^3p}{(2\pi)^3} - \frac{Ne^2}{mc} \mathbf{A}(\zeta_s, \mathbf{k}),$$

g being given by (96.16). As the vectors $\mathbf{j}$ and $\mathbf{A}$ are transverse to $\mathbf{k}$, we average in the integrand over the directions of the vector $\mathbf{p}_\perp$ in the plane perpendicular to $\mathbf{k}$. The functions $\mathcal{G}^{(0)}$ and $\bar{\mathcal{F}}^{(0)}$ in (96.16) do not depend on the direction of $\mathbf{p}_\perp$; averaging of the factor $\mathbf{p}_\perp(\mathbf{p}_\perp \cdot \mathbf{A})$ converts it to $\mathbf{A}p^2 \sin^2 \tfrac{1}{2}\theta$, where θ is the angle between $\mathbf{p}$ and $\mathbf{k}$. We thus have the following final expression for the Matsubara susceptibility:

$$Q_M(\zeta_s, \mathbf{k}) = \frac{Ne^2}{mc} + \frac{e^2T}{m^2c} \int \sum_{s'=-\infty}^{\infty} p^2 \sin^2\theta \times$$

$$\times [\mathcal{G}^{(0)}(P_+)\mathcal{G}^{(0)}(P_-) + \bar{\mathcal{F}}^{(0)}(P_+)\bar{\mathcal{F}}^{(0)}(P_-)]\, d^3p/(2\pi)^3. \tag{96.18}$$

Let us now make the analytical continuation of this function from the discrete series of points $\zeta_s = 2s\pi T$ to the whole of the right-hand half of the complex ζ-plane, i.e. to the upper half of the ω-plane ($\omega = i\zeta$). This amounts to the analytical continuation of the integrand in (96.18); let us consider its first term, for example:

$$J_M(\zeta_s) \equiv T \sum_{s'=-\infty}^{\infty} \mathcal{G}_+(\zeta'_{s'} + \tfrac{1}{2}\zeta_s)\mathcal{G}_-(\zeta'_{s'} - \tfrac{1}{2}\zeta_s)$$

$$= T \sum_{s'=-\infty}^{\infty} \mathcal{G}_+((2s'+1)\pi T)\mathcal{G}_-((2s'+1)\pi T - \zeta_s). \tag{96.19}$$

For brevity, we omit the index (0), and replace the arguments $\mathbf{p}_\pm = \mathbf{p} \pm \tfrac{1}{2}\mathbf{k}$ by suffixes $\pm$. This expression may be written as the integral

$$J_M(\zeta_s) = \frac{1}{4\pi i} \oint \mathcal{G}_+(z)\mathcal{G}_-(z - \zeta_s) \tan(z/2T)\, dz, \tag{96.20}$$

taken along the three closed contours C_1, C_2 and C_3 in Fig. 32, which together enclose the infinite set of poles of the factor $\tan(z/2T)$ at the points $z = (2s'+1)\pi T$ (marked by strokes in the diagram). The residues of the integrand at each pole give corresponding terms in the sum (96.19); at infinity, $\mathcal{G}(z) \propto 1/z$, and the integral therefore converges. In choosing the contours, we have used the fact that $\mathcal{G}(z)$ is analytic in each of the two half-planes:

$$\mathcal{G}(z) = \begin{cases} G^R(iz), & \text{re } z > 0; \\ G^A(iz), & \text{re } z < 0, \end{cases}$$

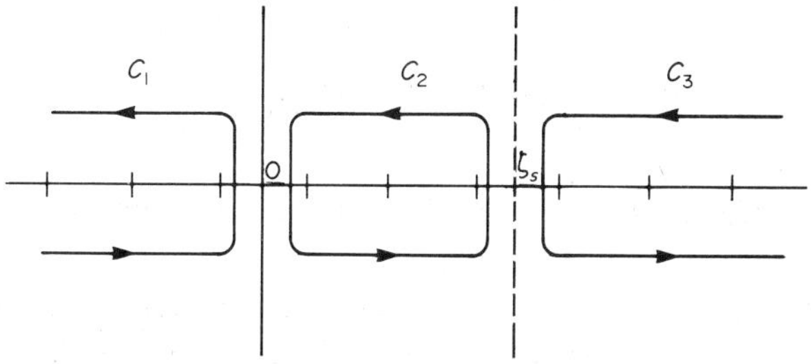

FIG. 32.

where G^R and G^A are analytic functions, the retarded and advanced Green's functions (see *SP* 2, § 37); the imaginary z-axis is in general a cut for the function $\mathcal{G}(z)$.

We now rotate the contours so as to pass vertically on either side of the cuts re $z = 0$ and re $z = \zeta_s$ (Fig. 33; the infinitely distant parts which close the contours

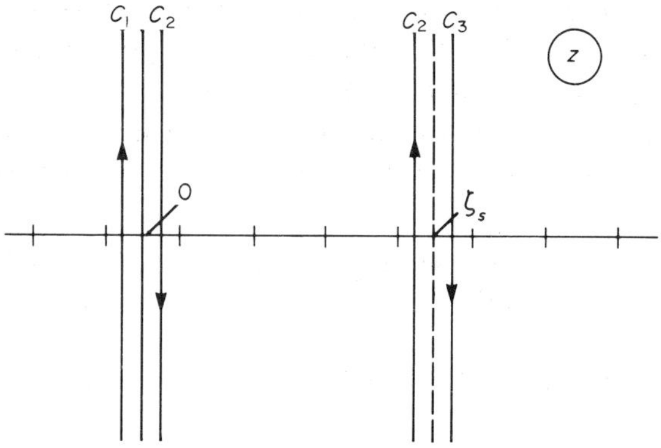

FIG. 33.

are not shown). On the pair of lines C_1, C_2 we change the variable of integration by putting $z = i\omega'$, and on C_2, C_3 we put $z - \zeta_s = i\omega'$. Then, when $\zeta_s > 0$,

$$J_M(\zeta_s) = -\frac{1}{4\pi}\int_{-\infty}^{\infty}\left\{\tan\frac{i\omega'}{2T}[G_+^R(\omega') - G_+^A(\omega')]G_-^A(\omega' - i\zeta_s)\right.$$

$$\left. + \tan\frac{i\omega' + \zeta_s}{2T}[G_-^R(\omega') - G_-^A(\omega')]G_+^R(\omega' + i\zeta_s)\right\} d\omega'. \qquad (96.21)$$

In the derivation of this equation, ζ_s has been fixed as $2\pi sT$. For such values,

$$\tan\frac{i\omega' + \zeta_s}{2T} = \tan\frac{i\omega'}{2T} = i\tanh\frac{\omega'}{2T}.$$

The fact that the expression (96.21) is analytic for all $\zeta_s > 0$ is then evident from the fact that G^A and G^R are analytic in the corresponding half-planes. Now putting $i\zeta_s = \omega$, we have for the analytically continued function†

$$J(\omega) \equiv J_M(-i\omega)$$
$$= -\frac{i}{4\pi}\int \tanh\frac{\omega'}{2T}\{[G^R_+(\omega') - G^A_+(\omega')]G^A_-(\omega'-\omega)$$
$$+ [G^R_-(\omega') - G^A_-(\omega')]G^R_+(\omega'+\omega)\}\, d\omega'. \tag{96.22}$$

The second term in the integrand in (96.18) is analytically continued in a similar way, and the result differs from (96.22) only in that G^R and G^A are replaced by F^{+R} and F^{+A}.‡ These functions are (see *SP* 2, § 41)

$$\left.\begin{aligned} G^R(\omega, \mathbf{p}) &= \frac{u_p{}^2}{\omega - \epsilon + i0} + \frac{v_p{}^2}{\omega + \epsilon + i0},\\ F^{+R}(\omega, \mathbf{p}) &= \frac{\Delta}{2\epsilon}\left[\frac{1}{\omega + \epsilon + i0} - \frac{1}{\omega - \epsilon + i0}\right], \end{aligned}\right\} \tag{96.23}$$

where

$$\left.\begin{matrix} u_p{}^2 \\ v_p{}^2 \end{matrix}\right\} = \tfrac{1}{2}(1 \pm \eta/\epsilon).$$

The functions G^A and F^{+A} are the same with the sign of $i0$ changed. Hence

$$G^R - G^A = 2\,\mathrm{im}\, G^R = -\pi[u_p{}^2\delta(\omega - \epsilon) + v_p{}^2\delta(\omega + \epsilon)],$$
$$F^{+R} - F^{+A} = (\pi\Delta/2\epsilon)[\delta(\omega - \epsilon) - \delta(\omega + \epsilon)],$$

and the integration in (96.22) amounts to the removal of the delta functions.

After some simple but laborious algebra we arrive at the final expression§

$$Q(\omega, \mathbf{k}) = \frac{Ne^2}{mc} - \frac{e^2}{4m^2c}\int p^2 \sin^2\theta \tanh\frac{\epsilon_+}{2T} \times$$
$$\times\left\{\left[1 + \frac{\eta_+\eta_- + \Delta^2}{\epsilon_+\epsilon_-}\right]\left[\frac{1}{\epsilon_+ - \epsilon_- - \omega - i0} + \frac{1}{\epsilon_+ - \epsilon_- + \omega + i0}\right]\right.$$
$$\left. + \left[1 - \frac{\eta_+\eta_- + \Delta^2}{\epsilon_+\epsilon_-}\right]\left[\frac{1}{\epsilon_+ + \epsilon_- - \omega - i0} + \frac{1}{\epsilon_+ + \epsilon_- + \omega + i0}\right]\right\}\frac{d^3p}{(2\pi)^3}, \tag{96.24}$$

†This method of analytical continuation is due to G.M. Éliashberg (1962).

‡The definition of the Green's function F^+ (corresponding to the temperature function $\bar{\mathscr{F}}$) is given in *SP* 2, §41. The definitions of F^{+R} and F^{+A} differ from that of F^+ in that the T product is replaced by the commutator, the relationship being similar to that between G^R, G^A and G.

§Mention has been made (*SP* 2, §51) of the need for caution in calculating sums and integrals of the form (96.18), because of the slowness of decrease of the integrand. With the order of operations used here, this difficulty is avoided, as is confirmed by the fact that the final expression (96.24) satisfies the necessary condition: $Q = 0$ when $\Delta = 0$ and $\omega = 0$ (a normal metal in a static field); see the second footnote to §97.

where

$$\eta_{\pm} = \frac{1}{2m}(\mathbf{p} \pm \tfrac{1}{2}\mathbf{k})^2 - \mu, \quad \epsilon_{\pm}^2 = \Delta^2 + \eta_{\pm}^2. \tag{96.25}$$

The two terms in the braces in (96.24) are quite different in origin and significance. The first is an odd function of **p**, and the integral of it is therefore zero for $T = 0$, when $\tanh(\epsilon_+/2T) = 1$. This part of Q is related to the collisionless dynamics of elementary excitations. Its imaginary part, which exists for all ω and **k**, is related to the collisionless Landau damping.

The integral of the second term is not zero even when $T = 0$. This part of Q is related to the formation or break-up of Cooper pairs. The poles of the integrand in this part are at $\epsilon_+ + \epsilon_- = \pm\omega$. For them to exist (and so for there to be dissipation because Q has an imaginary part), the frequency must exceed 2Δ, the Cooper pair binding energy.

§97. High-frequency properties of superconductors. Limiting cases

Let us now examine the general formula (96.24). The number of limiting cases here is very large on account of the presence of four independent parameters $\hbar\omega$, $\hbar k v_F$, Δ and T, which can be in various relationships to one another. Several of these limiting cases will be considered.

When $\hbar\omega \gg \Delta$, the gap in the superconductor spectrum is unimportant. Putting $\Delta = 0$ in the first approximation, we should obtain the formula for the transverse permittivity of a normal electron Fermi gas; we shall not pause to give the relevant calculations.†

London case

Let us take the London limiting case, in which

$$\hbar k v_F \ll \Delta_0, \tag{97.1}$$

where Δ_0 is the value of $\Delta(T)$ when $T = 0$. We shall assume that $\Delta \lesssim T$, thus excluding the range of very low temperatures. The frequency will be regarded as small, in the sense that $\omega \lesssim k v_F$.

As $k \to 0$,

$$1 - \frac{\eta_+\eta_- + \Delta^2}{\epsilon_+\epsilon_-} \propto k^2.$$

The second term in the braces in (96.24) is therefore small, and may be neglected. In the first term, the first square bracket is equal to 2; since the second square

†The relation between $Q(\omega, \mathbf{k})$ and the transverse permittivity $\epsilon_t(\omega, \mathbf{k})$ is ascertained as follows. Expressing the current density in terms of the polarization vector by $-i\omega\mathbf{P} = \mathbf{j}$, and the vector potential **A** in terms of the electric field by $\mathbf{E} = i\omega\mathbf{A}/c$, we can rewrite (96.4) as $\mathbf{P} = -c\omega^{-2}Q\mathbf{E}$. This shows that

$$-cQ/\omega^2 = (\epsilon_t - 1)/4\pi.$$

bracket is an odd function of $\mathbf{p}$, we can then write

$$Q(\omega, \mathbf{k}) = \frac{Ne^2}{mc} - \frac{e^2}{2m^2c} \int \left[\tanh \frac{\epsilon_+}{2T} - \tanh \frac{\epsilon_-}{2T}\right] \frac{p^2 \sin^2 \theta}{\epsilon_+ - \epsilon_- - \hbar\omega - i0} \frac{d^3p}{(2\pi\hbar)^3}.$$

We have $\tanh(\epsilon/2T) = 1 - 2n_0(\epsilon)$, where

$$n_0 = [e^{\epsilon/T} + 1]^{-1} \tag{97.2}$$

is the distribution function of elementary excitations in a superconducting Fermi gas (a Fermi distribution with zero chemical potential), and thus put

$$\begin{aligned} \tanh \frac{\epsilon_+}{2T} - \tanh \frac{\epsilon_-}{2T} &= -2[n_0(\epsilon_+) - n_0(\epsilon_-)] \\ &\approx -2\hbar\mathbf{k}\cdot\mathbf{v}\, \partial n_0/\partial\epsilon, \end{aligned}$$

where

$$\mathbf{v} = \partial\epsilon/\partial\mathbf{p} = \eta\mathbf{p}/m\epsilon.$$

Then

$$Q(\omega, \mathbf{k}) = \frac{Ne^2}{mc} + \frac{e^2}{m^2c} \int \frac{\partial n_0}{\partial\epsilon} \frac{\mathbf{k}\cdot\mathbf{v}p^2 \sin\theta}{\mathbf{k}\cdot\mathbf{v} - \omega - i0} \frac{d^3p}{(2\pi\hbar)^3}. \tag{97.3}$$

When $\omega = 0$, this expression agrees, as it should, with the London value N_se^2/mc, where $N_s(T)$ is the density of superconducting electrons.† We can therefore rewrite (97.3) in the equivalent form

$$Q(\omega, \mathbf{k}) = \frac{N_se^2}{mc} + \frac{\omega e^2}{m^2c} \int \frac{\partial n_0}{\partial\epsilon} \frac{p^2 \sin^2\theta}{\mathbf{k}\cdot\mathbf{v} - \omega - i0} \frac{d^3p}{(2\pi\hbar)^3}. \tag{97.4}$$

The second term in this expression represents the contribution to the permittivity from the elementary excitations in the Fermi gas.‡

When $\omega \ll kv$, we may neglect ω in the denominator of the integrand in (97.4):

$$Q(\omega, \mathbf{k}) = \frac{N_se^2}{mc} + \frac{\omega e^2}{4\pi^2 c\hbar^3 k} \int_{-1}^{1} \frac{\sin^2\theta \, d\cos\theta}{\cos\theta - i0} \int \frac{\partial n_0}{\partial\epsilon} \frac{p^4}{m^2 v} dp. \tag{97.5}$$

†This is easily shown by means of the formulae given in *SP* 2, §40, for the calculation of $\rho_s = mN_s$. The function $Q(0, \mathbf{k})$ tends to zero (as does N_s) when $T \to T_c$, as already mentioned in the last footnote to §96.

‡This may be seen by comparing (97.4) with formula (2) in §31, Problem 2, for the transverse permittivity of a collisionless electron plasma. In making the comparison, it must be noted that the London case corresponds to the quasi-classical limit, so that the formula for a degenerate gas differs from that for a Maxwellian plasma only in the form of the distribution function and the dispersion relation $\epsilon(p)$.

The integral with respect to $\cos\theta$ is calculated from the residue at the pole $\cos\theta = i0$, and is equal to $i\pi$. The integral with respect to p, written as

$$\int \frac{\partial n_0}{\partial \epsilon}\frac{p^2\epsilon}{\eta}\,d\eta,$$

diverges logarithmically when $|\eta| \ll \Delta$. With a cut-off at $|\eta| \sim \omega\Delta/kv_F$ (where $kv \sim \omega$), we find with logarithmic accuracy

$$\left[\frac{\partial n_0}{\partial \epsilon}\right]_{\epsilon=\Delta} p_F{}^2\Delta \,.\, 2\int_{\omega\Delta/kv_F}^{\Delta}\frac{d\eta}{\eta}.$$

Thus

$$Q(\omega, \mathbf{k}) = \frac{N_s e^2}{mc} - i\,\frac{e^2 p_F{}^2\Delta\omega\log(kv_F/\omega)}{2\pi c\hbar^3 Tk(e^{\Delta/T}+1)(e^{-\Delta/T}+1)}. \tag{97.6}$$

The imaginary part of Q determines the dissipation; a negative sign of this part corresponds to a positive imaginary part of the permittivity.

The expression (97.6) becomes invalid when $T \to T_c$ and N_s and Δ tend to zero. The principal contribution to the integral with respect to p in (97.5) here comes from the range $\eta \sim T \gg \Delta$, and in it we may put $\Delta = 0$. The result is

$$Q(\omega, \mathbf{k}) = -i\,.\,\tfrac{3}{4}\pi\,\frac{Ne^2}{mc}\,\frac{\omega}{kv_F},$$

where $N = p_F{}^3/3\pi^2\hbar^3$ is the electron density. This expression simply represents the anomalous skin effect in a normal metal, with the dispersion relation $\epsilon = p^2/2m$.†

PIPPARD CASE

In a static magnetic field, the Pippard limiting case corresponds to the inequality

$$\hbar kv_F \gg \Delta_0 \sim T_c. \tag{97.7}$$

To consider an alternating electromagnetic field, we add the further condition

$$kv_F \gg \omega. \tag{97.8}$$

The calculations in this case are considerably simplified by first subtracting from $Q(\omega, \mathbf{k})$ (96.24) its static value $Q(0, \mathbf{k})$; this is equivalent to omitting the constant term Ne^2/mc and subtracting from each term $(\epsilon_+ \pm \epsilon_- \pm \hbar\omega)^{-1}$ in the integrand a similar term with $\omega = 0$. The difference $Q(\omega, \mathbf{k}) - Q(0, \mathbf{k})$ is found to be proportional to $1/k$. The Pippard-case $Q(0, \mathbf{k})$ has a similar dependence on k:

$$Q(0, \mathbf{k}) = \frac{c\beta}{4\pi k}, \quad \beta = \frac{4\pi Ne^2}{mc^2}\frac{3\pi^2}{4\hbar v_F}\Delta\tanh\frac{\Delta}{2T}; \tag{97.9}$$

†See (86.16). In making the comparison, it is important that K in this case is independent of φ, and that Q relates $\mathbf{j}$ to $\mathbf{A}$, not to $\mathbf{E}$ as σ does in (86.16).

see *SP* 2 (51.21). We can therefore write $Q(\omega, \mathbf{k})$ in the form

$$Q(\omega, \mathbf{k}) = \frac{c}{4\pi k}[\beta + \gamma(\omega)], \tag{97.10}$$

where $\gamma(\omega)$ is a function that may be calculated and is zero when $\omega = 0$. Because of this dependence on k, the formula *SP* 2 (52.6) for the penetration depth δ remains valid, if we replace β by $\beta + \gamma(\omega)$. However, since $\gamma(\omega)$ is complex (see below), it is natural to use here not δ itself but the related surface impedance $\zeta(\omega) = -i\omega\delta/c$.

In the integral which gives the difference $Q(\omega, \mathbf{k}) - Q(0, \mathbf{k})$, the important range is that of small values of $\cos\theta$, as in the calculation of $Q(0, \mathbf{k})$ in *SP* 2, §51, and the integral converges rapidly as $\cos\theta$ increases; we can therefore put $\sin\theta = 1$, and extend the integration with respect to $\cos\theta$ from $-\infty$ to ∞.

The integral is transformed by means of

$$d^3p = 2\pi p^2\, dp\, d\cos\theta \approx 2\pi p_F^2 m\, d\eta\, d\cos\theta$$

($\eta = p^2/2m - \mu$), and new variables of integration are used:

$$x_1 = \epsilon_+/\Delta, \quad x_2 = \epsilon_-/\Delta.$$

We have

$$\eta_+ + \eta_- \approx 2\eta, \quad \eta_+ - \eta_- \approx \hbar k v_F \cos\theta.$$

The integration over $d\eta\, d\cos\theta$ can therefore be replaced by one over $d\eta_+ d\eta_-/kv_F$ from $-\infty$ to ∞ for each of the variables η_+ and η_-. All terms in the integrand which contain the product $\eta_+\eta_-$ and are therefore odd functions of these variables then give zero on integration. We can then change to integration from 1 to ∞ with respect to each of the variables x_1 and x_2, putting

$$d\eta\, d\cos\theta \to 4\frac{\epsilon_+\epsilon_-}{\hbar k v_F \eta_+\eta_-}\, d\epsilon_+ d\epsilon_-$$

$$= \frac{4\Delta^2 x_1 x_2\, dx_1 dx_2}{\hbar k v_F [(x_1^2 - 1)(x_2^2 - 1)]^{1/2}}.$$

These transformations lead to the result

$$\gamma(\omega) = -3\pi \frac{Ne^2}{mc^2}\frac{\Delta}{\hbar v_F} J,$$

$$J = \int_1^\infty \int_1^\infty \frac{dx_1 dx_2}{[(x_1^2 - 1)(x_2^2 - 1)]^{1/2}} \tanh\frac{x_1\Delta}{2T} \times$$

$$\times \left\{(x_1x_2 + 1)\left[\frac{1}{x_1 - x_2 - \bar\omega - i0} + \frac{1}{x_1 - x_2 + \bar\omega + i0} - \mathrm{P}\frac{2}{x_1 - x_2}\right]\right.$$

$$\left. + (x_1x_2 - 1)\left[\frac{1}{x_1 + x_2 - \bar\omega - i0} + \frac{1}{x_1 + x_2 + \bar\omega + i0} - \frac{2}{x_1 + x_2}\right]\right\}, \tag{97.11}$$

where $\tilde{\omega} = \hbar\omega/\Delta$. We shall consider only the imaginary part of this expression, which determines the absorption of energy from the field.

The imaginary part of the integrands in (97.11) is separated by means of the rule (29.8), and the delta functions are then eliminated by integrating with respect to one variable x_1 or x_2; it is necessary here to verify that the point at which the argument of the delta function is zero does in fact lie in the range of integration. A simple calculation gives, when $\omega > 0$,

$$J'' \equiv \operatorname{im} J = \pi \int_1^{\infty} \frac{x(x+\tilde{\omega})+1}{(x^2-1)^{1/2}[(x+\tilde{\omega})^2-1]^{1/2}} \left[\tanh\frac{(x+\tilde{\omega})\Delta}{2T} - \tanh\frac{x\Delta}{2T}\right] dx$$

$$+ \pi \int_1^{\tilde{\omega}-1} \frac{x(\tilde{\omega}-x)-1}{(x^2-1)^{1/2}[(x-\tilde{\omega})^2-1]^{1/2}} \tanh\frac{x\Delta}{2T}\, dx; \tag{97.12}$$

the second term occurs only when $\tilde{\omega} > 2$. Similarly, we can easily show that $J''(-\tilde{\omega}) = J''(\tilde{\omega})$. The integral (97.12) depends on two parameters Δ/T and $\hbar\omega/\Delta$, which may bear various relations to each other and to unity. Let us consider some of the possible limiting cases.

Let $T = 0$. Then the first integral in (97.12) is zero; the second is non-zero when $\omega > 2\Delta_0$; that is, there is an absorption threshold at the binding energy of the Cooper pairs. The presence of this threshold, which is a direct result of the gap in the spectrum, is a specific property of a superconductor.

Near the threshold, when $\tilde{\omega} - 2 \ll 1$, x is close to unity throughout the range of integration. Putting $\tilde{\omega} - 2 = \delta$, $x - 1 = z\delta$, we find

$$J'' \approx \tfrac{1}{2}\pi\delta \int_0^1 \frac{dz}{\sqrt{[z(1-z)]}} = \tfrac{1}{2}\pi^2\delta = \pi^2(\tfrac{1}{2}\tilde{\omega} - 1).$$

Collecting the above formulae, we have the following expression for the imaginary part of Q at $T = 0$ near the absorption threshold:

$$Q'' = -\frac{3\pi^2 Ne^2}{4mc}\frac{\Delta_0}{\hbar v_F k}\left(\frac{\hbar\omega}{2\Delta_0} - 1\right). \tag{97.13}$$

If the temperature is not zero, let us consider the case of low frequencies $\hbar\omega \ll \Delta$, and assume that $\Delta(T) \sim T$ (thus excluding both temperatures near zero and those near T_c). The second integral in (97.12) is then absent. In the first integral, the important range is $x - 1 \sim \tilde{\omega} \ll 1$. Expanding the difference of two tanh in the integrand in powers of $\tilde{\omega}$ and using the variable $x - 1 = u$, we find with logarithmic accuracy

$$J'' \approx \frac{\pi\hbar\omega}{2T}\cosh^{-2}\frac{\Delta}{2T}\int_0^{\sim 1}\frac{du}{\sqrt{[u(u+\tilde{\omega})]}} = \frac{\pi\hbar\omega}{2T}\cosh^{-2}\frac{\Delta}{2T}\log\frac{\Delta}{\hbar\omega}.$$

The result is then

$$Q'' = -\frac{3\pi}{8}\frac{Ne^2}{mc}\frac{\omega}{v_F k}\frac{\Delta}{T}\cosh^{-2}\frac{\Delta}{2T}\log\frac{\Delta}{\hbar\omega}. \tag{97.14}$$

§98. Thermal conductivity of superconductors

The physical nature of electronic thermal conduction in superconductors is similar to that of thermal conduction or viscosity in Bose superfluids. In both cases we are concerned with the transport coefficients of the normal component of a quantum liquid, which forms a set of elementary excitations therein. Here we shall consider this topic also in the BCS model (B. T. Geĭlikman 1958).

We start from the transport equation for the distribution function of quasi-particles in a superconductor where there is a temperature gradient:

$$\mathbf{v}\cdot\frac{\partial n}{\partial \mathbf{r}}-\frac{\partial \epsilon}{\partial \mathbf{r}}\cdot\frac{\partial n}{\partial \mathbf{p}}=C(n), \tag{98.1}$$

where $\mathbf{v}=\partial\epsilon/\partial\mathbf{p}$ is the quasi-particle velocity. The energy of a quasi-particle is

$$\epsilon=[v_F^2(p-p_F)^2+\Delta^2(T)]^{1/2}, \tag{98.2}$$

and itself depends on the temperature through the energy gap $\Delta(T)$. Hence, when a temperature gradient is present, the energy ϵ also becomes a function of the coordinates, and $-\partial\epsilon/\partial\mathbf{r}$ represents the force acting on a quasi-particle. This is the source of the second term on the left of (98.1).

As usual, we put $n=n_0(\epsilon)+\delta n(\mathbf{r},\mathbf{p})$, where

$$n_0(\epsilon)=(e^{\epsilon/T}+1)^{-1} \tag{98.3}$$

is the equilibrium distribution function. Retaining only the terms in n_0 on the left, we have as the equation for n_0

$$\mathbf{v}\cdot\frac{\partial n_0}{\partial \mathbf{r}}-\frac{\partial \epsilon}{\partial \mathbf{r}}\cdot\frac{\partial n_0}{\partial \mathbf{p}}=\left[\frac{\partial n_0}{\partial T}-\frac{\partial n_0}{\partial \epsilon}\frac{\partial \epsilon}{\partial T}\right]\mathbf{v}\,.\,\nabla T.$$

The difference of terms containing the derivative of Δ is zero in the square brackets, leaving

$$-\frac{\epsilon}{T}\frac{\partial n_0}{\partial \epsilon}\mathbf{v}\,.\,\nabla T=\frac{\mathbf{v}\,.\,\nabla T}{T^2}\frac{\epsilon}{(e^{\epsilon/T}+1)(e^{-\epsilon/T}+1)}.$$

The collision integral depends on the quasi-particle scattering mechanism. We shall consider the case where the principal mechanism is elastic scattering by impurity atoms at rest, and assume this scattering to be isotropic. Then the collision integral reduces (cf. (11.3)) to

$$C(n)=-\nu\,\delta n,$$

where $\nu=vN_{\text{imp}}\sigma_t$ is the effective collision frequency, N_{imp} the number density of impurity atoms, and σ_t the transport cross-section for the scattering of a quasi-particle by an impurity atom. The latter quantity is a constant, of the order of atomic dimensions.

The transport equation thus becomes

$$\frac{\mathbf{v}\,.\,\nabla T}{v}\frac{\epsilon}{T}\frac{\partial n_0}{\partial \epsilon} = \frac{\delta n}{l}, \tag{98.4}$$

where $l = 1/N_{\mathrm{imp}}\sigma_t$ is the constant mean free path.

The heat flux is calculated as the integral

$$\mathbf{q} = \int \epsilon \mathbf{v} \delta n\,.\,2d^3p/(2\pi\hbar)^3, \tag{98.5}$$

the factor 2 coming from the two directions of the quasi-particle spin. The distribution function $n = n_0 + \delta n$ is also related to the normal electric current in the superconductor, with density

$$\mathbf{j}_n = \mathbf{j} - \mathbf{j}_s = -\frac{e}{m}\int \mathbf{p}\,\delta n\,\frac{2\,d^3p}{(2\pi\hbar)^3} - e(N - N_s)\mathbf{v}_s;$$

in the model under consideration, $\mathbf{j} = -e\mathbf{i}/m$, with $\mathbf{i}$ given by (77.7).

The thermal conductivity is defined in terms of the heat flux with $\mathbf{j} = 0$. In the present case, however, this condition does not call for any change in equation (98.4). The reason is that the total current density in the superconductor is $\mathbf{j} = \mathbf{j}_n + \mathbf{j}_s$, the sum of the normal and superconducting currents. The current $\mathbf{j}_n$ that occurs in the presence of a temperature gradient is automatically balanced (in an open circuit) by the superconducting current $\mathbf{j}_s = -\mathbf{j}_n$. An important point here is that the movement of superconducting electrons does not involve any transfer of heat. The equilibrium distribution function of quasi-particles against the "background" of the superfluid flow with velocity $\mathbf{v}_s = -\mathbf{j}_s/eN_s$ differs from (98.3) in that ϵ is replaced by $\epsilon + \mathbf{p}\,.\,\mathbf{v}_s$ (cf. §77); this change would also have to be made in the transport equation (98.1). The velocity $\mathbf{v}_s$ is proportional to $\mathbf{j}_n$, and therefore to the small gradient ∇T; hence the above change would give rise only to second-order small terms on the left of the transport equation, and these would in any case have to be omitted in arriving at (98.4).

Substituting δn from (98.4) in (98.5), we find, after averaging over the directions of $\mathbf{p}$, the thermal conductivity

$$\kappa = -\frac{l}{3T}\int v\epsilon^2 \frac{\partial n_0}{\partial \epsilon}\frac{2.4\pi p^2\,dp}{(2\pi\hbar)^3},$$

or, with $v\,dp = d\epsilon$, $p^2 \approx p_F^2$,

$$\kappa = -\frac{lp_F^2}{3\pi^2\hbar^3 T}\cdot 2\int_\Delta^\infty \epsilon^2 \frac{\partial n_0}{\partial \epsilon}\,d\epsilon. \tag{98.6}$$

Finally, after some obvious substitutions,

$$\kappa = \frac{2lp_F^2\Delta^3}{3\pi^2\hbar^3 T^2}\int_1^\infty \frac{u^2\,du}{(e^{u\Delta/T}+1)(e^{-u\Delta/T}+1)}. \tag{98.7}$$

When $T \to 0$, $\Delta \to \Delta_0$, the conductivity tends to zero:

$$\kappa = \frac{2lp_F^2\Delta^2}{3\pi^2\hbar^3 T} e^{-\Delta/T}. \tag{98.8}$$

When $T \to T_c$, $\Delta \to 0$, it is seen from (98.6) to tend to the limit

$$\kappa = \frac{4lp_F^2 T}{3\pi^2\hbar^3} \int_0^\infty \epsilon n_0(\epsilon)\, d\epsilon = \frac{lp_F^2 T}{9\hbar^3},$$

corresponding to the case of a normal metal.

CHAPTER XII

KINETICS OF PHASE TRANSITIONS

§99. Kinetics of first-order phase transitions. Nucleation

THE basic ideas of the thermodynamic theory of nucleus formation in a phase transition are as follows (*SP* 1, § 162).

The change from a metastable to a stable phase occurs as the result of fluctuations in a homogeneous medium, which form small quantities of a new phase, or nuclei. The energetically unfavourable process of creation of an interface, however, has the result that when the nucleus is below a certain size it is unstable and disappears again. Only nuclei whose size a is above a definite value a_{cr} (for a given state of the metastable phase) are stable; this is called the *critical size*, and nuclei of this size will be called *critical nuclei*.† They are assumed to be macroscopic objects containing large numbers of molecules. The entire theory is therefore valid only for metastable states that are not too close to the limit of absolute instability of the phase; as this limit is approached, the critical size decreases to a value of the order of molecular dimensions.

With a purely thermodynamic approach, one can put only the problem of calculating the probability of occurrence in a medium of fluctuational nuclei of various sizes, the medium being regarded as in equilibrium. This is a point of fundamental importance. Since the state of the metastable phase does not actually correspond to complete statistical equilibrium, this treatment applies only to times much less than the critical nucleus formation time (reciprocal probability per unit time), after which the change to the new phase occurs in practice, and the metastable state ceases to exist. For the same reason, the thermodynamic calculation of the formation probability is feasible only for nuclei with size $a < a_{cr}$; larger nuclei develop into the new phase. That is, such large fluctuations are not among the group of microscopic states which correspond to the (metastable) macroscopic state under consideration.

Instead of the thermodynamic probability of nucleation, we shall refer to a quantity proportional to this, the "equilibrium" (in the sense mentioned) distribution function for nuclei of various sizes existing in the medium, denoted by $f_0(a)$; $f_0\,da$ is the number of nuclei per unit volume of the medium with sizes in the range da. According to the thermodynamic theory of fluctuations,

$$f_0(a) \propto \exp\{-R_{\min}(a)/T\}, \tag{99.1}$$

†In *SP* 1, § 162, only nuclei of the new phase which have just this critical size were considered.

where $R_{\min}$ is the minimum work needed to form a nucleus of a given size. This is made up of volume and surface parts, and for a spherical nucleus with radius a it is

$$R_{\min} = -\frac{8\pi a^3 \alpha}{3a_{\mathrm{cr}}} + 4\pi a^2 \alpha,$$

where α is the surface tension coefficient and the critical radius a_{cr} is expressed in terms of the thermodynamic quantities for the two phases; see *SP* 1, § 162, Problem 2. The value $a = a_{\mathrm{cr}}$ corresponds to the maximum of $R_{\min}(a)$, and near it

$$R_{\min} = \tfrac{4}{3}\pi\alpha a_{\mathrm{cr}}^2 - 4\pi\alpha(a - a_{\mathrm{cr}})^2. \qquad (99.2)$$

The maximum of $R_{\min}$ corresponds to an exponentially sharp minimum of the distribution function. Neglecting the much slower variation with a of the coefficient of the exponential, we have

$$f_0(a) = f_0(a_{\mathrm{cr}}) \exp\{4\pi\alpha(a - a_{\mathrm{cr}})^2/T\}, \qquad (99.3)$$

where†

$$f_0(a_{\mathrm{cr}}) = \mathrm{constant} \times \exp\{-4\pi\alpha a_{\mathrm{cr}}^2/3T\}.$$

From the above discussion, the value $a = a_{\mathrm{cr}}$ corresponds to the limit beyond which large quantities of the new phase begin to be formed. More precisely, we should refer not to a limit point $a = a_{\mathrm{cr}}$ but to a critical range of values of a near that point, with width $\delta a \sim (T/4\pi\alpha)^{1/2}$. The fluctuational development of nuclei in this size range can still, with appreciable probability, throw them back into the subcritical range, but nuclei beyond the critical range will inevitably develop into a new phase.

Since the thermodynamic theory is limited to the stage before the actual phase transition, it cannot provide information about the course of this process, for instance the rate of the process. That would require a kinetic analysis of the development of the nuclei, which ultimately merge into the new phase.‡

Let $f(t, a)$ be the required "kinetic" size distribution function of the nuclei. The "elementary process" which changes the size of a nucleus is the attachment to it, or the loss by it, of one molecule, and this is to be regarded as a small change, since in the present theory the nuclei are macroscopic objects. We may therefore describe the growth of the nuclei by a Fokker–Planck equation:

$$\partial f/\partial t = -\partial s/\partial a, \qquad (99.4)$$

†The coefficient of the exponential in $f_0(a_{\mathrm{cr}})$ cannot be expressed in terms of just the macroscopic properties of the phases. For a qualitative estimate, we may suppose that this factor is proportional to the particle number density N_1 in the main phase (1) and to the derivative $d\mathcal{N}/da$, where $\mathcal{N}$ is the number of particles in a nucleus of the new phase (2). Putting $N_1 \sim 1/v_1$, $\mathcal{N} \sim a_{\mathrm{cr}}^3/v_2$, where v_1 and v_2 are the volumes per molecule in the two phases, we obtain as an estimate of the constant $a_{\mathrm{cr}}^2/v_1 v_2$.

‡The theory given below is due to Ya. B. Zel'dovich (1942).

where s is the flux in "size space":

$$s = -B\,\partial f/\partial a + Af. \tag{99.5}$$

The quantity B is a "nuclear size diffusion coefficient"; A is connected with B by a relationship which follows from the fact that s is zero for the equilibrium distribution. With the latter in the form (99.1), and neglecting the slow variation of the coefficient of the exponential, we find

$$A = -BR'_{\min}(a)/T. \tag{99.6}$$

Let us now find the stationary solution corresponding to a continuous phase-transition process. Then s = constant, and this constant flux (in the direction of increasing size) is just the number of nuclei passing through the critical range per unit time and unit volume of the medium, i.e. it defines the rate of the process.

We can rewrite the expression (99.5) for the flux by expressing it, using (99.6), in terms of the ratio f/f_0 instead of f itself. Then the condition of constant flux becomes

$$-Bf_0\frac{\partial}{\partial a}(f/f_0) = s. \tag{99.7}$$

Hence

$$f/f_0 = -s\int\frac{da}{Bf_0} + \text{constant}.$$

The constant here, and s, are found from the boundary conditions for small and large a. The fluctuation probability increases rapidly with decreasing size, and small nuclei therefore have a high probability of occurrence. The stock of such nuclei may be regarded as made up so quickly that their number continues to have its equilibrium value, despite the constant depletion by the flux s. This is expressed by the boundary condition $f/f_0 \to 1$ as $a \to 0$. The boundary condition for large a can be established by noting that above the critical range the function f_0 defined by (90.1) (which is actually not valid there) increases without limit, whereas the true distribution function $f(a)$ of course remains finite. This situation is expressed by the condition $f/f_0 = 0$, imposed somewhere above the critical range; precisely where, is of no importance (see below), and we shall arbitrarily apply it as $a \to \infty$.†

The solution which satisfies both the above conditions is

$$f/f_0 = s\int_a^\infty da/Bf_0, \tag{99.8}$$

and s is determined by

$$1/s = \int_0^\infty da/Bf_0. \tag{99.9}$$

†Similar arguments have been used in solving a different problem (§24).

The integrand has a sharp maximum at $a = a_{cr}$. Using the expression (99.3) near that point, we can extend the integration with respect to $a - a_{cr}$ in (99.9) from $-\infty$ to ∞, regardless of precisely where (outside the critical range) the upper limit in (99.8) and (99.9) is taken, i.e. precisely where the boundary condition is imposed. The result is

$$s = 2\sqrt{(\alpha/T)}B(a_{cr})f_0(a_{cr}). \tag{99.10}$$

This is the number of "viable" nuclei (i.e. those that have passed through the critical range) formed in stationary conditions per unit time and per unit volume of the metastable phase, expressed in terms of the equilibrium number of critical nuclei given by the thermodynamic theory.

For the distribution function $f(a)$ itself, formula (99.8) in the subcritical range gives simply $f(a) \approx f_0(a)$. Above the critical range, (99.8) tells us only that $f \ll f_0$, in accordance with the boundary condition stated. It is evident from the physical picture of the process that in this range the distribution function is constant: having reached that point, the nucleus becomes steadily larger, with practically no change in the reverse direction. Accordingly, we can here neglect the term containing the derivative $\partial f/\partial a$ in the flux (99.5), writing $s = Af$. From the significance of the flux s, the coefficient A acts as a velocity in size space, da/dt. The growth of a nucleus beyond the critical range, however, takes place in accordance with the macroscopic equations, by means of which the derivative da/dt can be independently determined:

$$A = (da/dt)_{\text{macro}}, \tag{99.11}$$

the subscript indicating the result of such a calculation.†

From (99.6), we then find

$$\begin{aligned} B(a) &= -\frac{T}{R'_{\text{min}}(a)}\left(\frac{da}{dt}\right)_{\text{macro}} \\ &= \frac{T}{8\pi\alpha(a - a_{cr})}\left(\frac{da}{dt}\right)_{\text{macro}}. \end{aligned} \tag{99.12}$$

Strictly speaking, the function $B(a)$ thus calculated pertains to the range $a > a_{cr}$, whereas we are interested in the value of $B(a_{cr})$ for substitution in (99.10). However, since $B(a)$ has no singularity at $a = a_{cr}$, the function just found can be used at that point also. As $a \to a_{cr}$, the derivative $(da/dt)_{\text{macro}}$ tends to zero (the nucleus is in—unstable—equilibrium); division by $a - a_{cr}$ gives a finite result.

Formula (99.12) makes it possible in principle to calculate the coefficient $B(a_{cr})$ and hence the rate of formation of nuclei, without using a microscopic treatment.

†The question may arise of the correspondence between (99.11) and the "microscopic" definition (21.4), according to which the rate $\Sigma\, \delta a/\delta t$ (summed over elementary growth events) is not A itself but the sum $\tilde{A} = A + B'(a)$. But the derivative $B'(a)$ is small (outside the critical range) in comparison with the value (99.6), which includes the large factor R'_{min}/T, and must be omitted. Quantities of this order have already been neglected in deriving (99.6), when the coefficient of the exponential in (99.1) was regarded as constant.

For example, in the case of boiling we have to apply the hydrodynamic equations to consider the growth of a vapour bubble in the liquid; for precipitation of a solute from a supersaturated solution, we have to deal with the growth of a precipitated grain as a result of the diffusion of the substance to it from the surrounding solution.

PROBLEM

Determine the "size diffusion coefficient" for the precipitation of a substance from a supersaturated (but nevertheless weak) solution; the nuclei are assumed to be spherical.

SOLUTION. The thermodynamic formulae are as follows. The critical radius at which a nucleus is precipitated from a supersaturated solution is

$$a_{cr} = 2\alpha v'/(\mu' - \mu_0');$$

see *SP* 1, §162, Problem 2. In the present case, μ_0' and v' are the chemical potential and molecular volume of the substance of the nucleus, and μ' the chemical potential of the solute in solution: $\mu' = T \log c + \psi(P, T)$, where c is the concentration. With the concentration $c_{0\infty}$ of the saturated solution above a plane surface of solute: $T \log c_{0\infty} + \psi = \mu_0'$, we have

$$\mu' - \mu_0' = T \log(c/c_{0\infty}) \approx T(c - c_{0\infty})/c_{0\infty};$$

the latter equation is valid for weak solutions. The critical radius is therefore

$$a_{cr} = 2\alpha v' c_{0\infty}/T(c - c_{0\infty}). \tag{1}$$

The formula

$$\begin{aligned} c_{0a} &= c_{0\infty}(1 + 2\alpha v'/Ta) \\ &= c_{0\infty} + (a_{cr}/a)(c - c_{0\infty}) \end{aligned} \tag{2}$$

gives the saturation concentration c_{0a} above a spherical solute surface with radius a.

The substance reaches the nucleus as it grows, beyond the critical range, by diffusion from the surrounding solution. In a steady state, the spherically symmetric concentration distribution $c(r)$ round a nucleus of radius a is given by the solution of the diffusion equation

$$D\triangle c(r) = D\frac{1}{r}\frac{\partial^2}{\partial r^2} rc(r) = \frac{\partial c(r)}{\partial t} \equiv 0,$$

with the boundary conditions $c(\infty) = c$ (the given value of the concentration of the supersaturated solution) and $c(a) = c_{0a}$. Hence

$$c(r) = c - (c - c_{0a})a/r,$$

and the diffusional flux towards the nucleus is

$$\begin{aligned} I &= 4\pi r^2 D\, dc/dr \\ &= 4\pi Da(c - c_{0a}) \\ &= 4\pi D(c - c_{0\infty})(a - a_{cr}); \end{aligned}$$

in the last equation, formula (2) is used.

If the concentration is defined as the number of dissolved molecules per unit volume, then I is the number of molecules deposited on the surface of the nucleus per unit time. We have

$$\begin{aligned} (da/dt)_{\text{macro}} &= Iv'/4\pi a^2 \\ &= (Dv'/a^2)(a - a_{cr})(c - c_{0\infty}) \end{aligned}$$

and, from (99.12),

$$B(a_{cr}) = TDv'(c - c_{0\infty})/8\pi\alpha a_{cr}^2$$
$$= Dv'^2 c_{0\infty}/4\pi a_{cr}^3.$$

§ 100. Kinetics of first-order phase transitions. Coalescence

The treatment in § 99 of phase-transition kinetics relates only to the initial stage of the transition: the total volume of all nuclei of the new phase has to be so small that their formation and growth have no appreciable effect on the "degree of metastability" of the main phase, and the critical size of the nuclei, determined by the degree of metastability, may be regarded as a constant. In this stage there is a fluctuational formation of nuclei of the new phase, and the growth of each nucleus is independent of the behaviour of the others. We shall refer to the particular case of solute precipitation from a supersaturated solution; the degree of metastability is then the degree of supersaturation of the solution.

In the later stage when the supersaturation of the solution becomes very slight, the nature of the process is quite different. The fluctuational formation of new nuclei has now practically ceased, as the critical size is great. The increase in the critical size accompanying the steady decrease in the degree of supersaturation has the result that the smaller among the grains of the new phase already formed fall below the critical range and redissolve. Thus a decisive role at this stage is played by the "swallowing up" of small grains by large ones, which grow as the result of the dissolution of the small ones (*coalescence*). This stage will be discussed in the present section. It is assumed that the initial concentration of the solution is so small that the precipitated grains are far apart and their direct "interaction" may be neglected.†

We shall consider a solid solution in which the precipitated grains are at rest and grow only by diffusion from the surrounding solution. In order to illustrate the method and the basic qualitative features of the process, we shall also make a number of other simplifying assumptions, neglecting the elastic stresses round the precipitated grains, and assuming that these are spherical.

The equilibrium concentration of the solution at the surface of a grain with radius a is given by the thermodynamic formula

$$c_{0a} = c_{0\infty}(1 + 2\alpha v'/Ta), \qquad (100.1)$$

where $c_{0\infty}$ is the concentration of a saturated solution above a plane surface of the solute, α the surface tension coefficient at the phase interface, v' the molecular volume of the solute; see § 99, Problem. The concentration is defined in terms of the volume of the substance dissolved in unit volume of the solution. With this definition, the diffusive flux $i = D\partial c/\partial r$ at a grain surface is equal to the rate of

†The theory given here is due to I. M. Lifshitz and V. V. Slezov (1958).

change of the grain radius:

$$da/dt = D[\partial c/\partial r]_{r=a},$$

where D is the solute diffusion coefficient. Because the concentration is assumed small, this rate is so small that the concentration distribution round the grain can be regarded as equal, at each instant, to the steady distribution $c(r)$ corresponding to the relevant value of a:

$$c(r) = c - (c - c_{0a})a/r,$$

where c is the mean concentration of the solution. Hence the diffusive flux $i(r) = Da(c - c_{0a})/r^2$ and, with (100.1),

$$\begin{aligned} i(a) &= da/dt \\ &= D(c - c_{0a})/a \\ &= \frac{D}{a}\left(\Delta - \frac{\sigma}{a}\right), \end{aligned}$$

where the parameter $\sigma = 2\alpha v' c_{0\infty}/T$ and the quantity $\Delta = c - c_{0\infty}$ is the supersaturation of the solution. The quantity

$$a_{\mathrm{cr}}(t) = \sigma/\Delta(t) \tag{100.2}$$

is the critical radius: when $a > a_{\mathrm{cr}}$, the grain becomes larger ($da/dt > 0$), and when $a < a_{\mathrm{cr}}$ it dissolves ($da/dt < 0$). In the following analysis, up to the final result, we shall measure time in units of $a_{\mathrm{cr}}^3(0)/D\sigma$, where $a_{\mathrm{cr}}(0)$ is the critical radius at the point where coalescence begins. Thus we have the equation

$$\frac{da}{dt} = \frac{a_{\mathrm{cr}}^3(0)}{a}\left(\frac{1}{a_{\mathrm{cr}}} - \frac{1}{a}\right). \tag{100.3}$$

Next, let $f(t, a)$ be the grain size distribution function normalized so that the integral

$$N(t) = \int_0^\infty f(t, a)\,da$$

is the number of grains per unit volume. Regarding $v_a = da/dt$ as the rate of movement of the grain in size space, we can write the continuity equation in that space as

$$\frac{\partial f}{\partial t} + \frac{\partial}{\partial a}(f v_a) = 0. \tag{100.4}$$

Lastly, the conservation of the total quantity of solute is expressed by

$$\Delta + q = \text{constant} \equiv Q, \quad q(t) = \tfrac{4}{3}\pi \int a^3 f(t, a)\, da, \tag{100.5}$$

where Q is the total initial concentration and q the volume of precipitated grains per unit volume of the solution.

Equations (100.3)–(100.5) form a complete set of equations for the problem concerned. They can be transformed so as to involve variables that are more convenient for the analysis.

We use the dimensionless quantity

$$x(t) = a_{\text{cr}}(t)/a_{\text{cr}}(0). \tag{100.6}$$

As $t \to \infty$, the supersaturation $\Delta(t)$ tends to zero, and the critical radius correspondingly tends to infinity. Hence, as t varies from 0 to ∞, the quantity

$$\tau = 3 \log x(t) \tag{100.7}$$

also varies monotonically from 0 to ∞, and we shall take this as a new time variable. As the unknown function in (100.3) we take the ratio

$$u = a/a_{\text{cr}}(t). \tag{100.8}$$

The equation then becomes

$$du^3/d\tau = \gamma(u-1) - u^3, \tag{100.9}$$

where

$$\gamma = \gamma(\tau) = dt/x^2\, dx > 0. \tag{100.10}$$

Going on now to analyse the equations, we shall first show that as $\tau \to \infty$ the function $\gamma(\tau)$ must tend to a particular finite limit.

The right-hand side of (100.9) has a maximum at $u^2 = \frac{1}{3}\gamma$, where its value is $\gamma[\frac{2}{3}(\frac{1}{3}\gamma)^{1/2} - 1]$. Thus, depending on the value of γ, the rate $du^3/d\tau$ as a function of u may have any of the three forms shown in Fig. 34. When $\gamma = \gamma_0 = 27/4$, the curve touches the abscissa axis at $u = u_0 = 3/2$.

Each point on the abscissa axis representing the state of a grain moves to the right or to the left, according to the sign of the derivative $du^3/d\tau$. When $\gamma > \gamma_0$, all points to the left of u_1 move to the left and disappear on reaching the origin. The points with $u > u_1$ move to the point u_2, approaching it asymptotically from the right or from the left. This means that all grains with $u > u_1$, i.e. with radius $a > u_1 a_{\text{cr}}$, would reach asymptotically the size $a = a_{\text{cr}} u_2$, which tends to infinity with a_{cr}; the total volume q of precipitated grains would thus also tend to infinity, so that the equation of conservation of matter (100.5) could not be satisfied. When $\gamma < \gamma_0$, all points move to the left and disappear on reaching the origin after a finite time; in this case, $q(\tau) \to 0$, and equation (100.5) again cannot be satisfied.

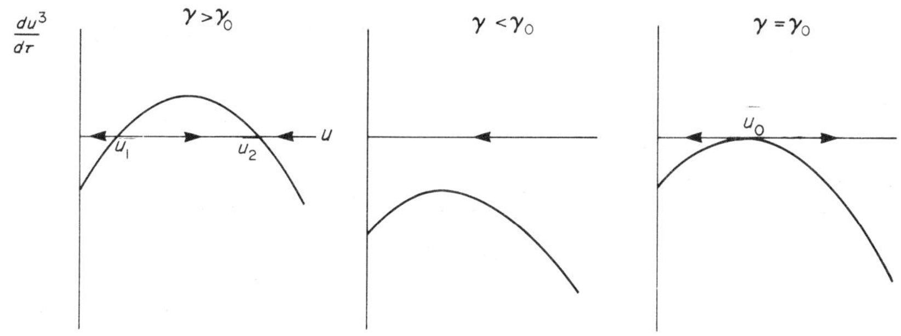

FIG. 34.

Thus the function $\gamma(t)$ must tend to the limit γ_0 and must do so from below: if it did so from above, or if $\gamma = \gamma_0$ exactly, all points with $u > u_0$, moving to the left, would still become "stuck" at $u = u_0$ (where $du^3/d\tau = 0$), and equation (100.5) could not be satisfied, as in the case $\gamma(\infty) > \gamma_0$. We must therefore have

$$\gamma(\tau) = \tfrac{27}{4}[1 - \epsilon^2(\tau)], \tag{100.11}$$

with $\epsilon \to 0$ as $\tau \to \infty$. The points approaching from the right pass more and more slowly through the "hold-up" point $u = u_0$. Their rate of passage is governed by the function $\epsilon(\tau)$, which must again be determined from the equation of motion (100.9) and the equation of conservation of matter (100.5).

Near the point $u = u_0$, equation (100.9), with γ from (100.11), is

$$\frac{du}{d\tau} = -\tfrac{2}{3}(u - \tfrac{3}{2})^2 - \tfrac{1}{2}\epsilon^2.$$

With a new unknown function, the ratio $z = (u - \tfrac{3}{2})/\epsilon$ of two small quantities, we can write this equation as

$$\frac{3}{2\epsilon}\frac{dz}{d\tau} = -z^2 - \frac{3}{4} + \frac{3}{2}z\eta \quad \left(\eta = \frac{d(1/\epsilon)}{d\tau}\right). \tag{100.12}$$

Analysis of this, similar to that of (100.9), leads to the conclusion that as $\tau \to \infty$ the function $\eta(\tau)$ must tend asymptotically to a finite limit $\eta_0 = 2/\sqrt{3}$, the value for which the right-hand side of (100.12) as a function of z touches the abscissa axis at the hold-up point $z_0 = \sqrt{3}/2$. The asymptotic equation $\eta = \eta_0$ gives the limiting form

$$\epsilon(\tau) = \sqrt{3}/2\tau. \tag{100.13}$$

When $\tau^2 \gg 1$, the correlation term in (100.11) may be neglected. The equation $1/\gamma = x^2\, dx/dt = 4/27$ then gives the limiting form of the time dependence of the critical radius:

$$x(t) = a_{cr}(t)/a_{cr}(0) = (4t/9)^{1/3}. \tag{100.14}$$

Since $\tau = \log x^3$, the condition for (100.14) to be applicable, expressed in terms of the actual time t, is $\log^2 t \gg 1$. It is noteworthy that, although the relative magnitude of the corrections to γ_0 decreases rapidly with increasing τ and the first approximation (100.14) becomes more and more nearly exact, the behaviour of the solution near the hold-up point is governed by just these corrections.

Let us now calculate the grain size distribution function. The distribution function in the variables u and τ is related to that in t and a by

$$\varphi(\tau, u)\, du = f(t, a)\, da, \quad f = \varphi / a_{\text{cr}}. \tag{100.15}$$

The continuity equation for this function is

$$\frac{\partial \varphi}{\partial \tau} + \frac{\partial}{\partial u}(v_u \varphi) = 0, \quad v_u = du/d\tau. \tag{100.16}$$

The rate v_u is given by (100.9), with $\gamma = 27/4$, everywhere except in a neighbourhood ($\sim \epsilon$) of u_0:

$$v_u = \frac{du}{d\tau} = -\frac{1}{3u^2}(u - \tfrac{3}{2})^2(u + 3). \tag{100.17}$$

The solution of equation (100.16) has the form

$$\varphi(\tau, u) = \chi[\tau - \tau(u)]/-v_u, \quad \tau(u) = \int_0^u du/v_u, \tag{100.18}$$

where χ is a function to be determined.

We have seen that all points representing grains move from right to left on the u-axis, pass through the neighbourhood of the hold-up point, and spend a longer time there if they arrive later. This neighbourhood thus acts as a sink for points with $u > u_0$ and as a source for the range $u < u_0$.

The distribution function to the right of u_0, as $\tau \to \infty$, is determined by the points arriving from infinity, which correspond to grains in the "tail" of their initial ($\tau = 0$) distribution. Since the number of grains in that distribution of course decreases rapidly (in practice exponentially) with increasing size, the distribution function in the range $u > u_0$ (outside the immediate neighbourhood of u_0) tends to zero as $\tau \to \infty$.

In the equation of conservation of matter (100.5), the term $\Delta(\tau) \to 0$ as $\tau \to \infty$. Expressing the integral q in terms of the variables τ and u, with $a^3 = u^3 x^3 a_{\text{cr}}^3(0) = u^3 e^\tau a_{\text{cr}}^3(0)$, we find

$$\kappa e^\tau \int_0^u u^3 \varphi(\tau, u)\, du = 1, \quad \kappa = 4\pi a_{\text{cr}}^3(0)/3Q; \tag{100.19}$$

here φ is to be substituted from (100.18), with v_u from (100.17).† It is immediately

†We shall not pause to prove that the relative contribution to the integral from the neighbourhood of u_0, where (100.17) is not valid, tends to zero as $\tau \to \infty$.

evident that the expression on the left of (100.19) can be independent of τ only if χ has the form

$$\chi[\tau-\tau(u)] = Ae^{-\tau+\tau(u)}.$$

The function $\tau(u)$ is calculated by elementary integration, and the result is

$$\varphi(\tau, u) = Ae^{-\tau}P(u), \tag{100.20}$$

where

$$P(u) = \frac{3^4 e}{2^{5/3}} \frac{u^2 \exp[-1/(1-\frac{2}{3}u)]}{(u+3)^{7/3}(\frac{3}{2}-u)^{11/3}}, \quad u < \tfrac{3}{2}, \tag{100.21}$$

$$P(u) = 0, \qquad u > \tfrac{3}{2}.$$

The constant A is determined by substituting (100.20) back into (100.19); numerical evaluation of the resulting integral gives $A = 0.9/\kappa$. The function $P(u)$ is necessarily normalized to unity:

$$\int_0^{u_0} P(u)\,du = \int_0^{3/2} \frac{e^{\tau(u)}}{-v_u}\,du = -\int_0^{-\infty} e^{\tau}\,d\tau = 1.$$

The number of grains per unit volume is therefore

$$N = \int_0^{u_0} \varphi(\tau, u)\,du = Ae^{-\tau} = 9A/4t. \tag{100.22}$$

It is easy to find also the value $\bar{u}$ averaged over the distribution (100.21). To do so, we consider the integral

$$\int_0^{u_0} P(u)(u-1)du = \int_0^{u_0} e^{\tau(u)}(u-1)\frac{du}{-v_u} = \int_{-\infty}^{0} e^{\tau}[u(\tau)-1]d\tau.$$

Substitution of $u(\tau)-1$ from (100.9) gives

$$\frac{4}{27}\int_{-\infty}^{0} e^{\tau}\left[u^3(\tau) + \frac{du^3(\tau)}{d\tau}\right] d\tau = \frac{4}{27}[u^3(\tau)e^{\tau}]_{-\infty}^{0} = 0.$$

Thus

$$\bar{u} = \int_0^{u_0} P(u)u\,du = \int_0^{u_0} P(u)\,du = 1,$$

i.e. $\bar{a} = a_{cr}(t)$, the mean size being equal to the critical size.

We can assemble the above formulae and rewrite the results in terms of the original variables—the grain radius a and the dimensional time t. The mean grain radius increases asymptotically with time according to

$$\bar{a} = (4\sigma Dt/9)^{1/3}. \tag{100.23}$$

The grain size distribution is given at any time by (100.21): the number of grains whose radius is in the range da is $P(a/\bar{a})\, da/\bar{a}$. The function $P(u)$ is non-zero only when $u < \frac{3}{2}$, and is shown graphically in Fig. 35. The asymptotic distribution is independent of the initial distribution at the start of coalescence. The total number of grains per unit volume decreases with time according to

$$N(t) = 0.5Q/D\sigma t. \tag{100.24}$$

The supersaturation of the solution tends to zero:

$$\Delta(t) = (9\sigma^2/Dt)^{1/3}. \tag{100.25}$$

To see the significance of these relations, note that in the above treatment the total volume of the solution is regarded as infinite, and the total amount of solute is therefore infinite also. In a finite volume the process is of course complete after a finite time, when the whole of the solute has been precipitated into one mass.

§101. Relaxation of the order parameter near a second-order phase transition

It is well known that the change in state of a body in a second-order phase transition (phase transition of the second kind) is described by the order parameter η, which is non-zero on one side of the transition point (in the "unsymmetrical" phase) and zero on the other side (in the "symmetrical" phase). The discussion in *SP* 1, Chapter XIV, related to the properties of bodies in thermodynamic equili-

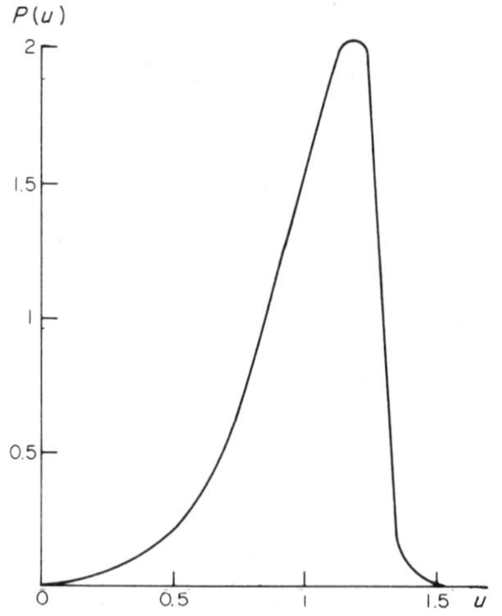

FIG. 35.

brium near transition points. Let us consider the relaxation of the order parameter in a system not in equilibrium.

The equilibrium value of the order parameter, denoted here by $\bar{\eta}$, is determined by minimizing the corresponding thermodynamic potential. In order to deal with cases of spatial homogeneity and inhomogeneity, we shall use the potential Ω, a function of the temperature T and the chemical potential μ (for a given total volume of the body); cf. *SP* 1, § 146.

In a spatially homogeneous body, the value of η is determined by the minimum of $\Omega(T, \mu, \eta)$ (the thermodynamic potential per unit volume) as a function of η with given T and μ:

$$\partial\Omega/\partial\eta = 0. \tag{101.1}$$

If this condition is not satisfied, a relaxation process occurs, in which η varies with time and tends to $\bar{\eta}$. In a state not far from equilibrium, i.e. when $\partial\Omega/\partial\eta$ is small but not zero, the relaxation rate (the derivative $d\eta/dt$) is also small. In the Landau theory, where fluctuations of the order parameter are neglected, we must suppose that the relation between these two derivatives amounts to a simple proportionality:

$$d\eta/dt = -\gamma\partial\Omega/\partial\eta, \tag{101.2}$$

with a constant coefficient γ (L. D. Landau and I. M. Khalatnikov, 1954).

In the Landau theory, the thermodynamic potential near the transition point has the form

$$\Omega = \Omega_0(T, \mu) + (T - T_c)\alpha\eta^2 + b\eta^4, \tag{101.3}$$

with a positive coefficient b. If the unsymmetrical phase corresponds to $T < T_c$, then $\alpha > 0$ also; see *SP* 1 (146.3). The equilibrium value of the order parameter in the unsymmetrical phase, i.e. the solution of equation (101.1), is

$$\bar{\eta} = [\alpha(T_c - T)/2b]^{1/2}. \tag{101.4}$$

The relaxation equation (101.2) becomes

$$d\eta/dt = -2\gamma[(T - T_c)\alpha\eta + 2b\eta^3],$$

or linearizing with respect to the small difference $\delta\eta = \eta - \bar{\eta}$,

$$d\delta\eta/dt = -\delta\eta/\tau_0, \tag{101.5}$$

where

$$\tau_0 = 1/4\gamma\alpha(T_c - T), \quad T < T_c. \tag{101.6}$$

As $t \to \infty$, the difference $\delta\eta$ must tend to zero; hence we must have $\tau_0 > 0$, and therefore $\gamma > 0$.

The relaxation in the range $T > T_c$ is treated similarly. Here $\bar{\eta} = 0$, and the linearized expression for the derivative is

$$\partial\Omega/\partial\eta = -2\alpha(T - T_c)\delta\eta.$$

Accordingly, (101.6) is replaced by

$$\tau_0 = 1/2\gamma\alpha(T - T_c), \quad T > T_c. \tag{101.7}$$

The quantity τ_0 is the relaxation time for the order parameter. We see that it tends to infinity as $T \to T_c$. This is of fundamental importance for the whole theory of phase transitions. As already noted in *SP* 1, § 143, it ensures the existence of macroscopic states corresponding to incomplete equilibrium for given non-equilibrium values of η. The theory given in §§ 101 and 102, which treats the relaxation of the order parameter independently of that of other macroscopic characteristics of the body, depends on this property for its significance.

In a spatially inhomogeneous system, we have to consider the total thermodynamic potential, given by the integral

$$\Omega_t = \int \{\Omega_0 + \alpha(T - T_c)\eta^2 + b\eta^4 + g(\nabla\eta)^2\}\, dV; \tag{101.8}$$

see *SP* 1 (146.5). The corresponding equilibrium condition is found by varying the integral with respect to η and equating the variation to zero. Integrating by parts in the gradient term, we get as the condition of equilibrium

$$2\alpha(T - T_c)\eta + 4b\eta^3 - 2g\Delta\eta \approx \delta\eta/\gamma\tau_0 - 2g\Delta\delta\eta = 0;$$

we have taken the particular case of the unsymmetrical phase with $T < T_c$. Correspondingly, there is an additional term in the relaxation equation:

$$\frac{\partial\delta\eta}{\partial t} = -\left\{\frac{\delta\eta}{\tau_0} - 2\gamma g\Delta\delta\eta\right\}. \tag{101.9}$$

For each of the spatial Fourier components of the function $\delta\eta(t, \mathbf{r})$, this gives

$$\frac{d\delta\eta_{\mathbf{k}}}{dt} = -\frac{\delta\eta_{\mathbf{k}}}{\tau_{\mathbf{k}}}, \quad \frac{1}{\tau_{\mathbf{k}}} = \frac{1}{\tau_0} + 2\gamma g k^2. \tag{101.10}$$

We see that the relaxation time for components with $\mathbf{k} \neq 0$ remains finite as $T \mapsto T_c$, but increases with decreasing k.

Lastly, if we include in Ω the term $-\eta h$ which describes the effect of an external field on the transition (see *SP* 1 (146.5)), the relaxation equation becomes

$$\frac{\partial\delta\eta}{\partial t} = -\frac{\delta\eta}{\tau_0} + 2\gamma g\Delta\delta\eta + \gamma h. \tag{101.11}$$

If the field is assumed periodic,

$$h \propto e^{i(\mathbf{k}\cdot\mathbf{r}-\omega t)},$$

we then obtain

$$\delta\eta_{\mathbf{k}} = \chi(\omega, \mathbf{k})/h,$$

with the generalized susceptibility

$$\chi(\omega, \mathbf{k}) = \gamma/(\tau_{\mathbf{k}}^{-1} - i\omega). \qquad (101.12)$$

This expression has a pole at $\omega = -i\tau_{\mathbf{k}}^{-1}$, in accordance with the general statement made at the end of §91. When $\omega = 0$ and $\mathbf{k} = 0$, it reduces to $\chi(0, 0) = 1/4\alpha(T_c - T)$, in agreement with *SP* 1 (144.8).

According to the fluctuation–dissipation theorem, the generalized susceptibility (101.12) determines the spectral correlation function of the fluctuations of the order parameter by the formula (in the classical limit $\hbar\omega \ll T$)

$$(\delta\eta^2)_{\omega\mathbf{k}} = (2T/\omega)\,\mathrm{im}\,\chi(\omega, \mathbf{k}) = 2\gamma T/(\omega^2 + \tau_{\mathbf{k}}^{-2}). \qquad (101.13)$$

This is the space–time Fourier component of the correlation function $\langle\delta\eta(0,0)\delta\eta(t,\mathbf{r})\rangle$; the mean values of the products of Fourier components of the fluctuations are related to $(\delta\eta^2)_{\omega\mathbf{k}}$ by

$$\langle\delta\eta_{\omega\mathbf{k}}\delta\eta_{\omega'\mathbf{k}'}\rangle = (2\pi)^4\delta(\omega+\omega')\delta(\mathbf{k}+\mathbf{k}')(\delta\eta^2)_{\omega\mathbf{k}}.$$

Integration of (101.13) over $d\omega/2\pi$ gives the spatial Fourier component of the single-time correlation function $\langle\delta\eta(0,0)\delta\eta(0,\mathbf{r})\rangle$:†

$$(\delta\eta^2)_{\mathbf{k}} = \int (\delta\eta^2)_{\omega\mathbf{k}}\, d\omega/2\pi$$

$$= T/[2gk^2 + 4\alpha(T_c - T)]. \qquad (101.14)$$

§102. **Dynamical scale invariance**

The theory in §101 does not take account of fluctuations of the order parameter. Its applicability is therefore restricted by the same conditions as for the Landau thermodynamic theory of phase transitions. These conditions are not satisfied in a neighbourhood of the transition point, the "fluctuation" region.

In this region, the kinetic properties of the body, like the purely thermodynamic properties (see *SP* 1, §148), can be described by a set of *critical indices* (or *critical*

†In comparing (101.14) with *SP* 1 (146.8), it must be remembered that the latter formula relates to components in the expansion as a Fourier series in a finite volume V, not as a Fourier integral.

exponents) which specify the manner of variation of quantities as the transition point is approached. It proves possible to derive certain relations between these indices by extending to include kinetic effects the hypothesis of scale invariance formulated for the thermodynamic properties in *SP* 1, § 149; this generalization is termed *dynamical scale invariance.*

The nature of the singularity of the thermodynamic quantities at the transition point depends on the number of components of the order parameter describing the transition, and on the structure of the effective Hamiltonian formed from them (see *SP* 1, § 147). For the kinetic quantities, the range of possible cases become more various because of the different possible forms of the "equations of motion" describing the relaxation. Let us first consider the simplest case, that of an order parameter having only one component (B. I. Halperin and P. C. Hohenberg 1969).†

A way to determine the relaxation behaviour that is possible in principle, but not in practice, is to calculate the exact (including fluctuations) generalized susceptibility $\chi(\omega, k; T)$ for the order parameter η under the action of the external field. The time variation of η during the relaxation is governed (as was explained in § 91) by the singularities of χ as a function of the complex variable ω. If the singularity nearest the real axis is the simple pole at $\omega = -i\tau^{-1}(k; T)$ on the imaginary axis, each Fourier component of the order parameter decays exponentially, with relaxation time $\tau(k; T)$. As well as the critical indices which determine the behaviour of the thermodynamic quantities, we use two indices y and z which describe the function $\chi(\omega, k; T)$:

$$\tau \propto |T - T_c|^{-y} \quad \text{when} \quad k = 0, \tag{102.1}$$

$$\tau \propto k^{-z} \quad \text{when} \quad T = T_c, \tag{102.2}$$

with $y > 0$, $z > 0$, since the relaxation time becomes infinite for $k = 0$, $T = T_c$.

It is plausible to assume that, near a second-order phase transition (in the fluctuation range), the relaxation time is independent of the temperature if it is measured in units of $\tau_0 \equiv \tau(0; T)$ and the lengths $1/k$ are measured in units of $r_c(T)$, the correlation radius for fluctuations of the order parameter. Thus $\tau(k; T)$ must take the form

$$\tau(k; T) = |T - T_c|^{-y} f(kr_c), \tag{102.3}$$

where f depends on the temperature only through $r_c(T)$ in the product kr_c, and $f(0) = \text{constant}$.

Since $r_c \to \infty$ when $T \to T_c$, in accordance with the definition of the critical index z we must have $f(\xi) \propto \xi^{-z}$ as $\xi \to \infty$. The temperature dependence of τ can then be separated as the product

$$|T - T_c|^{-y}\, |T - T_c|^{z\nu},$$

†This is the case, for instance, in the relaxation of the magnitude of the magnetization vector in a ferromagnet near its Curie point, where strong relativistic interactions fix the crystallographic direction of the vector.

where ν is the critical index for the correlation radius:†

$$r_c \propto |T - T_c|^{-\nu}. \tag{102.4}$$

But τ must remain finite as $T \to T_c$ (with $k \neq 0$). Hence it follows that we must have

$$y = z\nu. \tag{102.5}$$

Thus the assumption of scale invariance enables us to relate the two indices in (102.1) and (102.2).

As in the static case, there is good reason to suppose that the critical indices are the same on both sides of the transition point. This is because the spatial inhomogeneity ($k \neq 0$) blurs the phase transition, in the sense that it eliminates the singularities of all quantities at $T = T_c$, in this respect, the inhomogeneity influences the phase transition in the same way as an external field. In other words, the point $T = T_c$ is no longer distinctive, so that there is no reason to expect a difference between the values of z as T tends to T_c from above and from below. By virtue of the relation (102.5), the same is then true of the index y.

We can similarly relate z to the other critical indices. Let us consider, for instance, the dependence of the susceptibility χ on ω when $k = 0$, at the point $T = T_c$. According to scale invariance, the function $\chi(\omega, k; T_c)$ may be put in the form

$$\chi = |T - T_c|^{-\gamma} f(\omega\tau_0, kr_c), \quad f(0,0) = \text{constant},$$

where γ is the critical index for the susceptibility when $k = 0$ and $\omega = 0$. For $k = 0$ and $T \to T_c$, the susceptibility must tend to a finite limit (if $\omega \neq 0$). Since $\tau_0 \propto |T - T_c|^{-z\nu}$, we find that this implies

$$f(\xi, 0) \propto \xi^{-\gamma/\nu z} \quad \text{as} \quad \xi \to \infty.$$

The required dependence of χ on ω is therefore

$$\chi \propto \omega^{-\gamma/\nu z} \quad \text{for} \quad k = 0,\ T = T_c. \tag{102.6}$$

In the case considered, then, the demands of scale invariance enable us to establish a relation between the kinetic and thermodynamic critical indices, but not to determine the former entirely from the latter.

§103. Relaxation in liquid helium near the λ-point

Let us now consider "degenerate" systems in which the order parameter has n components η_i but the effective Hamiltonian depends (in a homogeneous system)

†The notation for the critical indices of the thermodynamic quantities here and below is the same as in *SP* 1, §148.

only on the sum of the squares of these components. That is, if the set of the η_i is regarded as an n-dimensional vector, the effective Hamiltonian is independent of the direction of the vector.

A typical example is a purely exchange ferromagnet, whose energy is independent of the direction of the magnetization vector. Another example is a superfluid (liquid helium), in which the order parameter is represented by the condensate wave function

$$\Xi = \sqrt{n_0} e^{i\Phi}; \tag{103.1}$$

see *SP* 2, §§26 and 27. This complex quantity is a set of two independent quantities, but the energy of a homogeneous liquid depends only on the squared modulus $|\Xi|^2 = n_0$, the density of the condensate.

The specific properties of degenerate systems are due to the presence in their vibrational spectra of a *soft mode*, a branch which results from variations in the direction of the "order parameter vector"; the frequency of these is zero at the phase transition point. Their dispersion relation can be found from the macroscopic equations of motion, and must satisfy the requirements of scale invariance. As we shall see, this allows the kinetic critical indices to be expressed entirely in terms of the thermodynamic ones. We shall do this for the case of liquid helium (R. A. Ferrell, N. Menyhárd, H. Schmidt, F. Schwabl and P. Szépfalusy 1967).

In this case, the soft mode is second sound. Near the transition point, it consists of combined oscillations of the superfluid velocity $\mathbf{v}_s$ and the entropy; the normal velocity oscillation amplitude in second sound is $v_n \sim v_s\rho_s/\rho_n$, and near the phase transition point (the λ-point) it is small, like ρ_s. The superfluid velocity is related to the phase of the condensate wave function, $\mathbf{v}_s = \hbar\nabla\Phi/m$, so that oscillations of $\mathbf{v}_s$ imply oscillations of the phase, i.e. of the direction of the order parameter vector. The dispersion relation for these oscillations is

$$\omega = u_2 k, \tag{103.2}$$

where

$$u_2 = \sqrt{(TS^2\rho_s/C_p\rho_n)} \approx \sqrt{(T_\lambda S_\lambda^2\rho_s/C_p\rho)} \tag{103.3}$$

is the speed of second sound (S being the entropy and C_p the specific heat per unit mass of the liquid); near the λ-point, T and S may be replaced by their values T_λ and S_λ at that point, and ρ_n (the density of the normal component of the liquid) by the total density ρ.†

†The speeds of first and second sound in liquid helium are calculated (see *FM*, § 130) as the roots of the dispersion relation

$$u^4 - u^2\left[\left(\frac{\partial P}{\partial \rho}\right)_S + \frac{\rho_s TS^2}{\rho_n C_v}\right] + \frac{\rho_s TS^2}{\rho_n C_p}\left(\frac{\partial P}{\partial \rho}\right)_S = 0.$$

Outside the immediate neighbourhood of the λ-point, the thermal expansion coefficient is small, and therefore so is the difference $C_p - C_v$, so that we can put $C_p \approx C_v$. As $T \to T_\lambda$, C_p becomes noticeably different from C_v, but $\rho_s \to 0$, and we then arrive at (103.3).

As $T \to T_\lambda$, the density ρ_s tends to zero according to

$$\rho_s \propto (T_\lambda - T)^{(2-\alpha)/3}, \tag{103.4}$$

where α is the critical index for the specific heat:

$$C_p \propto |T_\lambda - T|^{-\alpha}; \tag{103.5}$$

see *SP* 2 (28.4). The way in which u_2 tends to zero depends on the sign of α. If $\alpha > 0$, so that $C_p \to \infty$, we have

$$u_2 \propto (T_\lambda - T)^{(1+\alpha)/3}, \quad \alpha > 0.$$

If $\alpha < 0$, C_p tends to a finite limit (the critical index defines the behaviour of only the singular part of the specific heat near the transition point); then

$$u_2 \propto (T_\lambda - T)^{(2-\alpha)/6}, \quad \alpha < 0. \tag{103.6}$$

We shall suppose here that $\alpha < 0$, as in fact seems to be true for liquid helium ($\alpha \approx -0.02$).

The damping of second sound is governed by the imaginary part of the frequency. Far below the λ-point, this imaginary part is small, but it increases as the λ-point is approached, and in its immediate neighbourhood ($kr_c \sim 1$) becomes of the order of unity ($\operatorname{im}\omega \sim |\omega|$). At a sufficient distance above the λ-point, we have an ordinary damped thermal wave (the solution of the thermal conduction equation), with the dispersion relation

$$\omega = i\kappa k^2/\rho C_p, \tag{103.7}$$

where κ is the thermal conductivity.

We now apply the hypothesis of scale invariance, according to which the dispersion relation near the λ-point must have the form

$$\omega = k^z f(kr_c).$$

This may also be written as†

$$\omega = k^z f\left(\frac{T - T_\lambda}{k^{1/\nu}}\right), \tag{103.8}$$

with a different function f, ν being the critical index for the correlation radius.

The validity of the dispersion relations (103.2) and (103.7) is not restricted by any condition of distance from the λ-point, but at a given temperature it is limited by

†These relations must be valid in the fluctuation region, which means that the inequality $|T - T_\lambda| \ll T_\lambda$ must always be satisfied. There is evidence, however, that in liquid helium this inequality must actually be satisfied with plenty to spare, implying that the theory should involve some small numerical parameter.

the condition $kr_c \ll 1$: the wavelength must be much greater than the correlation radius, since otherwise the macroscopic equations on which these relations are based cease to be valid.

Let us first consider temperatures below the transition point. The requirement that for $kr_c \ll 1$ the dispersion relation be linear in k determines the limiting form of $f(\xi)$ in (103.8):

$$f(\xi) \propto (-\xi)^{\nu(z-1)} \quad \text{as} \quad \xi \to -\infty.$$

The temperature dependence of the dispersion relation is found similarly:

$$\omega \propto k(T_\lambda - T)^{\nu(z-1)}. \tag{103.9}$$

Comparison of this with (103.6) gives

$$\nu(z-1) = (2-\alpha)/6.$$

The critical indices ν and α are related by $3\nu = 2-\alpha$ (see *SP* 1 (149.2)); hence†

$$z = 3/2. \tag{103.10}$$

As $T \to T_\lambda$, the frequency must tend to a finite limit, and therefore $f(0) =$ constant. Thus the dispersion relation for second sound at the λ-point itself is

$$\omega \propto k^z. \tag{103.11}$$

The imaginary part of ω is of the same order of magnitude as the real part. When $T \neq T_\lambda$, the dispersion relation (103.11) is valid for short waves such that $kr_c \gg 1$.

Lastly, let us consider the temperature range $T > T_\lambda$. Here, when $kr_c \ll 1$, ω must be a quadratic function of k. This implies that

$$f(\xi) \propto \xi^{\nu(z-2)} \quad \text{as} \quad \xi \to +\infty.$$

Then

$$\omega \propto k^2(T - T_\lambda)^{\nu(z-2)}.$$

Comparison with (103.7), expressing ν in terms of α, gives the temperature dependence of the thermal conductivity:

$$\kappa \propto (T - T_\lambda)^{-(2-\alpha)/6}. \tag{103.12}$$

This tends to infinity as $T \to T_\lambda$, approximately as $(T - T_\lambda)^{-1/3}$.

Second sound involves oscillations of the phase Φ of the condensate wave function. Hence $1/\text{im}\,\omega$ also represents a phase relaxation time. When $k \to 0$ it of

†If $\alpha < 0$, $z = 3/(2-\alpha)$.

course tends to infinity: in a homogeneous liquid, the change in phase does not lead to a change in energy, and phase relaxation is therefore not possible.

The relaxation time for $|\Xi| = \sqrt{n_0}$, the condensate density, is not in general the same as the phase relaxation time. However, from the sense of scale invariance we can say that the two times agree in order of magnitude when $kr_c \sim 1$. According to (103.9), this time is

$$\tau \sim \frac{1}{\omega(1/r_c)} \propto r_c(T_\lambda - T)^{-\nu(z-1)} \propto (T_\lambda - T)^{-\nu z}.$$

With z from (103.10),

$$\tau \propto (T_\lambda - T)^{-1+\alpha/2}. \qquad (103.13)$$

The relaxation time for the condensate density remains finite as $k \to 0$, and does not tend to infinity like the phase relaxation time. The temperature dependence (103.13) of the condensate density relaxation therefore remains valid when $k = 0$ (V. L. Pokrovskiĭ and I. M. Khalatnikov 1969).†

†If $\alpha > 0$, we should obtain $\tau \propto (T_\lambda - T)^{-1}$, in exact agreement with the result (101.6) of the Landau theory. This agreement, however, is to some extent accidental.

INDEX